SECONDARY PLANT METABOLISM

SECONDARY PLANT METABOLISM

Margaret L. Vickery and Brian Vickery

UNIVERSITY PARK PRESS

Baltimore

First published 1981 by
The Macmillan Press Ltd
London and Basingstoke

Published in North America by
UNIVERSITY PARK PRESS
300 North Charles Street
Baltimore, Maryland 21201

Printed in Hong Kong

LCCCN 80-54786
ISBN 0-8391-1676-4

CONTENTS

Preface   xi

Chapter 1   *Introduction*   1
Primary and secondary metabolism   1
The function of secondary metabolites   4
The usefulness of secondary metabolites   6
Secondary metabolites in chemosystematics   11
Plant tissue cultures   12
The investigation of biosynthetic pathways by isotopic tracer analysis   13
Bibliography   17

Chapter 2   *Sugar metabolism*   20
  Introduction   20
Photosynthesis   20
Monosaccharide derivatives   23
  Myo-inositol   25
  Ascorbic acid   30
Glycosides   32
Storage carbohydrates   39
  Sucrose   39
  Starch   41
  Phytoglycogen   43
  Fructans   43
  Mannans   44
  Exudate gums   45
Structural carbohydrates   46
  Plant cell walls   46
  Cellulose   47
  Hemicelluloses   48
    D-Xylans   48
    D-Mannans   49
    L-Arabino-D-galactans   49
  Pectic substances   50
    Pectin   50
  Seaweed gums   51
    Alginic acid   52
    Agar   52
    Carrageenan   52
Carbohydrates in chemosystematics   53
Bibliography   54

Chapter 3   *The acetate–malonate pathway*   56
  Introduction   56
  Acetyl coenzyme A   56
The acetate–malonate pathway   58
  Fatty acid elongation systems   60
  The biosynthesis of hydroxy acids   62
Catabolism of fatty acids   62
  β-Oxidation   62
  α-Oxidation   65
  Lipoxygenase-catalysed oxidation   66
Fatty acids in plants   68
Plant lipids   73
The function of plant lipids   76
The economic importance of seed oils   78
Plant lipids in chemosystematics   79
Polyacetylenes and thiophenes   81
Polyacetylenes and thiophenes in chemosystematics   84
Bibliography   85

Chapter 4   *Polyketides*   88
  Introduction   88
The biosynthesis of polyketides   88
  Orsellinic acid   89
  6-Methylsalicylic acid   90
  Patulin and penicillic acid   91
  Griseofulvin   93
  Phloroglucinol derivatives   94
  Polyketides in lichens   96
  Quinones   97
  Aflatoxins   103
  Fungal tropolones   103
  Mycophenolic acid   104
Deviations in polyketide chain biosynthesis   105
The function and uses of polyketides   108
Polyketides in chemosystematics   109
Bibliography   110

Chapter 5   *The acetate–mevalonate pathway*   112
  Introduction   112
  Mevalonic acid   113
  Dimethylallyl pyrophosphate   115
Hemiterpenoids   116
Geranyl pyrophosphate and the monoterpenoids   117
  Essential oils   119
  Iridoids   119
  Monoterpenoids with irregular structures   122
Farnesyl pyrophosphate and the sesquiterpenoids   122
  Abscisic acid   122

Geranylgeranyl pyrophosphate and the diterpenoids   124
  Phytol   124
  Gibberellins   126
Squalene and the triterpenoids   127
  Triterpenoids   128
  Metabolism of triterpenoids   132
Cycloartenol and the steroids   133
  Insect moulting hormones   137
  Sapogenins   137
  Steroidal alkaloids   138
  C_{21} steroids   140
  Cardiac glycosides   141
  Steroid degradation   144
  Steroids in tissue cultures   145
Carotenoids   146
  Xanthophylls   148
Polyisoprenes   148
The function of terpenoids and steroids in plants   149
Terpenoids and steroids in chemosystematics   151
Bibliography   154

Chapter 6   *Shikimic acid pathway metabolites*   157
  Introduction   157
The shikimic acid pathway   157
  Cinnamic acid derivatives   161
  Benzoic acid derivatives   166
  Lignin   168
  Coumarins   171
  Furanocoumarins   173
  The hydrolysable tannins   175
  p-Aminobenzoic acid   177
The catabolism of shikimic acid pathway metabolites   177
The function of phenols in higher plants   178
Simple phenols, phenolic acids and coumarins in chemosystematics 180
Bibliography   181

Chapter 7   *Compounds with a mixed biogenesis*   183
  Introduction   183
Flavonoids   183
  Chalcones   184
  Aurones   186
  Flavanones   187
  Flavones   187
  Flavonols   190
  Flavanols   192
  Proanthocyanidins   192
  Anthocyanins   194
  Anthocyanin pigmentation   196

Isoflavonoids 197
Catabolism of flavonoids 202
The function and physiological effects of plant flavonoids 203
Flavonoids in chemosystematics 204
Xanthones and stilbenes 206
Quinones 207
Benzoquinones 208
Ubiquinones, plastoquinones and tocopherols 208
Naphthoquinones 212
Phylloquinones and menaquinones 215
Anthraquinones 215
Quinones in chemosystematics 216
Bibliography 217

Chapter 8   Compounds derived from amino acids 220
Introduction 220
Non-protein amino acids 221
3-Cyanoalanine 223
Mimosine 224
N^3-Oxalyldiaminopropionic acid 225
Canavanine 226
Hypoglycin A 227
Azetidine-2-carboxylic acid 227
Selenoamino acids 228
Non-protein amino acids in chemosystematics 230
Amines 231
3-Aminopropionitrile 232
Polyamines 232
Histamine 234
Protoalkaloids 234
Amines in chemosystematics 236
Colchicine 236
Cyanogenic glycosides and cyanolipids 238
Cyanogenic glycosides 238
Cyanolipids 241
Cyanogenic glycosides in chemosystematics 241
Glucosinolates 241
Glucosinolates in chemosystematics 244
Penicillins and cephalosporins 245
Sulphur compounds derived from cysteine 247
3-Indolylacetic acid 249
Ethylene 251
Bibliography 252

Chapter 9   Alkaloids 255
Introduction 255
Alkaloids biosynthesised from ornithine 257
Pyrrolidine and tropane alkaloids 257

The pyrrolizidine alkaloids   260
Alkaloids biosynthesised from lysine   263
  The piperidine alkaloids   263
  The lupin alkaloids   264
  Cytisine   266
Alkaloids biosynthesised from nicotinic acid   267
  The tobacco alkaloids   267
Alkaloids biosynthesised from tyrosine   269
  The benzylisoquinoline alkaloids   270
Alkaloids biosynthesised from tyrosine and phenylalanine 273
  The phenanthridine alkaloids   273
Alkaloids biosynthesised from tryptophan   275
  The simple indole alkaloids   275
    Physostigmine   276
    The ergot alkaloids   277
  The complex indole alkaloids   279
    Strychnine   283
    Quinine   284
Alkaloids in tissue cultures   284
Alkaloids in chemosystematics   284
Bibliography   287

Chapter 10   *Porphyrins, purines and pyrimidines* 289
  Introduction   289
Porphyrins   289
Purine and pyrimidine derivatives   293
  The biosynthesis of purines and pyrimidines   293
  Nucleotides   296
  Nucleic acids   300
    The catabolism of nucleic acids   302
Nucleic acids in plant chemosystematics   306
Bibliography   307

General Index   309
Chemical Index   314
Botanical Index   328

PREFACE

Considerable advances in the understanding of secondary metabolism in plants have been made over the past decade. Elucidation of the biosynthetic pathways of secondary metabolites is continuing apace, and, while their function is still open to controversy, it is now widely acknowledged that these compounds are not the metabolic waste products they were once thought to be.

While we have endeavoured to include the most recent results reported in the literature of plant biochemistry, such is the volume of research being undertaken in this subject that the reader is recommended to consult the continuing journals listed in the bibliographies at the end of each chapter to keep abreast of developments in secondary plant metabolism. In particular, subsequent volumes of the Chemical Society's *Biosynthesis* will, it is hoped, continue to give a good coverage of the subject.

Exciting discoveries are continually being made in the field of secondary plant metabolism. Since completion of the manuscript of this book, it has been reported that not only can plants biosynthesise vitamin D_3 (cholecalciferol), once assumed the exclusive preserve of vertebrates, but the biosynthetic pathways forming this compound are similar in both plants and animals.

The overall purpose of our book, *Secondary Plant Metabolism*, is to show that plants do not haphazardly produce a large number of chemical compounds, but that each metabolite is biosynthesised for a definite purpose (although we may not, as yet, have discovered this purpose), and that all products are interrelated according to a complex plan which conserves energy and scarce inorganic nutrients. The format of the book is based on five main biosynthetic routes, namely sugar metabolism, the acetate–malonate pathway, the acetate–mevalonate pathway, the shikimic acid pathway, and metabolites derived from amino acids. An introductory chapter attempts to define primary and secondary metabolism, although in the light of modern work, it is no longer logical to make a clear-cut separation of the two. This chapter also discusses possible functions of secondary metabolites and describes some of the more important uses that man has made of these compounds. A brief introduction to isotopic tracer analysis is included, as this is the tool which more than any other has led to the elucidation of many of the sequences of secondary metabolite biosynthesis. Enzymology is not included, although the isolation and characterisation of the enzymes catalysing a particular biosynthetic step are the only means of proving that such a step takes place. This subject is usually well covered in any good book of biochemistry, however, and it would be superfluous to repeat the information here.

Many attempts have been made using secondary metabolites in the classification of plants, which is traditionally based on morphology. Such attempts have had some success and brief descriptions of examples are included at the end of each chapter or section.

Although this book is mainly concerned with biosynthesis, no description of secondary metabolites would be complete without reference to the ways in which these compounds affect animals, insects and other plants. Such ecological properties are therefore briefly included wherever applicable.

University of Warwick, 1980   M.L.V.
                                                                           B.V.

1 Introduction

Animal life on Earth depends on plants, as, without their capacity for converting carbon dioxide and water to sugars, and nitrogen to amino acids, animals, including man, could not survive. Thus, it could be argued that green plants are the most important constituents of this planet. Their activities, however, are not confined to sugar and protein production, but include the biosynthesis of the numerous types of organic compound described in this book.

Primary and Secondary Metabolism

Historically, the processes generating plant compounds have been separated into primary and secondary metabolism. However, in the light of present-day knowledge, this distinction is arbitrary, as there is no sharp division between primary and secondary metabolites. The once-popular definition of a secondary metabolite — that it does not play an indispensable role in the plant and is not ubiquitous — is no longer applicable. Secondary metabolites are now known to be very necessary to plant life, many of them providing a defence mechanism against bacterial, viral and fungal attack analogous to the immune system of animals. The detection of a compound depends on the sensitivity of the analytical procedure, and many compounds that now seem to be confined to particular plants will no doubt be found to be widespread as analytical techniques advance. Normal analysis of sugar beet (*Beta vulgaris,* Chenopodiaceae), for example, cannot detect the presence of the amino acid, azetidine-2-carboxylic acid (1.1), but analysis of the huge quantities of sugar beet residue remaining after the extraction of sugar shows this compound to be a minor constituent of the plant. In fact, there is an accumulation of evidence indicating that azetidine-2-carboxylic acid, which was once considered restricted to the Liliaceae, is probably a ubiquitous plant constituent.

azetidine-2-carboxylic acid
(1.1)

Although primary and secondary metabolism are interrelated to the extent that an absolute distinction is meaningless, for the purpose of this book some division has had to be made and this has been based on biosynthetic pathways. Excluding the primary processes of sugar and protein biosynthesis, there are three main routes to the wealth of chemical compounds found in plants — the acetate–malonate, acetate–mevalonate and shikimic acid pathways (described in

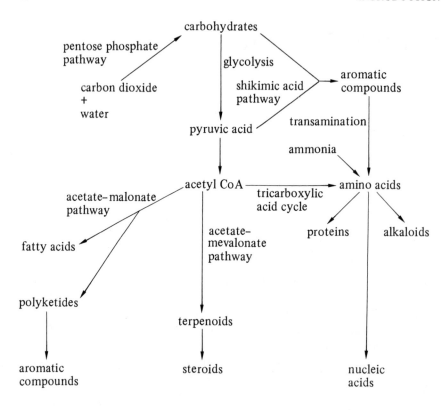

Figure 1.1 Biosynthetic pathways in plants.

Chapters 3, 5 and 6, respectively), which are interrelated as shown in *Figure 1.1*. These pathways are ubiquitous and the first products formed in quantity can be considered primary metabolites. Thus, palmitic acid (1.2) is the primary metabolite of the acetate–malonate pathway, other fatty acids and their derivatives being secondary metabolites.

$$CH_3(CH_2)_{14}COOH$$

palmitic acid
(1.2)

The terpenoids can be chemically and biochemically divided into classes depending on the number of C_5 isoprene units they contain, each class being derived from a primary metabolite precursor biosynthesised by the acetate–mevalonate pathway. Thus, geranyl pyrophosphate (1.3) is the primary metabolite from which the monoterpenoids are derived, farnesyl pyrophosphate (1.4) gives rise to the sesquiterpenoids and, through its conversion to squalene (1.5), the triterpenoids and steroids, while geranylgeranyl pyrophosphate (1.6) is the primary metabolite precursor of the diterpenoids and carotenoids (*Figure 1.2*).

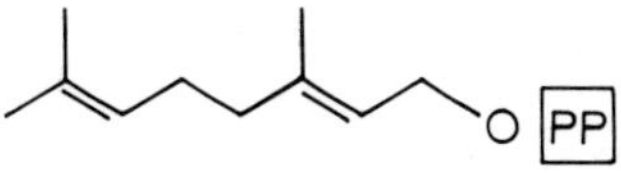

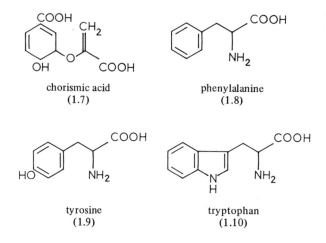

geranyl pyrophosphate
(1.3)

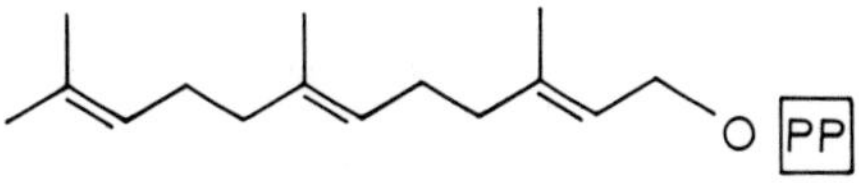

farnesyl pyrophosphate
(1.4)

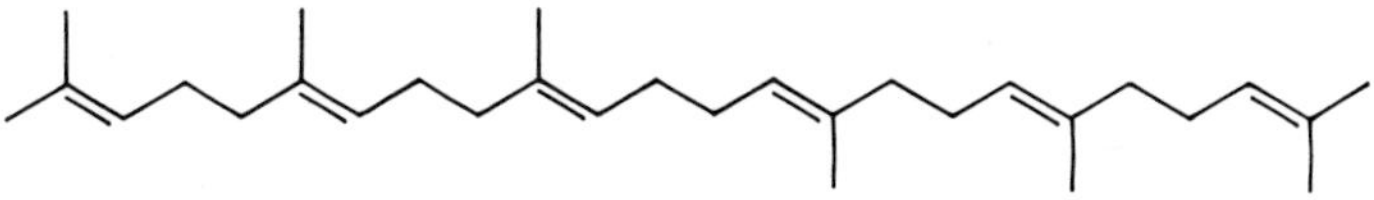

squalene
(1.5)

$$PP = H_3P_2O_6$$

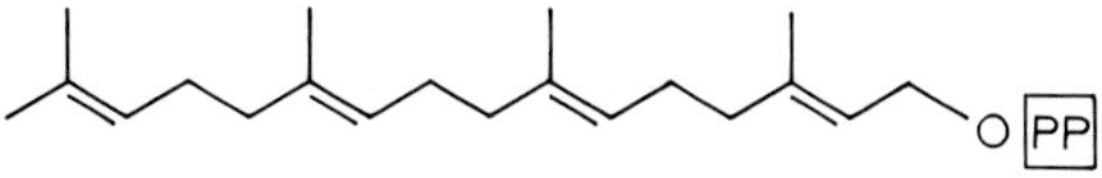

geranylgeranyl pyrophosphate
(1.6)

Figure 1.2 Primary metabolites of the acetate–mevalonate pathway.

It is less easy to separate the primary and secondary metabolites of the shikimic acid pathway, as there are several branch points. However, chorismic acid (1.7), the precursor of phenylalanine (1.8), tyrosine (1.9) and tryptophan (1.10) (*Figure 1.3*), can be considered the final primary metabolite.

chorismic acid
(1.7)

phenylalanine
(1.8)

tyrosine
(1.9)

tryptophan
(1.10)

Figure 1.3 Some metabolites of the shikimic acid pathway.

Variations in primary processes also lead to secondary metabolites. Thus, the pentose phosphate pathway producing the primary metabolites glucose (1.11) and fructose (1.12) (*Figure 1.4*), also generates the rare sugars found in the cardiac glycosides, while variations in the pathways synthesising the protein amino acids gives the non-protein amino acids.

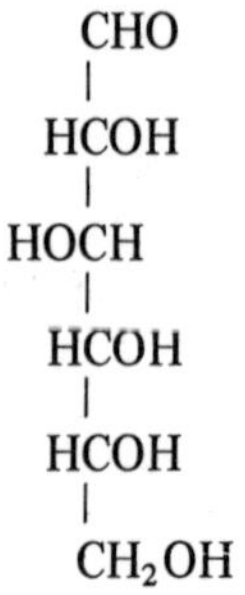

```
        CHO                              CH₂OH
         |                                 |
        HCOH                              C=O
         |                                 |
        HOCH                              HOCH
         |                                 |
        HCOH                              HCOII
         |                                 |
        HCOH                              HCOH
         |                                 |
        CH₂OH                             CH₂OH

       glucose                          fructose
       (1.11)                           (1.12)
```

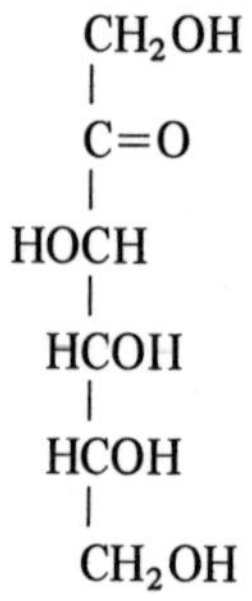

Figure 1.4 Primary metabolites of the pentose phosphate pathway.

The Function of Secondary Metabolites

Plant secondary metabolites are not waste products as was once concluded but have many useful functions. Auxin (indolylacetic acid (1.13)), the cytokinins (e.g. zeatin (1.14)), ethylene (1.15), abscisic acid (1.16) and the gibberellins (e.g. gibberellic acid (1.17)) are all hormones concerned with plant growth, while the wound hormone, traumatic acid (1.18), is responsible for the healing of damaged tissue (*Figure 1.5*).

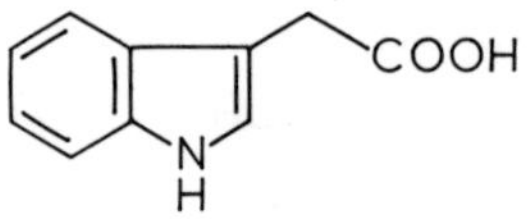

auxin
(1.13)

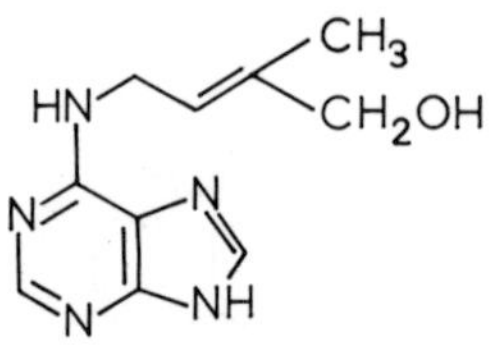

zeatin
(1.14)

$H_2C{=}CH_2$

ethylene
(1.15)

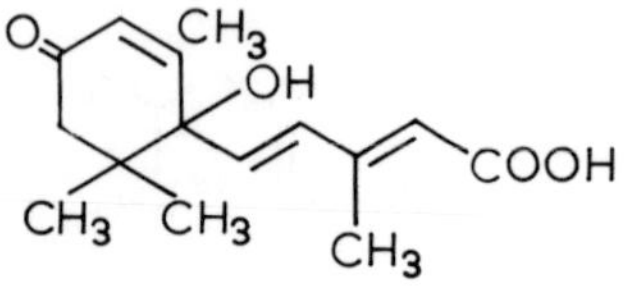

abscisic acid
(1.16)

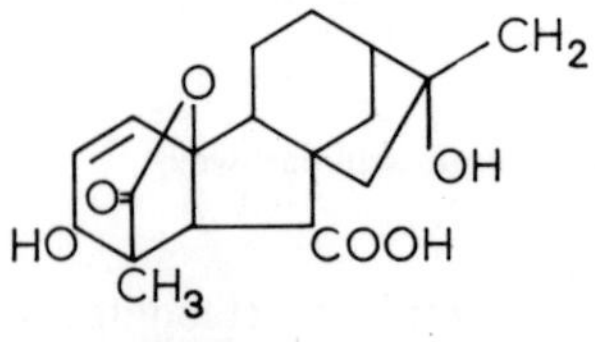

gibberellic acid
(1.17)

$HOOCCH{=}CH(CH_2)_8COOH$

traumatic acid
(1.18)

Figure 1.5 Plant hormones.

Many of the coloured flavonoids and carotenoids serve to attract insect and bird pollinators or seed dispersal agents, while other secondary metabolites, such as the volatile terpenoids, repel potential invaders. Numerous secondary metabolites are fungicides or antibiotics, protecting plants from fungal or bacterial invasion. Some, such as the pterocarpan, pisatin (1.19), and sesquiterpenoid, ipomeamarone (1.20), are phytoalexins (*Figure 1.6*) produced in response to fungal attack.

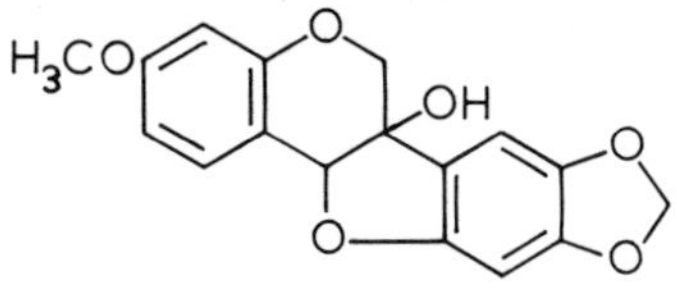

pisatin
(1.19)

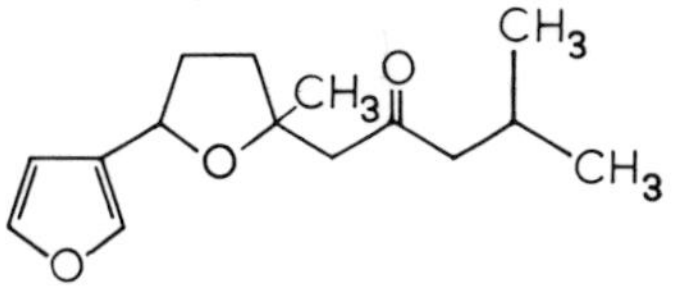

ipomeamarone
(1.20)

Figure 1.6 Phytoalexins.

A large number of secondary plant metabolites are toxic to animals or insects. Those, such as the alkaloids and cyanogenic glycosides, which also have a bitter taste act as feeding deterrents, and thus both repel invaders and prevent further damage to the plant species by killing the predator. Toxic plants are of great economic importance to farmers, as many losses are suffered every year through their agency. Herbaceous plants which grow where animals graze are the most dangerous, and probably more losses occur from the pyrrolizidine alkaloids of ragwort (*Senecio* species, Compositae) and *Crotalaria* species (Leguminosae), than from any other chemical compound. Other alkaloid-containing plants responsible for serious stock losses include yew (*Taxus baccata,* Taxaceae), *Delphinium* species (Ranunculaceae) and green potatoes (*Solanum tuberosum,* Solanaceae). Deaths amongst herbivores would be much greater, however, if these animals were not able to recognise and avoid poisonous plants within an area well known to them. Animals moved to a strange environment are unable to differentiate between poisonous and non-poisonous plants and are thus much more vulnerable. Unfortunately, an unpleasant taste is not always a deterrent. It is well known that once addicted to buttercups (*Ranunculus* species, Ranunculaceae), cattle will continue to eat the plants even after recovering from a serious bout of poisoning.

Poisonous plants are also dangerous to man, as they can be eaten by children or mistaken for edible plants by adults. Particularly dangerous to children are plants with attractive fruits, such as deadly nightshade (*Atropa belladonna,* Solanaceae), while hemlock (*Conium maculatum,* Umbelliferae) is easily mistaken for parsley (*Petroselinum crispum* syn *Apium petroselinum*) or other edible umbelliferous herbs. In the past, epidemics due to plant toxins have occurred, including St Anthony's fire caused by ergot infestation of rye (*Secale cereale,* Gramineae) and milk sickness due to *Eupatorium rugosum* (Compositae) ingested by cows.

The interactions between plants and insects are of particular ecological importance. Some insects, essential as pollinators, are attracted to plants by the colours or odours of flowers, while phytophages are repelled by leaf odours or tastes. Repellents include volatile compounds, such as certain monoterpenoids, or feeding deterrents, such as the cyanogenic glycosides. Saponins are insecticidal, which possibly explains their widespread occurrence in plants. However, almost all insecticidal compounds are specific feeding attractants for a particular insect. Thus, cabbage white butterflies are attracted by the mustard oil resulting from the hydrolysis of sinigrin (1.21), a glucosinolate characteristic of the Cruciferae family. The best insect repellents appear to be the condensed tannins, these compounds also deterring animals from feeding on plants containing them.

$$CH_2 = CHCH_2 - C \overset{\displaystyle \underset{\|}{N-O-SO_3^-}}{\underset{\displaystyle S-glucose}{\diagdown}}$$

sinigrin
(1.21)

Plants also produce steroids analogous to insect hormones, which interfere with metabolism and prevent the larvae reaching the adult stage and breeding. However, many insects are able to detoxify these compounds and, in fact, plant steroids are insect vitamins, insects being unable to synthesise the steroid ring system.

Plant secondary compounds are actively metabolised, especially in growing tissue. Evidence is accumulating to show that all types of compound are eventually degraded and enter the various metabolic pools or are completely oxidised to carbon dioxide. Flavonoids and phenolic compounds are degraded to aliphatic acids; terpenoids to isoprenoid units; and alkaloids to amino acids. Often, turnover is rapid, showing that secondary metabolites are not, in general, storage compounds, although this is the function of starch, some fatty acids, and non-protein amino acids found in high concentrations in seeds.

The Usefulness of Secondary Metabolites

Not only do plants provide the carbohydrates, proteins and fats necessary to the diet of man and other animals, but they also provide most of the essential vitamins (*Figure 1.7*). Vitamins C (ascorbic acid (1.22)), E (α-tocopherol (1.23)) and K (phylloquinone (1.24)) are biosynthesised by plants, while β-carotene (1.25), the precursor of vitamin A ('1.26), and ergosterol (1.27), the precursor of vitamin D (1.28), are also secondary plant metabolites.

For thousands of years, plants have been used as medicines, and many plant drugs are still prescribed today, although others have fallen into disrepute. Some higher plant drugs used in orthodox medicine are listed in *Table 1.1* together with their chemical identity, occurrence and main therapeutic use.

Plant secondary metabolites are also important as precursors for the synthesis of some drugs, particularly hormones and steroids.

Besides their medicinal value, plant poisons have been used for thousands of years to kill animals. Arrow poisons are generally prepared from plants containing cardiac glycosides or alkaloids. Cardiac glycoside-containing arrow poisons

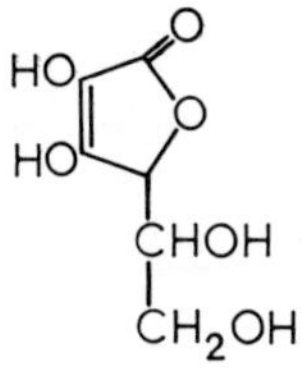

vitamin C
(1.22)

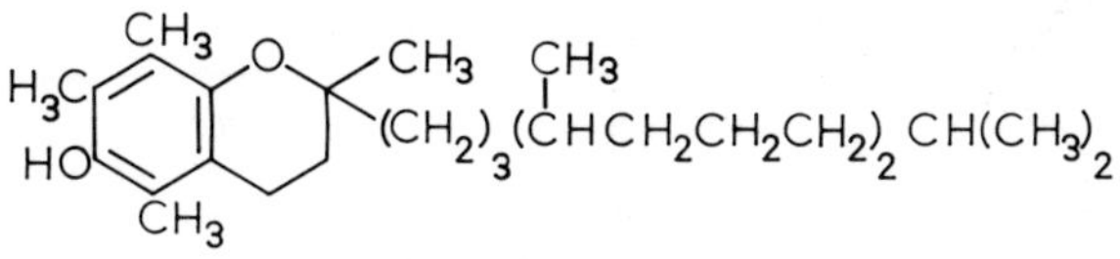

vitamin E
(1.23)

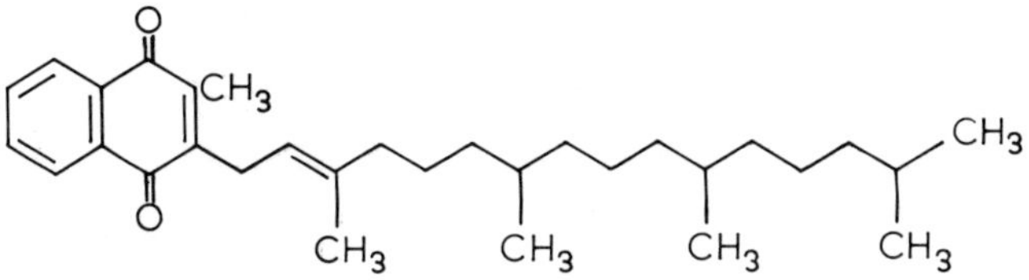

vitamin K
(1.24)

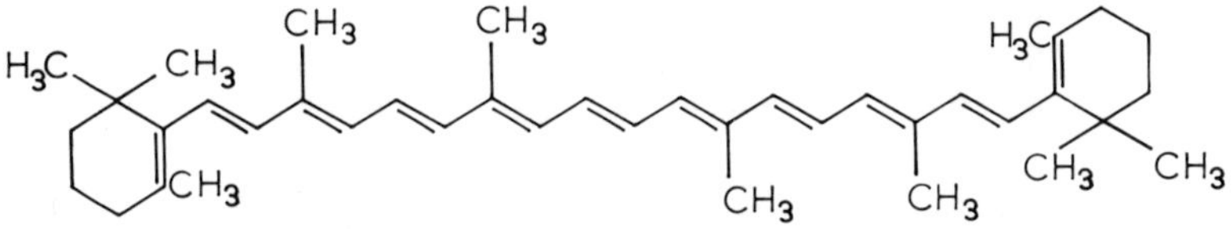

β-carotene
(1.25)

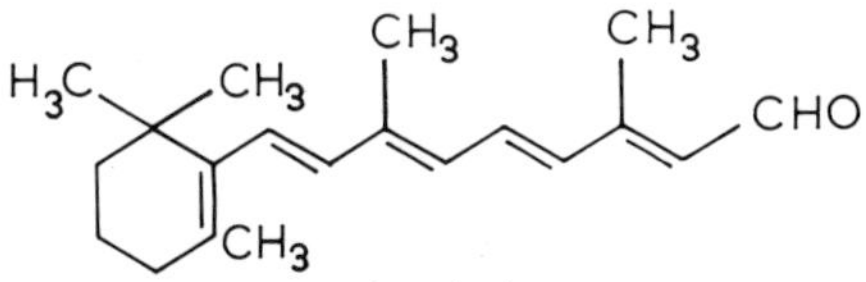

vitamin A
(1.26)

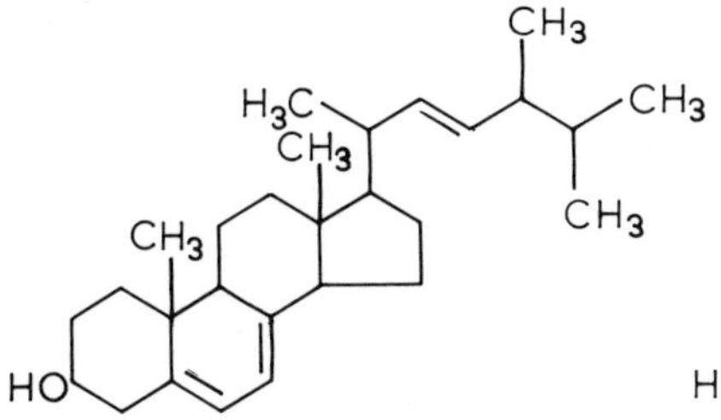

ergosterol
(1.27)

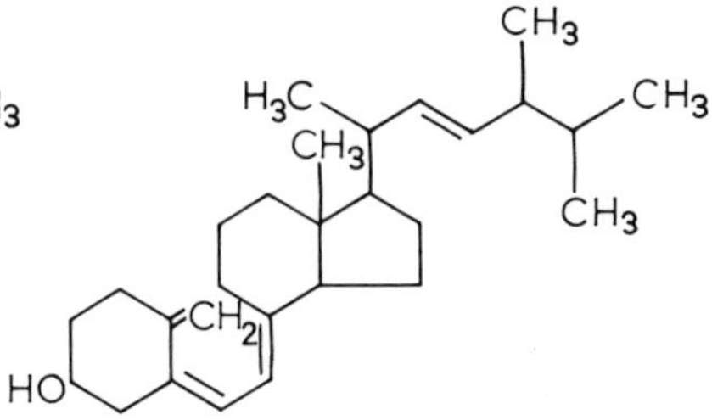

vitamin D
(1.28)

Figure 1.7 Vitamins.

include those from the Asian upas tree (*Antiaris toxicaria*, Moraceae) and African *Acocanthera* and *Strophanthus* species (Apocynaceae), while alkaloid-containing arrow poisons include the South American curare prepared from *Strychnos* species (Loganiaceae) and *Chondrodendron* species (Menispermaceae). Rat

Table 1.1 Plant drugs used in orthodox medicine.

Drug	Chemical identity	Occurrence	Use
aloes	anthraquinone	*Aloe* sp.	cathartic
caffeine	purine	many plants	stimulant
castor oil	lipid	*Ricinus communis*	cathartic
cocaine	alkaloid	*Erythroxylum coca*	local anaesthetic
codeine	alkaloid	*Papaver* sp.	analgesic
colchicine	alkaloid	Liliaceae	treatment of gout
digitoxin	cardiac glycoside	*Digitalis purpurea*	heart disease
digoxin	cardiac glycoside	*Digitalis lanata*	heart disease
emetine	alkaloid	*Cephaelis ipecacuanha*	emetic, amoebiasis
emodin	anthraquinone	several plants	cathartic
ephedrine	alkaloid	*Ephedra* sp.	bronchodilator, to prevent sleep
hyoscine	alkaloid	Solanaceae	sedative, to prevent secretions
hyoscyamine	alkaloid	Solanaceae	antispasmodic, mydriatic
lobeline	alkaloid	*Lobelia* sp.	respiratory stimulant
morphine	alkaloid	*Papaver somniferum*	narcotic
ouabain	cardiac glycoside	Apocynaceae	heart disease
papaverine	alkaloid	*Papaver* sp.	heart disease
physostigmine	alkaloid	*Physostigma venenosum*	glaucoma
quinine	alkaloid	*Cinchona* sp.	malaria
quinidine	alkaloid	*Cinchona* sp.	heart disease
reserpine	alkaloid	*Rauvolfia* sp.	tranquilliser
senna	anthraquinone	*Cassia* sp.	cathartic
strychnine	alkaloid	*Strychnos* sp.	stimulant
tubocurarine	alkaloid	*Strychnos* sp. *Chondrodendron* sp.	muscle relaxant
theophylline	purine	*Camellia sinensis*	stimulant
theobromine	purine	*Theobroma cacao*	stimulant
vinblastine	alkaloid	*Catharanthus roseus*	leukaemia, Hodgkin's disease
vincristine	alkaloid	*Catharanthus roseus*	leukaemia

poisons include ratsbane (*Dichapetalum toxicarium,* Dichapetalaceae) and squill (*Urginea maritima,* Liliaceae), while around 150 species have been recorded as fish poisons. The safest of these contain saponins which stun the fish without

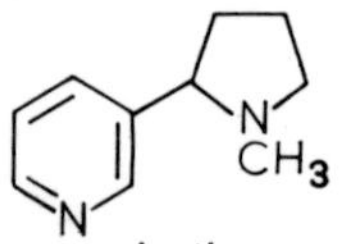

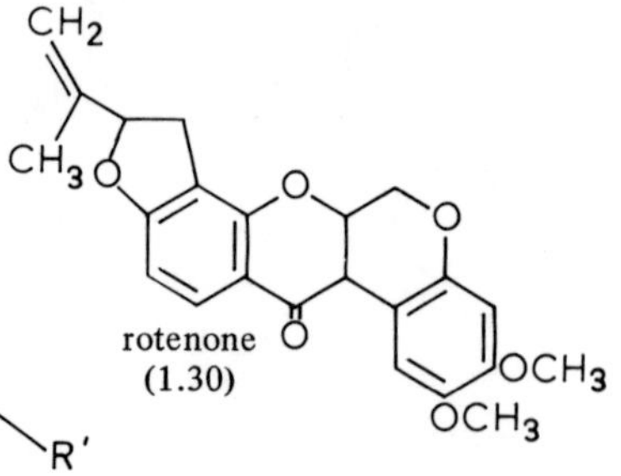

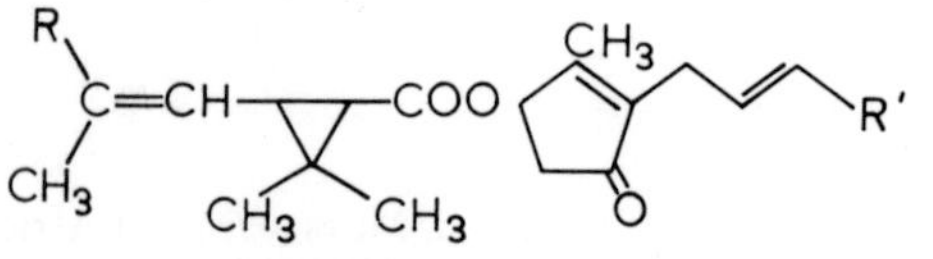

Figure 1.8 Commercial insecticides obtained from plants.

killing it, although *Mundulea sericea* (Leguminosae) has been reported to kill small crocodiles.

Although many plants contain insecticidal compounds, only a few have been used on a large scale *(Figure 1.8)*. These include nicotine (1.29) from *Nicotiana* species (Solanaceae), rotenone (1.30) from *Derris* and *Lonchocarpus* species (Leguminosae), and the pyrethrins (1.31) from pyrethrum *(Chrysanthemum cinerariaefolium,* Compositae).

Plant poisons are also notorious as homicidal agents. Poisoned arrows are still used for this purpose in Africa, although the method of justice known as trial by ordeal has almost vanished. Plants used to determine the guilt of a suspect usually contained highly toxic alkaloids and included the Calabar bean *(Physostigma venenosum)* and *Erythrophleum* species, both members of the Leguminosae family. Poisoning with plant extracts was rife in Europe in the Middle Ages, the plants most commonly used being deadly nightshade *(Atropa belladonna)* and mandrake *(Mandragora officinalis)*, both members of the Solanaceae family, and hemlock *(Conium maculatum,* Umbelliferae). All these plants contain alkaloids.

Secondary plant products find innumerable industrial uses and only a few can be mentioned here. Cellulose, in the form of wood pulp, is the raw material of the paper industry, while as cotton or rayon it is woven into fabrics. Fermentation of starch produces several important chemicals, particularly industrial alcohol. The bulk of plant oils not used in the food industry are solvents for paints and varnishes or are employed in the pharmaceutical and cosmetics industries. Waxes are the basis of polishes and large amounts are required for the production of carbon paper. Plant gums and mucilages are important as adhesives and thickening agents, while tannins are essential to the leather industry. Lignin has, as yet, found little application despite the large quantities obtained as a by-product of the paper industry. A small amount, however, is employed as a cheap precursor of vanillin (1.40).

Rubber finds many industrial applications, the bulk being used in the manufacture of tyres. Gutta percha and balata are no longer important industrially, as they have been superseded by synthetic polymers.

Since the advent of synthetic dyes, natural products *(Figure 1.9)* such as madder from *Rubia tinctorum* (Rubiaceae), which contains alizarin (1.32), are no longer used industrially. However, indigotin (1.33), from woad *(Isatis tinctoria,* Cruciferae) and *Indigofera tinctoria* (Leguminosae), has recently regained its popularity and is used to dye denim. Carthamin (1.34) from *Carthamus tinctorius* (Compositae), bixin (1.35) from annatto *(Bixa orellana,* Bixaceae), saffron from *Crocus sativa* (Iridaceae), which contains crocetin (1.36), and turmeric from *Curcuma longa* (Zingiberaceae), which contains curcumin (1.37), are still used to colour foods as they are considered safer than synthetic dyes.

Although expensive, plant perfumes and flavourings are preferred to synthetic mixtures, which can never exactly duplicate the natural product. Perfumes and flavourings are obtained from the essential oils of plants which contain many volatile components, including terpenoids, alcohols, aldehydes, ketones and esters. Species grown on a large scale for their perfumes include *Rosa* (Rosaceae), *Citrus* (Rutaceae), *Pelargonium* (Geraniaceae), *Viola* (Violaceae), *Lavandula* (Labiatae), *Cymbopogon* (Gramineae) and *Iris* (Iridaceae). Flavourings are obtained from

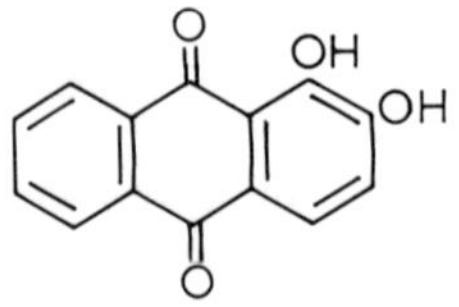

alizarin
(1.32)

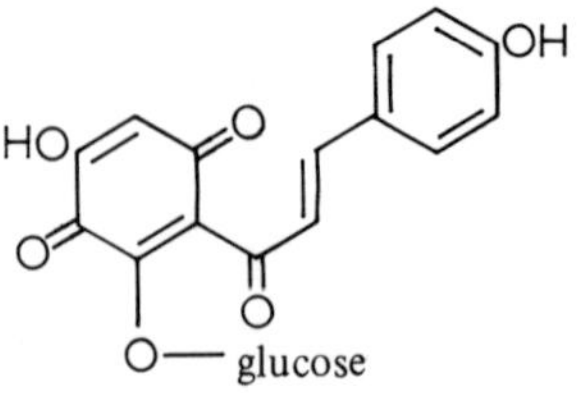

carthamin
(1.34)

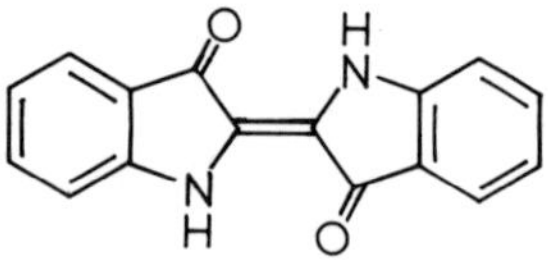

indigotin
(1.33)

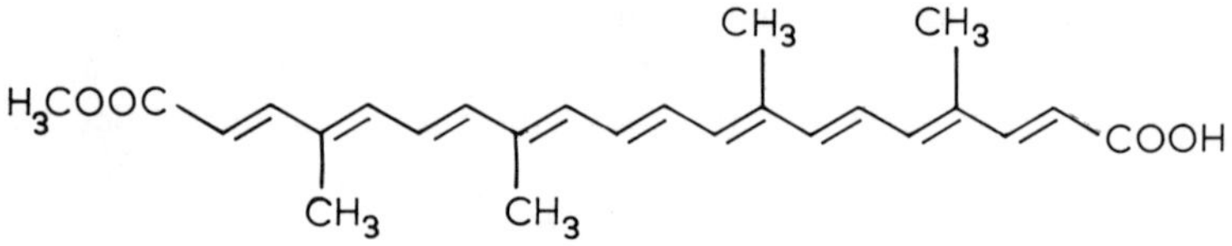

bixin
(1.35)

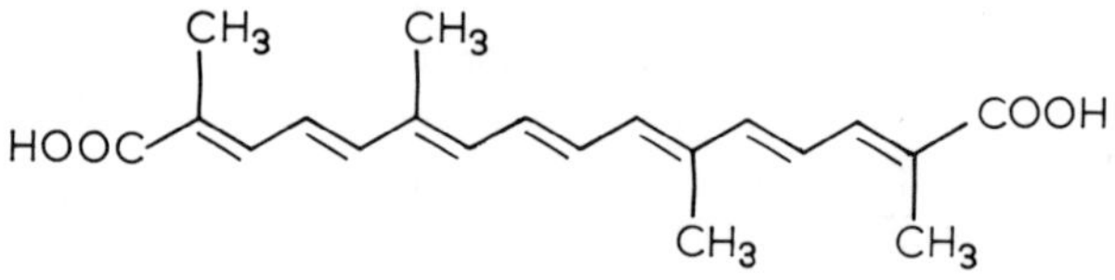

crocetin
(1.36)

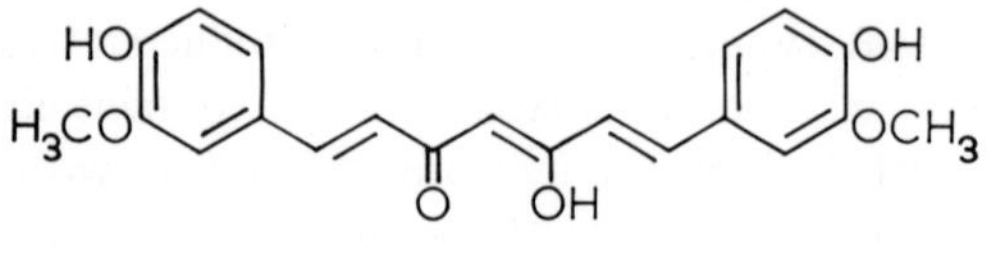

curcumin
(1.37)

Figure 1.9 Some naturally occurring dyes.

herbs belonging particularly to the Labiatae and Umbelliferae families, while the
spices (*Figure 1.10*) come from tropical plants. The characteristic flavour of
cloves (*Eugenia caryophyllata,* Myrtaceae) is due to eugenol (1.38), while
cinnamon (*Cinnamomum zeylanicum,* Lauraceae) contains the related cinnamaldehyde (1.39). Vanillin (1.40) gives the vanilla orchid (*Vanilla fragrans,* Orchid-

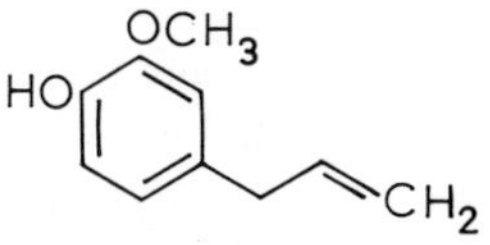

eugenol
(1.38)

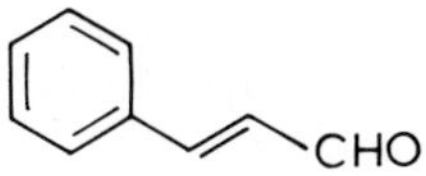

cinnamaldehyde
(1.39)

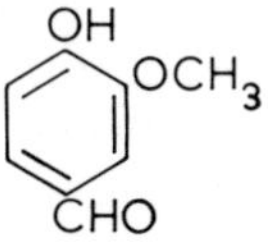

vanillin
(1.40)

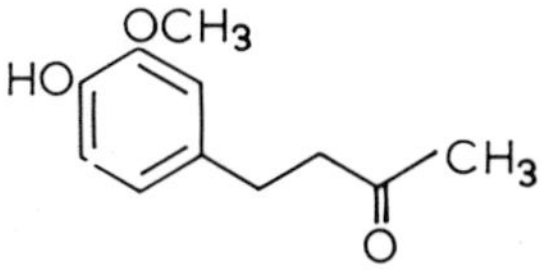

zingerone
(1.41)

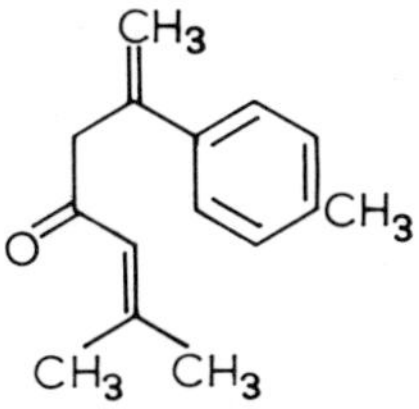

tumerone
(1.42)

Figure 1.10 Some naturally occurring spices.

aceae) its scent and flavour, while zingerone (1.41) is responsible for the pungent flavour of ginger (*Zingiber officinale,* Zingiberaceae) and tumerone (1.42) for that of turmeric (*Curcuma longa*).

Secondary Metabolites in Chemosystematics

Taxonomists are concerned with the description and classification of plants. Until recently, such work was confined to morphological features (i.e. the form, arrangement and interrelationships of plant structures), but it has now become evident that secondary metabolites can contribute much to plant taxonomy, systematics and the study of evolution. The evolution of chemical and morphological features are interrelated and thus most studies of secondary metabolites serve to confirm morphological classifications. However, in cases where morphological relationships are unclear, chemotaxonomic markers often prove of value. Several examples are given in the following chapters, including the use of secondary metabolites in the classification of the tribes of the Compositae (Chapters 3 and 5). This is a highly evolved family which is difficult to classify on morphological grounds, but the wealth of secondary products has already helped to solve some problems and should prove of increasing value as surveys are extended.

When considering chemosystematics and plant evolution, it is the biosynthetic pathway which is important, as the same type of metabolite can be the product of two quite different pathways. The enzymes catalysing metabolic reactions are of greater importance than the products of the reactions, as evolution depends on changes in enzyme characteristics. Unfortunately, only a few of the vast number of enzymes catalysing the production of secondary metabolites have so far been characterised. However, work in this field is progressing steadily.

One of the reasons for classifying plants into species, genera, families, etc., is

to trace their evolution, as plants which are similar morphologically and chemically should have the same ancestor. This is not always true, however, due to
convergent (parallel) evolution. For instance, both *Crotalaria* (Leguminosae) and
Senecio (Compositae) produce pyrrolizidine alkaloids by similar pathways, but
these genera are not related morphologically. It is thus important when classifying plants from biochemical data to consider all secondary metabolites and not
just one class, as has so often been done in the past. Unfortunately, such an
exercise is outside the scope of this book.

One of the difficulties encountered in determining the chemical features of a
plant species is the variation found in content of secondary metabolites. This can
be due to many factors, which are not necessarily genetic, including environment, the organ analysed, the stage of development of the plant and the particular plant population analysed. Thus, before any important conclusions are drawn
about the presence or absence of a particular compound, adequate sampling of a
large number of members of the species, at different stages of development and
growing in different environments, should be made. Also, when applying analytical
results to systematics, all species in a genus should be investigated. When such
criteria are applied to the results of chemical surveys obtained so far, it is found
that most fall far short of these ideals and chemosystematics is still in its infancy.

Plant Tissue Cultures

Plant tissue cultures (cell suspension cultures) are theoretically the ideal medium
for studying metabolism and the biosynthetic pathways producing secondary
metabolites. Such cultures are grown from callus tissue kept in a sterile, liquid
medium which contains the necessary nutrients. Growth is at first exponential
but eventually reaches a stationary phase, when batches of cells can be transferred to a new medium. Thus, the number of cells can be continually increased,
a distinct advantage if plant tissue cultures are used on an industrial scale.

It is often found that a compound characteristic of the intact plant is only
synthesised after differentiation of aggregated cells into developing roots, buds,
etc. Thus, with tissue cultures of deadly nightshade (*Atropa belladonna*, Solanaceae), initiation of root formation precedes tropane alkaloid synthesis, while in
tobacco (*Nicotiana tabacum*, Solanaceae) cultures bud formation precedes
nicotine (1.29) synthesis. However, suspension cultures of some plants produce
characteristic metabolites without such differentiation taking place. For example,

diosgenin
(1.43)

Dioscorea (Dioscoreaceae) and *Solanum* (Solanaceae) callus cells biosynthesise diosgenin (1.43), while those of rue (*Ruta graveolens,* Rutaceae) synthesise the volatile oil characteristic of the plant. However, rose (*Rosa* sp., Rosaceae) suspension cultures do not produce the terpenoids characteristic of rose oil.

It has been found that cultured cells rarely mimic the intact plant and often the secondary metabolites produced by these cells are different from those synthesised by the plant. However, the sterility and controlled environment, and the fact that only one type of cell is involved, makes interpretation of results more certain than when working with intact plants. It is also possible to obtain colonies originating from a single cell; such colonies often differ from one another in their biochemical properties. Mutants containing a blocked biosynthetic pathway are particularly valuable in studying metabolism.

Although not yet as useful as cultures of microorganisms for producing secondary metabolites on an industrial scale, plant tissue cultures are potential sources of such compounds, particularly steroids and alkaloids.

The Investigation of Biosynthetic Pathways by Isotopic Tracer Analysis

Although sequential analysis (analysis of the formation of secondary metabolites over a period of time), enzymology and the use of microbial auxotrophic mutants (organisms containing a blocked biosynthetic pathway) have contributed much to our knowledge of biosynthetic pathways, the most important method for such investigations today is isotopic tracer analysis. Until a few years ago, most work involved the use of precursors containing radioactive ^{14}C or tritium (^{3}H). Since the development of ^{13}C nuclear magnetic resonance (CMR) spectroscopy, however, the stable nuclide ^{13}C has featured prominently, while ^{18}O and ^{14}N have also been used.

No single isotope possesses all the desirable properties required of a tracer. Radioactive isotopes are preferred for quantitative analysis, or if the conversion of precursor to product is low. However, unless the labelled product can be readily degraded to an identifiable fragment, it is difficult to be precise about the position of the label. The advantage of CMR spectroscopy is that the distribution of labelled atoms within a molecule can be determined accurately without the necessity for degradation. This gives information on the mechanisms of precursor incorporation, the stereospecificity of the reaction and any randomisation (scrambling) of label that may have taken place due to degradation of the precursor. However, the labelled metabolite must be obtained in a pure form.

When administering labelled precursors to plants, it is important to ensure that the compound reaches the site of metabolism. Gases such as $^{14}CO_2$ are administered in sealed polythene bags, while solid substrates are dissolved in water or emulsified with the help of surface-active agents such as detergents. For plants rooted in soil, wick feeding is generally employed. This consists of threading non-mercerised cotton through the stem, the ends being immersed in a small volume of solution in a container taped to the stem. The substrate may also be given to individual parts of a plant by, for instance, dipping the cut ends of stems into the solution, by injecting the solution into the stem or seed pod or by adding the compound to the nutrient solution of tissue cultures.

Although well absorbed by the plant, the labelled precursor may fail to reach the biosynthetic site because of permeability barriers between and within cells. Phosphate esters, for example, can fail to penetrate cell walls, while both acetate and mevalonate are unable to cross certain permeability barriers. Thus, negative results in biosynthetic studies may only be due to the existence of such barriers.

The effectiveness of a precursor in a biosynthetic pathway may be measured by dilution value and degree of incorporation. The dilution value is defined as A_1/A_2, where A_1 is the specific activity mol^{-1} of the precursor and A_2 is the specific activity mol^{-1} of the product. In a reaction sequence A→B→C→D, pools of B, C and D may exist such that the label from A appearing in D will be diluted, the dilution depending on the respective sizes of the B, C and D pools. The longer the reaction sequence, the higher will be the dilution value. A disadvantage of using dilution value is that pool sizes may vary between organisms, making comparison difficult. However, the final product D does not have to be recovered quantitatively to determine A_2, as it is activity mol^{-1} which is measured.

In principle, the disadvantage of dilution value can be overcome by calculating the degree of incorporation. This is defined as $(A_2 M_2/A_1 M_1)$ x 100, where A_1 and A_2 are the specific activities as before and M_1 and M_2 are the molar quantities of precursor and product, respectively. However, to determine M_2 quantitative recovery is necessary, while radiochemical purity is essential to obtain an accurate value for A_2.

In theory, the position of a label should be the same in the product as in the starting material, but in practice some randomisation may occur due to degradation and reformation of the precursor before it is incorporated into the product. Thus, if methyl-labelled acetic acid (1.44) is the precurosr of squalene (1.45), scrambling of label occurs and some carboxyl-derived atoms are also labelled (*Figure 1.11*).

One of the most successful applications of ^{14}C to biosynthesis is that proving that both rings of the alkaloid, anatabine (1.46), are derived from nicotinic acid

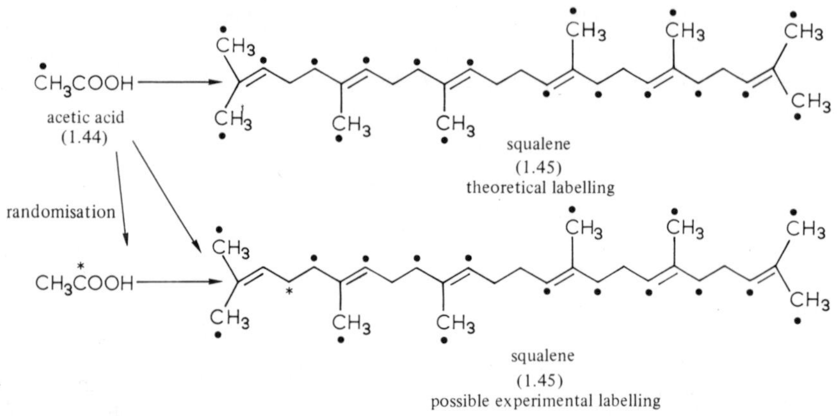

Figure 1.11 The labelling patterns of squalene.

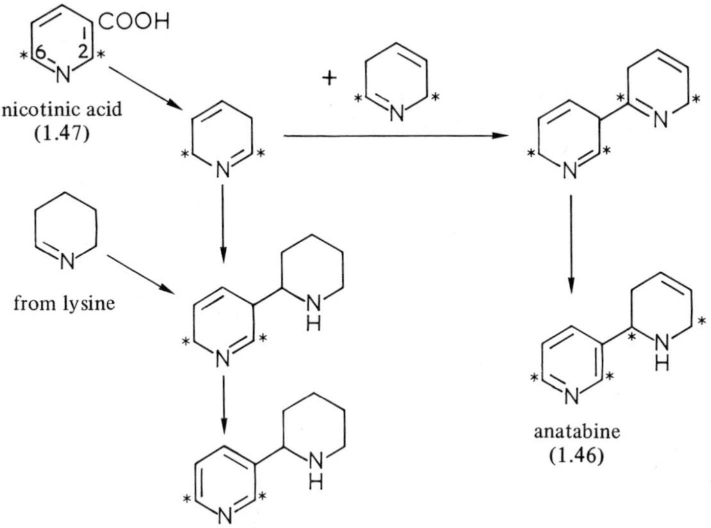

Figure 1.12 The biosynthesis of anatabine and anabasine.

(1.47). This is in contrast to the related anabasine (1.48), in which it had been previously established that the pyridine ring is derived from nicotinic acid, while the piperidine ring is formed from lysine (*Figure 1.12*). Three separate experiments involving three *Nicotiana* species, in which varying feeding times and weights of nicotinic acid were used, showed that both rings of anatabine were equally labelled. It was also found that no randomisation of labelling between C-2′ and C-6′ derived from [2-^{14}C] and [6-^{14}C] nicotinic acid took place. The former gave [2,2′-^{14}C], and the latter [6,6′-^{14}C] anatabine.

Multiple labelling of precursors with the same or different isotopes detects any *in vivo* cleavage between the labelled atoms. When applied to ^{14}C analysis, this method has the further advantage that the multiple labels need not be incorporated into the same molecule. Mixtures in which the precursor molecules are labelled singly but in different positions can be used. However, in the application of CMR spectroscopy, the two ^{13}C atoms must be adjacent atoms in the same molecule, so that resonance splitting can be detected.

Multiple labelling has revealed several inaccuracies in the interpretation of earlier work. Thus, in the synthesis of eugenol (1.49) in basil (*Ocimum basilicum*, Umbelliferae), it had been proposed that C-1 was lost and subsequently replaced by a carbon atom from methionine. However, work using doubly labelled glucoferulic acid (1.50), in which the methoxyl group was labelled with ^{3}H, and C-1 with ^{14}C, showed that eugenol (1.49) and methyleugenol (1.51) were synthesised with no loss of carbon at any stage (*Figure 1.13*). The ^{3}H:^{14}C ratios in the starting material and products were the same and 97% of the ^{14}C label was found at C-1 of eugenol. The feeding of [1-^{14}C^{3}H$_2$OH] coniferin (1.52) gave eugenol with no change in the ^{14}C:^{3}H ratio, indicating that both hydrogen atoms at C-1 of coniferin are retained in its conversion to eugenol.

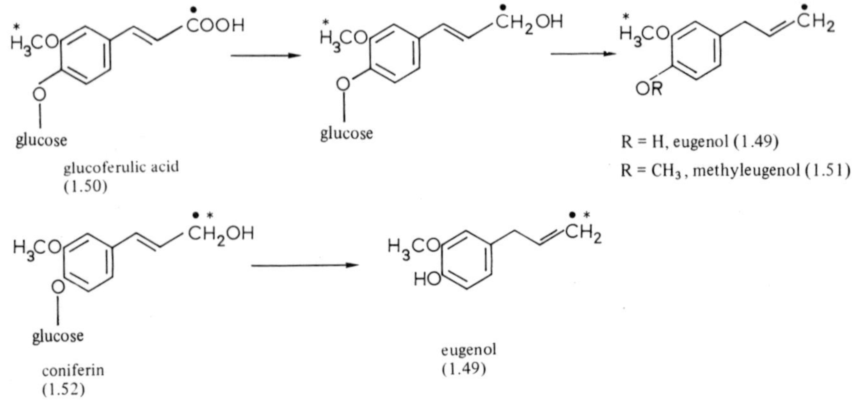

Figure 1.13 The biosynthetic pathway of eugenol.

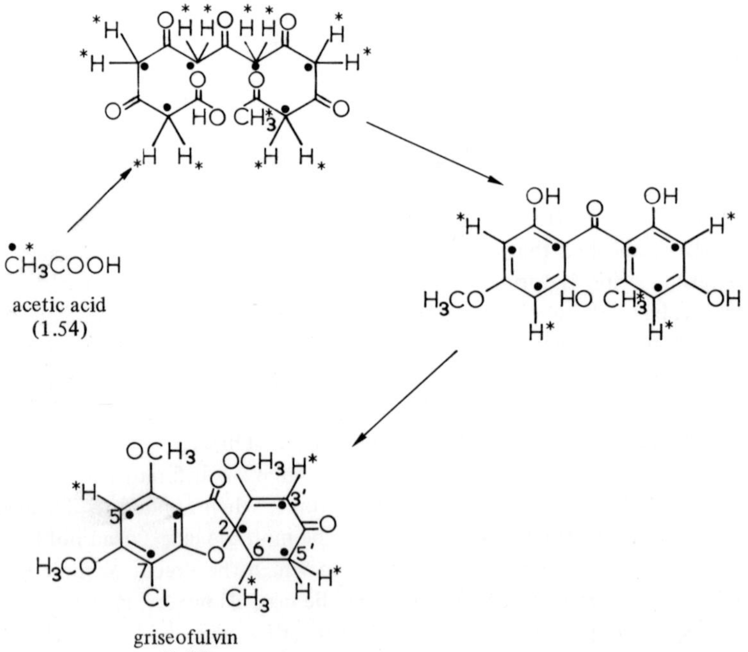

Figure 1.14 The formation of griseofulvin.

Multiple labelling of precursors also destroyed the hypothesis that griseofulvin (1.53) and other heptaketides were biosynthesised by the cleavage of a carbocyclic molecule rather than by folding of a carbon chain. When *Penicillium urticae* cultures were treated with [2-^{3}H,2-^{14}C] acetate (1.54), degradation of the griseofulvin formed showed that three of the ^{3}H atoms were located on the 6'-methyl group, while the remainder were at the 5', 3' and 5 positions, as would be expected if a polyketide chain was folded as shown in *Figure 1.14*.

NMR (nuclear magnetic resonance) spectroscopy employing ^{13}C (also known as CMR spectroscopy) has been used to study the biosynthesis of fungal and bacterial metabolites, as sufficient incorporation of ^{13}C and yields of products are obtained for the latter to be isolated in a pure state. Although incorporation and yields are much smaller in higher plants, CMR spectroscopy has the advantage that no degradation of product is necessary, and it will doubtless prove a useful tool in future investigations.

The principles underlying ^{13}C and ^1H (proton) NMR spectroscopy are similar and it is assumed that the reader is familiar with the latter (if not, see *Proton and Carbon-13 N.M.R. Spectroscopy*, listed in the bibliography). Only those nuclei which have an intrinsic nuclear spin give an NMR spectrum, and besides ^1H those commonly encountered include ^2H, ^{13}C, ^{14}N, ^{15}N, ^{17}O and ^{19}F. ^{12}C, ^{16}O and ^{32}S have zero spin and therefore cannot be used in NMR spectroscopy. The low abundance of ^{13}C (1.11%) in naturally occurring carbon and the poor sensitivity of the NMR technique to ^{13}C (about 5700 times less than to ^1H) has prevented the use of CMR in the past. However, techniques have now been devised to overcome these problems.

The first biosynthetic studies using ^{13}C utilised the ^{13}C satellites in the ^1H NMR spectrum, which arose through coupling of ^{13}C with ^1H. This method has several limitations however, as only carbon atoms to which one or more hydrogen atoms are attached can be studied. Often, complex spectra result in which the satellite bands are undetectable, and spinning sidebands may also obscure the satellite bands. However, the ^{13}C satellite method has been used in the study of griseofulvin (1.53) synthesis in cultures of *Penicillium urticae*. When griseofulvin biosynthesised from 55% enriched $[2\text{-}^{13}C]$ acetate was compared with the unenriched compound, it was found that the satellites from the C-6' methyl group were about three times as intense in the enriched metabolite. Similar increases were observed for the C-3' and C-5 proton signals. The proton satellite signals for the $-OCH_3$ groups showed no increase in intensity, as expected, since it had been previously shown that these CH_3 groups were derived from the single carbon pool. These results are in agreement with the work using ^{14}C described above. Later work employing CMR spectroscopy showed that griseofulvin is formed from seven acetate units with C-2 incorporated into the 2, 5, 7, 3', 5' and 6' positions and also into one of the ring junction carbon atoms (*Figure 1.14*).

Bibliography

Primary Metabolism

Primary Metabolism, J. Staunton, Oxford University Press, 1978
Plant Biochemistry, 3rd edn, J. Bonner and J. E. Varner (eds.), Academic Press, 1976
Cellular Biochemistry and Physiology, N. A. Edwards and K. A. Hassall, McGraw-Hill, 1971
Introduction to Modern Biochemistry, 4th edn, P. Karlson, Academic Press, 1975

Functions of Secondary Metabolites

Phytochemical Ecology, J. B. Harborne (ed.), Academic Press, 1972
Introduction to Chemical Ecology, M. Barbier, Longman, 1979
Introduction to Ecological Biochemistry, J. B. Harborne, Academic Press, 1977
Allelopathy, E. L. Rice, Academic Press, 1974

Uses of Secondary Metabolites

'Importance of Secondary Plant Constituents as Drugs', in *Phytochemistry*, vol. 3,
    L. P. Miller (ed.), Van Nostrand Reinhold Co., 1973
'Importance of Plant Chemicals in Human Affairs', *ibid.*
Plant Products of Tropical Africa, M. L. Vickery and B. Vickery, The Macmillan
    Press, 1979

Chemosystematics

'Methods of Classical Plant Taxonomy', in *Chemical Plant Taxonomy,* T. Swain
    (ed.), Academic Press, 1963
'History of Chemical Taxonomy', *ibid.*
Chemistry in Botanical Classification, G. Bendz and J. Santesson (eds.), Academic
    Press, 1974
The Chemotaxonomy of Plants, P. M. Smith, Edward Arnold, 1976
'Biochemical Evolution of Plants', in *Comprehensive Biochemistry,* vol. 29A,
    M. Florkin and E. H. Stotz (eds.), Elsevier, 1974
Chemistry in Evolution and Systematics, T. Swain (ed.), Butterworths, 1973
'Recent Developments in Molecular Taxonomy', in *Perspectives in Phytochemistry,*
    J. B. Harborne and T. Swain (eds.), Academic Press, 1969
'Chemical Evidence for the Classification of Some Plant Taxa', *ibid.*
'Phytochemistry and Taxonomy', in *Comparative Phytochemistry,* T. Swain
    (ed.), Academic Press, 1966
'Biogenetic Classification of Plant Constituents', *ibid.*
'Molecular Taxonomy', in *Phytochemistry, ibid.*
'Biochemical Systematics', in *Plant Taxonomy,* 2nd edn, V. H. Heywood, Edward
    Arnold, 1976

Tissue Cultures

Tissue Cultures and Plant Science, H. E. Street (ed.), Academic Press, 1974
'Tissue Cultures', in *Recent Advances in Phytochemistry,* vol. 2, M. K. Seikel
    and V. C. Runeckles (eds.), Appleton (N.Y.), 1969
'Plant Cell Cultures: Their Potential for Metabolic Studies', in *Biosynthesis and
    its Control in Plants,* B. V. Milborrow (ed.), Academic Press, 1973

Isotopic Tracer Analysis

'Methodology', in *Biosynthesis,* vol. 1, Specialist Periodical Reports, T. Swain
    (ed.), The Chemical Society, 1972
'Stable Isotopes in Biosynthetic Studies', *ibid.,* vols. 2 and 3, 1973 and 1975
'N.M.R. with Stable Isotopes in Biosynthetic Studies', *ibid.,* vol. 4, 1976

Proton and Carbon-13 N.M.R. Spectroscopy, R. J. Abraham and P. Loftus, Heyden, 1978

'Methods for Investigation of Biosynthesis in Higher Plants', in *Progress in Phytochemistry,* vol. 3, L. Reinhold and Y. Liwschitz (eds.), Interscience, 1972

2 Sugar Metabolism

Introduction

Sugar metabolism in plants is such a vast subject that we cannot hope to include every aspect in this chapter. In order to cover as wide a field as possible, carbohydrate types, rather than individual compounds, will be discussed. However, exceptions are made for the more important carbohydrates such as fructose, glucose, sucrose, ascorbic acid, starch and cellulose. More detailed information can be obtained from the books listed in the bibliography.

Plants convert carbon dioxide and water into sugars by the primary process of photosynthesis. Animals are unable to synthesise sugars (except in small amounts by the process of gluconeogenesis, which reconverts to glucose products resulting from the degradation of sugars). Thus, animals depend entirely on plants for their energy requirements and the photosynthetic biosynthesis of sugars is the primary process on which the lives of all higher plants and animals depend.

Photosynthesis

The photosynthetic process takes place in two stages — the light reactions and the dark reactions. The light reactions are concerned with the absorption of energy from sunlight by chlorophyll a molecules (see Chapter 10 for a description of chlorophyll), and the transfer of this energy to ATP (adenosine triphosphate) and NADPH (reduced nicotinamide adenine dinucleotide phosphate). These important compounds are also further discussed in Chapter 10.

The dark reactions can take place in the absence of light. Plants have two pathways of carbon dioxide assimilation — the reductive pentose phosphate, or Calvin, cycle and the pyruvate–malate, or Slack and Hatch, pathway. The Calvin cycle is the more important as this takes place in all higher plants, whereas the Slack and Hatch pathway has only been detected in some tropical plants. The energy stored in ATP and NADPH is used in the dark reactions to convert carbon dioxide to fructose-6-phosphate (in the Calvin cycle) or malic acid (in the Slack and Hatch pathway).

The key intermediate of the Calvin cycle (*Figure 2.1*) is ribulose-1,5-diphosphate (2.1), which is regenerated at the end of the cycle. As only one molecule of carbon dioxide is reduced during each complete cycle, theoretically it takes six cycles, 18 molecules of ATP and 12 molecules of NADPH to form a completely new fructose-6-phosphate molecule. Further details of the light reactions, the Calvin cycle and the Slack and Hatch pathway can be found in such advanced textbooks of biochemistry as *Plant Biochemistry* listed in the bibliography.

It can be seen from *Figure 2.1* that the sugars involved in the Calvin cycle — ribulose, glycerate, glyceraldehyde, fructose, erythrose, sedoheptulose, ribose and xylulose — contain phosphate groups. Free sugars are unreactive compounds and

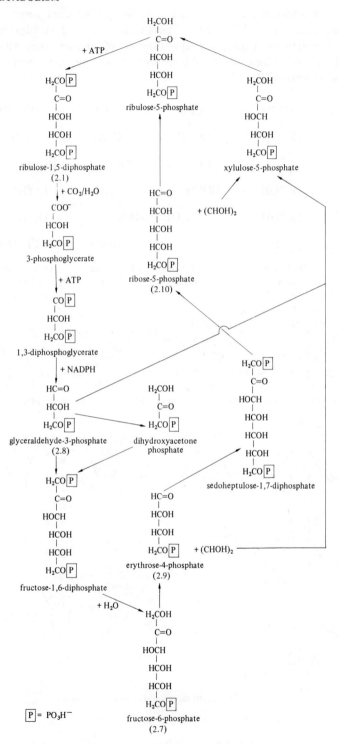

Figure 2.1 The reductive pentose phosphate cycle (Calvin cycle).

only small amounts are present in plants, the predominant compounds being glucose (2.2), fructose (2.3) and the disaccharide, sucrose (2.6). Sugar metabolism in plants is active and there is a rapid turnover of intermediates. All sugars concerned in metabolism are combined with phosphate, the phosphate groups originating from ATP.

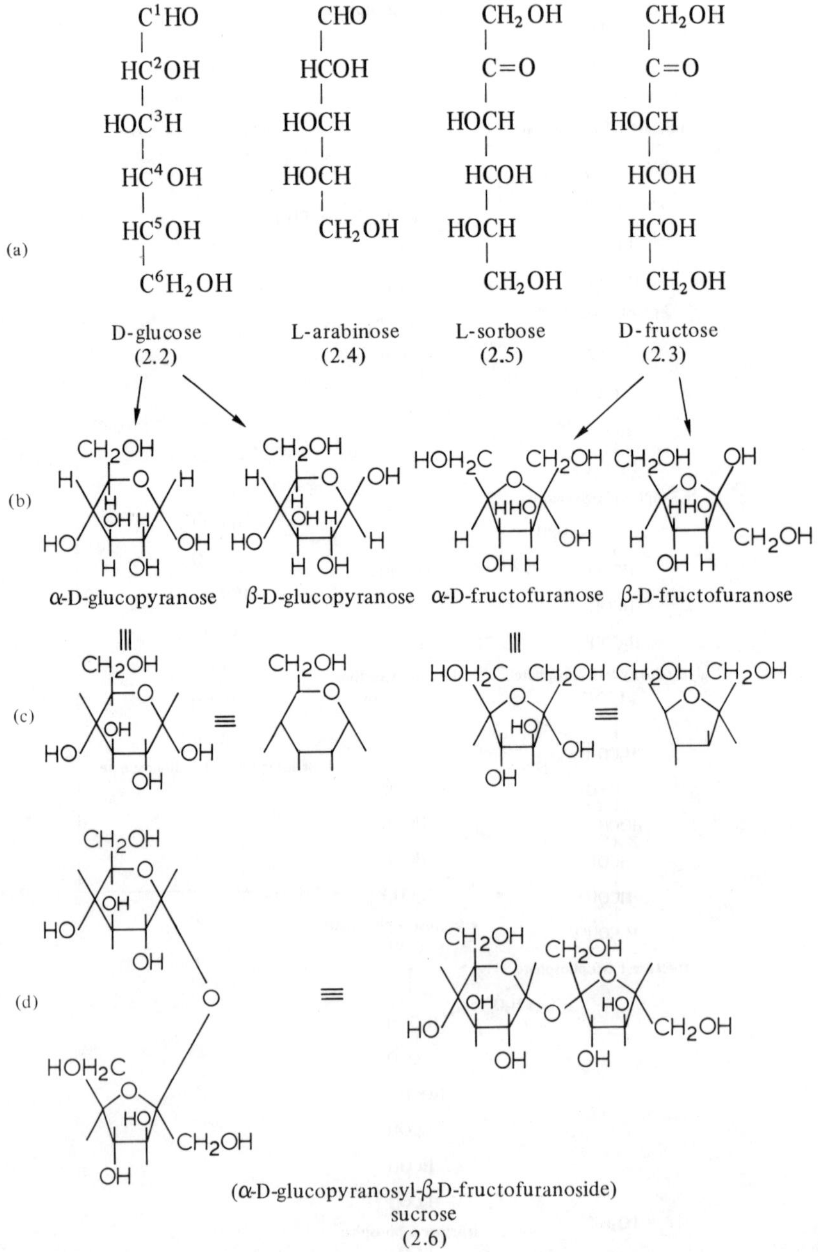

Figure 2.2 Sugar structures.

The monosaccharides, glucose and fructose, can be written in their straight-chain forms as shown in *Figure 2.2*. Conventionally, all D-sugars are written with the hydroxyl group on the final asymmetric (chiral) carbon atom (C-5 in glucose and fructose) to the right. L-sugars (e.g. L-arabinose (2.4) and L-sorbose (2.5)) have this hydroxyl group to the left. Sometimes (+) or (–) is added to the name of a sugar to indicate the optical rotation. Thus, we have D(+)glucose and D(–)fructose. In plants, sugars usually exist in their pyranose or furanose cyclic forms, glucose generally forming a pyranose ring and fructose a furanose ring (*Figure 2.2*). In their cyclic forms, these sugars are properly called glucopyranose and fructofuranose, respectively, but this convention is not always adhered to, and, in this book, sugars will be referred to by their trivial names unless the ring system is of especial importance.

The hydroxyl group on the carbon atom adjacent to the ring oxygen atom of the cyclic sugars can be in either the α or β position (*Figure 2.2b*). As we shall see later, the α or β position of this hydroxyl group is important when monosaccharides link up through these hydroxyl groups to form polysaccharides (e.g. sucrose (2.6) is α-D-glucopyranosyl-β-D-fructofuranoside). If the α or β position of the hydroxyl group is not known, then it is usually written –H,OH or ∿OH. To simplify the drawing of sugar formulae, two shorthand notations are in common use (*Figure 2.2c*). Either the hydrogen atoms are represented by single lines, or the hydrogen atoms are omitted altogether and the hydroxyl groups written as single lines. We shall use the first representation. (The single lines should not be confused with those used to represent $-CH_3$ in terpenoid compounds).

Glucose-6-phosphate (2.11) is derived from fructose-6-phosphate (2.7), the end-product of photosynthesis and it can be seen from *Figure 2.3* that glucose-6-phosphate is the precursor of all secondary metabolites in plants, as, in every case, the biosynthetic pathway can be traced back to this sugar. In this chapter, we shall consider the various types of carbohydrates which can be derived from glucose-6-phosphate. These can be divided into the monosaccharide derivatives, the oligosaccharides, which contain two or more monosaccharide units, and the polysaccharides, high-molecular-weight polymers containing many monosaccharide units. The polysaccharides are further divided into storage carbohydrates and structural carbohydrates. Sugars occurring as glycosides can be either mono-saccharides or oligosaccharides and these are considered separately.

Monosaccharide Derivatives

Glucose-6-phosphate (2.11) can be converted to fructose-6-phosphate (2.7), glycer-aldehyde-3-phosphate (2.8), erythrose-4-phosphate (2.9) and ribose-5-phosphate (2.10) by the oxidative pentose phosphate cycle (hexose monophosphate shunt), which is the opposite of the Calvin cycle (*Figure 2.1*). Erythrose-4-phosphate is a precursor of the shikimic acid pathway to the formation of aromatic compounds, described in Chapter 6, while ribose-5-phosphate forms part of the ribonucleic acids and is a precursor of the deoxyribonucleic acids and the purine derivatives described in Chapter 10.

Over 100 monosaccharides have been found to occur in plants, most as

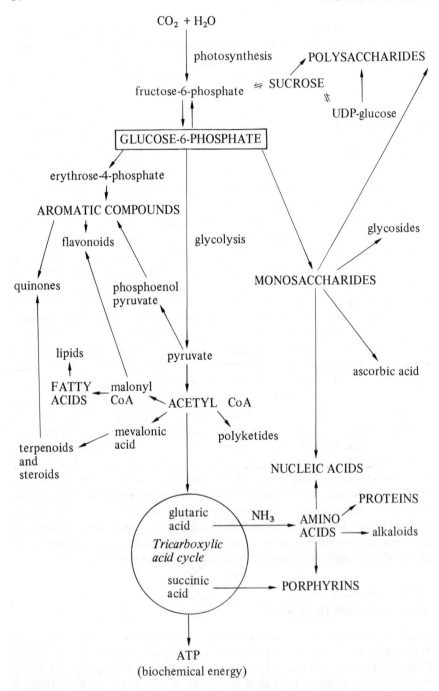

Figure 2.3 Glucose-6-phosphate as a precursor of secondary metabolites.

constituents of glycosides (see below). The more common of these sugars are known to be derived from glucose-6-phosphate, as shown in *Figure 2.4*, and it is probable that all are ultimately derived from this compound. Before interconversion of monosaccharides can occur, they must be converted to their nucleotide glycosides, the most common being UDP (uridine diphosphate) and GDP (guanine diphosphate) derivatives. The nucleotide moieties of these glycosides are described in Chapter 10.

An enzyme preparation from tobacco (*Nicotiana tabacum*, Solanaceae) catalyses the conversion of UDP-D-glucose to UDP-L-rhamnose.

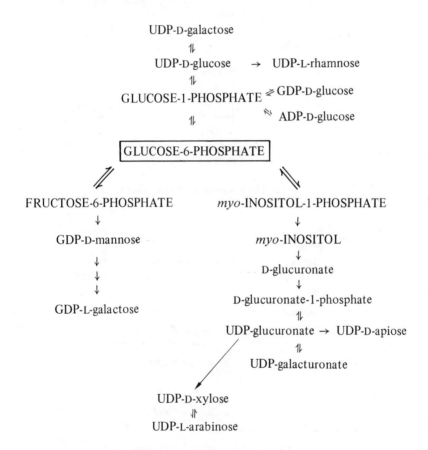

Figure 2.4 Monosaccharides derived from glucose-6-phosphate.

Myo-inositol

One of the most important derivatives of glucose-6-phosphate (2.11) is the cyclitol *myo*-inositol (2.12), a sugar alcohol found in all plants. The enzyme catalysing the conversion of D-glucose-6-phosphate to *myo*-inositol-1-phosphate (2.13), *myo*-inositol-1-phosphate synthase, has been isolated from sycamore (*Acer pseudoplatanus*, Aceraceae) cell cultures. The cyclisation is an oxidation–reduction reaction dependent on NAD^+ (*Figure 2.5*).

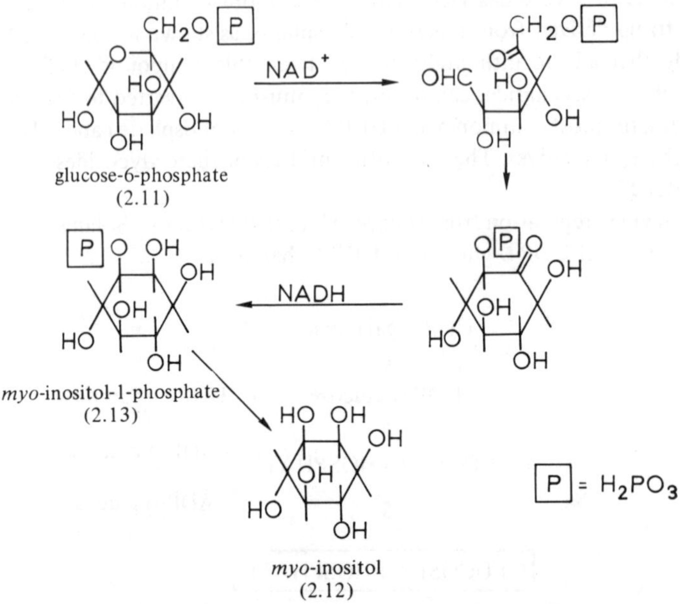

Figure 2.5 The biosynthesis of *myo*-inositol.

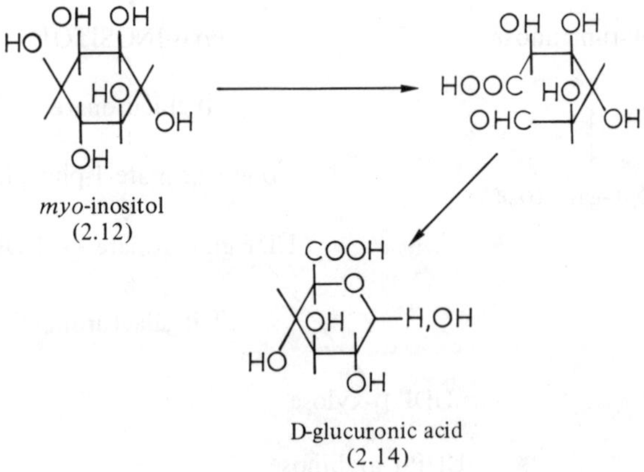

Figure 2.6 The biosynthesis of D-glucuronic acid.

The oxidation of *myo*-inositol to D-glucuronic acid (2.14), a reaction catalysed by oxygenase, is unique to plants (*Figure 2.6*), the more usual pathway to the formation of glucuronate being the oxidation of UDP-glucose. However, it has been clearly shown in several plants that the pentose units of cell wall poly-saccharides, which are derived from UDP-D-glucuronate, incorporate activity from isotopically labelled *myo*-inositol.

Through its conversion to glucuronic acid, *myo*-inositol is also an intermediate in the biosynthesis of the branched sugar, D-apiose (2.15), from D-glucose-6-phosphate. The conversion of the UDP derivative of glucuronic acid (2.14a) to

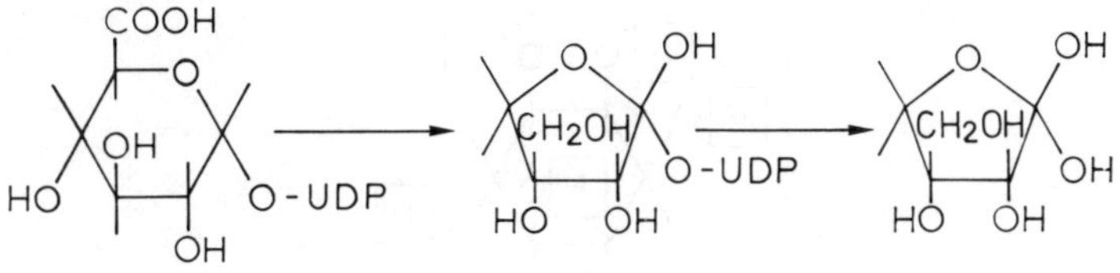

UDP-glucuronic acid
(2.14a)

apiose
(2.15)

Figure 2.7 The biosynthesis of apiose.

UDP-apiose (*Figure 2.7*) is catalysed by UDP-glucuronate cyclase, which has been isolated from duckweed (*Lemna minor*, Lemnaceae), a plant containing apiose in its cell wall polysaccharides.

Apiose was first isolated from apiin, a glycoside occurring in parsley (*Petroselinum crispum* syn. *Apium petroselinum*, Umbelliferae) and this sugar was once considered rare. However, it is now known to be widespread, occurring in the cell wall polysaccharides of many plants.

In contrast to apiose, the branched sugar, hamamelose (2.16), does not have *myo*-inositol as an intermediate in its biosynthesis. Tracer work has shown this sugar to be closely connected with photosynthesis and it has been suggested that it is biosynthesised from glyceraldehyde-3-phosphate (2.8), as shown in *Figure 2.8*.

Hamamelose was first isolated from witch hazel (*Hamamelis virginiana*, Hamamelidaceae), where it occurs in tannin. Like apiose, it was once thought to be a rare sugar but has since been isolated from many plants. Some *Primula* species (Primulaceae), in particular, accumulate derivatives of hamamelose.

An important derivative of *myo*-inositol, the hexaphosphate, phytic acid (2.17), is a storage phosphate compound in plants. Phytic acid is biosynthesised from *myo*-inositol-1-phosphate (2.13), the phosphate groups being derived from ATP and added one at a time. The reaction is catalysed by phosphoinositol kinase. The reverse process takes place when phosphate is required for the biosynthesis of other compounds, the stepwise hydrolysis being catalysed by phytases.

It was shown in germinating wheat (*Aestivum sativum*, Gramineae) that phytic acid gradually disappeared over 14 days. However, an increase in *myo*-inositol was

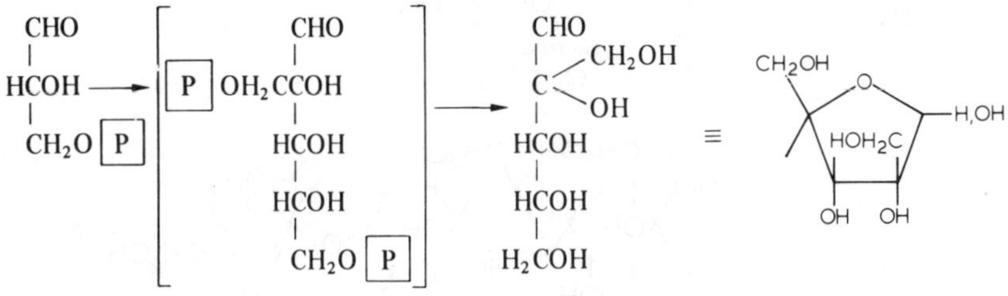

glyceraldehyde-
3-phosphate
(2.8)

hamamelose
(2.16)

Figure 2.8 The biosynthesis of hamamelose.

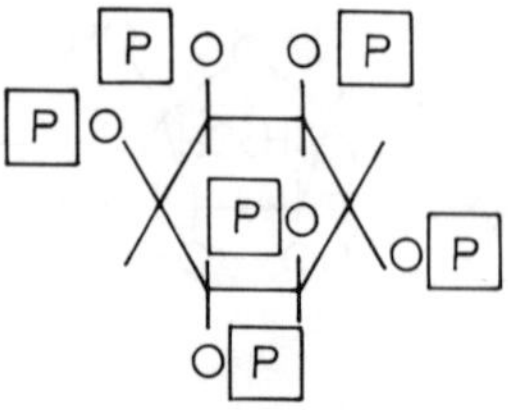

phytic acid
(2.17)

not detected, indicating that this compound was actively metabolised to other carbohydrates.

Myo-inositol synthesised in the chloroplasts is required to transport sugar residues through the chloroplast membrane into the cytoplasm. This sugar alcohol also forms esters with the plant growth hormone 3-indolylacetic acid (auxin), which are possibly concerned in the modification of cell walls (see Chapter 8). Myo-inositol readily forms methylethers which are possibly the precursors of

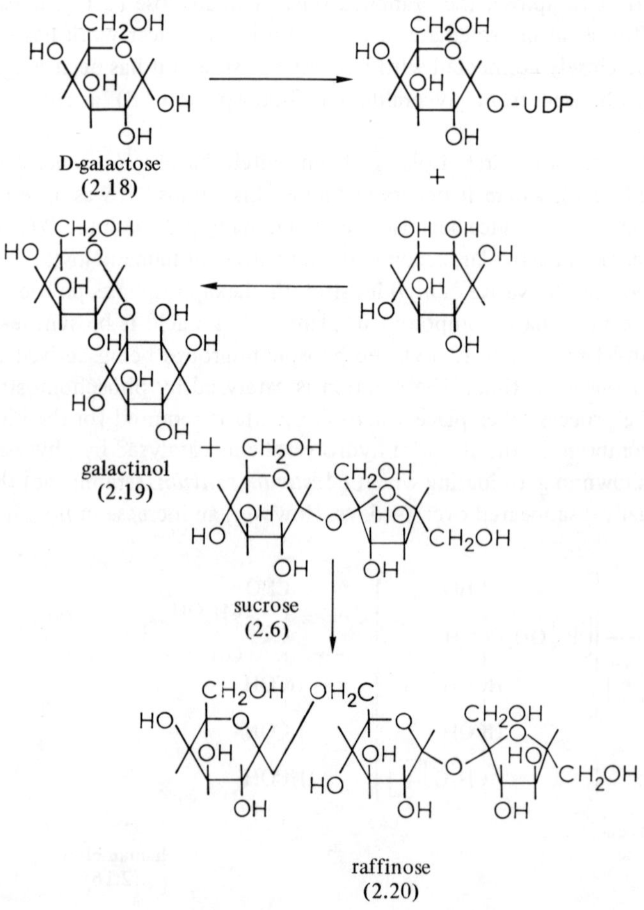

D-galactose
(2.18)

galactinol
(2.19)

sucrose
(2.6)

raffinose
(2.20)

Figure 2.9 The biosynthesis of raffinose.

some methylated polysaccharides, although it has been established that, in general, polysaccharides are methylated after formation by *S*-adenosylmethionine. *Myo*-inositol is also concerned in the biosynthesis of the complex glycolipids known as phosphoinositides. In combination with galactose (2.18), *myo*-inositol forms the complex galactinol (1-*O*-α-D-galactopyranosyl-*myo*-inositol) (2.19), which is found in plants accumulating sugars of the raffinose (2.20) series. Galactinol transfers galactosyl residues to sucrose (2.6), raffinose and higher homologues of raffinose (*Figure 2.9*). The reaction, which is reversible, is catalysed by trans-ferases.

Isomeric inositols, derived from *myo*-inositol, occur in plants, but in general they are of lesser importance. (Sugar alcohols in general are discussed in the section on carbohydrates in chemosystematics, p.53.) An amino derivative of the isomer *scyllo*-inositol (scyllitol) occurs in the antibiotic streptomycin (2.21), a trisaccharide obtained from the fungus *Streptomyces griseus*. The streptidine (2.22) portion of the antibiotic is biosynthesised from *myo*-inositol (2.12) and arginine (2.23) through the intermediate formation of streptamine (2.24), as shown in *Figure 2.10*.

Many other antibiotics containing amino sugars have been isolated from fungi.

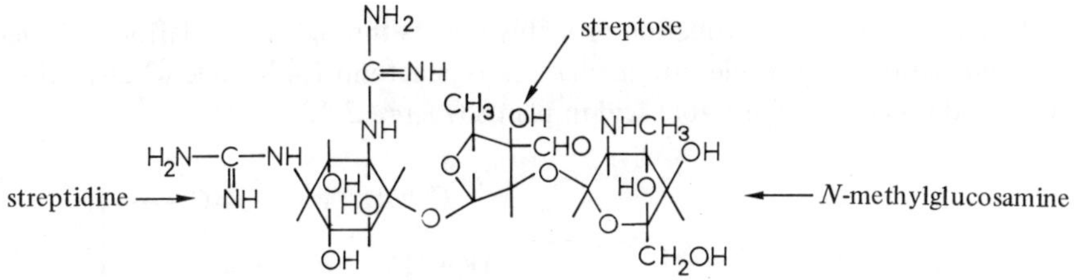

streptomycin
(2.21)

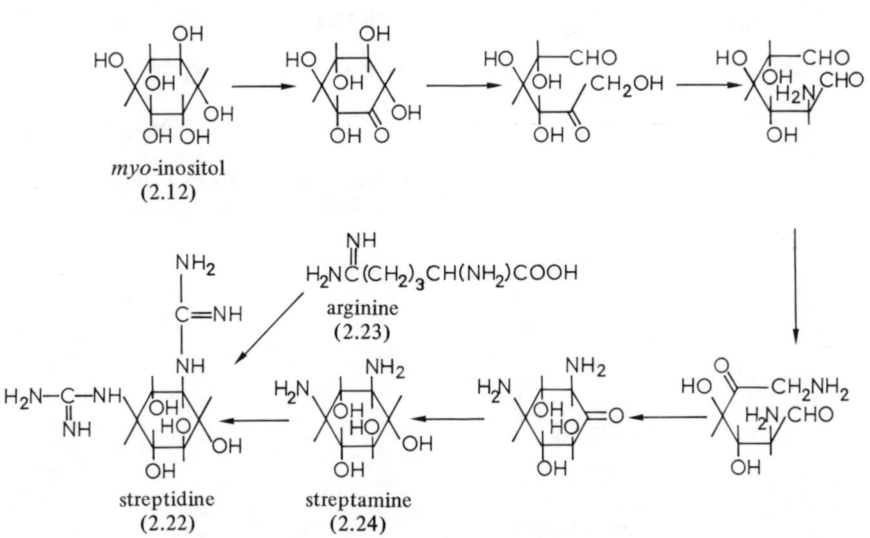

myo-inositol
(2.12)

streptidine
(2.22)

streptamine
(2.24)

Figure 2.10 The biosynthesis of streptidine.

It seems likely that these plants produce such compounds as a protection against bacterial attack.

Streptomycin and the other amino sugar antibiotics inhibit protein synthesis by acting on ribosomes; in overdoses, therefore, they are toxic to animals as well as bacteria. Streptomycin is only used nowadays to treat tuberculosis, but gentamicin is a broad-spectrum agent effective against several *Staphylococcus* infections.

Ascorbic acid

Animals convert D-glucose (2.2) to L-ascorbic acid (2.25) through the intermediates, D-glucuronic acid (2.14), L-gulonic acid (2.26) and 2-oxo-L-gulonolactone (2.27) (*Figure 2.11*). Although plants are able to metabolise these uronic acid derivatives to ascorbic acid, work with isotopically labelled intermediates indicates that such derivatives are not the true intermediates when glucose or galactose are the precursors.

The detailed pathway to ascorbic acid synthesis in plants is not known, but any postulated scheme must take into account the oxidation of C-1 of glucose to the lactone group of ascorbic acid and the change from D to L configuration at C-5. The C-2=C-3 diol formation probably occurs through an oxidation followed by enolisation. A possible intermediate is L-galactono-1,4-lactone (2.28), which is a good precursor of ascorbic acid in plants (*Figure 2.12*).

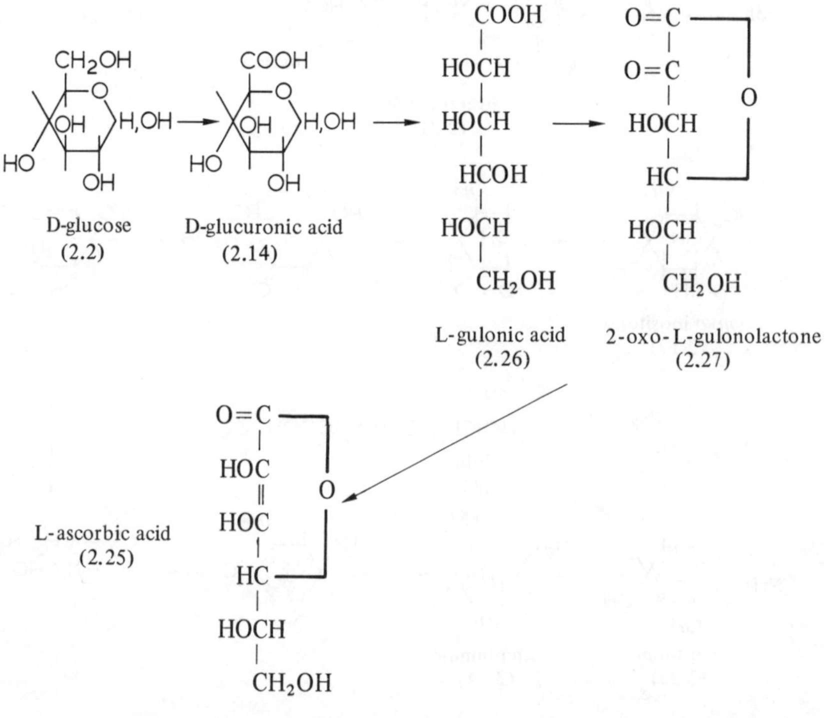

Figure 2.11 The biosynthesis of ascorbic acid in animals.

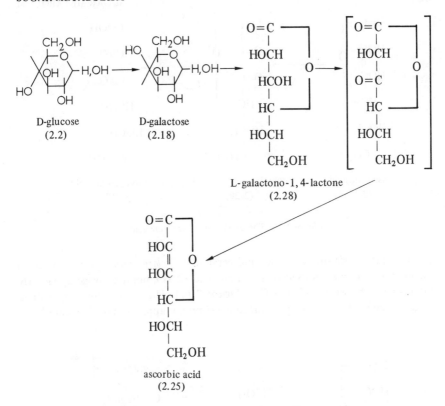

L-galactono-1, 4-lactone
(2.28)

ascorbic acid
(2.25)

Figure 2.12 The possible biosynthesis of ascorbic acid in plants.

Ascorbic acid, also known as vitamin C, is an important intermediate in carbo-hydrate metabolism in plants and animals. Man, monkey and the guinea pig are the only animals unable to synthesise ascorbic acid from L-gulonolactone, pre-sumably through a gene-controlled enzyme deficiency. Thus, these animals suffer scurvy unless exogenous vitamin C is provided. The fundamental effect of lack of vitamin C is the disappearance of collagen fibres, which causes a weakening of the intercellular cement. This leads to the symptoms of scurvy (haemorrhage, weakening of bones and teeth, lack of healing of wounds, etc.). Vitamin C is also needed to fight infections, and large amounts are needed to overcome the trauma of shock.

In plants, the largest concentration of ascorbic acid is found in the chloroplasts, which accords with the suggestion that this sugar acid is an intermediate of carbo-hydrate metabolism.

Ascorbic acid is easily oxidised to dehydroascorbic acid (2.29) in plants and animals. This oxidation is reversible and thus dehydroascorbic acid has all the properties of vitamin C. In animals, dehydroascorbic acid is converted irreversibly to 2,3-dioxogulonic acid (2.30) (*Figure 2.13*), this compound being subsequently metabolised to glucose and other sugars. Man and some other animals also degrade ascorbic acid to oxalic acid, which is excreted in the urine.

$$
\begin{array}{ccc}
\text{O=C} \longrightarrow & \text{O=C} \longrightarrow & \text{COOH} \\
| & | & | \\
\text{HOC} \;\;\; \text{O} & \text{O=C} \;\;\; \text{O} & \text{C=O} \\
|| & | & | \\
\text{HOC} & \text{O=C} & \text{C=O} \\
| & | & | \\
\text{HC} \longleftarrow & \text{HC} \longleftarrow & \text{HCOH} \\
| & | & | \\
\text{HOCH} & \text{HOCH} & \text{HOCH} \\
| & | & | \\
\text{CH}_2\text{OH} & \text{CH}_2\text{OH} & \text{CH}_2\text{OH}
\end{array}
$$

ascorbic acid	dehydroascorbic acid	2,3-dioxogulonic acid
(2.25)	(2.29)	(2.30)

Figure 2.13 The oxidation of ascorbic acid.

In grapes (*Vitis* sp., Vitaceae) and geraniums (*Pelargonium* sp., Geraniaceae) ascorbic acid is metabolised to L-tartaric acid (2.31), bond cleavage apparently taking place between C-4 and C-5 (*Figure 2.14*). The two-carbon fragment is metabolised to glucose in grapes, but in *P. crispum* it appears as oxalic acid (2.32).

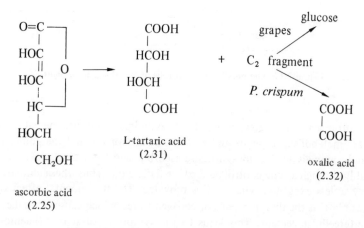

Figure 2.14 The degradation of ascorbic acid.

Glycosides

When sugars combine or react with non-carbohydrates, the resulting compound is a glycoside. However, compounds consisting of sugars only are generally referred to as disaccharides, trisaccharides, etc., and the term glycoside restricted to compounds in which one or more sugars are combined with a non-sugar, known as the genin or aglycon. The most common are *O*-glycosides where the sugar is linked through the oxygen atom of an alcohol or phenol. Phenolic glycosides are the most widespread of all such compounds. The sugar can also be linked directly to a carbon atom, when the compound is known as a *C*-glycoside, or to a nitrogen atom (*N*-glycosides) or sulphur atom (*S*-glycosides).

Many different types of compound provide the genins of glycosides, and such compounds appear throughout this book; some examples are phenols (Chapters 6 and 7), terpenoids and steroids (Chapter 5), and various nitrogen- and sulphur-containing compounds (Chapters 8 and 10).

Several suggestions have been made as to the function of glycosides in plants. They undoubtedly increase the solubility of the genins, and thus their main purpose may be to transport the genins in the plant sap. Glycoside formation also decreases the reactivity and/or toxicity of some genins, thus allowing such compounds to be stored or transported without harm to the plant.

D-Glucose (2.2) is the most common sugar to be found in glycosides. It is the only sugar occurring in the *S*-glycosides known as glucosinolates (Chapter 8) and is the major sugar found in the cyanogenic glycosides (Chapter 8). Other common sugars are D-galactose (2.18), D-xylose (2.33), L-rhamnose (2.34) and L-arabinose (2.4), but D-fructose (2.3) is a rare constituent of glycosides.

<pre>
      CHO   CHO
       | |
      HCOH   HCOH
       | |
      HOCH   HCOH
       | |
      HCOH   HOCH
       | |
      CH₂OH   HOCH
                               |
                              CH₃

    D-xylose   L-rhamnose
     (2.33)   (2.34)
</pre>

Some unusual monosaccharides are found amongst the cardiac glycosides including 6-deoxy (e.g. 6-deoxy-D-glucose (2.35)) and 2-deoxy (e.g. 2-deoxy-D-arabinohexose (2.36)) derivatives. The 2,6-dideoxy-sugar methylethers (deoxy-methylpentoses) which are common amongst the cardiac glycosides are not found

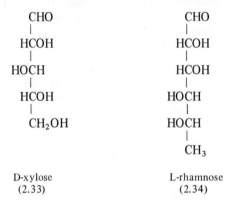

<pre>
   CHO   CHO   CHO   CHO   CHO
    | | | | |
   HCOH   CH₂   CH₂   CH₂   CH₂
    | | | | |
   HOCH   HOCH   HCOCH₃   H₃COCH   HCOCH₃
    | | | | |
   HCOH   HCOH   HCOH   HCOH   HOCH
    | | | | |
   HCOH   HCOH   HCOH   HCOH   HOCH
    | | | | |
   CH₃   CH₂OH   CH₃   CH₃   CH₃

 6-deoxy-D-   2-deoxy-D-   D-cymarose   D-oleandrose   L-oleandrose
  glucose   arabinohexose   (2.37)   (2.38)
  (2.35)   (2.36)
</pre>

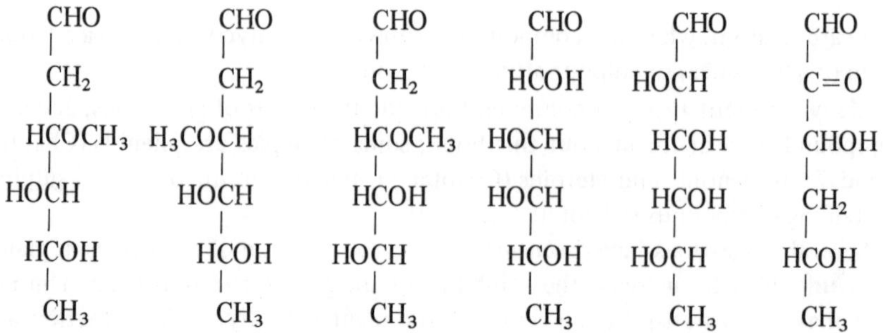

CHO	CHO	CHO	CHO	CHO	CHO
CH_2	CH_2	CH_2	HCOH	HOCH	C=O
$HCOCH_3$	H_3COCH	$HCOCH_3$	HOCH	HCOH	CHOH
HOCH	HOCH	HCOH	HOCH	HCOH	CH_2
HCOH	HCOH	HOCH	HCOH	HOCH	HCOH
CH_3	CH_3	CH_3	CH_3	CH_3	CH_3
D-sarmentose (2.39)	D-diginose (2.40)	L-diginose	D-fucose (6-deoxy-D-galactose) (2.41a)	L-fucose (2.41b)	4,6-dideoxy-hexosone (2.42)

in other types of glycoside. Some examples are D-cymarose (2.37), D- and L-oleandrose (2.38), D-sarmentose (2.39) and D- and L-diginose (2.40). D-Diginose occurs in African *Strophanthus* species (Apocynaceae) whereas the L-isomer occurs in Asian species. D-Fucose (2.41a) has been found only in cardiac glycosides, but L-fucose (2.41b) is widespread. Some interesting sugars found in the cardiac gly-

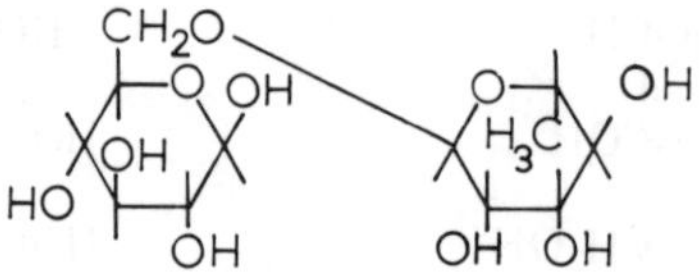

rutinose

(6-*O*-α-L-rhamnopyranosyl-β-D-glucopyranose)
(2.43)

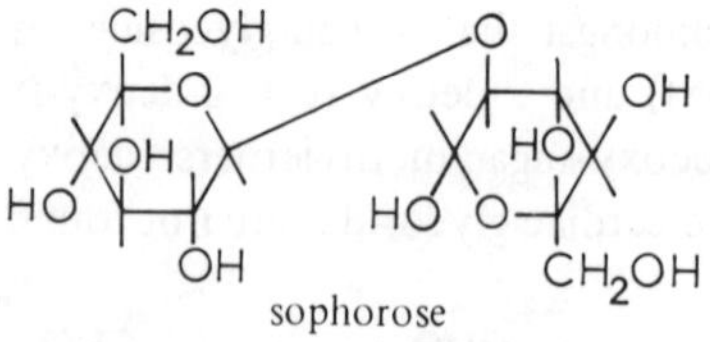

sophorose

(2-*O*-β-D-glucopyranosyl-β-D-glucopyranose)
(2.44)

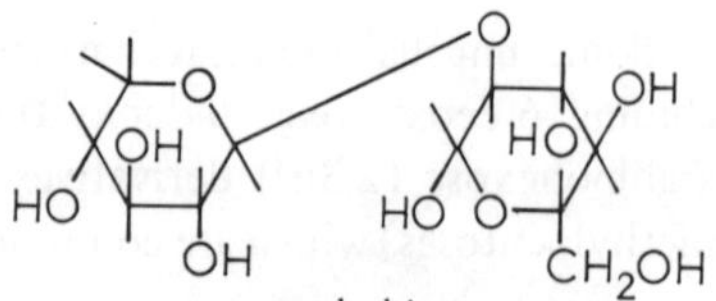

sambubiose

(2-*O*-β-D-xylopyranosyl-β-D-glucopyranose)
(2.45)

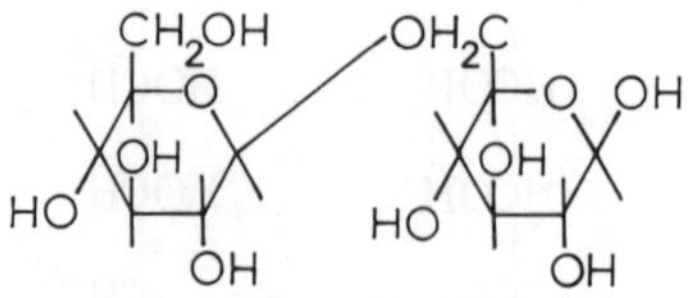

gentiobiose

(6-*O*-β-D-glucopyranosyl-β-D-glucopyranose)
(2.46)

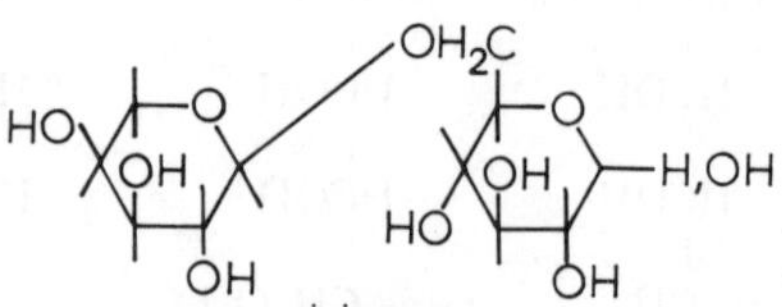

vicianose

(6-*O*-α-L-arabinopyranosyl-D-glucopyranose)
(2.47)

cosides of members of the Asclepiadaceae family arise from 4,6-dideoxyhexosone (2.42). The cardiac genins are described in Chapter 5.

If a genin contains more than one hydroxyl group, diglycosides, in which sugars are linked to two different hydroxyl groups of the genin, may be formed. However, much more common are the oligosaccharides, in which two or more sugars are linked together in a chain, the end sugar being bonded to the genin. Oligoglycosides are particularly common linked to terpenoid genins, while glycosides in which the sugars are linked to different hydroxyl groups are common amongst the flavonoids.

Rutinose (2.43), sophorose (2.44) and sambubiose (2.45) are common disaccharides, while gentiobiose (2.46) and vicianose (2.47) are rare. Gentiobiose has been found in the flavonols and anthocyanins of *Primula sinensis* (Primulaceae) and in the cyanogenic glycoside, amygdalin, which occurs in several species belonging to the Rosaceae. Vicianose occurs in violutoside, a salicylic ester of *Viola cornuta* (Violaceae), and in vicianin, the cyanogenic glycoside occurring in *Vicia angustifolia* (Leguminosae).

Oligoglycosides are most common amongst the saponins, which are terpenoid derivatives. Unlike the cardiac glycoside sugars, the monosaccharides making up the saponins are common and include D-glucose (2.2), D-galactose (2.18), D-galacturonic acid (2.48), D-glucuronic acid (2.14), D-xylose (2.33), L-arabinose (2.4), L-fucose (2.41b) and L-rhamnose (2.34).

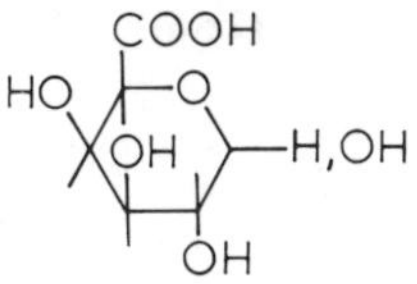

D-galacturonic acid
(2.48)

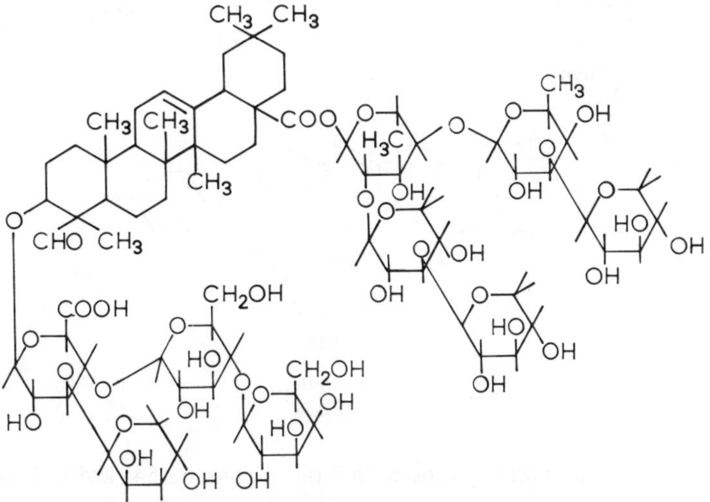

gypsoside A
(2.49)

The sugars are usually attached to the 3-hydroxyl group of the genin, but complex glycosides in which oligosaccharides containing branched chains are attached to two hydroxyl groups are known, such as gypsoside A (2.49) from *Gypsophila pacifica* (Caryophyllaceae). In the saponins, it is generally found that glucose links to the genin, and pentoses or methylpentoses are found at the ends of the chain. In the cardiac glycosides, however, the rare sugars are linked to the genin and glucose is found at the end of the chain.

C-glycosides are less common than *O*-glycosides and usually contain glucose. Some examples are mangiferin (2.50), a glucoside first isolated from the roots of *Mangifera indica* (Anacardiaceae), but since found in plants belonging to many different families. Barbaloin (2.51) occurs in *Aloe* species (Liliaceae), while flavone *C*-glycosides which have been characterised include vitexin (2.52), first isolated from the wood of *Vitex lucens* (Verbenaceae), and violanthin (2.53), from *Viola tricolor* (Violaceae), which contains both glucose and rhamnose in *C*-glycoside linkages.

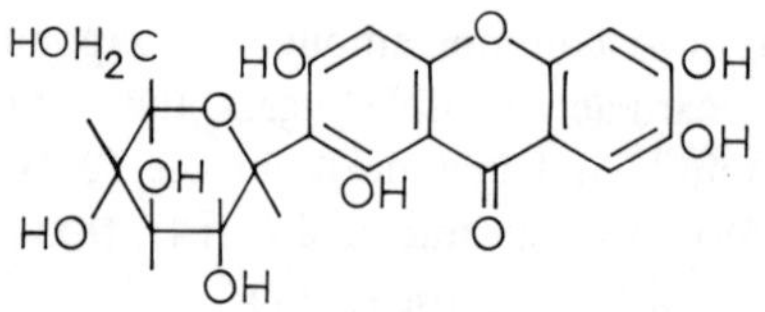

mangiferin
(2.50)

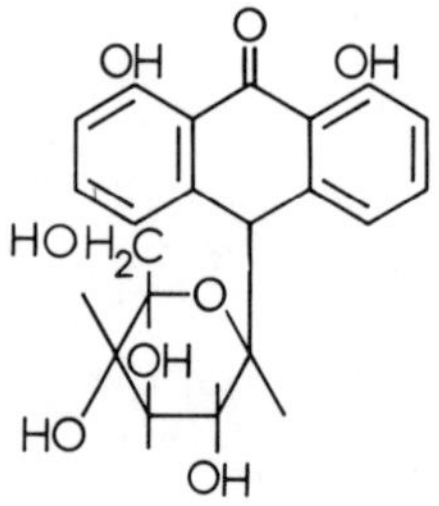

barbaloin
(2.51)

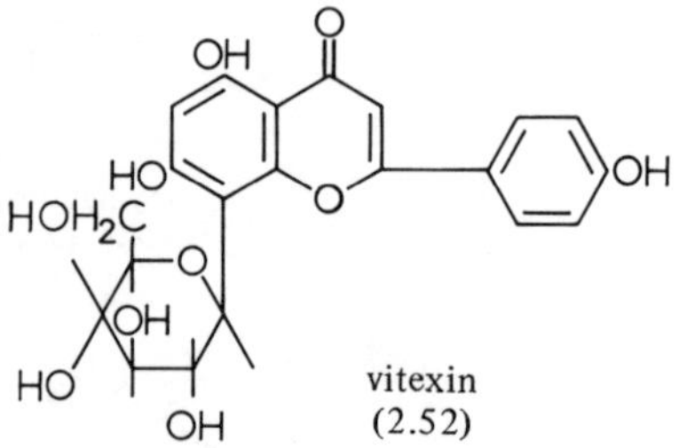

vitexin
(2.52)

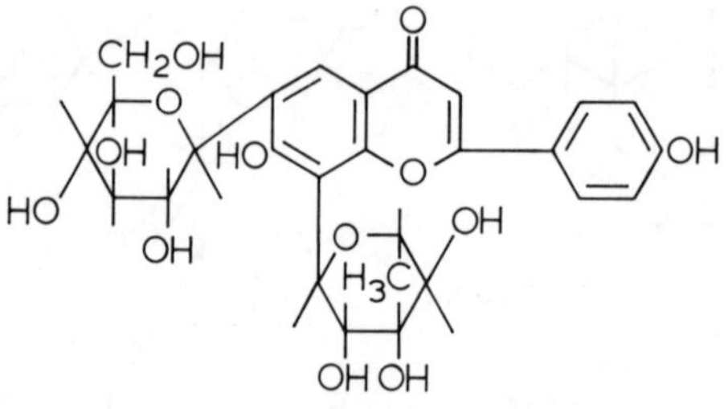

violanthin
(2.53)

The most important *N*-glycosides are the D-ribose (2.54) and D-deoxyribose (2.55) derivatives of purines and pyrimidines which constitute the nucleosides and nucleotides. These are described in Chapter 10. Other sugars have been found in combination with purine and pyrimidine bases, including D-glucose, D-galactose, L-arabinose, D-xylose, D-fructose and D-mannose (2.56). The purpose of these N-glycosides is unknown, as they do not form nucleic acids or coenzymes as do the ribose and deoxyribose derivatives.

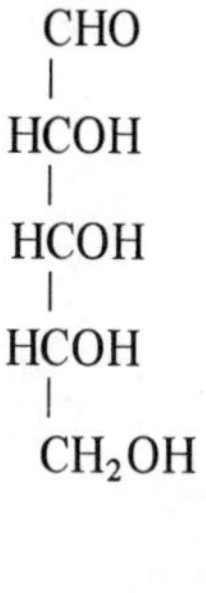

CHO	CHO	CHO
HCOH	CH_2	HOCH
HCOH	HCOH	HOCH
HCOH	HCOH	HCOH
CH_2OH	CH_2OH	HCOH
		CH_2OH
D-ribose	D-deoxyribose	D-mannose
(2.54)	(2.55)	(2.56)

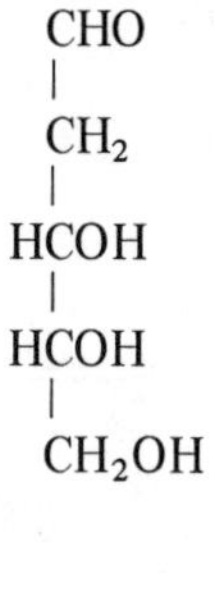

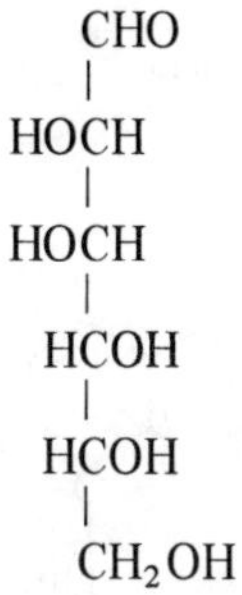

It has been clearly established that if free genins are introduced into plants they are glycosylated by nucleotide glycosides, usually the UDP derivatives. Oligosaccharides are formed in a series of reactions in which the monosaccharides are donated one at a time by nucleoside glycosides. Two enzyme systems have been separated from wheatgerm; one catalyses the formation of β-D-glucosides from various phenols and the other catalyses the formation of gentiobiosides from phenolic glucosides. Free disaccharides such as gentiobiose or rutinose are not found in plants, as hydrolysis of glycosides also takes place stepwise, the most common hydrolytic enzymes being β-glucosidases. Such enzymes also catalyse transglycosylation, in which a sugar is transferred from a glycoside of low energy to a genin. In *Daphne* sp. (Thymelaeaceae), the 8-glucoside of daphnetin (2.57) is formed by the transfer of glucose from daphnin (2.58), the 7-glucoside (*Figure 2.15*).

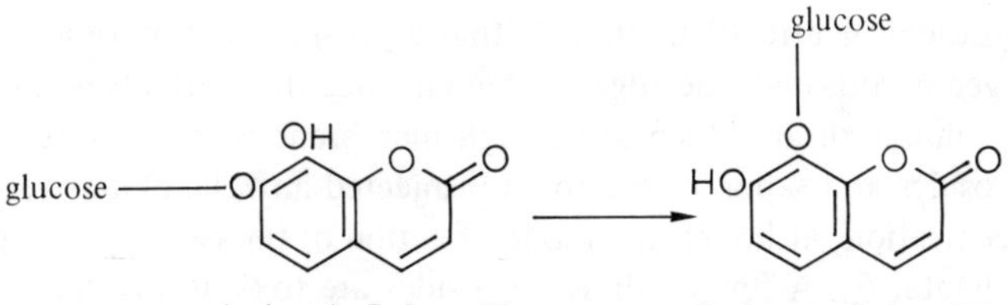

daphnetin-7-glucoside
(daphnin)
(2.58)

daphnetin-8-glucoside
(2.57)

Figure 2.15 Transglucosylation, forming daphnetin-8-glucoside.

In the biosynthesis of the coumarin, umbelliferone (2.59), glucosylation occurs at the 4-hydroxycinnamic acid (2.60) stage (*Figure 2.16*), the glucoside, skimmin (2.61), being formed before the free coumarin. Glucosylation of the 2-hydroxyl group of *trans*-cinnamic acid derivatives appears to be necessary to effect conversion to the *cis*-isomer and occurs in the biosynthesis of all coumarin derivatives.

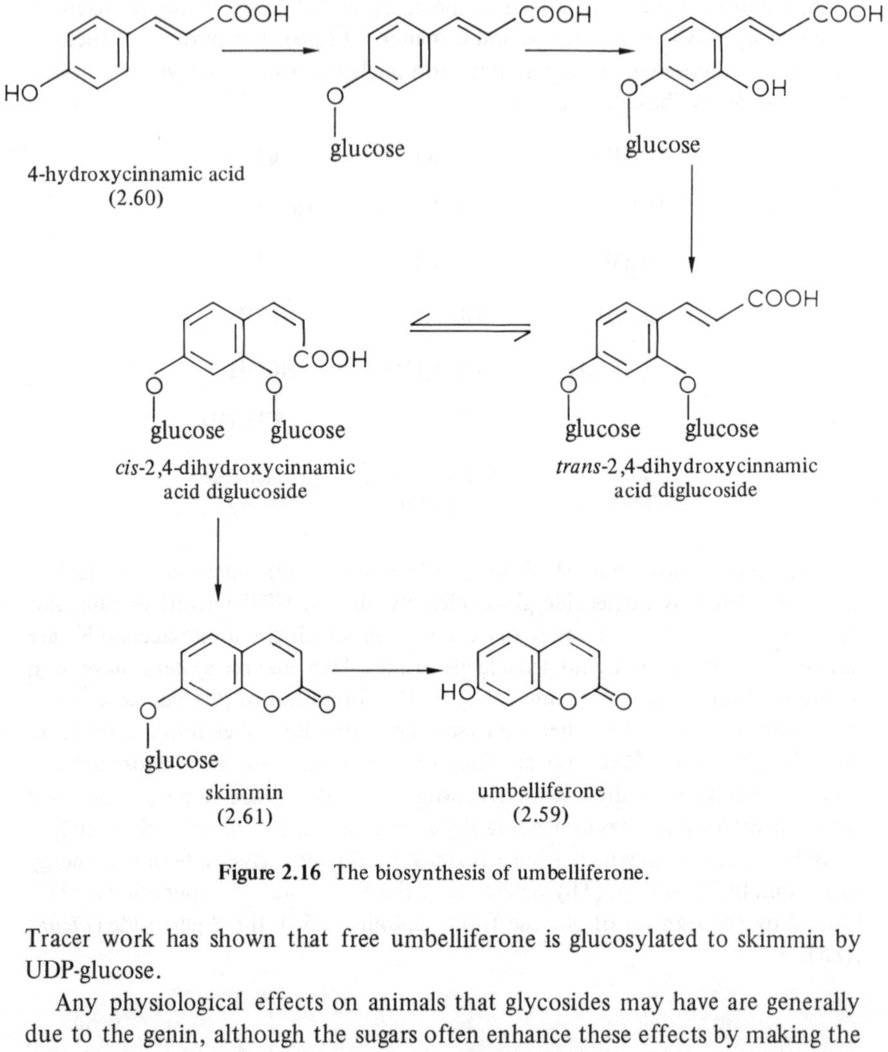

Figure 2.16 The biosynthesis of umbelliferone.

Tracer work has shown that free umbelliferone is glucosylated to skimmin by UDP-glucose.

Any physiological effects on animals that glycosides may have are generally due to the genin, although the sugars often enhance these effects by making the genin more soluble, or able to pass through membrane barriers more easily. All cardiac glycosides and saponins are toxic if injected into the bloodstream in sufficient concentration, although the mode of action of the two types of genin are different (Chapter 5). A few of these glycosides are toxic if ingested, but in the digestive tract the sugars are readily hydrolysed and the insoluble, bulky genins are, in general, unable to pass through the stomach wall into the bloodstream.

Cyanogenic glycosides are toxic to animals through their production of hydrogen cyanide (Chapter 8), while the glucosinolates produce irritant mustard oils on hydrolysis (Chapter 8). The glycoside, ranunculin (2.62), from buttercups and other *Ranunculus* species (Ranunculaceae), is a lactone glycoside which is enzymatically degraded to the highly irritant oil, protoanemonin (2.63), when buttercup plants are damaged. Dimerisation of protoanemonin (*Figure 2.17*) gives the harmless anemonin (2.64). Thus, fresh buttercups are harmful to grazing animals, but buttercups dried in hay are quite safe.

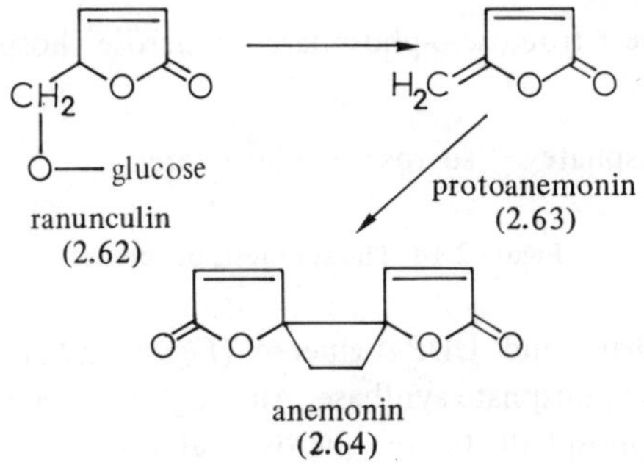

Figure 2.17 The degradation of ranunculin.

Phloridzin (2.65) is an interesting dihydrochalcone glycoside occurring in the bark and leaves of apples (*Malus* sp., Rosaceae) and some members of Ericaceae. In animals, this glycoside causes glycosuria (excess glucose in the urine), thus mimicking the symptoms of diabetes. This effect is due to interference with absorption of glucose from the intestine and with the reabsorption of this sugar from the kidneys.

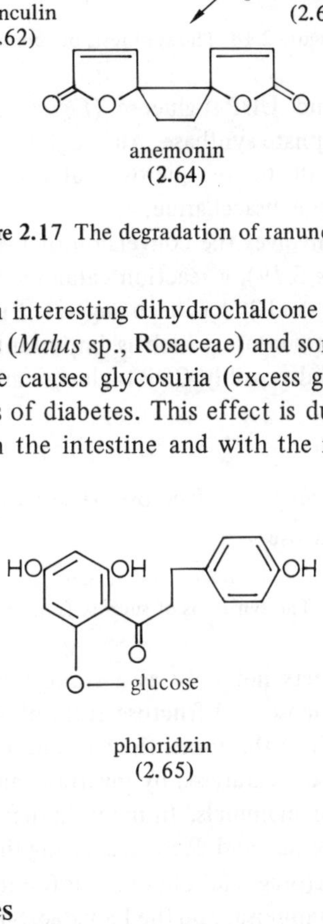

phloridzin
(2.65)

Storage Carbohydrates

The storage carbohydrates vary considerably in chain length, ranging from the disaccharide, sucrose, to the polysaccharides, starch and the fructosans. The storage carbohydrates are generally biosynthesised directly from photosynthetic monosaccharides and their derivatives. They are metabolised in the dark or stored in the seeds for use on germination, or in the roots or stems as a safeguard against adverse conditions. The storage carbohydrates are eventually metabolised to provide energy or the building blocks of cell wall polysaccharides.

Sucrose

Sucrose (2.6), one of the most common of the reserve carbohydrates, is ubiquitous in the plant world. It is particularly abundant in sugar cane (*Saccharum officinarum,* Gramineae) and sugar beet (*Beta vulgaris,* Chenopodiaceae). This sugar is biosynthesised in plants by two pathways involving either D-fructose-6-phosphate (2.7) or D-fructose (2.3). In the chloroplasts, sucrose phosphate is synthesised from

UDP-glucose + fructose-6-phosphate $\rightleftharpoons$ sucrose phosphate + UDP

sucrose phosphate $\rightarrow$ sucrose + phosphate

Figure 2.18 The synthesis of sucrose.

D-fructose-6-phosphate and UDP-D-glucose (*Figure 2.18*), the reaction being catalysed by sucrose phosphate synthase. Although this reaction is reversible, the hydrolysis of the phosphate to free sucrose, catalysed by sucrose phosphatase, favours the formation of the disaccharide.

The second pathway involves the conversion of D-fructose and nucleotide-D-glucose to sucrose (*Figure 2.19*), a reaction catalysed by sucrose synthase. This reaction is also reversible and it has been suggested that it is the formation of nucleotide-D-glucose from sucrose which is important. UDP-D-glucose, GDP-D-glucose, ADP-D-glucose and possibly CDP-D-glucose are involved in the synthesis of various polysaccharides.

NDP-D-glucose   + fructose $\rightleftharpoons$ sucrose +   NDP
nucleotide-D-glucose   nucleotide

Figure 2.19 The synthesis of sucrose from NDP-glucose.

In the plant, sucrose acts not only as a storage carbohydrate but also as a means of transporting glucose and fructose from photosynthetic tissue to non-photosynthetic tissue such as the roots, flowers and seeds. Hydrolysis of sucrose to D-glucose and D-fructose is catalysed by invertase, an enzyme present in plants and the small intestine of mammals. In many herbage species, sucrose concentration increases during the day and decreases during the night.

The raffinose family of storage carbohydrates is found in many plants, especially those belonging to the Leguminosae and the Labiatae. Sucrose and galactinol (2.19) are the precursors of the trisaccharide, raffinose (2.20) (*Figure 2.9*); the tetrasaccharide, stachyose, and the pentasaccharide, verbascose, being formed by stepwise addition of galactosyl units from galactinol. Stachyose is characteristic of the Labiatae family, while verbascose has been isolated from several plants including *Verbascum* species (Scrophulariaceae).

It has been shown in many plants that a large proportion of isotopically labelled carbon dioxide incorporated into sugars by photosynthesis is converted to galactose and thence to the raffinose family of sugars. These sugars are subsequently metabolised in the dark. In broad beans (*Vicia faba*, Leguminosae), the reaction hydrolysing raffinose to sucrose and galactose is catalysed by α-galactosidase, and the galactose is catalytically phosphorylated by galactokinase. In legumes, it has been found that raffinose and its homologues disappear during germination.

Gentianose (2.66) is a trisaccharide storage carbohydrate found in the rhizomes of many *Gentiana* species (Gentianaceae). It is biosynthesised by the glucosylation of sucrose, this reaction being reversed by emulsin, which hydrolyses gentianose to sucrose and glucose. Partial acid hydrolysis, however, yields gentiobiose (2.46) and fructose.

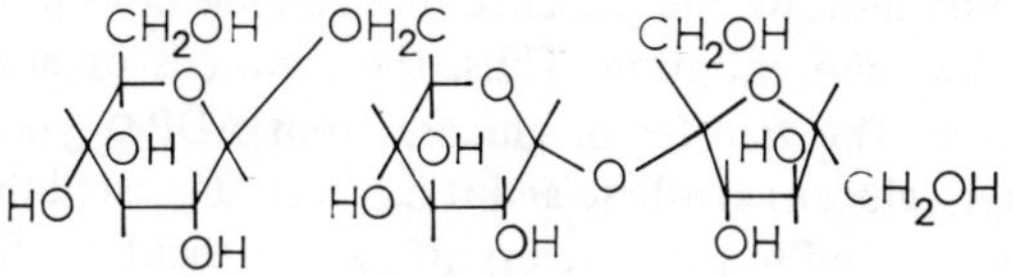

gentianose
(2.66)

Starch

Starch is yet another storage carbohydrate ubiquitous in higher plants. Starch granules, which vary in shape and size, are heterogeneous, containing both amorphous and crystalline regions. The amorphous regions are more susceptible to enzymatic and chemical attack than the crystalline regions, whose chains are held together by hydrogen bonds.

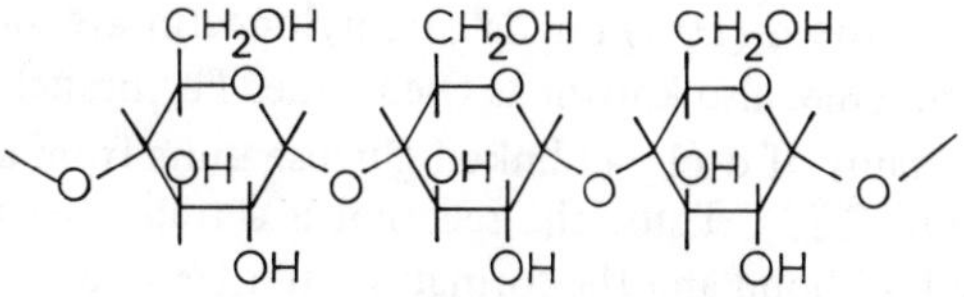

(a) Part of an amylose molecule

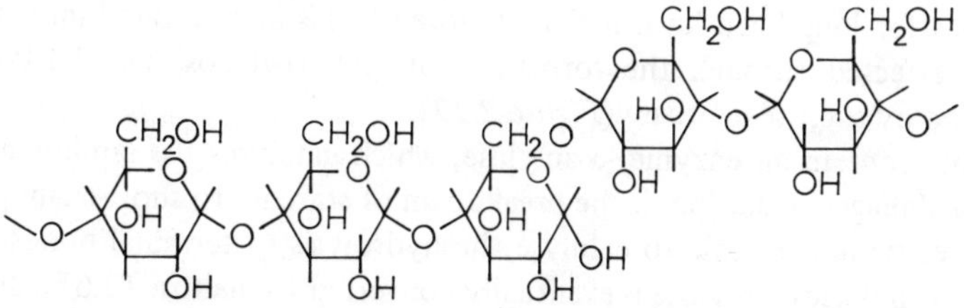

(b) Part of an amylopectin molecule

Figure 2.20 The molecular structures of starch.

Most starch granules contain a mixture of two α-D-glucose polymers known as amylose and amylopectin. Amylose (*Figure 2.20*) is a linear molecule in which the glucose units are joined by α (1→4) links. It has a relatively low molecular weight and is easily soluble in hot water. Amylopectin is a branched polymer (*Figure 2.20*) containing glucose units linked both α (1→4) and α (1→6). It has a high molecular weight and is insoluble in water. Amylose appears to be the predominant type of starch found in the leaves of such cereals as maize (*Zea mays*) and sorghum (*Sorghum bicolor*), while the seeds contain 70-90% amylopectin. However, several maize hybrids have been bred with amylose contents in the grains of from 0 to 60%.

Amylose is biosynthesised from ADP-D-glucose which transfers a glucose unit to a primer, an α-glucan molecule which already contains several α (1→4) linked glucose units. Although the *de novo* synthesis of primer has been suggested, most experimental results indicate the presence of an endogenous primer, which need only be in very low concentration. Thus, the synthesis of amylose is a chain-lengthening process. The transfer of glucose from ADP-D-glucose to primer is catalysed by starch (glucan) synthase and it has been suggested that primer glucan from parent cells is bound to the starch synthase of daughter cells.

Soluble starch synthase exhibits absolute specificity towards ADP-D-glucose, but an insoluble, granule-bound starch synthase can also utilise UDP-D-glucose, although the ADP glucoside is the more efficient donor.

$$
\begin{array}{cc}
 & G\text{--}G\text{--}G\text{--} \\
 & | \\
G\text{--}G\text{--}G\text{--}G\text{\textbarbelow{}}G\text{--}G\text{--}G \quad \rightarrow \quad & G\text{--}G\text{--}G\text{--}G \\
\text{α-glucan unit} & \text{amylopectin unit}
\end{array}
$$

Figure 2.21 The formation of amylopectin.

The formation of the α (1→6) bonds of amylopectin are catalysed by 1,4-α-glucan branching enzyme, also known as Q-enzyme. The branching is effected by transfer to short chains of α (1→4) linked glucose units from a 1→4 linkage to a 1→6 linkage (*Figure 2.21*). Thus, the reaction is a transglucosylation involving the breaking of a 1→4 bond and the formation of a 1→6 bond. There is evidence that the enzyme(s) catalysing the formation of 1→4 linkages of amylopectin may not be identical with the starch synthase of amylose.

It has for long been known that sucrose (2.6) is incorporated into starch and this is effected through the formation of ADP-D-glucose and UDP-D-glucose catalysed by sucrose synthase (*Figure 2.19*).

Plants contain an enzyme, α-amylase, which catalyses the random hydrolysis of 1→4 linkages, resulting in the breakdown of starches to short-chain polymers. As the enzyme is unable to catalyse the hydrolysis of terminal bonds or bonds near 1→6 linkages, amylose is eventually converted to maltose (2.67), and amylo-pectin to α-limit dextrins. The number of units contained in an α-limit dextrin depends on the source of the enzyme. Thus, malt α-amylase produces panose

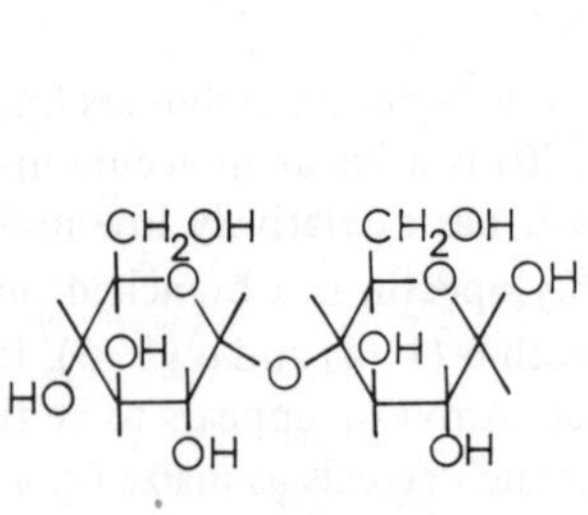

maltose
(2.67)

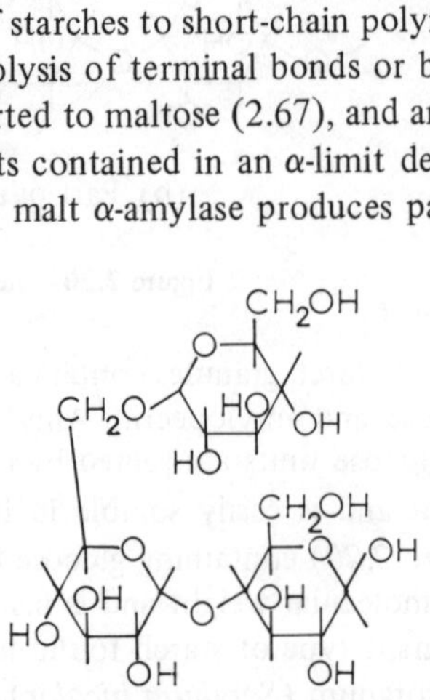

panose
(2.68)

(2.68), but other α-amylases produce molecules with more glucose units. It has been found in germinating cereals that α-amylases are synthesised in response to the plant hormone, gibberellic acid (Chapter 5).

The degradation of amylose catalysed by α-amylase also produces small amounts of glucose and maltotriose, but, if the reaction is catalysed by β-amylase, maltose is produced exclusively. This enzyme catalyses the hydrolysis of alternate 1→4 linkages. β-Amylase also catalyses the partial hydrolysis of amylopectin, but only of the outer chains. This useful distinction has led to an estimation of the average chain lengths for the inner (i.e. those chains of glucose units between each 1→6 linkage) and outer chains of amylopectin. These range from 5–9 units for the inner chains to 10–18 units for the outer chains. β-Amylase does not have the widespread occurrence of α-amylase and has only been found in cereals, sweet potatoes (*Ipomoea batatas*, Convolvulaceae) and soybeans (*Glycine max*, Leguminosae).

Phosphorylase is widely distributed in plants and occurs in both the leaves and the storage organs. This enzyme catalyses the conversion of starch to glucose-1-phosphate. The conversion of amylose is complete but only the outer chains of amylopectin are degraded.

Plants contain a number of enzymes which catalyse the hydrolysis of the 1→6 linkages of amylopectin. These include R-enzyme and limit dextrinase. Thus both amylose and amylopectin are eventually degraded to glucose which is metabolised in the various processes described earlier in this chapter. The extraction of starch from high-yielding plants such as cereals and tubers is an important industry in many countries. Apart from its use in food, starch is made into gels, adhesives, etc., and high-molecular-weight polymers. Chemically, starch can be converted to several compounds used industrially, including ethanol.

Phytoglycogen

The storage carbohydrate, phytoglycogen, has only been found in maize. It has a higher degree of branching than amylopectin and is not affected by the hydrolytic R-enzyme. Enzymes which catalyse the formation of phytoglycogen and those which catalyse its degradation have been isolated from maize.

Fructans

Some plants, notably those belonging to the Compositae and Gramineae families, contain storage carbohydrates which are polymers of fructose (2.3) linked to a single sucrose molecule. These polymers, known as fructans, have a relatively low molecular weight and are soluble in water. Unlike starch, they do not give a blue colour with iodine. Several types of fructan occur, including inulin in *Dahlia* tubers and *Inula* species (Compositae) and levans in grasses. Inulin (*Figure 2.22*) is a linear polymer containing β (2→1) linked fructose units, while levans (*Figure 2.22*) are short-chain, linear β (2→6) linked polymers. Levans are the storage carbohydrates found in the leaves of temperate grasses, while starch accumulates in the leaves of tropical grasses. In some plants, highly branched fructans containing both 2→1 and 2→6 linkages occur.

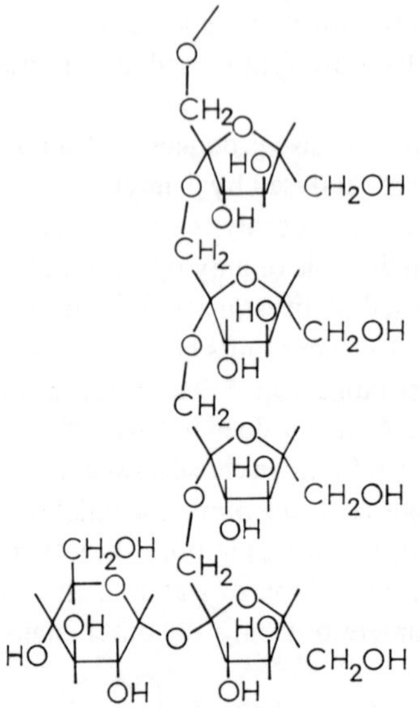

(a) Part of an inulin molecule

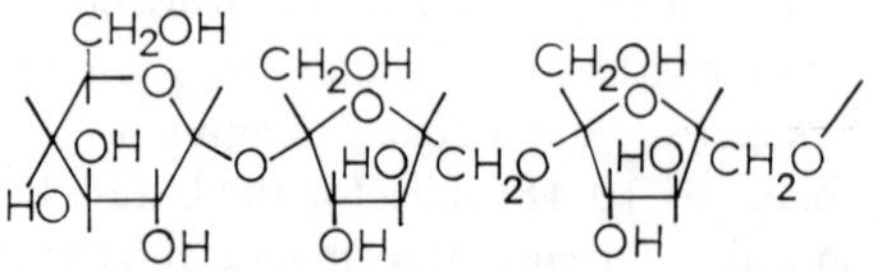

(b) Part of a levan molecule

Figure 2.22 Fructans.

Fructans are synthesised from sucrose, one molecule of which starts the synthesis by donating a fructose unit to another sucrose molecule, forming a trisaccharide (*Figure 2.23*). Chain lengthening takes place by the donation of further fructose units either from sucrose or from previously formed fructans (*Figure 2.23*).

glucose–fructose + glucose–fructose → glucose–fructose–fructose + glucose

glucose–fructose–(fructose)$_n$ + glucose–fructose–(fructose)$_m$

↓

glucose–fructose–(fructose)$_{n+1}$ + glucose–fructose–(fructose)$_{m-1}$

Figure 2.23 The biosynthesis of fructans.

Mannans

Much less is known of the structure, biosynthesis or degradation of the reserve mannans. Glucomannans have been isolated from a few tropical plants and some members of the Orchidaceae family, while galactomannans occur in the seeds of legumes.

The galactomannans present in guar seeds (*Cyamopsis tetragonolobus*) and carob or locust bean (*Ceratonia siliqua*) are extracted commercially and used as thickening or emulsifying agents. In the guar galactomannan, the ratio of mannose units to galactose units is 2 : 1, while that of locust bean varies from 3 : 1 to 6 : 1.

Both types of polymer consist of linear β (1→4) linked D-mannose (2.56) units with 1→6 linked D-galactose (2.18) units attached to the main chain. In guar gum, the galactose units occur in a regular manner (*Figure 2.24*), but in locust bean gum they are random.

```
         galactose          galactose
             |                  |
mannose—mannose—mannose—mannose—mannose
```

Figure 2.24 Guar gum.

Exudate gums

A number of plants, especially those of arid regions, exude gums in response to wounding. These gums consist of polysaccharides and harden on exposure to air, thus sealing the wound, preventing infection and dehydration. The deliberate wounding of some plants produces such exudates in sufficient quantity to warrant collection and commercial exploitation.

One of the best known of these exudates, gum arabic from *Acacia senegal* and *A. verek* (Leguminosae), has been an article of commerce for over 4000 years. Gum arabic is a highly branched polymer containing D-galactose (2.18), L-arabinose (2.4), D-glucuronic acid (2.14) and L-rhamnose (2.34), which occurs naturally as the mixed potassium, calcium and magnesium salt. In solution, the polysaccharide molecules are thought to exist as rigid spirals. A part of the gum arabic molecule is shown in *Figure 2.25*.

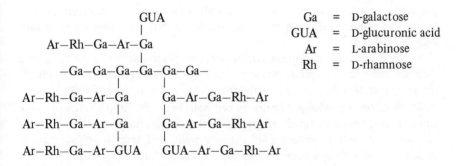

Ga	= D-galactose
GUA	= D-glucuronic acid
Ar	= L-arabinose
Rh	= D-rhamnose

Figure 2.25 Part of the gum arabic molecule.

The main commercial use of gum arabic is as an emulsifying agent for oil/water systems, especially in food, as this polysaccharide is totally non-toxic.

Some members of the *Prunus* genus (Rosaceae), including cherry (*P. cerasus* and *P. virginiana*), apricot (*P. armeniaca*) and damson (*P. insitia*), produce exudate gums which are similar to gum arabic.

Gum ghatti, or Indian gum, occurs as an exudate of the tree *Anogeissus latifolia* (Combretaceae), which grows in India and Sri Lanka. This polysaccharide contains the sugars L-arabinose (2.4), D-galactose (2.18), D-mannose (2.56), D-glucuronic acid (2.14) and D-xylose (2.33) and occurs naturally as the calcium salt.

Other commercially important gums include gum karaya from *Sterculia urens* (Sterculiaceae), gum tragacanth from *Astragalgus* species (Leguminosae), cholla gum from the white cactus (*Opuntia bulgida*, Cactaceae), mesquite gum from *Prosopis juliflora* (Leguminosae) and Khaya gums from *Khaya* species (Meliaceae).

Plants also exude sugars into the nectaries of their flowers to attract insects. Melibiose (6-*O*-α-D-galactosyl-β-D-glucopyranose) (2.69) has been found in the nectaries of several plants.

melibiose
(2.69)

Structural Carbohydrates

Plant cell walls

Plant cell walls contain protein, lignin and the polysaccharides, cellulose, the hemicelluloses and the pectic substances, as their major components. Young, undifferentiated cells form a primary cell wall which does not contain lignin. When the cell stops growing, a thicker, secondary wall, which does contain lignin, is laid down. The primary cell walls of all cells are similar, but the secondary walls vary with the cell type.

The cell walls are important constituents of plants, as they provide both a skeleton and a barrier against invasion. Many fungal pathogens are able to invade plants by secreting enzymes which degrade components of the cell walls.

Much of our knowledge of primary cell walls has come from a study of tissue cultures. Suspension-cultured cells are admirable for this purpose as they produce homogeneous cells possessing only primary walls, and secrete polysaccharides similar to cell wall components into the culture medium. A description of the complicated molecular structure of primary cell walls deduced from work with tissue cultures can be found in *Plant Biochemistry*, p. 254, listed in the bibliography.

The gross structure of cell walls is a matrix of pectin, hemicelluloses, lignin and protein in which cellulose microfibrils are embedded. Cellulose constitutes about 65% of perennial plant tissue and 50% of annual tissue and is the most important cell wall component. Pectic substances are largely concentrated in the primary wall, while the hemicelluloses occur throughout. Although once thought to be intermediates in the synthesis of cellulose, the hemicelluloses are now known to be a distinct group of polysaccharides whose biosynthesis is independent of that of cellulose.

Cellulose

Cellulose is a linear polymer of β (1→4) linked glucose units (*Figure 2.26*), and thus it is only in the stereochemistry of the linkages between glucose units that cellulose differs from starch. However, the physical properties of the two isomers are quite different.

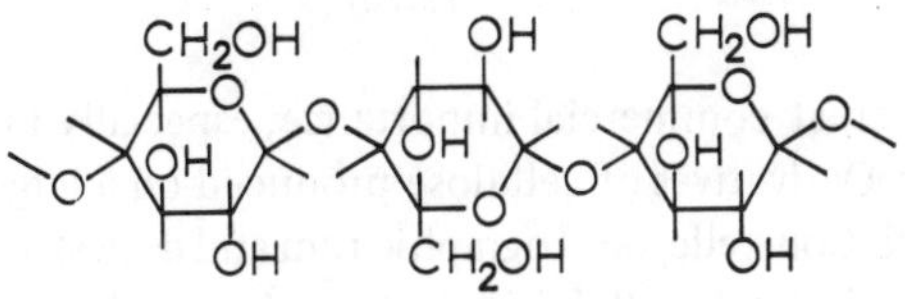

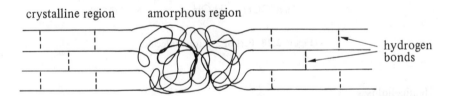

Figure 2.26 Part of a cellulose molecule.

Cellulose has a high molecular weight and is present in plant tissues as microfibrils, which consist mainly of parallel glucan chains held tightly together by hydrogen bonds. Such a regular arrangement gives the rigid crystalline cellulose constituting the greater part of the microfibrils. In cotton fibres, which are the purest form of natural cellulose, the regions of crystalline cellulose are interspersed with amorphous cellulose (*Figure 2.27*). It is these latter regions which give cotton fibres their elasticity.

crystalline region   amorphous region

hydrogen
bonds

Figure 2.27 Part of a cotton fibre.

The crystalline regions of cellulosic materials such as cotton and wood pulp can be separated from the amorphous regions by hydrolysis, the hydrogen bonding of the former making them much less susceptible to hydrolysis. Dry cellulose is inflexible and brittle, but in a humidity of 60–80% it absorbs 12% water and becomes flexible. The water molecules are held in the amorphous regions. Thus, any process which increases the amorphous regions of a cellulose fibre also increases its flexibility. Such amorphous regions are greatly increased when natural cellulose (usually wood pulp) is regenerated as viscose or rayon. The natural cellulose is dissolved in a solution of tetrammine copper(II) hydroxide in aqueous ammonia and regenerated by spinning into an acid bath.

Unlike starch, cellulose does not undergo extensive degradation in plants, but numerous microorganisms produce enzymes which readily catalyse the hydrolysis of the glucan. Cellobiose (2.70), an isomer of maltose (2.67), results from the hydrolysis of alternate linkages of cellulose. The degradation appears to involve phosphorylation, as the microorganisms concerned contain high levels of the enzyme, cellobiose phosphorylase.

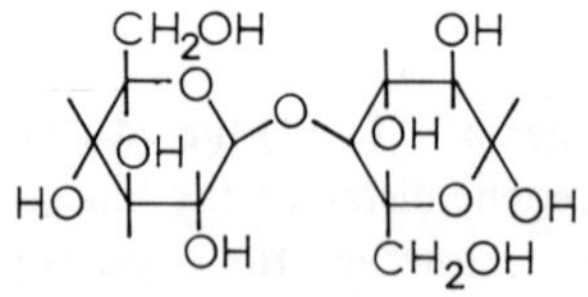

cellobiose
(2.70)

Cellulose is of great commercial importance, especially in the manufacture of textiles and paper. Derivatives of cellulose produced on a large scale include cellulose acetate (Tricel, Courtelle, photographic film and dopes), cellulose acetate butyrate (unbreakable plastics), cellulose trinitrate (gun cotton) and carboxymethylcellulose (washing powders giving dirt resistance to fabrics; thickeners and dispersants). Cotton/polyester fibres are produced by growing synthetic polymers onto cellulose chains, producing cross-linked molecules. Fabrics woven from such fibres are crease-resistant and 'easy care'.

In fungi, cellulose is largely replaced as a structural component by chitin (*Figure 2.28*), which is linked to protein to form a complex glycoprotein.

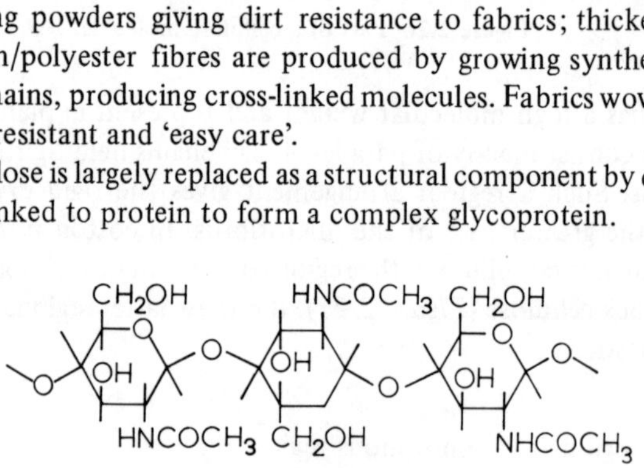

Figure 2.28 Part of a chitin molecule.

Hemicelluloses

The hemicelluloses are polysaccharides occurring throughout the cell wall matrix. They are usually branched-chain polymers containing two or more of the sugars: D-xylose (2.33), D-mannose (2.56), D-glucose (2.2), D-galactose (2.18), L-arabinose (2.4), D-glucuronic acid (2.14) and its 4-O-methyl derivative and D-galacturonic acid (2.48). The sugars are often partly acetylated.

The hemicelluloses are biosynthesised from their respective UDP-sugars, the reactions being catalysed by polysaccharide synthases. Many microorganisms contain enzymes which catalyse the degradation of plant hemicellulose. The hemicelluloses are generally divided into three families: the D-xylans, D-mannans and L-arabino-D-galactans.

D-XYLANS

The D-xylans are pentosans based on D-xylose. They form the bulk of the hemicelluloses present in annual and non-woody perennial plants and are also present in softwoods. In land plants, the xylans are based on linear chains of β (1→4) linked xylose units. Attached to these chains are L-arabinose, D-glucuronic acid or its 4-O-methyl derivative, D-galactose and possibly D-glucose (*Figure 2.29*).

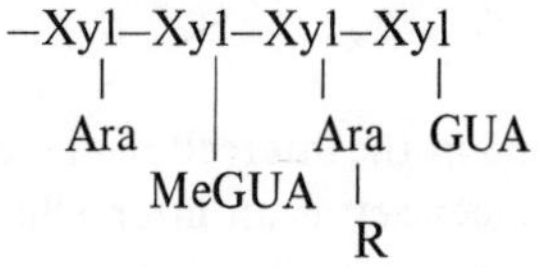

–Xyl–Xyl–Xyl–Xyl
Ara
Ara
GUA
MeGUA
R

Xyl = D-xylose
Ara = L-arabinose
MeGUA = D-4-*O*-methylglucuronic acid
GUA = D-glucuronic acid
R = side chain

Figure 2.29 Generalised xylan structure.

These sugars may form single-unit side chains or terminal units of the xylan chains, but some, particularly L-arabinose, have longer chains attached. D-Xylose is in the pyranose form and L-arabinose is in the furanose form.

D-MANNANS

The structural D-mannans are the main non-cellulosic hexosans in plants, and are either glucomannans or galactoglucomannans (the storage mannans are discussed above). The mannans are based on chains of β (1→4) linked D-glucose and D-mannose units. In the galactoglucomannans, varying amounts of galactose are α linked as single-unit side chains (*Figure 2.30*).

The glucomannans are found in both hardwoods and softwoods, but the galactoglucomannans occur only in softwoods.

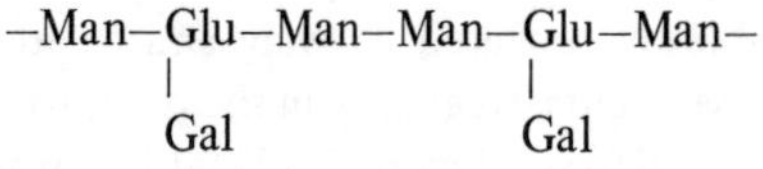

–Man–Glu–Man–Man–Glu–Man–
Gal
Gal

Man = D-mannose
Glu = D-glucose
Gal = D-galactose

Figure 2.30 Generalised structure of galactoglucomannans.

L-ARABINO-D-GALACTANS

Hemicelluloses containing L-arabinose and D-galactose units are found in the Coniferales, especially larches (*Laris* species), where they may constitute up to 25% of the wood. The arabinogalactans are pentosans with chains of β (1→3) linked galactose units containing side chains of various lengths (*Figure 2.31*). These side chains contain galactose and arabinose, the latter sugar occurring in both the pyranose and furanose forms. Unlike the other hemicelluloses, the arabinogalactans are soluble in water.

–Gal–Gal–Gal–Gal–Gal–
Gal
Gal–Gal
Gal
Gal
Ar
Ar
Gal
Ar
Ar
Ar
Ar

Gal = D-galactose
Ara = L-arabinose

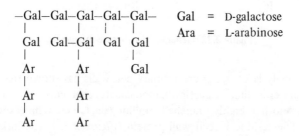

Figure 2.31 Generalised structure of arabinogalactans.

Pectic Substances

The pectic substances occur in the primary cell wall and in the intercellular layers but not in the secondary cell wall. The pectic substances act as an intercellular cement, binding individual cells together to form plant tissue. The abscission of leaves and fruits and the softening of fruits on ripening is attributed to a loss of cementing power of the pectic substances. In most plant tissues, the pectic substances are present as water-insoluble protopectin, but, when fruits ripen, this is converted to water-soluble pectins.

The pectic substances include polyuronide (pectin), galactans, arabans and galactoarabans. Only pectin will be discussed in detail.

PECTIN

Pectin consists of chains of α $(1\rightarrow4)$ linked D-galacturonic acid (2.48) units with many of the carboxyl groups esterified with methyl groups or in the form of calcium or magnesium salts. Pectins with a low degree of esterification are known as pectic acids, while those with a high degree of esterification are called pectinic acids. Attached to the polyuronide chains of pectins are varying amounts of other sugars including L-arabinose (2.4), D-galactose (2.18), L-rhamnose (2.34), D-xylose (2.33), L-fucose (2.41b) and D-glucose (2.2). The 2-O-methyl derivatives of D-xylose and L-fucose have also been detected. Not all these sugars are covalently linked, as removal of a proportion is possible by repeated washing and re-precipitation of the pectin. Some of the L-rhamnose occurs irregularly in the polyuronide chain, distorting its linearity and forming a zig-zag (*Figure 2.32*) which probably has better cementing powers.

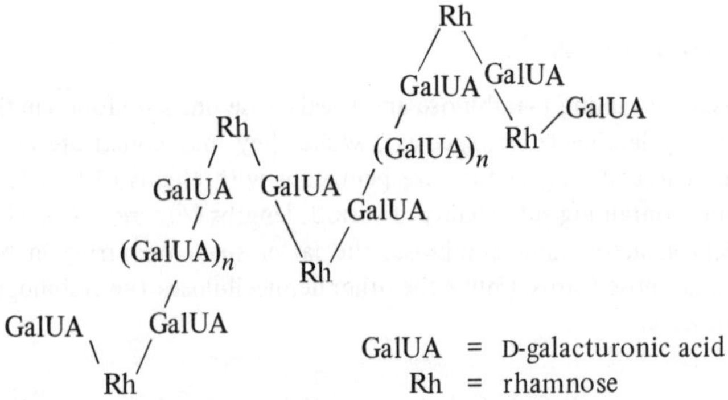

Figure 2.32 Generalised structure of a pectin.

Detailed analysis of sycamore primary cell walls has shown that an arabinogalactan chain cross-links a pectic rhamnogalacturonan chain to a hemicellulosic xyloglucan. Another, highly branched, arabinogalactan cross-links the pectin to a hydroxyproline residue in cell wall protein (*Figure 2.33*). The linking of these different types of polymers gives a rigid matrix which is strengthened by the cellulose microfibrils.

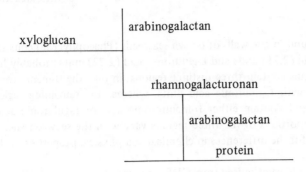

Figure 2.33 Part of a sycamore primary cell wall.

An enzyme system catalysing the formation of a polygalacturonic acid has been isolated from mung beans. The D-galacturonosyl donor is UDP-D-galacturonic acid and methylation takes place after polymerisation, the donor being S-adenosyl-L-methionine.

Enzymes catalysing the degradation of pectic substances are present in higher plants, fungi and microorganisms. Two types of enzymes have been identified — the hydrolases or glycosidases and the lyases or transeliminases. The hydrolases catalyse the hydrolysis of the α (1→4) glycosidic bonds, while the lyases effect the removal of a hydrogen atom at C-5, giving an unsaturated compound (*Figure 2.34*).

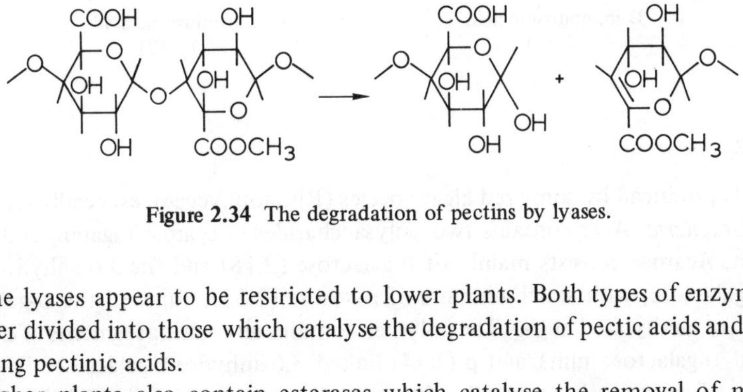

Figure 2.34 The degradation of pectins by lyases.

The lyases appear to be restricted to lower plants. Both types of enzyme are further divided into those which catalyse the degradation of pectic acids and those utilising pectinic acids.

Higher plants also contain esterases which catalyse the removal of methyl groups from pectins.

Due to their jelly-forming properties, pectins are commercially extracted from fruit waste and used in the manufacture of jams and jellies and as thickening agents.

Seaweed Gums

The seaweed gums occur as cell wall components or as reserve carbohydrates. In general, they consist of long chains of two different sugar units which often alternate in segments. Some of the seaweed gums have been exploited commercially and these are considered below.

ALGINIC ACID

Alginic acid found in the walls of brown seaweeds (Phaeophyceae) consists of D-mannuronic acid (2.71) units and L-guluronic acid (2.72) units probably linked β (1→4). The chains contain three distinct regions; in one, the D-mannuronic acid units alternate with L-guluronic acid units, whereas the remaining regions are homogeneous and contain either D-mannuronic acid or L-guluronic acid. The lengths and proportions of the three regions vary with the seaweed species and are responsible for the differences in chemical and physical properties of the various alginates.

Alginic acid is biosynthesised from GDP-D-mannuronic acid and GDP-guluronic acid. It seems probable that guluronic acid is derived from its isomer, mannuronic acid, the reaction being catalysed by an epimerase.

Monovalent salts of alginic acid are soluble in water, but polyvalent salts are either insoluble or form gels. Solutions of alginates are very viscous due to the high molecular weight and random-coil formation of the polymers. The jellying and thickening properties of alginates are widely used commercially in foodstuffs, etc.

D-mannuronic acid
(2.71)

L-guluronic acid
(2.72)

AGAR

Agar is produced by some red algae species (Rhodophyceae), especially *Gelidium* and *Gracilaria*. Agar contains two polysaccharides — agarose (agaran) and agaropectin. Agarose consists mainly of D-galactose (2.18) and the 3,6-anhydro form of L-galactose, with small amounts of D-xylose (2.33). Some of the D-galactose units are methylated at C-6. The polymer contains alternating segments of α (1→3) linked D-galactose units and β (1→4) linked 3,6-anhydro-L-galactose. The main chains of agaropectin are similar, but contain D-glucuronic acid (2.14) and small amounts of other sugars, including sulphate esters.

Agar is not attacked by microorganisms and its strong gel-forming properties make it an ideal matrix for microbial cultures.

CARRAGEENAN

The carrageenans are sulphate gums found in members of the Rhodophyceae. They consist of mixtures of polysaccharides which are mainly composed of D-galactose (2.18) units linked 1→3 and 1→4. A high proportion of the units are substituted with sulphate half-ester groups.

The carrageenans are useful stabilisers in foods.

Carbohydrates in Chemosystematics

The common sugars, such as fructose, glucose, sucrose, etc., are either ubiquitous in the plant world or so widespread as to have little taxonomic significance. Quantitative differences between taxa have been reported, but these are due more to environment than to genetic variation. Conversely, the rare sugars found in glycosides have such a restricted distribution that they contribute little to the understanding of phylogenetic relationships. In general, it is the genin of a glycoside, rather than the sugar, which has taxonomic significance, although the deoxy-methylpentoses encountered in the cardenolides appear to be restricted to these compounds, and hence their distribution follows that of the glycosides. Cardenolides containing such sugars are particularly abundant in the related Apocynaceae and Asclepiadaceae, although they appear spasmodically in various unrelated families (see Chapter 5).

Amygdalin, the cyanogenic glycoside characteristic of the Rosaceae family, contains the uncommon sugar, gentiobiose, while the trisaccharide, gentianose, appears to be restricted to the *Gentiana* genus (Gentianaceae).

Some flavonoid glycosides have proved useful taxonomic markers. For example, in the Zygophyllaceae family, quercetin-3-gentiobioside is restricted to members of the Tribuleae tribe, a fact which lends support to the suggestion made on morphological grounds that this tribe should be treated as a separate family. Flavonoid C-glycosides, such as vitexin, are considered primitive chemical characters which have often been lost in the more highly developed taxa. Within the sub-family Spiraeoideae of the Rosaceae, Only *Quillaja* contains C-glycosides, and it has been suggested that this genus is a relict of the now extinct ancestors of the Pomoideae. Within the latter subfamily, the occurrence of C-glycosides in *Aronia* and their absence in *Sorbus* argues against cytological and morphological evidence that the two genera should be united.

Of the mono- and oligosaccharides, probably the most useful from a chemo-systematic viewpoint are the polyols. The presence or absence of the sugar alcohol, sorbitol, in the Rosaceae has helped to define the subfamilies. Thus sorbitol has been found in all species belonging to Spiraeoideae, Pomoideae and Prunoideae examined, but in no species belonging to Rosoideae. *Ullmaria* does not contain the polyol, and the absence of this chemotaxonomic marker confirms the removal, on morphological grounds, of this genus from Spiraeoideae to Rosoideae. The three genera *Rhodotypos, Kerria* and *Neviusia*, whose placement on morphological grounds is unclear, contain sorbitol and are therefore chemically related to Spiraeoideae rather than to Rosoideae.

D-Quercitol is present in all *Quercus* species (Fagaceae) and serves to differen-tiate these from the closely related *Fagus* and *Castanea*. This cyclitol is also wide-spread in the Menispermaceae and has been detected in plants belonging to many different families.

The rare cyclitol, L-leucanthemitol, has only been found in three species of *Chrysanthemum* (Compositae), all of which belong to the Pyrethrum section of the genus. It may thus prove to be a distinctive characteristic for this section. Another cyclitol, L-viburnitol, is characteristic of the Compositae tribe, Anthem-ideae, although it has been found in plants belonging to three other families. The fact that L-inositol occurs in several totally unrelated members of the large Compositae family shows it to be a primitive feature which has been lost during the evolution of some species.

It is interesting that all members of Gymnospermae contain either D-pinitol or sequoyitol. D-Pinitol is also characteristic of the Caryophyllaceae and has been found in many families belonging to the angiosperms. Sequoyitol, however, has only been detected in the gymnosperms, whereas no methyl ethers of inositol have been found in the Monocotyledones. It seems, therefore, that the biosynthesis of these cyclitols is a primitive characteristic which has been partly retained by many of the Dicotyledones but lost during the evolution of the Monocotyledones.

As the end-products of sugar metabolism, the polysaccharides would be expected to provide information on phylogenetic relationships, but at present the detailed knowledge of their molecular structure is too sparse to draw more than tentative conclusions. Whereas starch is ubiquitous, the storage carbohydrate, inulin, is characteristic of the Compositae family, while members of the Gramineae store fructans. The grasses fall into two distinct groups based on their storage carbohydrate content. Thus, tropical and subtropical species accumulate starch in their leaves, but temperate grasses belonging to the tribes, Hordeae, Aveneae and Festuceae, accumulate fructans.

A comparison of D-xylans isolated from members of the Papilionoideae subfamily of the Leguminosae and the Gramineae show considerable differences between the two, but much smaller differences within each family. It would seem, therefore, that these polysaccharides are potentially useful taxonomic markers, but too little is known of their detailed structure at present to draw any definite conclusions.

The arabinogalactans appear to be characteristic of the Coniferales and are found in both the wood and in exudate gums. They are also present in green seaweeds (Chlorophyceae).

Analysis of *Acacia* exudates has shown distinct differences between the sections Gummiferae and Vulgares. However, the gums of the gymnosperm *Araucaria* (Araucariaceae) showed only small variations.

Bibliography

General

The Carbohydrates, 2nd edn, vols. 1 and 2, W. Pigman and D. Horton (eds.), Academic Press, 1970–72

'Carbohydrates and Nucleotides', in *Cellular Biochemistry and Physiology*, N. A. Edwards and K. A. Hassall, McGraw-Hill, 1971

'Simple Sugars, Monosaccharides', in *Introduction to Modern Biochemistry*, 4th edn, P. Karlson, Academic Press, 1975

'Glycosides, Oligosaccharides, Polysaccharides', *ibid*.

Advances in Carbohydrate Chemistry and Biochemistry (continuing series, vol. 1, 1945), Academic Press

Carbohydrate Chemistry, Specialist Periodical Reports (continuing series, vol. 1, 1967), The Chemical Society

Photosynthesis

'Photosynthesis: The Path of Carbon', in *Plant Biochemistry*, 3rd edn, J. Bonner and J. E. Varner (eds.), Academic Press, 1976
'Photosynthesis: The Path of Energy', *ibid*.
'Photosynthesis and the Pentose–Phosphate Cycle', in *Cellular Biochemistry and Physiology, ibid*.

Biosynthesis

'Mono- and Oligosaccharides', in *Plant Biochemistry,* 3rd edn, J. Bonner and J. E. Varner (eds.), Academic Press, 1976
'Polysaccharides', *ibid*.
'Cell Wall Biogenesis', *ibid*.
'The Primary Cell Wall', *ibid*.
Plant Carbohydrate Biochemistry, J. B. Pridham (ed.), Academic Press, 1974
Phytochemistry, vol. 1, L. P. Miller (ed.), Van Nostrand Reinhold Co., 1973
'The Biochemistry of *Myo*-inositol in Plants', in *Recent Advances in Phytochemistry*, vol. 8, V. C. Runeckles and E. E. Conn (eds.), Academic Press, 1974
'The Nonstructural Carbohydrates', in *Chemistry and Biochemistry of Plant Herbage*, vol. 1, G. W. Butler and R. W. Bailey (eds.), Academic Press, 1973
'Structural Carbohydrates', *ibid*.
Biogenesis of Plant Cell Wall Polysaccharides, F. Loewus (ed.), Academic Press, 1973
'Chemistry and Biochemistry of Algal Cell-wall Polysaccharides', in *Plant Biochemistry*, Biochemistry Series One, vol. 11, D. H. Northcote (ed.), Butterworths, 1974

Metabolism and Function

'Glycolysis and Other Pathways of Carbohydrate Metabolism', in *Metabolic Pathways,* 3rd edn, vol. 1, B. Axelrod (ed.), Academic Press, 1967
'Carbohydrate Metabolism', R. Caputto *et al.*, in *Annual Review of Biochemistry*, vol. 36, 1967
'Biochemistry of Plant Pollination', in *Introduction to Ecological Biochemistry*, J. B. Harborne, Academic Press, 1977

Chemosystematics

'The Distribution of Plant Glycosides', in *Chemical Plant Taxonomy*, T. Swain (ed.), Academic Press, 1963
'The Distribution of Aliphatic Polyols and Cyclitols', *ibid*.
'The Natural Distribution of Plant Polysaccharides', in *Comparative Phytochemistry*, T. Swain (ed.), Academic Press, 1966
'Comparative Biosynthetic Pathways in Higher Plants', section: Branch-chain Sugars, in *Chemistry in Evolution and Systematics,* T. Swain (ed.), Butterworths, 1973
'Polysaccharides of the Leguminosae', in *Chemotaxonomy of the Leguminosae*, J. B. Harborne *et al.* (eds.), Academic Press, 1971

3 The Acetate-Malonate Pathway

Introduction

The main products of the acetate-malonate pathway are the fatty acids, both those primary metabolites which occur universally and the more unusual compounds with a restricted distribution. This pathway also makes an important contribution to plant aliphatic and aromatic compounds, which are biosynthesised through the formation of polyketides. In this chapter, we shall describe the fatty acids and their metabolites, while in Chapter 4 the polyketides and their derivatives are discussed.

Acetyl Coenzyme A

Acetyl coenzyme A (3.1), the precursor of the acetate-malonate pathway, is a

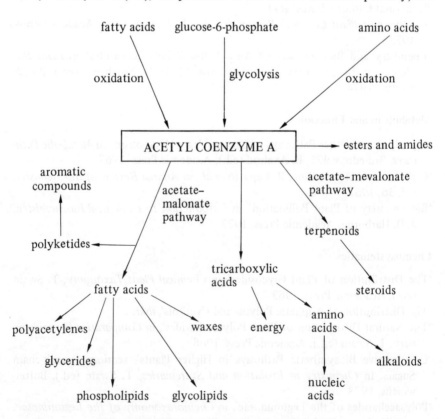

Figure 3.1 The acetyl coenzyme A metabolic pool.

metabolite of extreme importance in both primary and secondary metabolism. In every living organism, there exists a metabolic pool of acetyl CoA which is continually replenished and depleted. Glycolysis and the catabolism of fatty acids and amino acids produce acetyl CoA, while this compound is the precursor of a host of primary and secondary metabolites, including fatty acids, terpenoids, steroids, polyketides, aromatic compounds and acetyl esters and amides (*Figure 3.1*). The conversion of acetyl CoA to citrate and other tricarboxylic acids leads to the formation of the amino acids and their products, such as the nucleic acids and alkaloids.

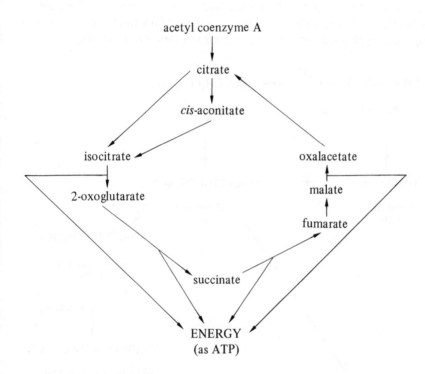

Figure 3.2 The tricarboxylic acid cycle.

Acetyl CoA is also central to the catabolism of glucose and the fatty acids, which, through the tricarboxylic acid cycle (*Figure 3.2*), are the main reaction sequences producing the energy necessary for metabolic processes.

In plants, an enzyme, acetyl CoA synthetase, catalyses the formation of the thiol ester (3.1) from acetate (3.2) (*Figure 3.3*).

$$CH_3COO^- + HS.CoA \rightarrow CH_3COS.CoA + OH^-$$

acetate   acetyl CoA
(3.2)   (3.1)

Figure 3.3 The formation of acetyl CoA.

The Acetate–Malonate Pathway

A thorough study of the acetate-malonate pathway (*Figure 3.4*), by which fatty acids are biosynthesised, has been made in plants, especially in spinach (*Spinacia oleracea*, Chenopodiaceae) chloroplasts, and the enzymes involved have been well characterised. The initial stage is the formation of malonyl CoA (3.3) by carboxylation of acetyl CoA, a reaction catalysed by acetyl CoA carboxylase.

Malonyl CoA is also formed from malonate (3.4) by a reaction analogous to that shown in *Figure 3.3*. Malonate is a product of the catabolism of oxalacetate (3.5) (*Figure 3.5*) and the production of malonyl CoA from malonate may be important in plants, as oxalacetate is formed in a variety of reactions.

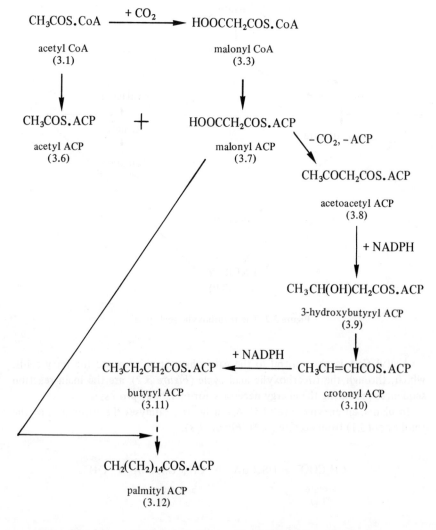

Figure 3.4 The acetate–malonate pathway to the formation of fatty acids.

$$^-OOCCOCH_2COO^- \xrightarrow{\frac{1}{2}O_2} {}^-OOCCH_2COO^- + CO_2$$

oxalacetate   malonate
(3.5)   (3.4)

Figure 3.5 The catabolism of oxalacetate.

Before the acetyl and malonyl radicals can react, they must be transferred from CoA esters to acyl carrier protein (ACP). Like coenzyme A (see Chapter 10), ACP is a phosphopantetheine derivative, but protein replaces the ADP moiety of the coenzyme. The enzymes catalysing the transfers are (ACP)acetyl transferase and (ACP)malonyl transferase. Each enzyme is specific to a small number of substrates and the acetyl transferase will not accept malonyl CoA as substrate and vice versa. This is possibly one way in which a control over fatty acid synthesis is exercised.

Acetyl ACP (3.6) and malonyl ACP (3.7) react to form acetoacetyl ACP (3.8) and carbon dioxide (*Figure 3.4*), the enzyme catalysing the reaction being 3-oxoacyl ACP synthase (condensing enzyme). Reduction of acetoacetyl ACP is catalysed by 3-oxoacyl ACP reductase, an enzyme requiring NADPH, while dehydroxylation of 3-hydroxybutyryl ACP (3.9) is catalysed by 3-hydroxyacyl[ACP] hydratase. The final step in the sequence is the reduction of crotonyl ACP (3.10) to butyryl ACP (3.11), a reaction catalysed by enoyl ACP reductase and also requiring NADPH.

The overall process of the first stage of the acetate–malonate pathway to the formation of fatty acids is therefore the addition of acetate to malonate with the evolution of carbon dioxide and reduction of the product to give butyrate. The enzymes catalysing the various reactions are known as the fatty acid synthetase complex.

This sequence of reactions is repeated until six further malonyl units have been added to butyryl ACP, forming the C_{16} palmityl ACP (3.12), when a termination reaction occurs. Three types of terminal reaction have been observed in plants – transfer of the palmityl radical to ACP which is not bound to the *de novo* fatty acid synthetase complex, transfer to CoA, or hydrolysis to the free acid. Palmitic acid, therefore, is the primary product of fatty acid synthesis, although termination can occur at shorter chain lengths to form acids with fewer carbon atoms.

A shorthand notation has been developed for fatty acids whereby palmitic acid is represented as 16:0 and oleic acid as 18:1(9c). The first number denotes the number of carbon atoms in the chain, and the second denotes the number of unsaturated linkages in the molecule. The number(s) in brackets shows the position of the ethylenic bonds, while c or t represent *cis* or *trans*. When calculating the position of unsaturation, the carbon atoms are numbered from the carboxyl group, which is C-1. Occasionally, numbering takes place from the other end of the chain, when the position of the double bond in oleic acid would be denoted as $\omega 9$.

Fatty Acid Elongation Systems

It has been clearly established that the *de novo* synthesis of fatty acids ends at palmitic acid (16:0) and that stearic acid (18:0) and higher fatty acids are formed by elongation systems, whose enzyme complex characteristics differ from those of the *de novo* complex. There appear to be two elongation systems; in type 1, palmitic acid is elongated to stearic acid, while in type 2, stearic acid is converted to the very-long-chain fatty acids found in waxes. In the type 1 system, palmityl ACP is the substrate, while in type 2 it is stearyl ACP. Both systems derive the extra C_2 units from malonyl ACP. The elongation enzyme complexes do not contain (ACP)acyltransferases, and thus neither the CoA esters nor the free acids can act as substrates for elongation.

The elongation systems will also accept unsaturated acyl ACPs as substrates.

Oleic acid (18:1(9c)), the most common of the monoenoic fatty acids, is biosynthesised in plants as oleoyl ACP (3.13) from stearyl ACP (3.14). The enzyme catalyst, stearyl ACP desaturase, requires oxygen, NADPH and ferredoxin, and, although several reaction mechanisms have been postulated, none have been proved to occur *in vivo*. The desaturation results in an unsaturated acid with *cis* configuration (*Figure 3.6*).

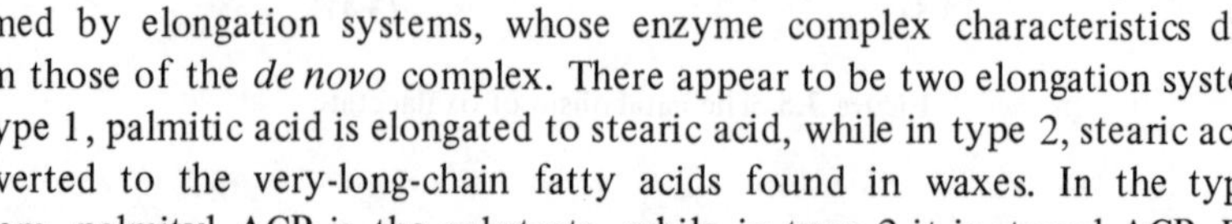

Figure 3.6 The biosynthesis of monoenoic fatty acids.

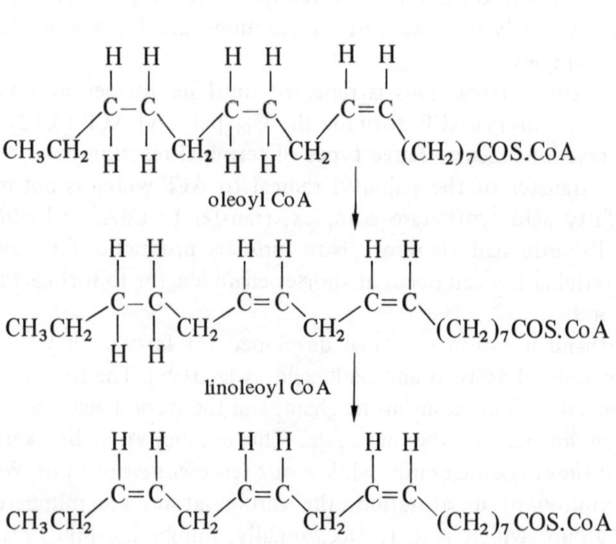

Figure 3.7 The biosynthesis of polyenoic fatty acids.

Oleic acid can also be formed by desaturation of shorter-chain saturated fatty acids followed by elongation.

Sequential desaturation of monoenoic acids gives first the dienoic and then the trienoic derivatives, the most common being linoleic acid (18:2(9c, 12c)) and linolenic acid (18:3(9c, 12c, 15c)). In contrast to the monoenoic system, the polyenoic systems require the CoA esters as substrates (*Figure 3. 7*).

Linolenic acid is also synthesised in plants by desaturation of lauric acid (12:0) to the trienoic derivative followed by elongation.

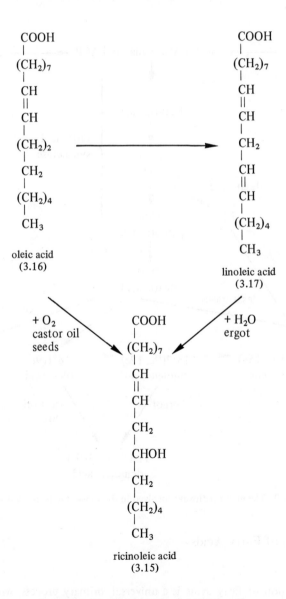

Figure 3.8 The biosynthesis of ricinoleic acid.

The Biosynthesis of Hydroxy Acids

Hydroxyl derivatives of fatty acids are biosynthesised by a number of pathways which include incorporation of oxygen into a saturated linkage, or addition of water to a double bond. Ricinoleic acid $(12\text{-}OH\text{-}18:1(9c))$ (3.15) is biosynthesised by both pathways, as in developing castor oil seeds (*Ricinus communis*, Euphorbiaceae) it is derived from oleic acid (3.16), while in the fungus, ergot (*Claviceps purpurea*), linoleic acid (3.17) is the precursor (*Figure 3.8*).

The main biosynthetic pathways leading to plant fatty acids are summarised in *Figure 3.9*.

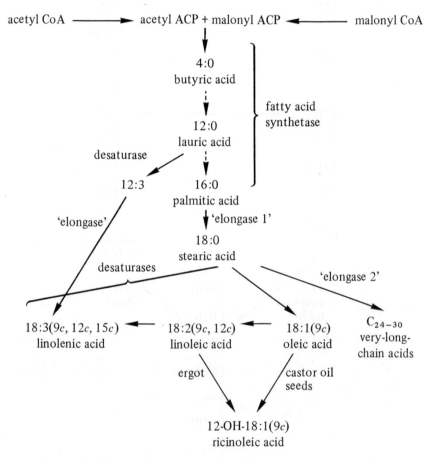

Figure 3.9 The major pathways involved in the biosynthesis of plant fatty acids.

Catabolism of Fatty Acids

β-Oxidation

The β-oxidation of fatty acids is a universal, primary process, which in *Figure 3.10* is illustrated for a saturated acid with an even number of carbon atoms.

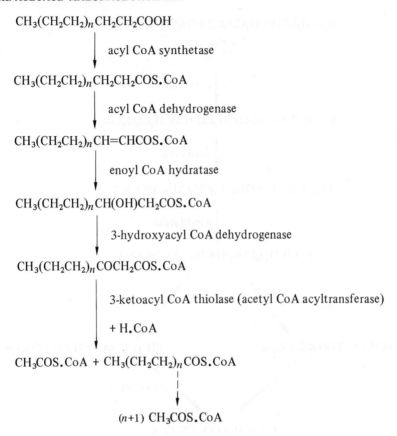

$$CH_3(CH_2CH_2)_nCH_2CH_2COOH$$

acyl CoA synthetase

$$CH_3(CH_2CH_2)_nCH_2CH_2COS.CoA$$

acyl CoA dehydrogenase

$$CH_3(CH_2CH_2)_nCH=CHCOS.CoA$$

enoyl CoA hydratase

$$CH_3(CH_2CH_2)_nCH(OH)CH_2COS.CoA$$

3-hydroxyacyl CoA dehydrogenase

$$CH_3(CH_2CH_2)_nCOCH_2COS.CoA$$

3-ketoacyl CoA thiolase (acetyl CoA acyltransferase)

+ H.CoA

$$CH_3COS.CoA + CH_3(CH_2CH_2)_nCOS.CoA$$

$$(n+1)\ CH_3COS.CoA$$

Figure 3.10 The β-oxidation of fatty acids.

Fatty acids with an odd number of carbon atoms give propionyl CoA as the final product of β-oxidation. This is further oxidised to acetyl CoA and carbon dioxide.

Unsaturated acids also act as substrates and oxidation proceeds normally until a 2-alkenoyl derivative is formed. *Trans* derivatives are natural intermediates and continue the cycle, but the more common *cis* derivatives undergo isomerisation reactions which convert them to *trans* compounds before oxidation continues. These reactions are catalysed by epimerases and isomerases.

Little detailed knowledge of the catabolism of substituted fatty acids has yet been acquired, but a mechanism (*Figure 3.11*) explaining the experimental observations of the oxidation of ricinoleic acid (3.15) in castor oil beans has been proposed.

The intermediates of β-oxidation do not accumulate, and thus short-chain fatty acids are formed by the *de novo* biosynthetic pathway (*Figure 3.4*) rather than by catabolism of long-chain acids.

In plants, acetyl CoA resulting from β-oxidation is concerned in two processes. In the mitochondria, it is oxidised to carbon dioxide by the

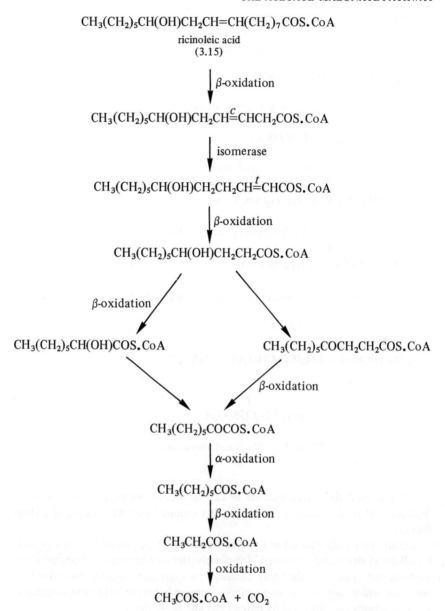

$$CH_3(CH_2)_5CH(OH)CH_2CH=CH(CH_2)_7COS.CoA$$

ricinoleic acid

(3.15)

$\downarrow$ β-oxidation

$$CH_3(CH_2)_5CH(OH)CH_2CH\overset{c}{=}CHCH_2COS.CoA$$

$\downarrow$ isomerase

$$CH_3(CH_2)_5CH(OH)CH_2CH_2CH\overset{t}{=}CHCOS.CoA$$

$\downarrow$ β-oxidation

$$CH_3(CH_2)_5CH(OH)CH_2CH_2COS.CoA$$

β-oxidation

$$CH_3(CH_2)_5CH(OH)COS.CoA \qquad\qquad CH_3(CH_2)_5COCH_2CH_2COS.CoA$$

β-oxidation

$$CH_3(CH_2)_5COCOS.CoA$$

$\downarrow$ α-oxidation

$$CH_3(CH_2)_5COS.CoA$$

$\downarrow$ β-oxidation

$$CH_3CH_2COS.CoA$$

$\downarrow$ oxidation

$$CH_3COS.CoA + CO_2$$

Figure 3.11 A proposed mechanism for the oxidation of ricinoleic acid.

tricarboxylic acid cycle (*Figure 3.2*). This process releases energy in the form of ATP. In the organelles known as glyoxysomes, however, acetyl CoA is converted to glucose through the glyoxalate pathway (*Figure 3.12*).

The succinate generated by this pathway is transported to the mitochondria where it joins the tricarboxylic acid cycle.

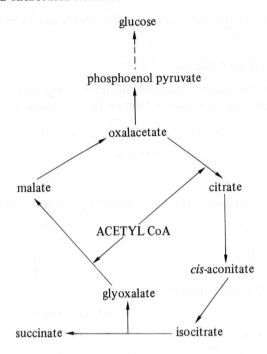

Figure 3.12 The glyoxalate pathway.

α-Oxidation

The α-oxidation of fatty acids (*Figure 3.13*) occurs without conversion of the acid to its CoA ester and this process probably produces the 2-hydroxy acids which are important constituents of the sphingolipids (see below). A 2-hydroperoxy acid is first generated by reaction of the fatty acid with oxygen, catalysed by a flavoprotein. This hydroperoxy derivative can form either a 2-hydroxy acid or, through the intermediate formation of an aldehyde, a fatty acid with one less carbon atom. The role of α-oxidation in plants has not yet

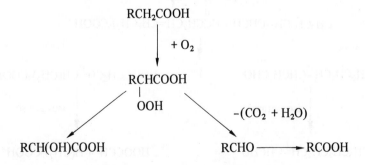

R = long saturated or unsaturated carbon chain

Figure 3.13 The α-oxidation of fatty acids.

been established, but evidence to date indicates that it may be as important as β-oxidation.

Lipoxygenase-catalysed Oxidation

Fatty acids with the linoleic pattern of unsaturation add on oxygen to form hydroperoxy derivatives, as shown in *Figure 3.14*. This reaction is catalysed by

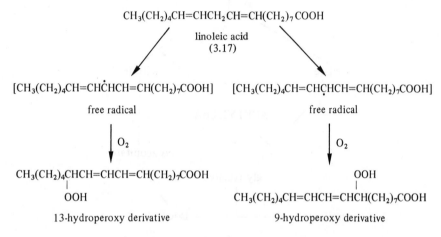

Figure 3.14 The lipoxygenase-catalysed formation of hydroperoxy derivatives.

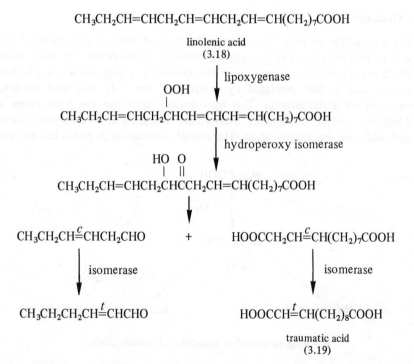

Figure 3.15 The formation of traumatic acid.

lipoxygenase. Hydroperoxy acids are cell toxins and are therefore metabolised as soon as they are formed. In damaged tissue, one of the main products of the catabolism of linolenic acid (3.18) is traumatic acid (3.19) (*Figure 3.15*), while, in potato tuber extracts, hydroperoxy compounds are broken down to aldehydes through the intermediate formation of divinylether derivatives (*Figure 3.16*).

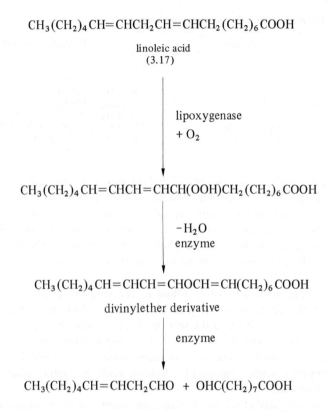

$$CH_3(CH_2)_4 CH=CHCH_2 CH=CHCH_2 (CH_2)_6 COOH$$

linoleic acid
(3.17)

lipoxygenase
+ O_2

$$CH_3(CH_2)_4 CH=CHCH=CHCH(OOH)CH_2 (CH_2)_6 COOH$$

$-H_2O$
enzyme

$$CH_3(CH_2)_4 CH=CHCH=CHOCH=CH(CH_2)_6 COOH$$

divinylether derivative

enzyme

$$CH_3(CH_2)_4CH=CHCH_2CHO \ + \ OHC(CH_2)_7COOH$$

Figure 3.16 The formation of aldehydes from linoleic acid in potato tuber extracts.

Other volatile breakdown products have been detected in plants, including ethylene and various aldehydes. The aldehydes, non-2-enal (3.20) and nona-2,4-dienal (3.21), which give cucumbers (*Cucumis sativa,* Cucurbitaceae) their characteristic odour, are derived from linoleic acid through the formation of hydroperoxides. It seems possible that common fatty acid hydroperoxides are also the precursors of many of the unusual fatty acids including hydroxy and epoxy derivatives.

$$CH_3(CH_2)_5CH=CHCHO \qquad\qquad CH_3(CH_2)_3CH=CHCH=CHCHO$$

non-2-enal   nona-2, 4-dienal
(3.20)   (3.21)

Fatty Acids in Plants

Fatty acids occurring in plants can be divided into major (*Table 3.1*), minor (*Table 3.2*) and unusual compounds. The major and minor acids are ubiquitous and are classed according to their concentrations, most minor acids only occurring in trace amounts.

Table 3.1 The major fatty acids occurring in plants.

Acid	Symbol	Molecular structure
lauric	12:0	$CH_3(CH_2)_{10}COOH$
myristic	14:0	$CH_3(CH_2)_{12}COOH$
palmitic	16:0	$CH_3(CH_2)_{14}COOH$
stearic	18:0	$CH_3(CH_2)_{16}COOH$
oleic	18:1(9c)	$CH_3(CH_2)_7CH=CH(CH_2)_7COOH$
linoleic	18:2(9c, 12c)	$CH_3(CH_2)_3(CH_2CH=CH)_2(CH_2)_7COOH$
linolenic	18:3(9c, 12c, 15c)	$CH_3(CH_2CH=CH)_3(CH_2)_7COOH$

Fatty acids with an odd number of carbon atoms are rare and margaric acid should, perhaps, be included with the unusual compounds. However, this acid occurs in trace amounts in many vegetable oils and is thus classed as a minor acid. Another ubiquitous compound, *trans*-3-hexadecenoic acid (16:1(3t) has an unusual configuration. It occurs in photosynthetic tissue as the phosphatidyl ester, while the seed oils of some members of the Compositae family are rich in this acid.

Unusual fatty acids are mainly found in the seed oils of plants, each compound generally being specific to a genus or family. Thus such compounds are useful taxonomic markers (see the section on plant lipids in chemosystematics, p. 79). Most of the unusual fatty acids occurring in plants are C_{18} derivatives which can be assigned to one of four main groups of compounds – non-conjugated ethylenic acids, conjugated ethylenic acids, acetylenic acids and substituted compounds.

The non-conjugated ethylenic acids contain double bonds in unusual positions or with a *trans* configuration. It is possible that the enzymes catalysing such common desaturations are mutants of the ubiquitous desaturases. Petroselinic

Table 3.2 The minor fatty acids occurring in plants.

Acid	Symbol	Molecular structure
caproic	6:0	$CH_3(CH_2)_4COOH$
caprylic	8:0	$CH_3(CH_2)_6COOH$
capric	10:0	$CH_3(CH_2)_8COOH$
margaric	17:0	$CH_3(CH_2)_{15}COOH$
arachidic	20:0	$CH_3(CH_2)_{18}COOH$
behenic	22:0	$CH_3(CH_2)_{20}COOH$
lignoceric	24:0	$CH_3(CH_2)_{22}COOH$
palmitoleic	16:1(9c)	$CH_3(CH_2)_5CH=CH(CH_2)_7COOH$
erucic	22:1(13c)	$CH_3(CH_2)_7CH=CH(CH_2)_{11}COOH$
γ-linolenic	18:3(6c, 9c, 12c)	$CH_3(CH_2)_3(CH_2CH=CH)_3(CH_2)_4COOH$
arachidonic	20:4(5c, 8c, 11c, 14c)	$CH_3(CH_2)_3(CH_2CH=CH)_4(CH_2)_3COOH$

acid $(18:1(6c))$, with the double bond at position 6 rather than position 9, as in oleic acid, is characteristic of the related Umbelliferae and Araliaceae families. *Trans* analogues of linoleic acid have been found in the seed oils of several plants, while a *trans*-2-enoic acid has been detected in pollen, where it apparently acts as a food marker for bees. *Trans*-2-enoic acids are important metabolites in bees, as the queen substance, 9-oxo-*trans*-2-decenoic acid $(9\text{-}CO\text{-}10:1(2t))$, is the hormone produced by the queen bee which is largely responsible for the functioning of the hive.

The conjugated ethylenic acids do not have methylene groups separating the double bonds as in linoleic and linolenic acids. Such acids occur widely and most can be assigned to one of four classes denoted *cis*-9, *cis*-12, *trans*-9 and *trans*-12. α-Eleostearic acid $(18:3(9c, 11t, 13t))$ (3.22), the main constituent of tung oil

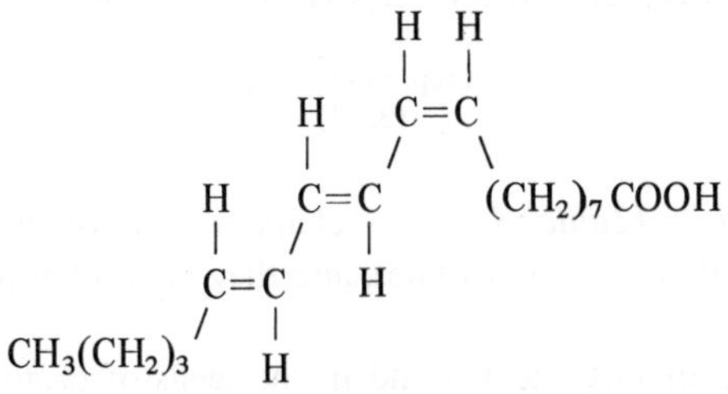

α-eleostearic acid
(3.22)

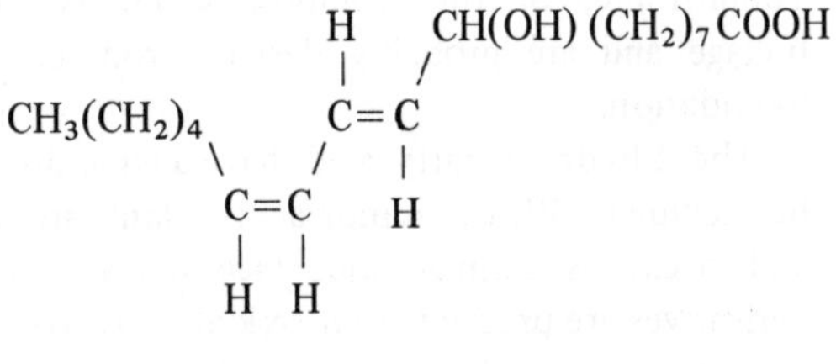

α-dimorphecolic acid
(3.23)

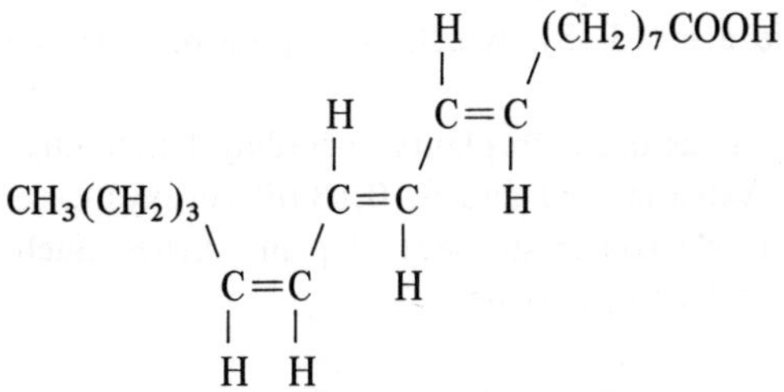

catalpic acid
(3.24)

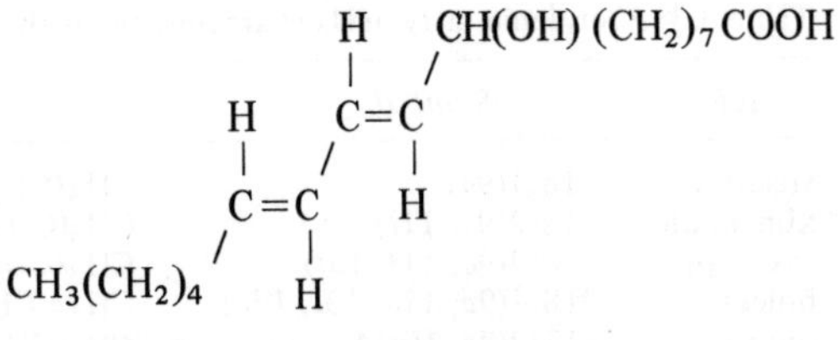

β-dimorphecolic acid
(3.25)

(*Aleurites montana,* Euphorbiaceae), is an example of a *cis*-9 compound, while
α-dimorphecolic acid (9-OH-18:2(10*t,* 12*c*)) (3.23), which occurs in the seed
oils of several plants, belongs to the *cis*-12 class. The seeds of *Catalpa ovata*
(Bignoniaceae) contain the trans-9 compound, catalpic acid (18:3(9*t,* 11*t,* 13*c*))
(3.24), while β-dimorphecolic acid (9-OH-18:2(10*t,* 12*t*)) (3.25), from
Dimorphotheca sinuata (Compositae), belongs to the *trans*-12 class.

A large number of fatty acids with acetylenic bonds have been isolated from
seed oils. One of the simplest, octadec-6-ynoic acid (18:1(6*a*), where *a* indicates
an acetylenic linkage) isolated from the seed oils of *Picramnia* species (Simaroubaceae), is the only naturally occurring compound with an acetylenic linkage
at position 6. The majority of the acetylenic acids are derived from either
stearolic acid (18:1(9*a*)) (3.26) or crepenynic acid (18:2(9*c,* 12*a*)) (3.27).

$$CH_3(CH_2)_7C{\equiv}C(CH_2)_7COOH \qquad CH_3(CH_2)_4C{\equiv}CCH_2CH{=}CH(CH_2)_7COOH$$

stearolic acid   crepenynic acid

(3.26)   (3.27)

Stearolic acid derived compounds (*Table 3.3*) are characteristic of the
Santalaceae and Olacaceae families of the order Santales, and they are probably
all biosynthesised from oleic acid.

Crepenynic acid (3.27) is the major fatty acid found in the seeds of *Crepis
foetida* (Compositae). Shorter-chain acids are characteristic of the subfamily,
Tubuliflorae, of the Compositae family. They all contain ω6 acetylenic
linkage and are probably derived from crepenynic acid or its derivatives by
β-oxidation.

The 2-hydroxy fatty acids have a wide distribution and should more properly
be included with the minor acids. Many are derivatives of very-long-chain acids
and occur as sphingolipids (see below). It seems probable that 2-hydroxy
derivatives are products of α-oxidation (*Figure 3.13*).

3-Hydroxy acids are intermediates in both the synthesis and degradation of
fatty acids, but they do not accumulate to any extent. Probably the best known
of the hydroxy acids is ricinoleic acid (3.15), which is responsible for the
purgative effect of castor oil.

A number of polyhydroxy fatty acids occur in plants, including 10,16-dihy-
droxypalmitic acid, 10,18-dihydroxystearic acid and 9,10,18-trihydroxystearic
acid, which are all found in the cuticle (outer surface) of plant tissues. Such
compounds are probably derived from hydroperoxides.

Table 3.3 Acetylenic fatty acids characteristic of the Santalaceae and Olacaceae families.

Acid	Symbol	Molecular structure
stearolic	18:1(9*a*)	$CH_3(CH_2)_7C{\equiv}C(CH_2)_7COOH$
ximenynic	18:2(9*a,* 11*t*)	$CH_3(CH_2)_5CH{=}CHC{\equiv}C(CH_2)_7COOH$
exocarpic	18:3(9*a,* 11*t,* 13*t*)	$CH_3(CH_2)_3(CH{=}CH)_2C{\equiv}C(CH_2)_7COOH$
bolekic	18:4(9*a,* 11*a,* 13*c,* 17*e*)	$CH_2{=}CH(CH_2)_2CH{=}CH(C{\equiv}C)_2(CH_2)_7COOH$
pyrulic	17:2(8*a,* 10*t*)*	$CH_3(CH_2)_5CH{=}CHC{\equiv}C(CH_2)_6COOH$

*Other C_{17} acids also occur, many are analogues of the C_{18} acids.

The oxo analogues of some hydroxy derivatives of fatty acids with conjugated ethylenic bonds have been detected in plants, but such compounds are rare.

Epoxy acids are characteristic of the seed oils of many plants belonging to the Euphorbiaceae, Onagraceae, Dipsaceae, Olacaceae, Valerianaceae and Compositae families. They appear to be biosynthesised from common unsaturated acids, probably through the intermediate formation of hydroperoxides. Vernolic acid (epoxyoleic acid, 12,13-epoxy-18:1(9c)) (3.28) has been found in the seed oils of a number of plants and it is the major constituent of the oil from *Vernonia anthelmintica* (Compositae). Epoxy acids are formed as artifacts when seed oils are stored under aerobic conditions.

$$CH_3(CH_2)_4 \overset{\displaystyle O}{\overset{\displaystyle / \ \backslash}{CH-CH}} CH_2 CH = CH(CH_2)_7 COOH$$

vernolic acid
(3.28)

Dicarboxylic acids are rare, but sumach fruits (*Rhus succedanea*, Anacardiaceae) contain several with the general formula $HOOC(CH_2)_n COOH$. Traumatic acid (2-dodecendioic acid) (3.19) is important in cellular repair mechanisms. It is biosynthesised from linoleic or linolenic acids in damaged tissue by oxidative degradation.

Cyclopropene derivatives such as sterculic acid (3.29) are characteristic of families belonging to the order Malvales.

$$CH_3(CH_2)_7 \overset{\displaystyle CH_2}{\overset{\displaystyle / \ \backslash}{C = C}} (CH_2)_7 COOH$$

sterculic acid
(3.29)

Although rare, other acids besides acetic acid can act as starter units for long-chain fatty acid biosynthesis. Thus, the acids which give rise to the 2- and 3-methylalkanes found in plant waxes are biosynthesised from simple branched compounds such as angelic acid (3.30), found in the roots of angelica (*Angelica archangelica*, Umbelliferae).

angelic acid
(3.30)

The cyclopentene compounds, such as hydnocarpic acid (3.31), characteristic of the Flacourtiaceae family, are biosynthesised from aleprolic acid (3.32), a compound which also occurs in this family.

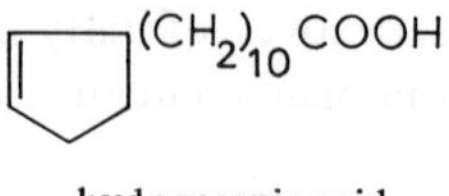

$(CH_2)_{10}COOH$

hydnocarpic acid
(3.31)

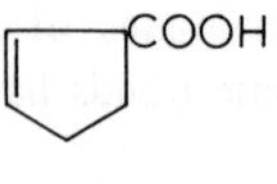

$COOH$

aleprolic acid
(3.32)

H_2COOCR^1
R^2COOCH
CH_2
CH_2OH

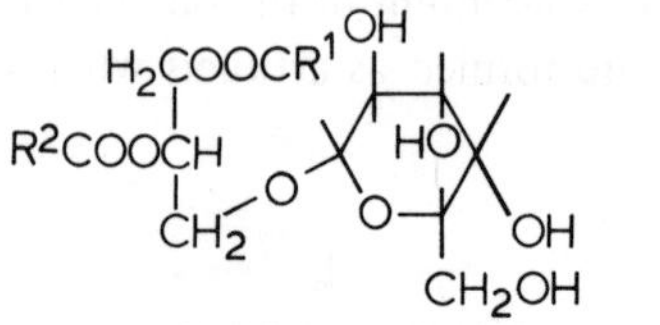

monogalactosyldiglyceride (MGDG)
(3.33)

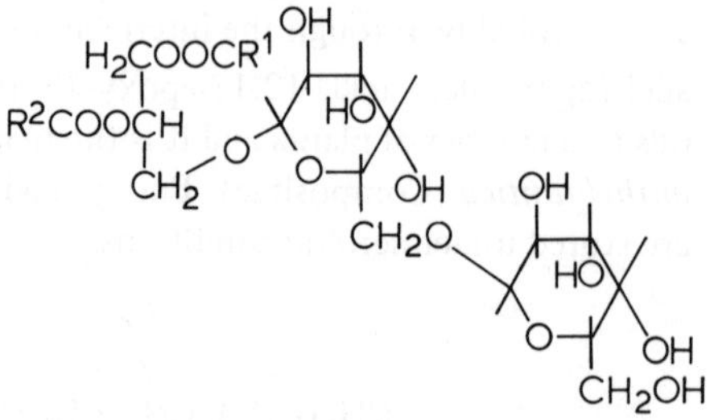

H_2COOCR^1
R^2COOCH
CH_2
CH_2O
CH_2OH

digalactosyldiglyceride (DGDG)
(3.34)

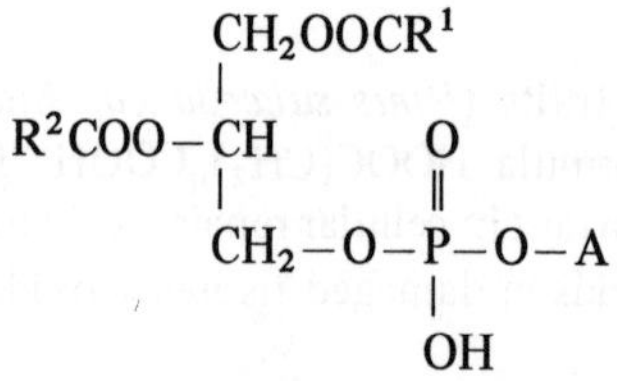

CH_2OOCR^1
$R^2COO-CH$
$CH_2-O-P-O-A$
OH

phosphatidyl lipids
(3.35)

$A = -CH_2CH_2NH_2$

phosphatidylethanolamine (PE)
(3.35c)

$A = -CH_2CH_2N(CH_3)_3OH$

phosphatidylcholine (PC)
(3.35a)

$A = -CH_2CH(NH_2)COOH$

phosphatidylserine (PS)
(3.35d)

$A = \begin{matrix} -CH_2 \\ | \\ CHOH \\ | \\ CH_2OH \end{matrix}$

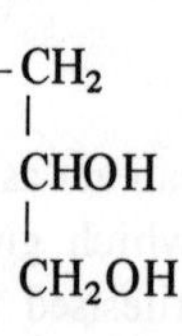

phosphatidylglycerol (PG)
(3.35b)

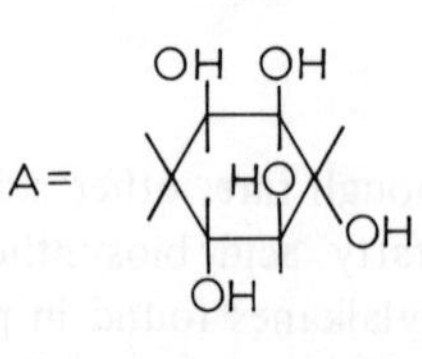

$A =$

phosphatidyl-*myo*-inositol (PI)
(3.35e)

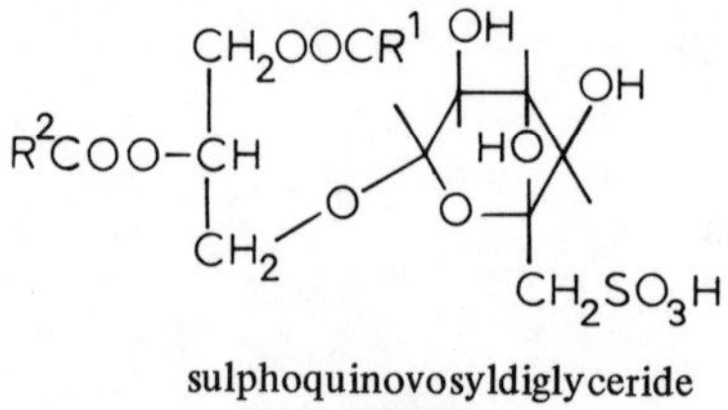

CH_2OOCR^1
$R^2COO-CH$
CH_2
CH_2SO_3H

sulphoquinovosyldiglyceride
(3.36)

$R^1, R^2 =$ long chains

Figure 3.17 The major leaf lipids.

The 18-fluorooleic and 16-fluoropalmitic acids found in the seeds of ratsbane (*Dichapetalum toxicarium*, Dichapetalaceae) would appear to be derived from a fluoroacetyl CoA starter unit, as fluoroacetic acid has also been found in the plant.

Plant Lipids

Free fatty acids are antimetabolites, reacting with protein to destroy enzymes necessary to cell processes. Thus, in all living systems, only minute amounts of the free acids exist and most are esterified in the form of lipids. Although all plant tissue contains lipids, those compounds which have received the most attention occur in the leaves and seeds. Leaf lipids are concentrated in the chloroplast lamellae and four types are generally found (*Figure 3.17*): the glycolipids (monogalactosyldiglyceride (MGDG) (3.33) and digalactosyl-diglyceride (DGDG) (3.34)), phosphatidyl lipids (3.35) and sulpholipid (sul-phoquinovosyldiglyceride (3.36)). The major components of the phosphatidyl lipids are phosphatidylcholine (PC) (3.35a), phosphatidylglycerol (PG) (3.35b), phosphatidylethanolamine (PE) (3.35c), phosphatidylserine (3.35d) and phos-phatidylinositol (PI) (3.35e).

The major fatty acids are the main acyl components of leaf lipids, MGDG and DGDG being rich in unsaturated acids, especially linolenic. The unusual acid, *trans*-3-hexadecenoic acid (16:1(3*t*)), is only found in PG lipids, where it is esterified to the 2 position of glycerol. There is some evidence that this acid is formed from palmitic acid (16:0) after synthesis of the lipid. Palmitic acid is the main fatty acid component of sulpholipid.

Seed lipids differ from leaf lipids in containing a high concentration of tri-glyceride esters (3.37), although glycolipids and phosphatidyl lipids also occur. Most of the unusual fatty acids are found in the triglycerides, where saturated acids are usually esterified to the 1 position of glycerol and unsaturated acids to the 2 position.

$$CH_2OOCR^1 \qquad R^1, R^2, R^3 = \text{long chains}$$
$$R^2COOCH$$
$$CH_2OOCR^3$$

triglycerides
(3.37)

The first stage in lipid synthesis is the transport of fatty acid ACP or CoA esters from the site of their biosynthesis to the site of lipid synthesis, a process catalysed by fatty acid transferases. At the site of lipid synthesis, the fatty acid is transferred to glycerol-3-phosphate (3.38) to give monoacylglycerol-3-phosphate (3.39). Addition of a second fatty acid gives diacylglycerol-3-phosphate (phos-phatidic acid) (3.40) (*Figure 3.18*).

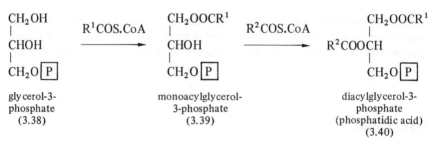

Figure 3.18 The biosynthesis of phosphatidic acid.

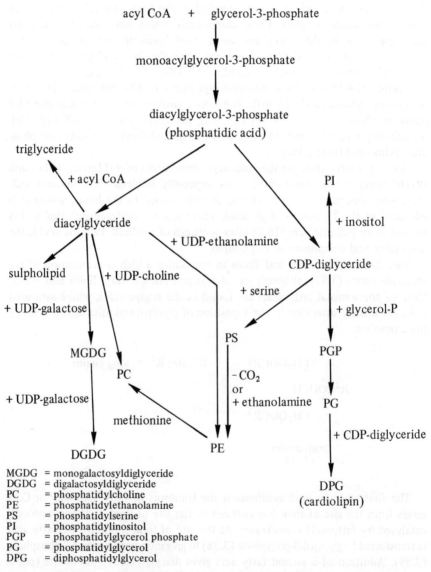

MGDG = monogalactosyldiglyceride
DGDG = digalactosyldiglyceride
PC   = phosphatidylcholine
PE   = phosphatidylethanolamine
PS   = phosphatidylserine
PI   = phosphatidylinositol
PGP   = phosphatidylglycerol phosphate
PG   = phosphatidylglycerol
DPG   = diphosphatidylglycerol

Figure 3.19 The biosynthesis of plant lipids.

Phosphatidic acid is central to the biosynthesis of most, if not all, lipids (*Figure 3.19*). Three alternative pathways for the biosynthesis of this compound have been demonstrated in animals, but they do not appear to be of importance in plants.

The phosphatidyl derivatives are biosynthesised from CDP-diglyceride, which is formed from CTP (cytosine triphosphate) and phosphatidic acid. Reaction of the nucleotide derivative with inositol, glycerol-3-phosphate, or serine gives phosphatidylinositol (3.35e), phosphatidylglycerol (3.35b) and phosphatidylserine (3.35d), respectively. The further reaction of phosphatidylglycerol with CDP-diglyceride gives diphosphatidylglycerol (cardiolipin) (3.41).

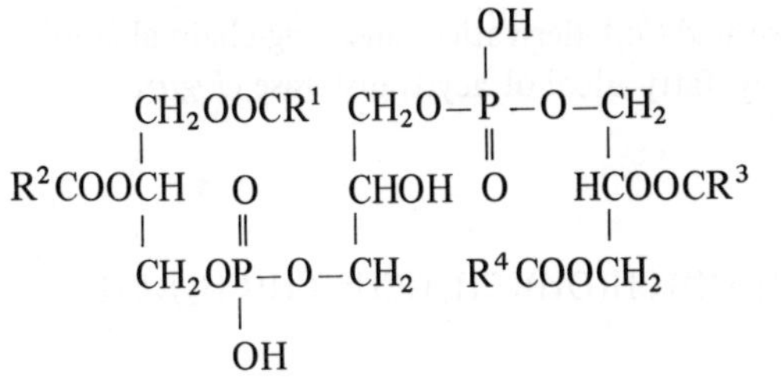

$$R^1, R^2, R^3, R^4 = \text{long chains}$$

$$
\begin{array}{l}
\qquad\qquad\quad OH \\
\qquad\qquad\quad | \\
CH_2OOCR^1\ \ CH_2O-P-O-CH_2 \\
|\qquad\qquad\ \ |\qquad \| \qquad\quad | \\
R^2COOCH\quad O\qquad CHOH\ O\quad HCOOCR^3 \\
|\qquad\ \|\qquad\ \ |\qquad\qquad\ \ | \\
CH_2OP-O-CH_2\quad R^4COOCH_2 \\
|\\
OH
\end{array}
$$

diphosphatidylglycerol
(cardiolipin)
(3.41)

Phosphatidylethanolamine (3.35c) can be synthesised in plants by two pathways, as both decarboxylation of PS catalysed by a decarboxylase and reaction of diglyceride with CDP-ethanolamine have been shown to take place in spinach leaves. The latter reaction is catalysed by ethanolamine phosphotransferase. There is evidence that an exchange reaction between PS and ethanolamine takes place, but this seems to be only a minor pathway.

Phosphatidylcholine (3.35a) is also biosynthesised from diglyceride, the nucleotide derivative being CDP-choline. Sequential methylation of PE by S-adenosylmethionine is a second, but probably less active, pathway.

Both triglycerides and glycolipids are biosynthesised from diglyceride and it is probable that this compound is also the precursor of sulpholipid. The

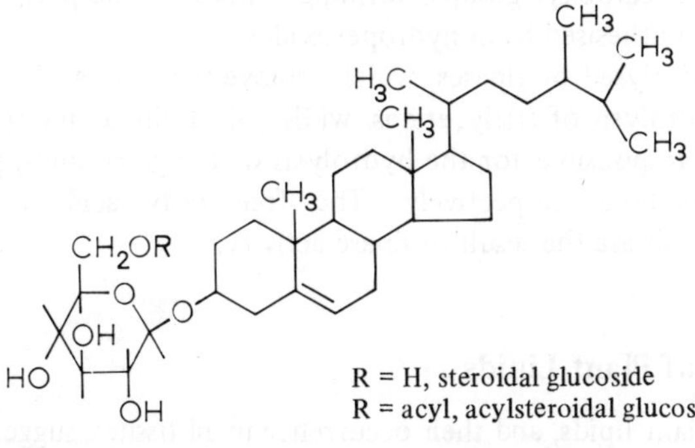

R = H, steroidal glucoside
R = acyl, acylsteroidal glucoside

Figure 3.20 Plant steroidal glucoside lipids.

addition of galactose units, in the form of UDP-galactose, to diglyceride forms first MGDG (3.33) and then DGDG (3.34).

Other types of lipid present in plants include the steroidal esters, sphingolipids (of which only the cerebrosides and phytoglycolipid are important), wax esters and cutin The steroidal glucoside lipids are unique to plants and consist of two types – those in which a steroid is glucosylated only and those in which both glucosylation and acylation has taken place to form an acylsteroidal glucoside (*Figure 3.20*).

The cerebrosides are derivatives of C_{18}-dehydrosphingosine containing glucose and 2-hydroxy fatty acids (*Figure 3.21*), while phytoglycolipid is a complex compound containing several sugar units.

Wax esters are formed from fatty acid CoA derivatives and long-chain alcohols, the esterification being catalysed by fatty alcohol acyltransferase (*Figure 3.22*).

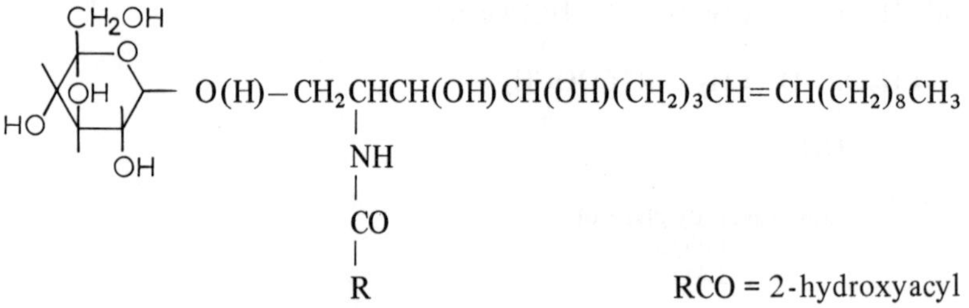

$$O(H) - CH_2CHCH(OH)CH(OH)(CH_2)_3CH{=}CH(CH_2)_8CH_3$$

$$NH$$

$$CO$$

$$R \qquad\qquad RCO = 2\text{-hydroxyacyl}$$

Figure 3.21 Cerebrosides.

$$R^1COS.CoA + R^2OH \;\rightarrow\; R^1COOR^2 + HS.CoA$$

acyl CoA   fatty   wax ester
              alcohol

Figure 3.22 The formation of wax esters.

Cutin occurs in the thick, outer covering of epidermal cells (cuticle), and contains polyhydroxy acids which are cross-linked in a random manner by esterification with carboxyl groups, forming a matrix. The polyhydroxy acids are probably biosynthesised from hydroperoxides.

Lipids are hydrolysed by lipases, which remove the fatty acids. Lipase, itself, catalyses the hydrolysis of triglycerides, while galactolipase, phospholipase and sulpholipase are responsible for the hydrolysis of the glycolipids, phosphatidyl lipids and sulpholipid, respectively. The free fatty acid components of commercial seed oils are the result of lipase activity.

The Function of Plant Lipids

The variety of plant lipids, and their occurrence in all tissues, suggests that these compounds have several functions. Although experimental observations indicate that lipids play a number of roles in plants, proof is lacking in many instances. The most firmly established function of plant lipids, especially in seed oils, is that of energy and carbon storage. Most mature seeds contain a high concentration of lipids which provide energy and a source of carbon to the germinating seedlings. These lipids are concentrated in the oil bodies (spherosomes, oleosomes) and are mostly triglyceride esters. On germination, the esters are hydrolysed and the fatty acids degraded through oxidation to acetyl CoA. This activated compound is either oxidised by the tricarboxylic acid cycle, releasing energy in the form of ATP, or it is converted to sugars through the glyoxalate pathway. The glyoxalate pathway is highly active in the germinating seedling, but ceases as photosynthesis begins.

Lipids are also important constituents of plant membranes, which are essential to the efficient transport of substrates and products of enzymatic reactions. Membranes are characterised by their semipermeability, which allows only certain compounds to pass through, acting as a barrier to others. The type of lipid present in a membrane appears to influence its permeability, although how it does so is not known. Membranes are also the site of many enzymatic reactions.

All membranes contain protein and lipid which are loosely held together by bonds which are non-ionic. The molecules are arranged with their hydrophilic groups at the aqueous interfaces and their hydrophobic groups within the membrane. Such an arrangement influences membrane semipermeability, but a detailed discussion of this property is outside the scope of this book (see 'Role of lipids in water and ion transport' in *Recent Advances in the Chemistry and Biochemistry of Plant Lipids* listed in the bibliography).

Many enzymatic reactions take place in membranes or on their surfaces and the presence of certain lipids appears to be necessary to the functioning of the majority of these enzymes. Although the exact mechanisms are not understood, there are indications that lipids act as cofactors in some reactions. It has been suggested, for example, that high concentrations of unsaturated fatty acids in the digalactosyl lipids of the chloroplasts are necessary as cofactors in the photosynthetic electron transport system. Similarly, phosphatidyl lipids are essential to the respiratory system of the mitochondria. The presence of linolenic acid also appears to be essential for the degradation of methionine to ethylene in apples (*Malus* sp., Rosaceae).

The unsaturated linoleic and linolenic acids are concerned in the healing of damaged plant tissue through their degradation to the wound hormone, traumatic acid. This compound causes enlargement of cells around the wound, a process which leads to eventual healing.

Some fatty acids and lipids act as attractants to insects. The unsaturated acid 18:3(2t, 9c, 12c) present in the pollen of clover (*Trifolium* spp., Leguminosae) acts as a food marker for bees, while the 1,3-diester of glycerol and oleic acid attracts flies to the mushroom, fly agaric (*Amanita muscaria*).

As with many secondary metabolites, a number of unusual fatty acids are poisonous to other organisms. The fluoro fatty acids are highly toxic, ratsbane (*Dichapetalum toxicarium*) having been used in West Africa to kill rats for centuries.

The external tissues of plants contain a protective layer known as the cuticle, which contains polyhydroxy fatty acids, wax esters, hydrocarbons, and fungicides such as flavonoids. It acts as a protective coating preventing fungal and bacterial attack and controlling transpiration The hydrocarbons present in plant cuticle are formed by decarboxylation of very-long-chain fatty acids.

The Economic Importance of Seed Oils

Seed oils are important to both the food and the non-food industries, as they are used in the manufacture of margarine, cooking oils, soap, glycerine, detergents, paints, varnishes, resins, cosmetics, pharmaceuticals and waterproofing agents. The most important oil-producing plants are listed in Table 3.4, together with their major fatty acid contents.

The oil palm produces two types of oil – palm oil from the fruit and palm kernel oil from the seed. Palm oil contains a high concentration of β-carotene, the precursor of vitamin A, and is thus an important item of diet in those countries which use the unrefined oil for cooking. Both types of oil are used for the manufacture of margarine, when the unsaturated acids are completely or partially hydrogenated. Unlike the natural oils, margarine does not go rancid and is solid at room temperature. Rancidity is due to oxidation of unsaturated acids. However, most medical authorities are now of the opinion that unsaturated acids are more healthful, especially for those suffering from heart disease, and thus oils such as corn, sunflower and safflower, with a high concentration of poly-unsaturated fatty acids, have become popular. Linoleic and linolenic acids are essential fatty acids which must be included in the diet or deficiency diseases develop.

Table 3.4 The main oil-producing plants of the world with their major fatty acids.

Common name	Botanical name	Family		Major fatty acids
oil palm	*Elaeis guineensis*	Palmae	*fruit:*	oleic, palmitic
			kernel:	lauric
coconut	*Cocos nucifera*	Palmae		lauric
corn (maize)	*Zea mays*	Gramineae		linoleic
groundnuts	*Arachis hypogaea*	Leguminosae		oleic, linoleic
olive	*Olea europaea*	Oleaceae		oleic
sunflower	*Helianthus annuus*	Compositae		linoleic
soybean	*Glycine max*	Leguminosae		linoleic
sesame	*Sesamum indicum*	Pedaliaceae		oleic, linoleic
safflower	*Carthamus tinctorius*	Compositae		linoleic
cotton	*Gossypium* sp.	Malvaceae		oleic, linoleic
cocoa	*Theobroma cacao*	Sterculiaceae		oleic, stearic
shea	*Butyrospermum paradoxum*	Sapotaceae		stearic
avocado	*Persea americana*	Lauraceae		palmitic, stearic
rapeseed	*Brassica napus*	Cruciferae		erucic (22:1)
rapeseed (erucic acid free)				oleic
linseed	*Linum usitatissimum*	Linaceae		linolenic
tung	*Aleurites montana*	Euphorbiaceae		eleostearic (18:3)
castor oil	*Ricinus communis*	Euphorbiaceae		ricinoleic (12-OH-18:1)

Palm oil and coconut oil are used in the manufacture of soap, when the tri-glyceride esters are hydrolysed by alkali (saponification), the fatty acid salts being used in soaps, while glycerol has many industrial uses.

Oils used in the manufacture of paints, varnishes, resins, etc., have a high concentration of unsaturated fatty acids. On exposure to air and light, these oils are oxidised and polymerised to form a hard, waterproof solid. Tung oil and linseed oil are mainly used as solvents for paints, etc., and in the manufacture of linoleum and oilcloth and for waterproofing fabrics.

Besides its use in the manufacture of paints, waxes, etc., castor oil is purified for administration as a purge. In the digestive tract, the ester is hydrolysed and ricinoleic acid irritates the intestinal walls causing evacuation. Other fatty acids used medicinally include hydnocarpic and chaulmoogric acids and their derivatives from members of the Flacourtiaceae family. Such compounds damage the envelope of the leprosy bacterium, *Bacillus leprae*, and thus effect a slow cure of this disfiguring disease.

Plant Lipids in Chemosystematics

Fatty acids occurring in bacterial lipids are characteristic of these organisms and are not generally found in higher plants. Most bacteria (except Mycobacteria) contain *cis*-vaccenic acid ($18:1(11c)$) in place of the oleic acid ($18:1(9c)$) of higher plants.

Some algal lipids are characterised by their content of unsaturated fatty acids and attempts have been made to classify algae according to their fatty acid composition. Blue-green algae (Cyanophyta) can be divided into three types – those which contain α-linolenic acid ($18:3(9c, 12c, 15c)$), those which contain

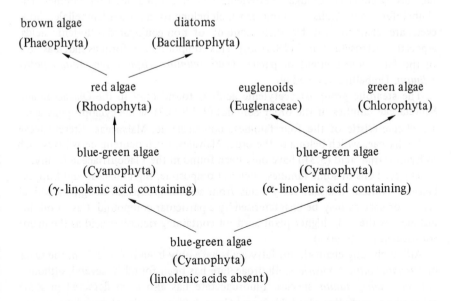

Figure 3.23 The evolution of algae according to fatty acid content.

γ-linolenic acid (18:3(6c, 9c, 12c)) and those which contain neither. Algae can be classified (*Figure 3.23*) according to fatty acid biosynthesis if it is assumed that blue-green algae without either α- or γ-linolenic acid are the most primitive.

In higher plants, the major and minor fatty acids are ubiquitous and are therefore of little use in chemosystematics. Attempts have been made to classify seed oils according to their quantitative content of fatty acids, but such classifications bear little relationship to classical taxonomy. Although most leaf lipids occur in the majority of plants, a few are specific to one or two families. Thus, malvalic acid (3.42) occurs in the leaf lipids of some species belonging to the Malvaceae family, while γ-linolenic acid and stearidonic acid (18:4(6c, 9c, 12c, 15c)) have been found in the leaf lipids of some members of the unrelated Boraginaceae and Caryophyllaceae families.

$$CH_3(CH_2)_7 \underset{\underset{CH_2}{\diagdown\ \diagup}}{C=C}(CH_2)_6 COOH$$

malvalic acid
(3.42)

It is the unusual fatty acids found in seed oils, however, which are potentially the most useful taxonomic markers, although, as yet, there has been little attempt to screen large numbers of species for these compounds. Plants seldom, if ever, biosynthesise only one unusual fatty acid, and thus these compounds are best considered in groups according to their substitution patterns. Some groups or individual acids are too widespread to have phylogenetic significance; the epoxy derivative, vernolic acid (3.28), for instance, occurs in five unrelated families, while *cis*-9 conjugated ethylenic acids occur in at least ten families. The Umbelliferae and Araliaceae families which belong to the order Umbellales, however, are characterised by their content of non-conjugated ethylenic acids, especially petroselinic acid (18:1(6c)). This compound can form as much as 76% of the fatty acid content of parsley (*Petroselinum crispum* syn *Apium petroselinum*, Umbelliferae) seed oil.

The stearolic group of acetylenic acids is found only in the Olacaceae and Santalaceae families of the order Santales (Table 3.3), while cyclopropene acids are characteristic of the four families, Bombacaceae, Malvaceae, Sterculiaceae and Tiliaceae, which belong to the order Malvales. Cyclopentene derivatives such as hydnocarpic acid (3.31) have only been found in the Flacourtiaceae family.

The seed oils of some families, such as Compositae, Bignoniaceae and Euphorbiaceae, contain unusual fatty acids from several groups, although individual genera or species may be characterised by a particular compound. Castor oil, for instance, is the only higher plant seed oil containing ricinoleic acid as the major constituent (up to 90%).

Although long-chain fluoro fatty acids have only been isolated from the seeds of *Diehapetalum toxicarium*, fluoroacetate has been found in several, although not all, *Dichapetalum* species. Fluoroacetate has also been detected in some other plants including *Oxylobium* and *Gastrolobium* species (Leguminosae), and it seems probable that a detailed examination of the seed oils of these plants would reveal the presence of long-chain fluoro fatty acids.

Polyacetylenes and Thiophenes

A large number of polyacetylenes and thiophenes have been isolated from plants since dehydromatricaria ester (3.43) and carlina oxide (3.44) were obtained from the roots of *Artemisia vulgaris* and *Carlina acaulis*, respectively, in the last

$$H_3C-C≡C-C≡C-C≡C-CH=CH-COOCH_3$$

dehydromatricaria ester
(3.43)

carlina oxide
(3.44)

century. Both these plants belong to the Compositae family, which, together with the Umbelliferae, contains the majority of naturally occurring acetylenes and thiophenes. Polyacetylenes with antibiotic properties also occur in species of moulds belonging to Basidiomycetes.

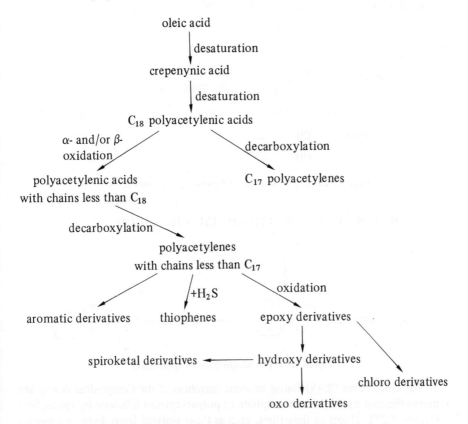

Figure 3.24 The biosynthesis of polyacetylenes and derivatives.

The range of unsaturation and substitution found in polyacetylenes and thiophenes is wide, and thus only a general discussion can be included in this chapter. A full account can be found in *Naturally Occurring Acetylenes* listed in the bibliography.

It has been firmly established that oleic acid is the primary precursor of polyacetylenes and thiophenes. Thus, these compounds are products of the acetate–malonate pathway. Crepenynic acid $(18:2(9c, 12a))$ appears to be the first acetylenic compound to be biosynthesised in plants, other compounds being formed by desaturation, degradation through α- and β-oxidation, decarboxylation and oxidation, as shown in *Figure 3.24*.

Aromatic polyacetylenes are formed by the cyclisation of the end six carbon atoms of a straight-chain compound (*Figure 3.25*), while spiroketals, characteristic of some genera of the Compositae family, are derived from alcohols as shown in *Figure 3.26*.

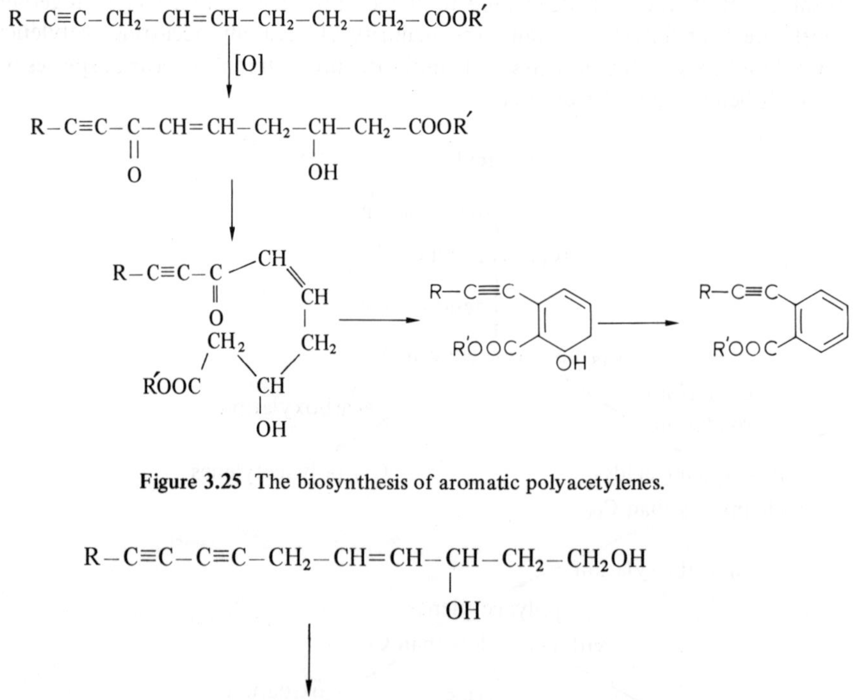

Figure 3.25 The biosynthesis of aromatic polyacetylenes.

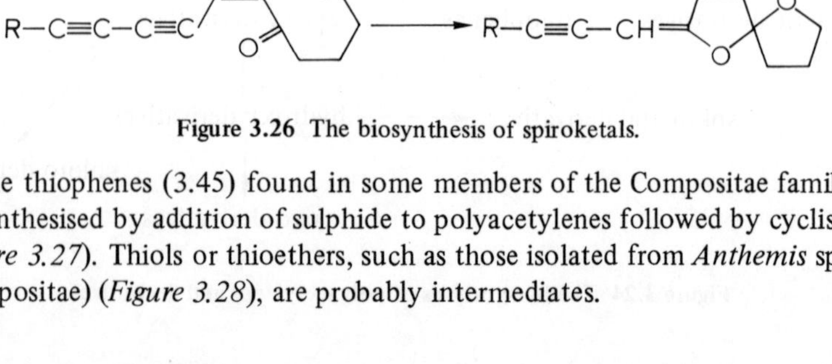

Figure 3.26 The biosynthesis of spiroketals.

The thiophenes (3.45) found in some members of the Compositae family are biosynthesised by addition of sulphide to polyacetylenes followed by cyclisation (*Figure 3.27*). Thiols or thioethers, such as those isolated from *Anthemis* species (Compositae) (*Figure 3.28*), are probably intermediates.

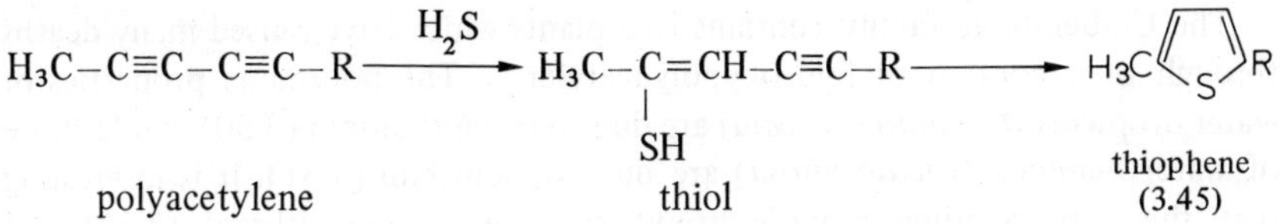

$$H_3C-C\equiv C-C\equiv C-R \xrightarrow{H_2S} H_3C-C=CH-C\equiv C-R \longrightarrow H_3C\underset{S}{\overset{}{\bigcirc}}R$$

polyacetylene   SH   thiophene

thiol   (3.45)

Figure 3.27 The biosynthesis of thiophenes.

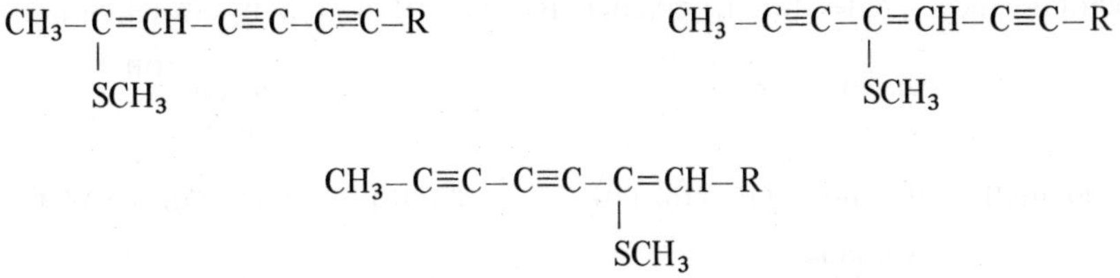

$$CH_3-C=CH-C\equiv C-C\equiv C-R \qquad\qquad CH_3-C\equiv C-C=CH-C\equiv C-R$$
$$\qquad\quad SCH_3 \qquad\qquad\qquad\qquad\qquad\qquad\qquad\qquad SCH_3$$

$$CH_3-C\equiv C-C\equiv C-C=CH-R$$
$$\qquad\qquad\qquad\qquad SCH_3$$

Figure 3.28 Thioethers isolated from *Anthemis* species.

Organochlorine compounds are rarely found in higher plants, but a number of chlorinated acetylenes have been isolated from members of the Compositae family, including (3.46) from *Centaurea ruthenica*, (3.47) from *Eclipta erecta* and (3.48) and (3.49) from *Helichrysum* and *Anaphalis* species.

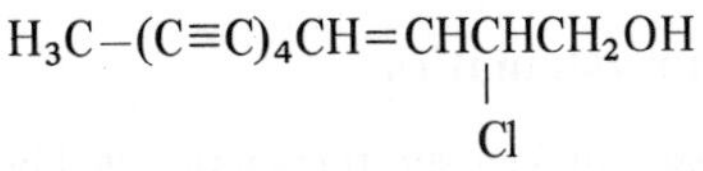

$$H_3C-(C\equiv C)_4CH=CHCHCH_2OH$$
$$\qquad\qquad\qquad\qquad\quad Cl$$

(3.46)

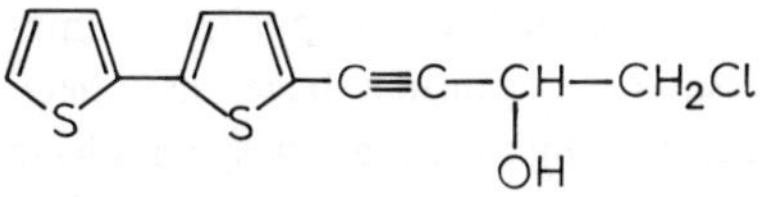

(3.47) 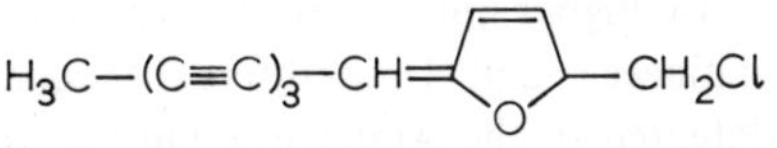   (3.48)

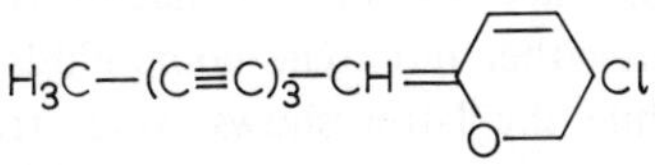

(3.49)

Polyacetylenes occurring in the Umbelliferae family can be distinguished from the Compositae compounds by their saturated end-groups (*Figure 3.29*).

$$-CH=CH_2 \qquad\qquad\qquad\qquad\qquad -CH_2CH_3$$

(a)   Characteristic vinyl end-group   (b)   Characteristic saturated end-group of
      of Compositae polyacetylenes   Umbelliferae polyacetylenes

Figure 3.29 The end-groups of Compositae and Umbelliferae polyacetylenes.

The Umbelliferae family contains two plants which have caused many deaths through their content of hydroxypolyacetylenes. The poisonous properties of water dropwort (*Oenanthe crocata*) are due to oenanthetoxin (3.50), while those of water hemlock (*Cicuta virosa*) are due to cicutoxin (3.51). It is interesting that the corresponding monohydroxyl compounds, oenanthetol (3.52) and cicutol (3.53), are relatively non-poisonous.

$$HOCH_2CH=CH(C\equiv C)_2\ (CH=CH)_2(CH_2)_2CH(CH_2)_2CH_3 \qquad HO(CH_2)_3(C\equiv C)_2(CH=CH)_3CH(CH_2)_2CH_3$$

$$\overset{\textstyle |}{OH} \qquad\qquad\qquad\qquad\qquad \overset{\textstyle |}{OH}$$

oenanthetoxin   cicutoxin

(3.50)   (3.51)

$$HOCH_2CH=CH(C\equiv C)_2(CH=CH)_2(CH_2)_5CH_3 \qquad HO(CH_2)_3(C\equiv C)_2(CH=CH)_3(CH_2)_3CH_3$$

oenanthetol   cicutol

(3.52)   (3.53)

The antibiotic or toxic properties of polyacetylenes confer an advantage on plants synthesising such compounds. Thus the polyacetylenes and thiophenes are examples of secondary metabolites whose function is one of protection against invasion by other organisms.

Polyacetylenes and Thiophenes in Chemosystematics

As the acetylenic fatty acids have been discussed previously, this section will be restricted to compounds without carboxyl groups.

In higher plants, polyacetylenes are widespread in the Compositae, Campanulaceae and Umbelliferae families, and such compounds have also been detected in the Araliaceae and Pittosporaceae, although only a few species have been examined. These five families fall into two natural groups — Compositae and Campanulaceae; and Umbelliferae, Araliaceae and Pittosporaceae. The relationship between Compositae and Campanulaceae is not close, however, as the former family contains either monoyne compounds or tetrahydropyranyl-ethers such as (3.54), while the latter shows wide structural variations, the pentayne (3.55) being the most common compound. Thiophenes have only been found in the Compositae.

$$ROCH_2CH=CH(C\equiv C)_2CH=CH \text{—} \qquad\qquad H_3C(C\equiv C)_5CH=CH_2$$

(3.54)   (3.55)

There is a closer chemical relationship between the Umbelliferae and Araliaceae as in both families the highly toxic falcarinone (3.56) is a common constituent. Thus chemosystematics agrees with classical taxonomy in placing the two families in the order Umbellales.

$$H_2C=CHCO(C\equiv C)_2CH_2CH=CH(CH_2)_6CH_3$$

falcarinone
(3.56)

Although too few species have been examined to draw any definite conclusions, the polyacetylenes of Pittosporaceae appear to be similar to those of Umbelliferae and Araliaceae.

The Compositae is a large family which is generally divided into thirteen tribes. Attempts have been made to characterise the tribes and subtribes by their polyacetylene content, and, despite overlapping, some success has been achieved in this direction. The tribes considered to be the most primitive, Vernonieae and Eupatorieae, contain only minute amounts of polyacetylenes, indicating that the ability to synthesise these compounds is an advanced characteristic. Other tribes in which these compounds are only minor constituents are Senecioneae, Calenduleae, Mutisieae and Cichorieae. It is interesting that the seeds of *Crepis foetida* (Cichorieae) contain crepenynic acid, considered to be the precursor of all the polyacetylenes and thiophenes, but no other acetylenes are found in this plant. Thiophenes occur sporadically in most tribes, but are particularly abundant in Heliantheae and the related Heleniceae.

Astereae is characterised by its content of C_{10} acetylenes, but such compounds are also found in the chemically and botanically complex Anthemideae. Spiroketals, such as (3.57), are abundant in the latter tribe, especially in the *Chrysanthemum* genus. The Inuleae tribe contains a number of chlorinated compounds with oxygen-containing rings, such as (3.58).

(3.57)   (3.58)

Polyacetylenes have also been isolated from fungi, particularly Basidiomycetes, and from some higher plant families other than those discussed above, including Cupressaceae, Annonaceae, Sapindaceae and Leguminosae. However, the number of species investigated is too few for these results to be of chemosystematic significance.

Bibliography

Lipid Biosynthesis and Metabolism

'Biosynthesis of Fatty Acids in Bacteria and Plants', in *Biosynthesis of Acetate-derived Compounds*, N. M. Packter, Wiley, 1973

'Compounds Formed by Linear Combination of C_2 Units', in *Organic Chemistry of Secondary Plant Metabolism*, T. A. Geissman and D. H. G. Crout, Freeman, Cooper and Co., 1969

'Fatty Acid Biogenesis in Plants', in *Perspectives in Phytochemistry*, J. B. Harborne and T. Swain (eds.), Academic Press, 1969

'Secondary Metabolites Derived from Acetate: Fatty Acids and Polyketides', in *Secondary Metabolism*, J. Mann, Oxford University Press, 1978

'The Control of Fatty Acid Biosynthesis in Plants', in *Biosynthesis and its Control in Plants*, B. V. Milborrow (ed.), Academic Press, 1973

'Unusual Fatty Acids in Plants', in *Recent Advances in Phytochemistry*, vol. 8, V. C. Runeckles and E. E. Conn (eds.), Academic Press, 1974

Plant Lipid Biochemistry, C. Hitchcock and B. W. Nichols, Academic Press, 1971

Recent Advances in the Chemistry and Biochemistry of Plant Lipids, T. Galliard and E. I. Mercer (eds.), Academic Press, 1975

Lipid Biochemistry: An Introduction, 2nd edn, M. I. Gurr and A. T. James, Chapman and Hall, 1975

'Lipid Metabolism' in *Cellular Biochemistry and Physiology*, N. A. Edwards and K. A. Hassall, McGraw-Hill, 1971

'Lipids', in *Chemistry and Biochemistry of Plant Herbage*, vol. 1, G. W. Butler and R. W. Bailey (eds.), Academic Press, 1973

Aliphatic Chemistry, Specialist Periodical Reports (continuing series, vol. 1, 1970–71), The Chemical Society

Aliphatic and Related Natural Product Chemistry, Specialist Periodical Reports (continuing series, vol. 1, 1979), The Chemical Society

'Fatty Acid Metabolism in Plants', in *Lipid Metabolism*, S. J. Wakil (ed.), Academic Press, 1970

'Cutins and Suberins', in *Phytochemistry*, vol. 3, L. P. Miller (ed.), Van Nostrand Reinhold Co., 1973

'Surface Waxes', *ibid.*

The Uses of Lipids

'Vegetable Oils', in *Plant Products of Tropical Africa*, M. L. Vickery and B. Vickery, The Macmillan Press, 1979

Chemosystematics of Lipids

'Distribution of Fatty Acids in Plant Lipids', in *Chemical Plant Taxonomy*, T. Swain (ed.), Academic Press, 1963

'Lipids of the Leguminosae', in *Chemotaxonomy of the Leguminosae*, J. B. Harborne *et al.* (eds.), Academic Press, 1971

'Vernonieae – Chemical Review', in *The Biology and Chemistry of the Compositae*, vol. 1, V. H. Heywood *et al.* (eds.), Academic Press, 1977

'Cynareae – Chemical Review', *ibid.*, vol. 2

'Lactuceae – Chemical Review', *ibid.*

Acetylenes and Thiophenes

Naturally Occurring Acetylenes, F. Bohlmann *et al.*, Academic Press, 1973

'Naturally Occurring Acetylenes', in *Phytochemistry*, vol. 3, L. P. Miller (ed.), Van Nostrand Reinhold Co., 1973

'Chemical Taxonomy of Acetylenic Compounds', in *Chemical Plant Taxonomy, ibid.*

'Acetylenic Compounds in the Umbelliferae', in *The Biology and Chemistry of the Umbelliferae,* V, H. Heywood (ed.), Academic Press, 1971

The Biology and Chemistry of the Compositae, vols. 1 and 2, *ibid.* (polyacetylenes are discussed in several chapters)

4 Polyketides

Introduction

Polyketides form a large class of compounds found extensively in fungi and also, to a lesser extent, in higher plants. The majority are aromatic derivatives, but alicyclic compounds also occur. Although also biosynthesised by the acetate-malonate pathway, true polyketides are distinguished from fatty acids and their derivatives by their mode of biogenesis. This involves a hypothetical chain of alternating keto and methylene groups (*Figure 4.1*), whereas fatty acid precursors contain only methylene groups, the malonyl keto group being reduced after each condensation (Chapter 3).

$$-CO-CH_2-CO-CH_2-CO-CH_2-CO-CH_2-$$

Figure 4.1 Polyketide chain.

Polyketide chains are derived from malonyl CoA units and a starter unit which is usually acetyl CoA, but can be other CoA esters. Reduction of some of the keto groups often occurs during the biosynthesis of derivatives, but this takes place after the formation of the polyketide chains. The great structural variety of polyketide-derived compounds is due to the numerous secondary modifications, which can include the cleavage of aromatic rings.

The Biosynthesis of Polyketides

The intermediate polyketide chain has never been isolated, and its existence is deduced from the nature of intermediates which have been characterised. Such a chain would be highly reactive and is therefore likely to remain enzyme-bound throughout its formation and subsequent cyclisation. It has been suggested that the chain exists in its enol form and that the hydroxyl groups are hydrogen-bonded to amino acid residues and/or chelated to metal atoms in the enzyme, as represented schematically in *Figure 4.2*.

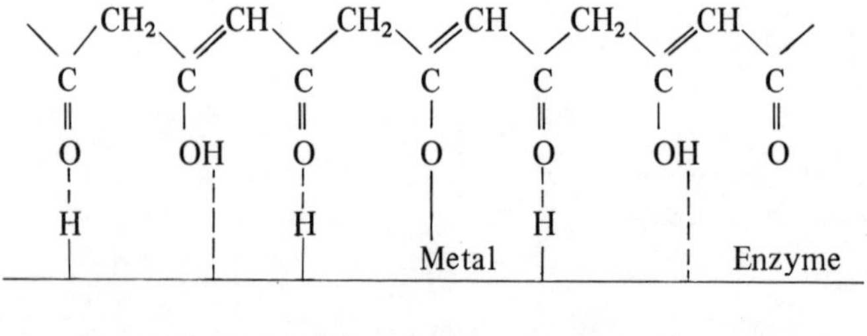

Figure 4.2 The possible stabilisation of a polyketide chain.

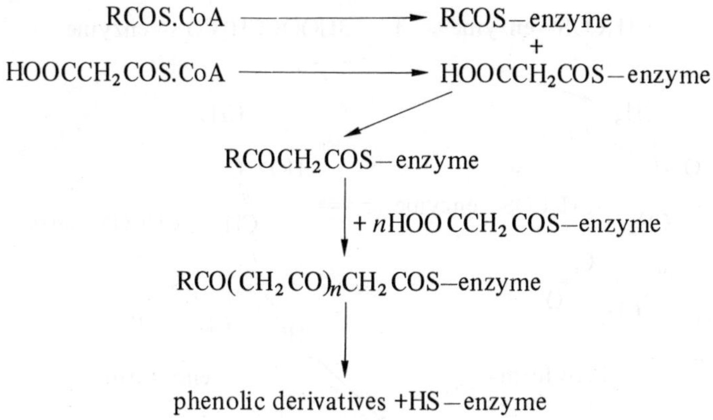

$$RCOS.CoA \longrightarrow RCOS-enzyme$$
$$+$$
$$HOOCCH_2COS.CoA \longrightarrow HOOCCH_2COS-enzyme$$

$$RCOCH_2COS-enzyme$$

$$\downarrow + nHOO\,CCH_2\,COS-enzyme$$

$$RCO(CH_2\,CO)_nCH_2\,COS-enzyme$$

$$\downarrow$$

phenolic derivatives $+HS-$enzyme

Figure 4.3 The biosynthesis of polyketides.

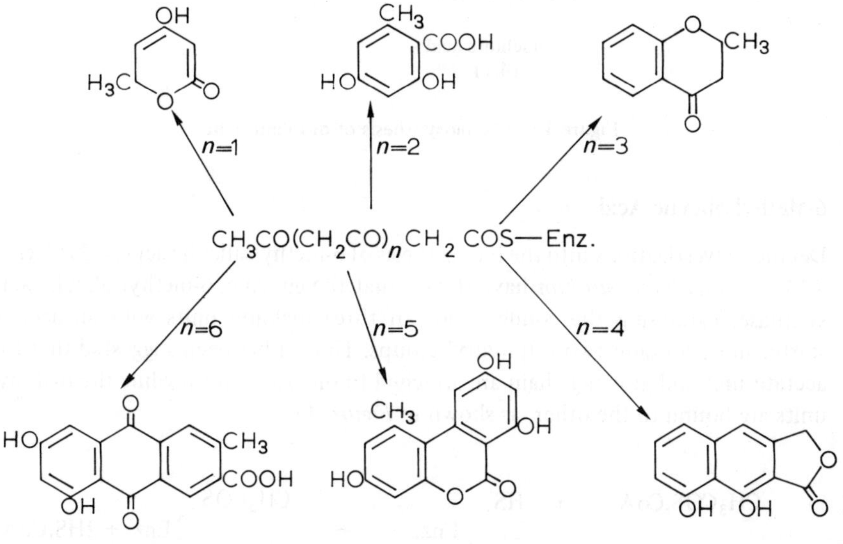

Figure 4.4 Compounds derived from polyketide chains.

Both the starter unit and the malonyl units are bound through enzyme sulphydryl groups before condensation and the product remains enzyme-bound until cyclisation occurs (*Figure 4.3*).

The various types of compound which are derived from polyketide chains are shown in *Figure 4.4*.

Orsellinic Acid

One of the simplest polyketide derivatives, orsellinic acid (4.1) is widely distributed in fungi. It is biosynthesised from an acetate starter unit and three malonyl units, as shown in *Figure 4.5*. Intramolecular condensation of the polyketide chain gives orsellinic acid without further modification.

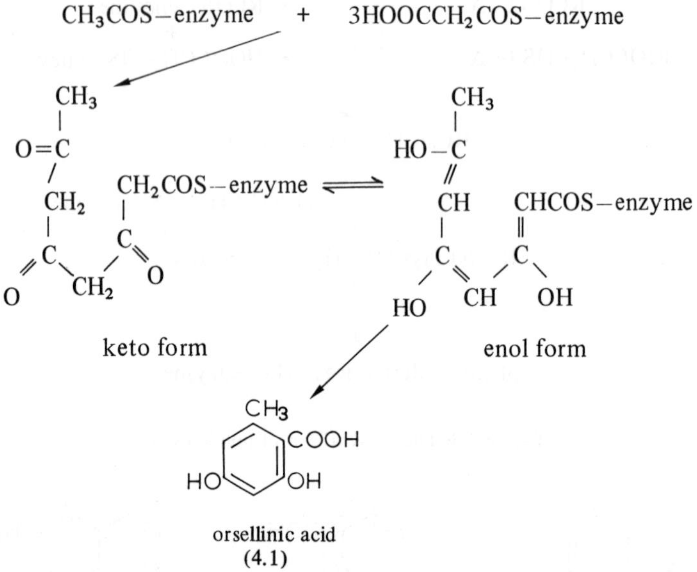

$$CH_3COS-enzyme \quad + \quad 3HOOCCH_2COS-enzyme$$

keto form   enol form

orsellinic acid
(4.1)

Figure 4.5 The biosynthesis of orsellinic acid.

6-Methylsalicylic Acid

Detailed investigations into the biosynthesis of 6-methylsalicylic acid (4.2) (*Figure 4.8*) in *Penicillium patulum* have shown that the enzyme, 6-methylsalicylic acid synthase, catalysing the condensation of three malonyl units with an acetate starter unit, contains two sulphydryl groups. Thus, it has been suggested that the acetate unit and growing chain are attached to one such group while the malonyl units are bound to the other, as shown in *Figure 4.6*.

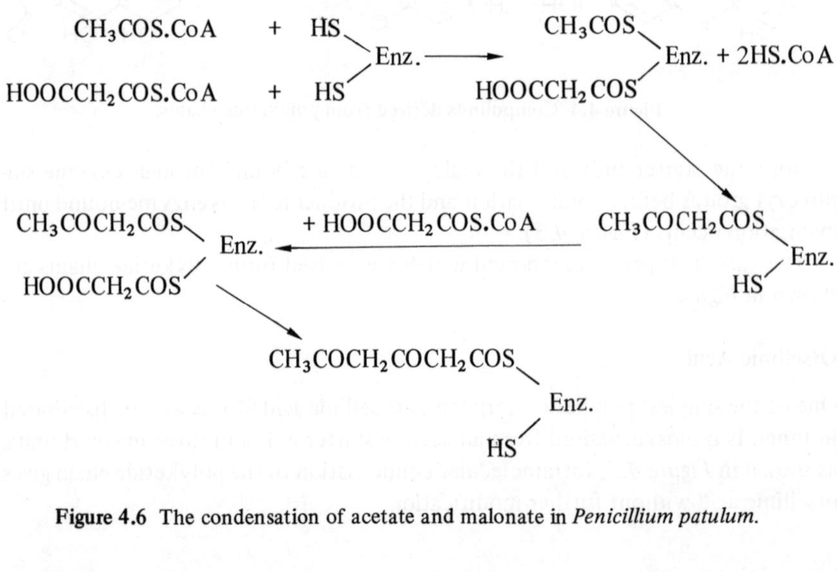

Figure 4.6 The condensation of acetate and malonate in *Penicillium patulum*.

One of the compounds isolated from *P. patulum*, triacetic acid lactone (4.3), is formed by direct cyclisation of the C_6 polyketide chain (*Figure 4.7*). It therefore seems probable that it is at this stage that the modifications to the polyketide chain occur which are necessary for the eventual synthesis of 6-methylsalicylic acid. The possibility that reduction and dehydration occur before condensation of the final malonyl unit (*Figure 4.8*) is also indicated by the behaviour of acetylenic inhibitors of 6-methylsalicylic acid synthase. The reductase catalysing the reduction of the polyketide chain is NADPH-dependent.

6-Methylsalicylic acid is also biosynthesised by the acetate–malonate pathway in higher plants, although salicylic acid, itself, is biosynthesised by the shikimic acid pathway (Chapter 6).

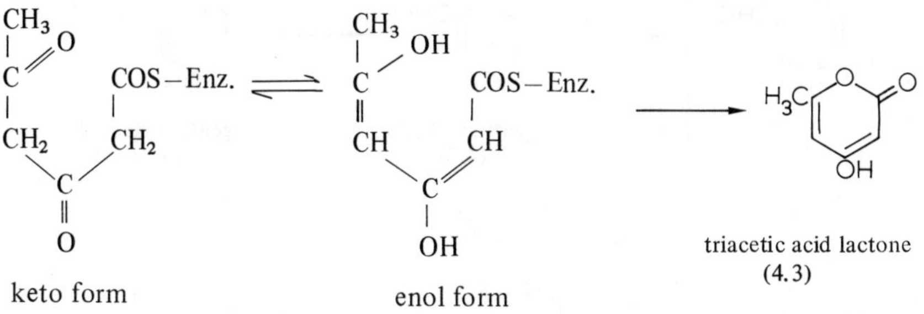

keto form   enol form   triacetic acid lactone
(4.3)

Figure 4.7 The formation of triacetic acid lactone.

6-methylsalicylic acid
(4.2)

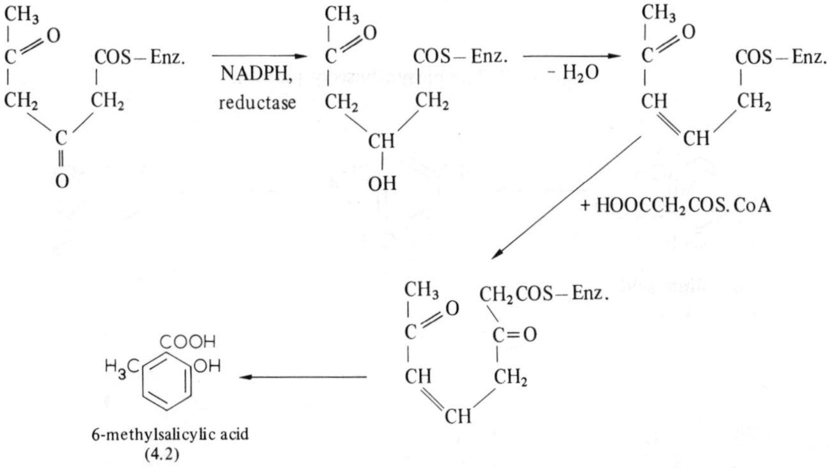

Figure 4.8 The probable final stages in the biosynthesis of 6-methylsalicylic acid.

Patulin and Penicillic Acid

In *P. patulum* 6-methylsalicylic acid (4.2) is converted to patulin (4.4) through the intermediate formation of *m*-cresol (4.5), 3-hydroxybenzyl alcohol (4.6), gentisyl alcohol (4.7) and gentisaldehyde (4.8), as shown in *Figure 4.9*. Enzyme

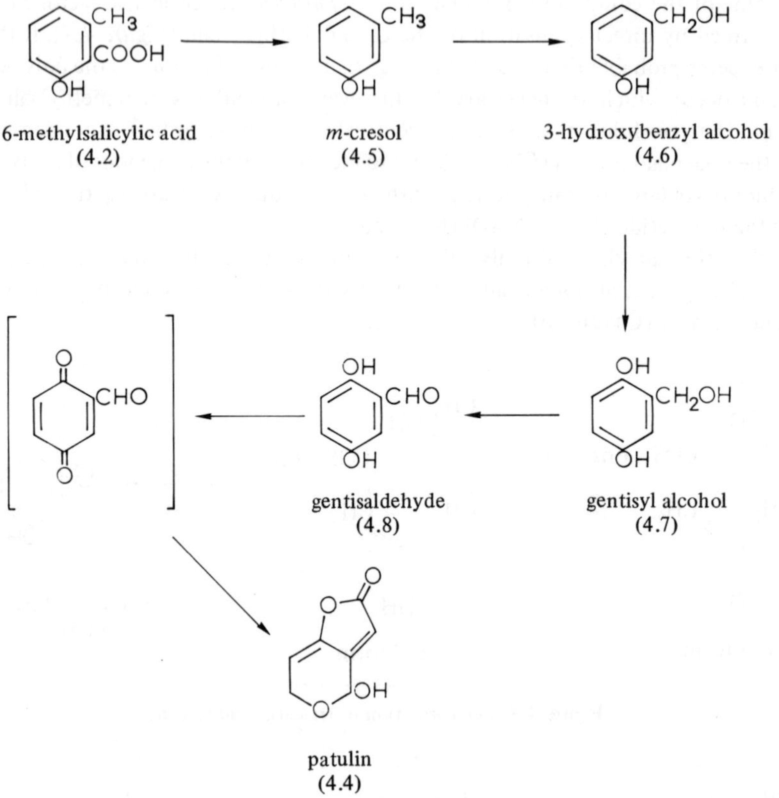

6-methylsalicylic acid
(4.2)

m-cresol
(4.5)

3-hydroxybenzyl alcohol
(4.6)

gentisaldehyde
(4.8)

gentisyl alcohol
(4.7)

patulin
(4.4)

Figure 4.9 The biosynthesis of patulin.

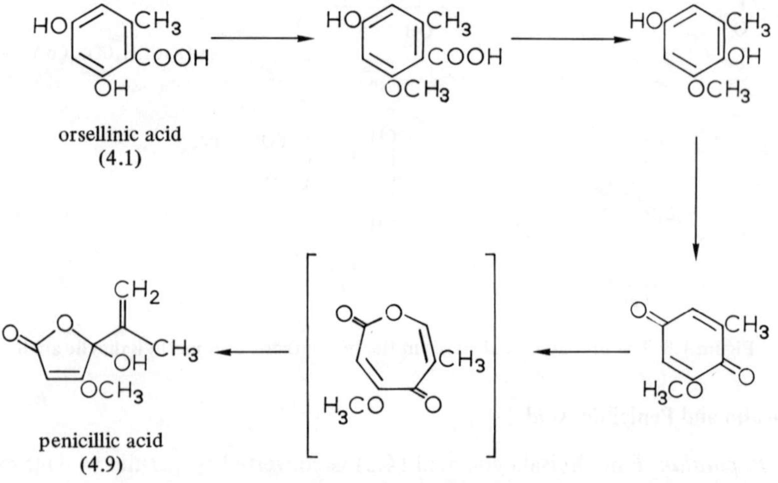

orsellinic acid
(4.1)

penicillic acid
(4.9)

Figure 4.10 The biosynthesis of penicillic acid.

preparations catalysing the hydroxylation of *m*-cresol and 3-hydroxybenzyl alcohol have been isolated from the mould. It is probable that the cleavage of the aromatic ring takes place via a quinonoid form. A similar transformation occurs in *P. cyclopium*, where orsellinic acid (4.1) is converted to penicillic acid (4.9) (*Figure 4.10*). An enzyme preparation catalysing the oxidative ring cleavage has been isolated from this mould.

Penicillic acid, produced by a variety of fungi, is both antibiotic and carcinogenic and is a possible contaminant of spoilt food. Patulin, also carcinogenic, is present in mouldy apples.

Griseofulvin

Griseofulvin (4.10) is an important fungicide isolated from *Penicillium* species.

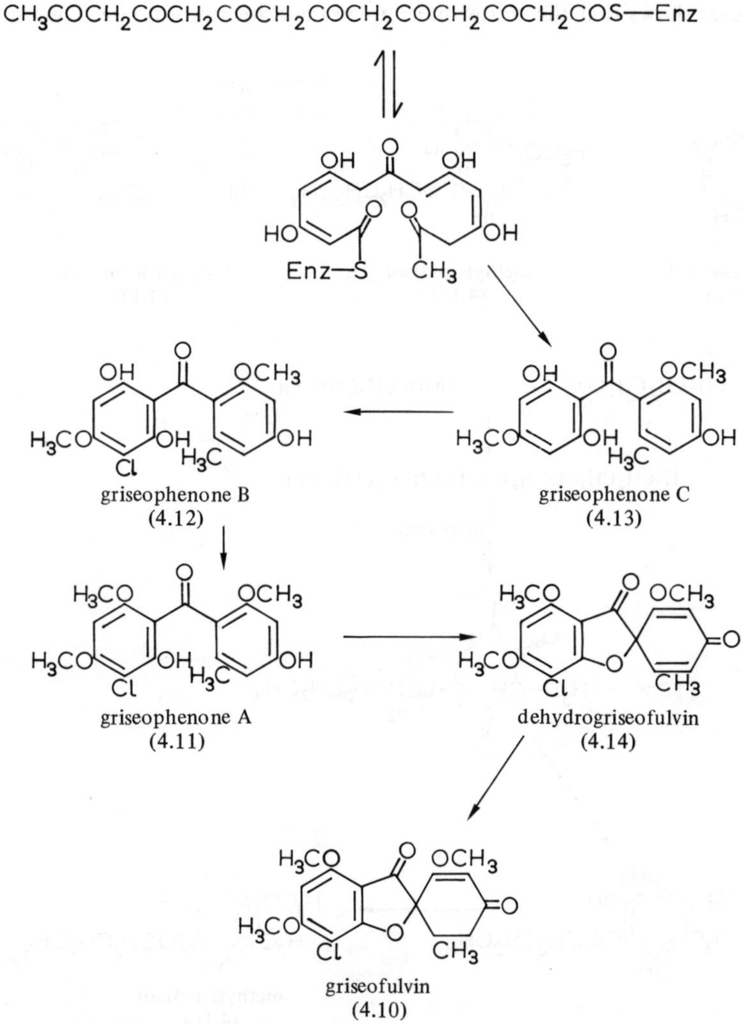

$$CH_3COCH_2COCH_2COCH_2COCH_2COCH_2COCH_2COS—Enz$$

Figure 4.11 The probable biosynthetic pathway to the formation of griseofulvin in *Penicillium* species.

Tracer experiments (see Chapter 1) with *P. patulum* have shown this compound to be derived from a C_{14} polyketide chain, while the isolation of griseophenones A, B and C (4.11–4.13) and dehydrogriseofulvin (4.14), all of which are acetate-derived, indicates a biosynthetic pathway similar to that shown in *Figure 4.11*.

If *P. patulum* is grown in a medium deficient in chloride, griseophenone C accumulates at the expense of the B and A compounds. Griseofulvin is used medicinally to treat fungal infections such as ringworm and athlete's foot.

Phloroglucinol Derivatives

A large number of compounds based on the phloroglucinol nucleus (4.15) have been isolated from plants. Some of the simplest, such as methylaspidinol (4.16) and butyrylfilicinic acid (4.17), are found in ferns, and examination of their structures shows that the starter units are not acetate but butyrate. Thus, in the

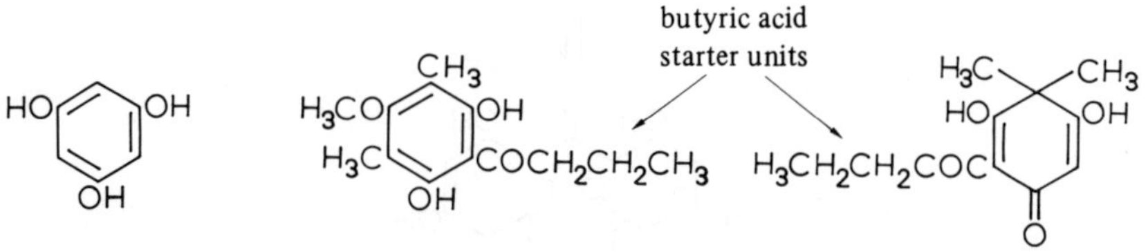
butyric acid
starter units
HO
OH
OH
CH3
H3CO
OH
H3C
COCH2CH2CH3
OH
H3C
CH3
HO
OH
H3CH2CH2COC
O

phloroglucinol (4.15) | methylaspidinol (4.16) | butyrylfilicinic acid (4.17)

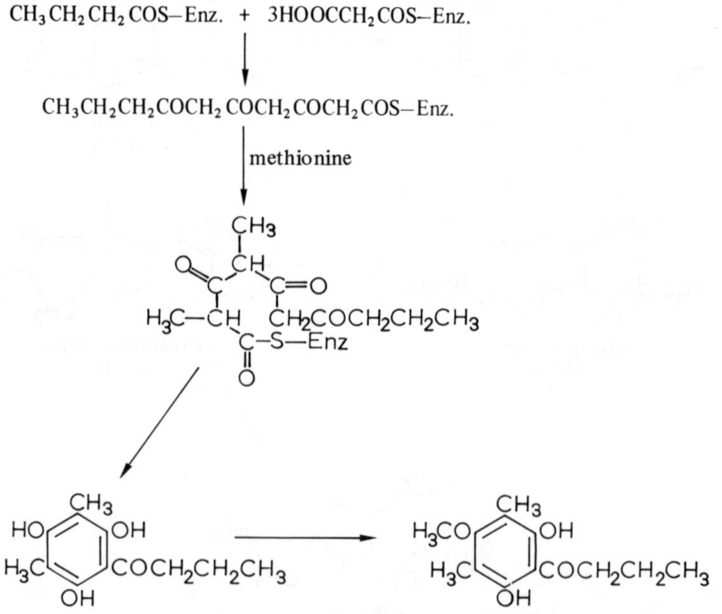
CH3CH2CH2COS–Enz. + 3HOOCCH2COS–Enz.
CH3CH2CH2COCH2COCH2COCH2COS–Enz.
methionine
CH3
CH
O
C
C=O
H3C–CH
CH2COCH2CH2CH3
C–S–Enz
O
CH3
HO
OH
H3C
COCH2CH2CH3
OH
CH3
H3CO
OH
H3C
COCH2CH2CH3
OH

methylaspidinol
(4.16)

Figure 4.12 The biosynthesis of methylaspidinol.

formation of these compounds, both the fatty acid and the polyketide acetate–malonate pathways are operative. The extra methyl groups are derived from methionine, ring methylation taking place at the polyketide stage (*Figure 4.12*).

The methylene-*bis*-phloroglucinol derivatives characteristic of some ferns are biosynthesised by linkage of two phloroglucinol derivatives through a methionine-derived methylene group. Thus, methylaspidinol (4.16) and aspidinol (4.18) give methylene-*bis*-aspidinol (4.19), as shown in *Figure 4.13*.

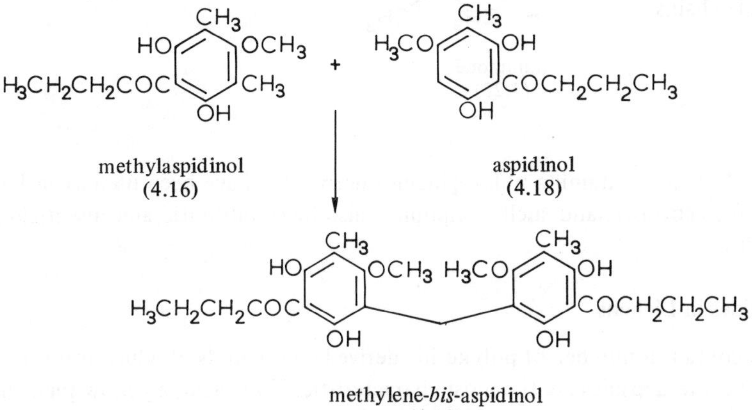

methylaspidinol
(4.16)

aspidinol
(4.18)

methylene-*bis*-aspidinol
(4.19)

Figure 4.13 The formation of methylene-*bis*-phloroglucinol derivatives.

Phloroglucinol derivatives also occur in higher plants and here, too, the starter units are often other than acetate. Branched-chain acids are common, such as iso-butyric which acts as the starter unit for tasmanone (4.20) from the essential oil of *Eucalyptus risdoni* (Myrtaceae) and the methylene-*bis*-phloroglucinol derivative, α-kosin (4.21), from the flowers of *Hagenia abyssinica* (Rosaceae). Eugenone (4.22), however, which occurs in the essential oil of cloves (*Eugenia caryophyllata*, Myrtaceae), has an acetate starter unit, the side chain being derived from one acetate and one malonate unit.

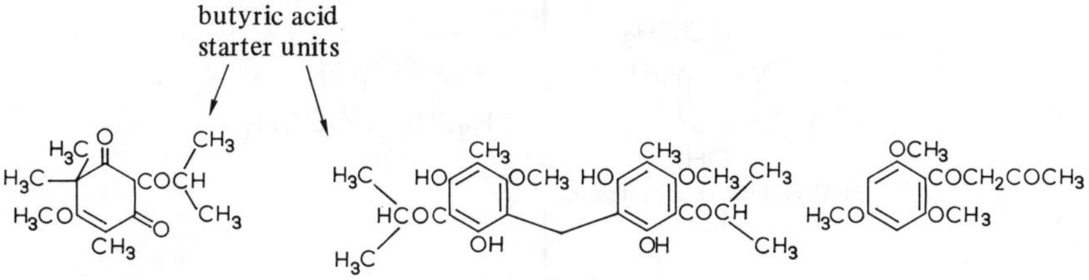

butyric acid
starter units

tasmanone
(4.20)

α-kosin
(4.21)

eugenone
(4.22)

Humulone (4.23), found in hops (*Humulus lupulus*, Cannabinaceae), is a more complex phloroglucinol derivative containing isoprenoid side chains. The starter unit for this compound is isovaleric acid. During the production of beer, humulone and other compounds found in hops are oxidised to complex structures which have a characteristic bitter flavour.

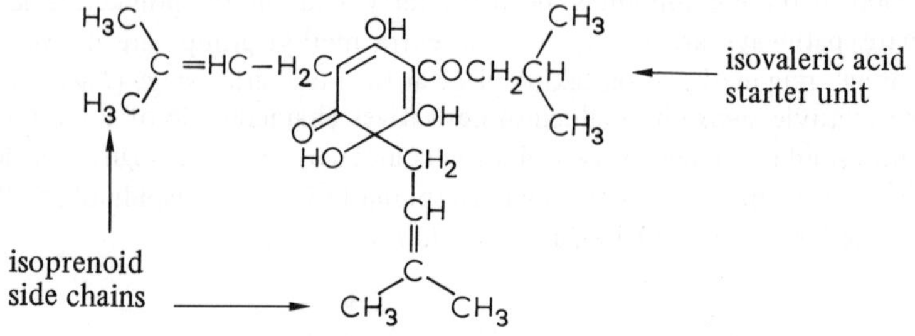

humulone
(4.23)

Several plants containing phloroglucinol derivatives have been used as anthelmintics for centuries, and such compounds also have antibiotic and insecticidal properties.

Polyketides in Lichens

Lichens contain a number of polyketide-derived compounds of which usnic acid (4.24) and the depsides are the most characteristic. Usnic acid, a yellow pigment with useful antibiotic properties, occurs widely in lichens. It is biosynthesised from two polyketide chains via methylphloroacetophenone (4.25) intermediates which are linked by oxidative coupling (*Figure 4.14*). Before cyclisation, the polyketide chains are methylated by methionine.

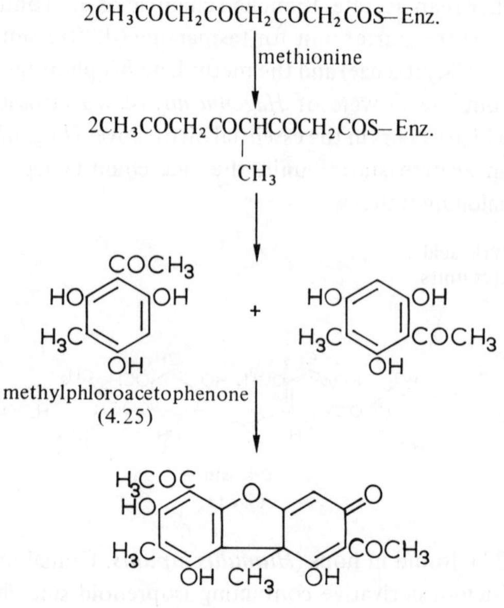

Figure 4.14 The biosynthesis of usnic acid.

A large number of depsides (hydroxyl compounds containing an ester linkage between a phenol and an acid) have been isolated from lichens. Such compounds often contain ring methyl groups, or their oxidised derivatives such as carboxyl, which originate from methionine. Such groups are always added at the polyketide stage and the methylated polyketide chains forming usnic acid are also the precursors of the depside, diffractic acid (4.26), in *Usnea diffracta* (*Figure 4.15*).

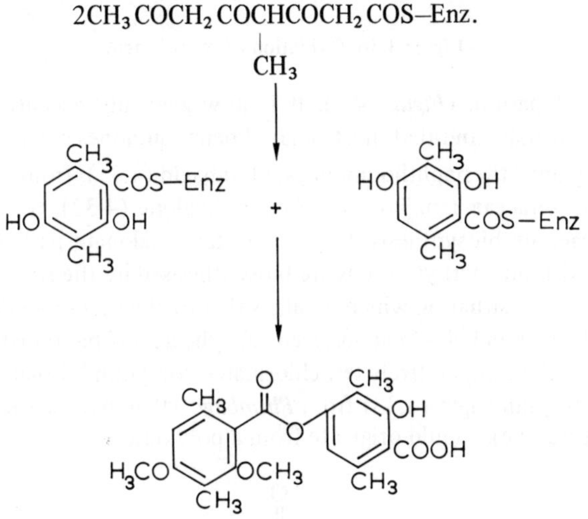

$$2CH_3 COCH_2 COCHCOCH_2 COS-Enz.$$
$$|$$
$$CH_3$$

diffractic acid
(4.26)

Figure 4.15 The biosynthesis of depsides.

Chlorinated compounds have been found in several lichens, including the depside, tumidulin (4.27), from *Ramalina ceruchis*, the depsidone, diploicin (4.28), from *Buellia canescens* and *Lecides carpathica* and the xanthone, thiophanic acid (4.29), from *Lecanora* species. All these compounds are presumably biosynthesised from polyketide precursors.

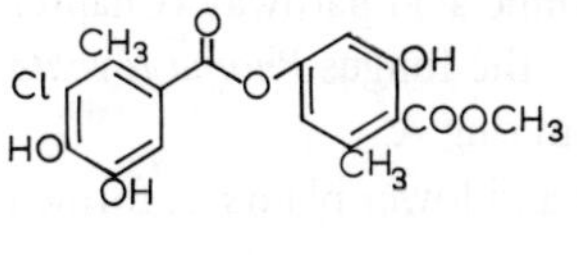

tumidulin
(4.27)

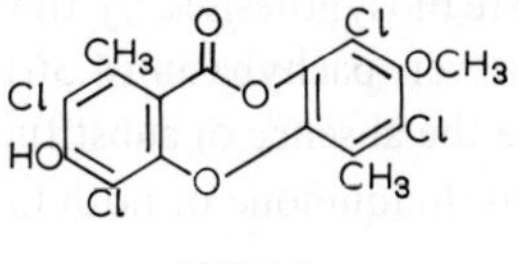

diploicin
(4.28)

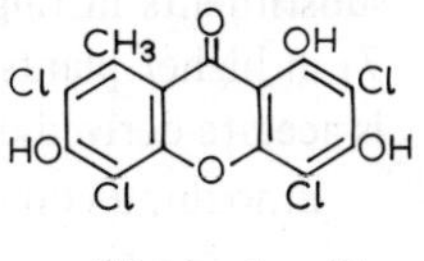

thiophanic acid
(4.29)

Quinones

Benzoquinones occurring in fungi can be biosynthesised by either the acetate–malonate or shikimic acid (Chapter 6) pathways. Thus, gentisic acid (4.30) (the quinol form, *Figure 4.16*) is biosynthesised from acetate in *Penicillium patulum* but from shikimate in *Polyporus tumulosus*. In *P. patulum* this compound is probably a side-product formed by oxidation of gentisaldehyde (4.8), in the

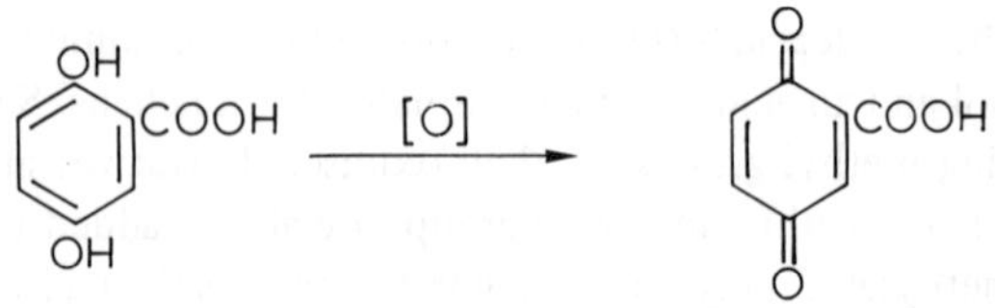

gentisic acid   quinone form
(4.30)

Figure 4.16 Oxidation of gentisic acid.

biosynthesis of patulin (*Figure 4.9*). It is now generally accepted that quinol forms occur naturally and that most isolated benzoquinones are artifacts.

In higher plants, the naphthoquinones, plumbagin (4.31), from the Plumbaginaceae and Droseraceae families, and 7-methyljuglone (4.32), from Droseraceae and Ebenaceae, are biosynthesised by the acetate–malonate pathway. However, compounds without methyl groups are biosynthesised by the shikimic acid pathway (Chapter 7), a situation which parallels that of the acetate-derived 6-methylsalicylic acid (4.2) and shikimate-derived salicylic acid (Chapter 6). It is interesting that one of the comparatively few chlorinated compounds isolated from higher plants, 3-chloroplumbagin (4.33) from *Plumbago* (Plumbaginaceae) and *Drosera* species (Droseraceae), should originate from a polyketide.

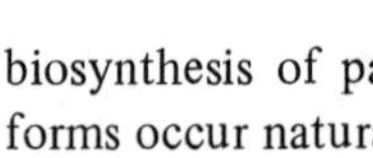

plumbagin   7-methyljuglone   3-chloroplumbagin
(4.31)   (4.32)   (4.33)

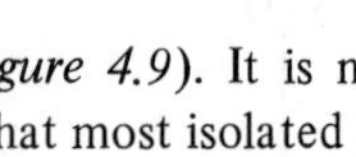

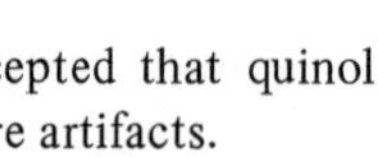

Anthraquinones are also biosynthesised by two pathways; those of the emodin (4.34) type with substituents in both rings A and C are usually acetate-derived in both higher and lower plants, while alizarin (4.35) and its derivatives, without substituents in ring A, are biosynthesised by the shikimic acid pathway (Chapter 7) in higher plants. However, pachybasin (4.36) from the fungus *Phoma foveata* is acetate-derived despite the absence of substituents in ring A.

Emodin, a common anthraquinone of both higher and lower plants, is formed

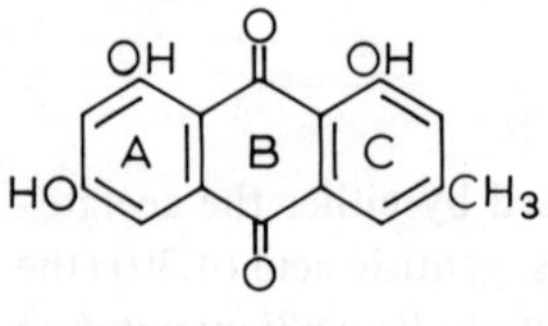

emodin   alizarin   pachybasin
(4.34)   (4.35)   (4.36)

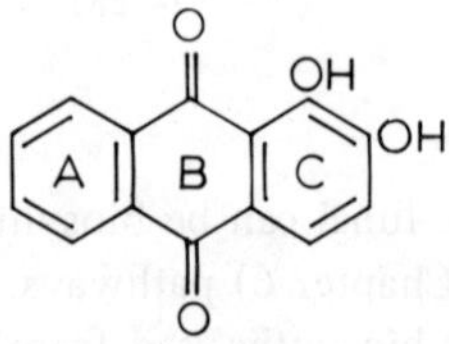

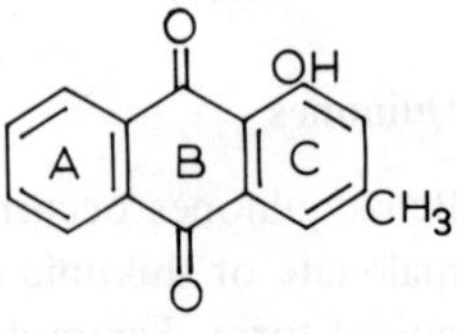

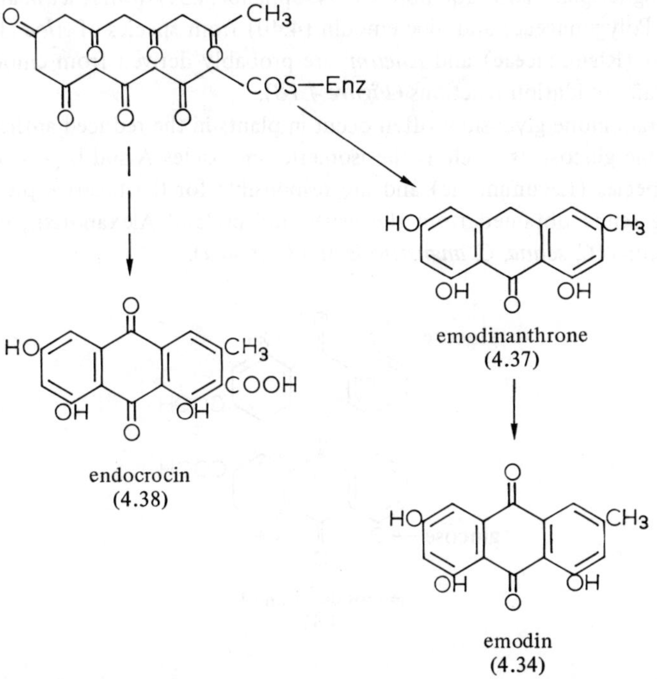

Figure 4.17 The probable biosynthesis of emodin and endocrocin.

from a C_{16} polyketide chain through the probable intermediate formation of an anthrone (4.37), as shown in *Figure 4.17*.

Endocrocin (4.38) (emodin-2-carboxylic acid) is derived from a similar polyketide chain but the carboxyl group is retained. It has been clearly shown in the mushroom, *Dermocybe sanguinea*, that endocrocin is not a precursor of emodin. Thus, decarboxylation of the polyketide chain must take place before cyclisation.

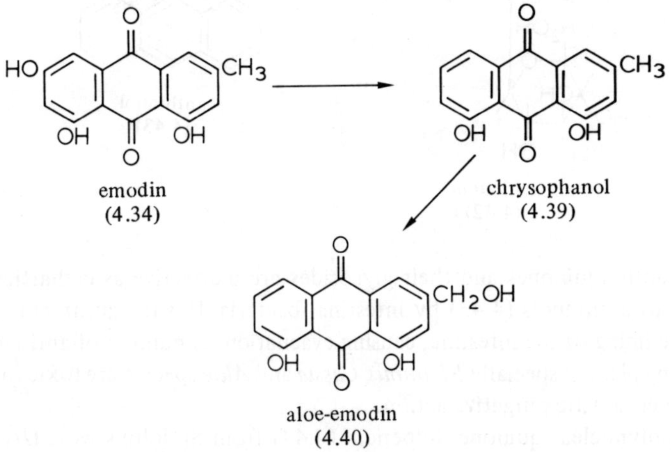

Figure 4.18 The probable derivation of chrysophanol and aloe-emodin.

The higher plant anthraquinones, chrysophanol (4.39) from *Rheum* and *Rumex* species (Polygonaceae) and aloe-emodin (4.40) from species of *Aloe* (Liliaceae), *Rhamnus* (Rhamnaceae) and *Rheum*, are probably derived from emodin by reduction and oxidation reactions (*Figure 4.18*).

Anthraquinone glycosides often occur in plants in the reduced anthrone form. Dianthrone glucosides, such as the isomeric sennosides A and B (4.41), occur in *Cassia* species (Leguminosae) and are responsible for the laxative properties of the drug senna obtained from the leaves and pods of Alexandrian, Indian and Italian senna (*C. senna, C. angustifolia* and *C. italica*).

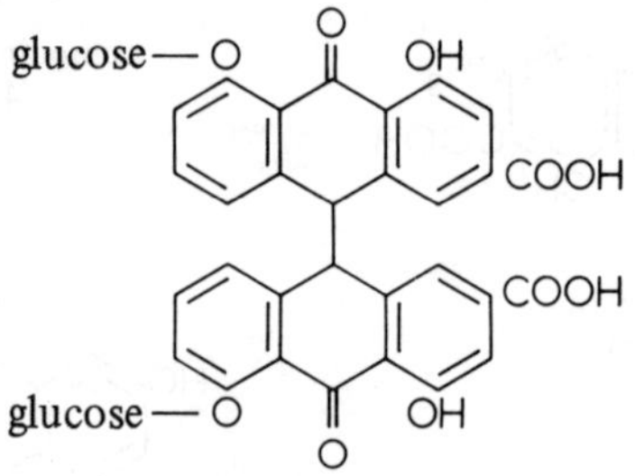

sennosides A and B
(4.41)

The laxative properties of cascara sagrada obtained from the sacred bark (*Rhamnus frangula*, Rhamnaceae) are due to aloe-emodin (4.40), which is probably a degradation product of a dianthrone glycoside. The purgative aloes, obtained from the leaf sap of *Aloe* species, contain anthraquinones and anthrone derivatives, including aloe-emodin and the C-glycoside, barbaloin (4.42).

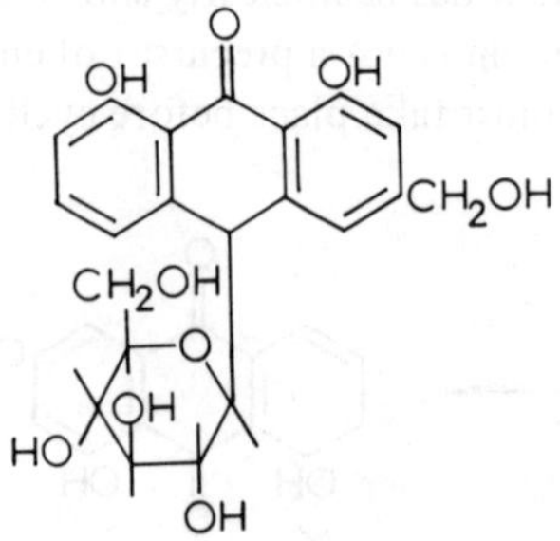

barbaloin
(4.42)

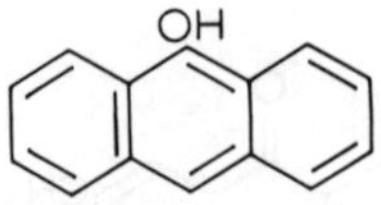

anthranol
(4.43)

Free anthraquinones and their glycosides are ineffective as cathartics but are reduced to anthranols (4.43) by intestinal bacteria. It is the anthranols which irritate the lining of the intestine, causing evacuation. A number of anthraquinone-containing plants, especially *Rhamnus, Cassia* and *Aloe* species, are toxic to animals, due to their drastic purgative action.

The polynuclear quinone, hypericin (4.44) from St John's wort (*Hypericum perforatum*, Hypericaceae) and other *Hypericum* species, would appear to be bio-

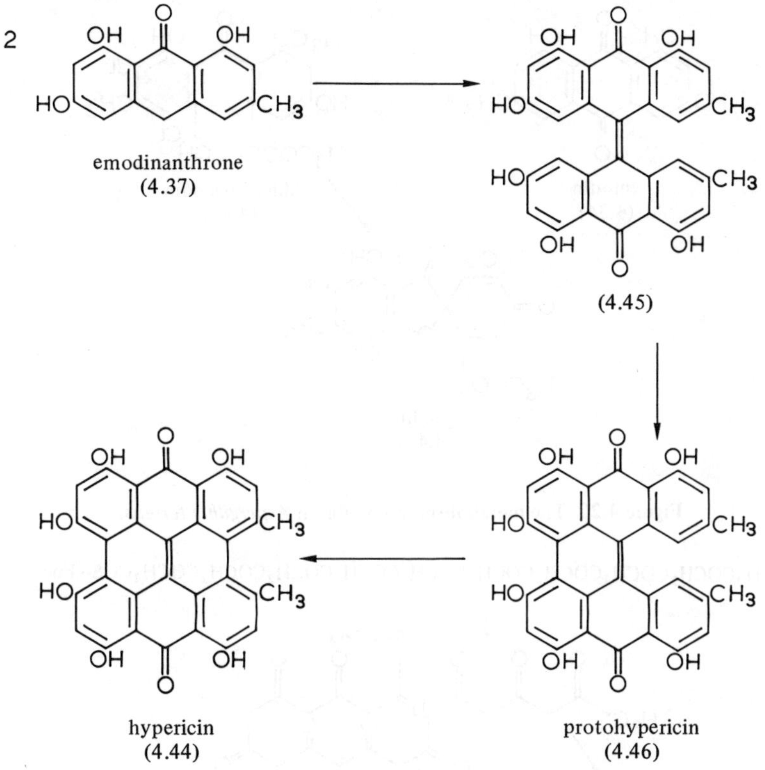

Figure 4.19 The probable biosynthesis of hypericin.

synthesised from two molecules of emodinanthrone (4.37), a compound which occurs as a minor constituent of these plants. If *Hypericum* species are grown in the dark, the dehydrodianthrone (4.45) accumulates and this is converted to hypericin through the intermediate formation of protohypericin (4.46) on exposure to light. Thus, the biosynthesis of hypericin would seem to follow the pathway shown in *Figure 4.19*.

Animals eating *Hypericum* species suffer photosensitisation from the effects of hypericin, a red compound which fluoresces in ultraviolet light. Only unpigmented, hairless animals or parts of animals are affected, for example pigs, and the udders, teats, muzzles and eyelids of cattle. Skin which is not protected from sunlight by pigmentation or hair dies and flakes off leaving wounds which are slow to heal. The liver, also, is often affected, giving rise to jaundice. Animals with these conditions become unthrifty and are often an economic loss to farmers, although some recovery can be achieved by keeping the animal in the dark. Hypericin is a feeding stimulant for the beetle *Chrysolina brunsvicensis*.

In fungi, anthraquinones are actively metabolised and ring cleavage has been observed in several instances. Emodin (4.34), for example, is the precursor of the chlorinated geodin (4.47) in cultures of *Aspergillus terreus*, dihydrogeodin (4.48) being an intermediate (*Figure 4.20*). In higher plants, anthraquinones are utilised during the early stages of fruit development.

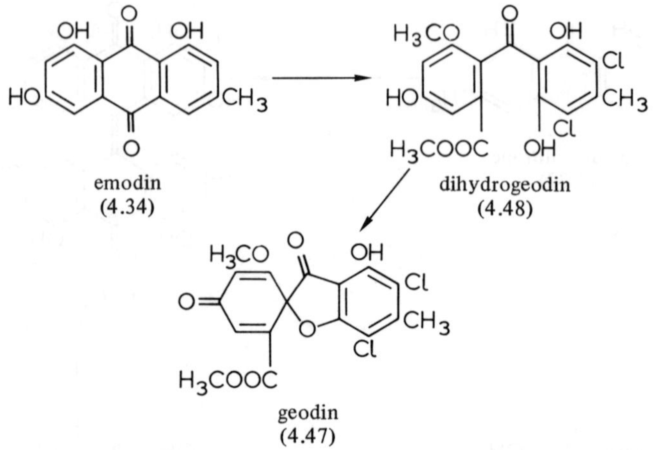

Figure 4.20 The metabolism of emodin in *Aspergillus terreus*.

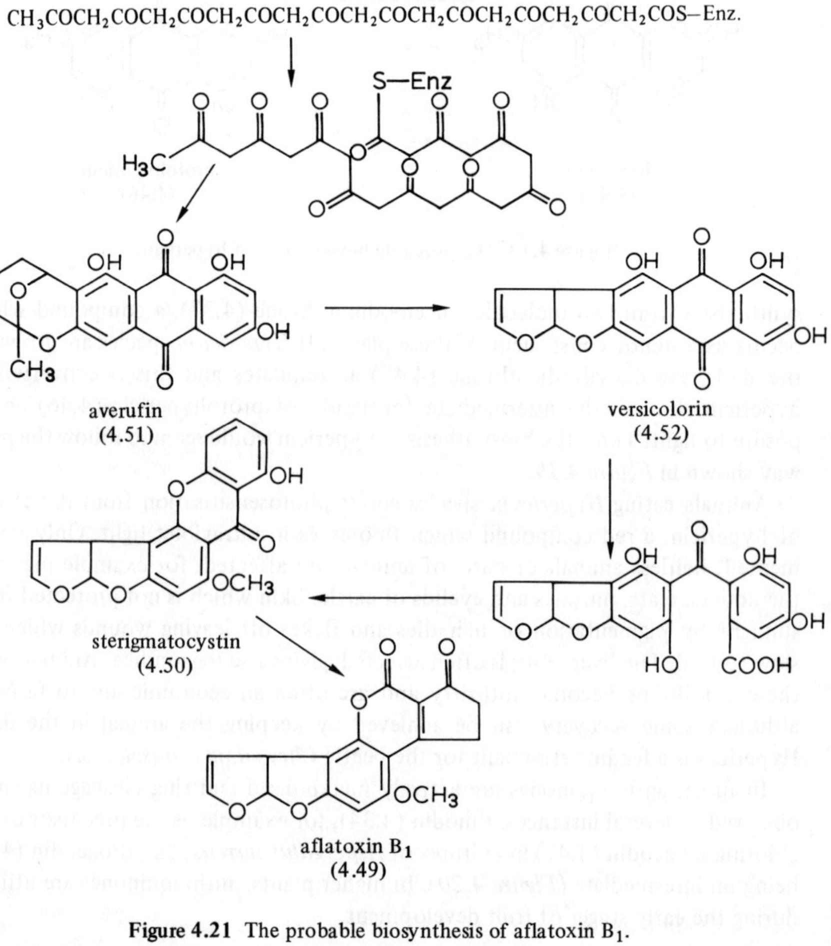

$$CH_3COCH_2COCH_2COCH_2COCH_2COCH_2COCH_2COCH_2COCH_2COS-Enz.$$

Figure 4.21 The probable biosynthesis of aflatoxin B_1.

Aflatoxins

Aflatoxins are poisonous furanocoumarin derivatives produced by *Aspergillus* species and *Penicillium puberulum*. They are of importance commercially, as contamination of food by such substances can lead to serious consequences. Aflatoxins, produced by *A. flavus* contamination of stored groundnuts (*Arachis hypogaea*, Leguminosae), cause fatal Turkey X disease, while the carcinogenic effects of these compounds are thought to be responsible for the high incidence of liver cancer amongst Africans whose diet depends on groundnuts. Dogs are also susceptible to the effects of aflatoxins, which can contaminate cereals, and many deaths have occurred due to contaminated cereal-based dog foods.

Aflatoxin B_1 (4.49) has been shown to be acetate-derived in *A. flavus* and *A. parasiticus*, and studies with isotopically labelled compounds fed to *A. parasiticus* have shown that sterigmatocystin (4.50) is an intermediate. In *A. versicolor*, sterigmatocystin is derived from averufin (4.51) through the intermediate formation of versicolorin (4.52). Thus, a scheme for the biosynthesis of aflatoxin B_1 from a C_{20} polyketide chain can be postulated as shown in *Figure 4.21*.

It can be seen from *Figure 4.21* that the formation of sterigmatocystin from versicolorin involves cleavage of an anthraquinone ring system similar to that observed in the formation of dihydrogeodin (4.48) from emodin (4.34) (*Figure 4.20*).

Fungal Tropolones

Tropolones contain a seven-membered ring system, and experiments with *Penicillium stipatum* have shown this ring to be biosynthesised from an aromatic

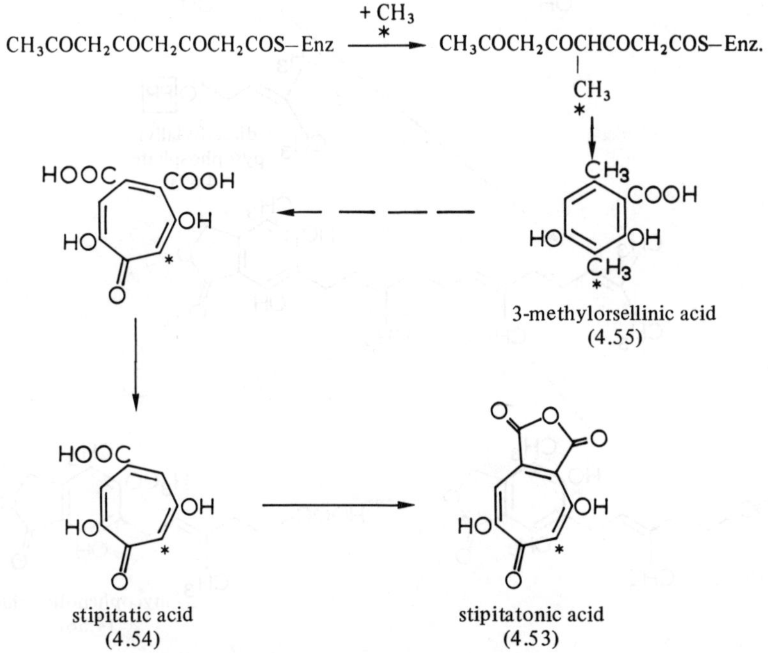

Figure 4.22 The derivation of the fungal tropolone ring system.

precursor, the extra carbon atom being donated by methionine. Thus, stipitatonic acid (4.53) and stipitatic acid (4.54) are derived from a methylated C_8 polyketide chain which cyclises to 3-methylorsellinic acid (4.55). Isotopic labelling of the methionine-derived carbon atom has shown that it becomes the seventh carbon atom of the tropolone ring (*Figure 4.22*).

Mycophenolic Acid

The antibiotic, mycophenolic acid (4.56), biosynthesised by *Penicillium brevi-compactum*, is unusual in that it is derived from a C_8 polyketide chain and three dimethylallyl pyrophosphate molecules (4.57) as shown in *Figure 4.23*.

Mycophenolic acid has useful antitumour properties and efforts have been made to prepare derivatives with enhanced or modified activity. It is of interest,

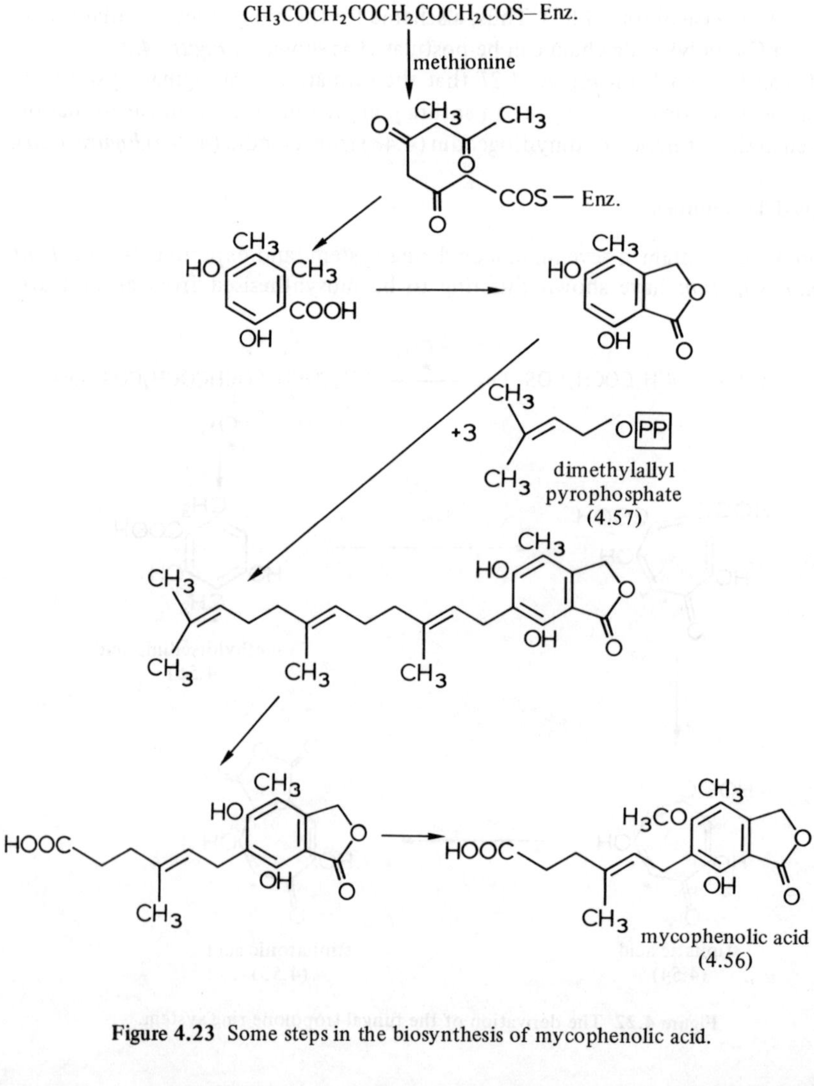

Figure 4.23 Some steps in the biosynthesis of mycophenolic acid.

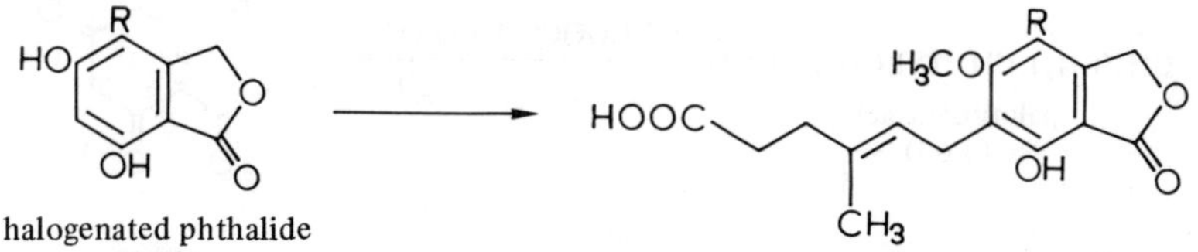

halogenated phthalide
(4.58)

halogenated mycophenolic acid
(4.59)

Figure 4.24 The conversion of halogenated phthalides into mycophenolic derivatives by *Penicillium brevicompactum.*

therefore, that cultures of *P. brevicompactum* can convert halogenated phthalides (4.58; R = Cl or Br) into the corresponding mycophenolic acid derivatives (4.59) (*Figure 4.24*).

Deviations in Polyketide Chain Biosynthesis

We have already seen in the section on phloroglucinol derivatives that polyketide chains can be formed from starter units other than acetate. The widespread plant flavonoids utilise cinnamic acid as a starter unit, as do the stilbenes and xanthones (see Chapter 7), while ferulic acid (4.60) or *p*-coumaric acid (4.61) appear to be the starter units for the biosynthesis of curcumin (4.62), the pigment from the spice plant, turmeric (*Curcuma longa*, Zingiberaceae) (*Figure 4.25*).

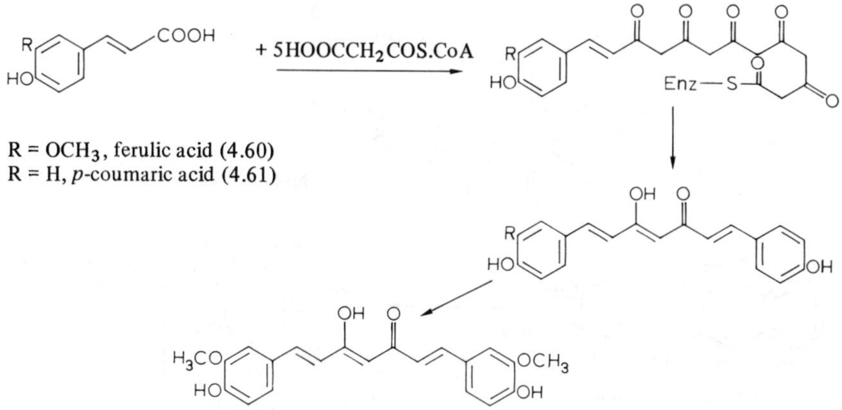

$R = OCH_3$, ferulic acid (4.60)
R = H, *p*-coumaric acid (4.61)

curcumin
(4.62)

Figure 4.25 Some steps in the biosynthesis of curcumin.

Other compounds with unusual starter units include the phenols with long side chains isolated from members of the Anacardiaceae. Here the starter units are long-chain fatty acids, palmitoleic acid (4.63), for instance, giving rise to the mixtures of compounds known as urushiol (4.64) found in poison ivy (*Rhus toxicodendron*) and cardol (4.65) and anacardic acid (4.66) found in cashew (*Anacardium occidentale*). The derivation of these compounds is shown in *Figure 4.26*.

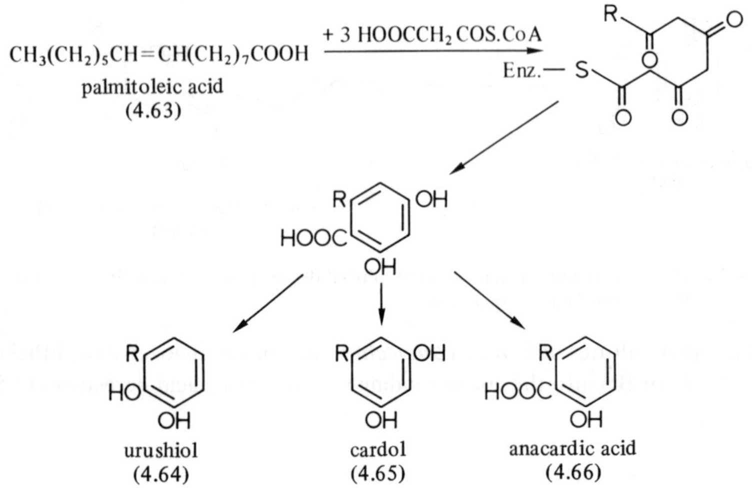

$$CH_3(CH_2)_5CH=CH(CH_2)_7COOH \xrightarrow[\text{Enz.}- S]{+ 3\ HOOCCH_2\ COS.CoA}$$

palmitoleic acid
(4.63)

urushiol
(4.64)

cardol
(4.65)

anacardic acid
(4.66)

R = C$_{15}$, with unsaturation at ω8; ω8, 11; or ω8, 11, 14

Figure 4.26 The derivation of the Anacardiaceae phenols.

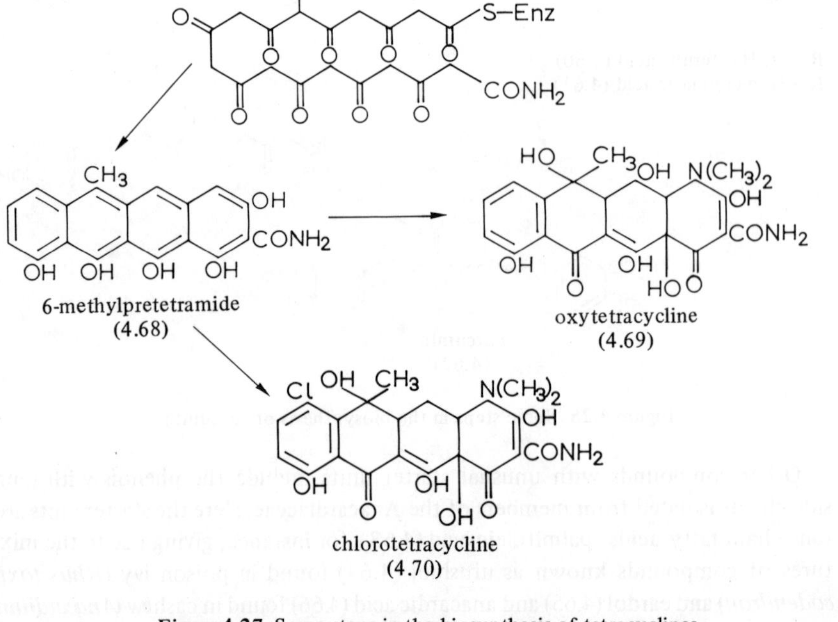

$$H_2NCOCH_2COOH \xrightarrow{+ 9HOOCCH_2COS.CoA}$$

malonamide
(4.67)

methionine

6-methylpretetramide
(4.68)

oxytetracycline
(4.69)

chlorotetracycline
(4.70)

Figure 4.27 Some steps in the biosynthesis of tetracyclines.

The Anacardiaceae phenols are highly toxic and powerful vesicants which produce severe dermatitis when in contact with the skin. Cashew shell oil, which is 90% anacardic acid and 10% cardol, is used industrially as a waterproofing agent or polymerised to form a plastic.

The tetracyclines elaborated by *Streptomyces* species contain both an amide and an amine group. Although the detailed biosynthetic pathway to the formation of these compounds is not known, the amide group has been shown to be derived from a malonamide starter unit (4.67), the first product to be isolated being 6-methylpretetramide (4.68). This is derived from a methylated C_{19} polyketide chain, as shown in *Figure 4.27*. Modification of the pretetramide, including transamination, results in oxytetracycline (4.69), while chlorination by inorganic chloride gives chlorotetracycline (aureomycin) (4.70).

The tetracyclines are broad-spectrum antibiotics which today are mainly used to treat penicillin-resistant cases, venereal diseases and cholera. Unfortunately, these compounds are liable to cause allergic reactions such as skin irritation and gastrointestinal upsets.

The antibiotic macrolides of *Streptomyces* species also contain non-acetate starter units and the chain-lengthening units are often other than malonyl esters. These compounds consist of a large lactone ring glycosylated with unusual sugars. In erythronolide A (4.71a), the lactone part of erythromycin A, which is elaborated by *S. erythreus*, propionic acid (4.72) initiates the polyketide chain, while its carboxylated derivative, 2-methylmalonic acid (4.73), is the chain-lengthening unit, as shown in *Figure 4.28*.

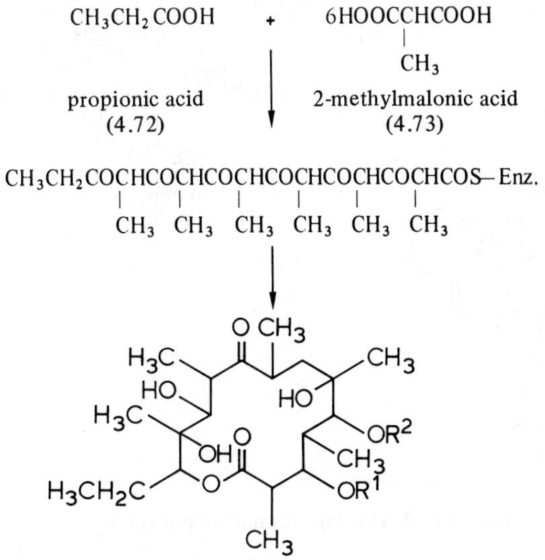

$R^1 = R^2 = H$, erythronolide A (4.71a)
R^1 = cladinose, R^2 = desosamine, erythromycin A (4.71b)

Figure 4.28 Some steps in the biosynthesis of erythronolide A.

Erythromycin A (4.71b) is active against many gram positive and gram negative bacteria, as it interferes with protein synthesis by binding to ribosomal structures. In particular, this antibiotic is used to treat penicillin-resistant staphylococci infections and pneumonia due to mycoplasma.

The ansamycins contain an aromatic nucleus attached to a large, cyclic ansa chain (*Figure 4.29*). They are important antibacterial, antiviral and antitumour agents which act by inhibiting the synthesis of bacterial and viral RNA. In rifamycin S (4.74), the ansa chain is initiated by 3-aminobenzoic acid (4.75) and lengthened by both malonyl and 2-methylmalonyl units (*Figure 4.29*).

Rifampicin (4.76), a derivative of rifamycin S, is an excellent mycobactericidal agent for the treatment of tuberculosis and leprosy.

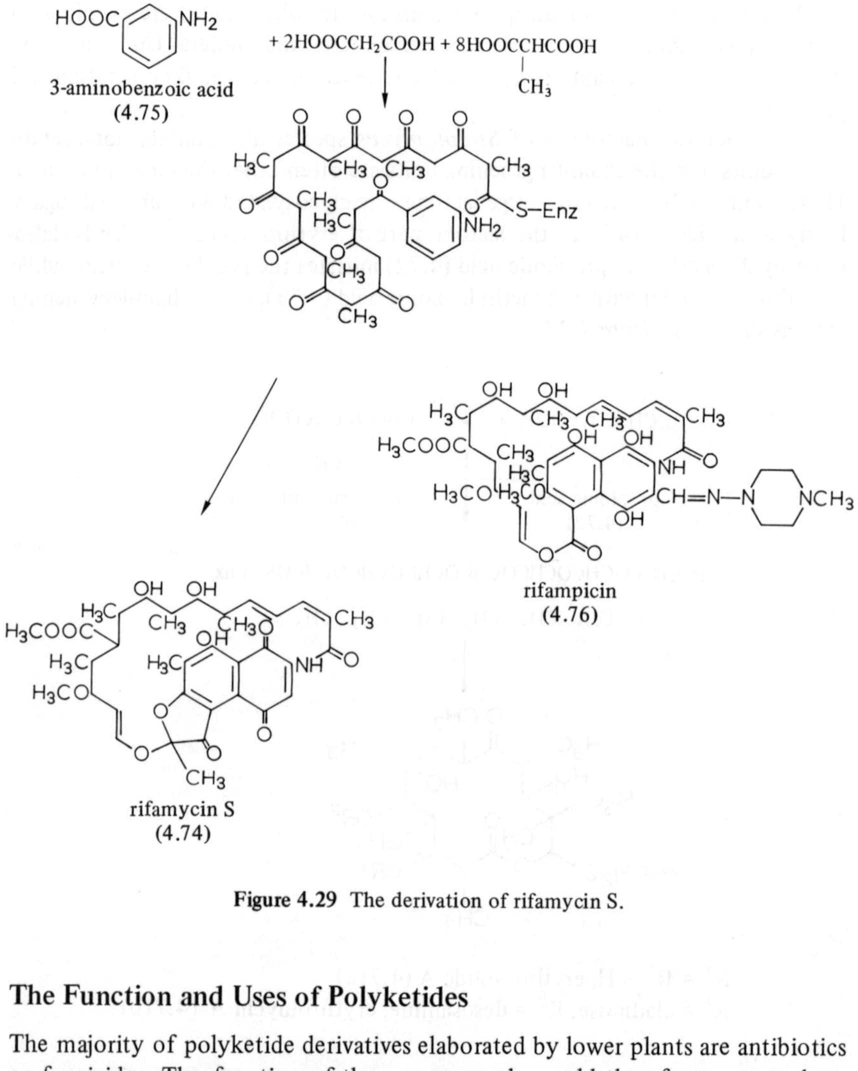

Figure 4.29 The derivation of rifamycin S.

The Function and Uses of Polyketides

The majority of polyketide derivatives elaborated by lower plants are antibiotics or fungicides. The function of these compounds would therefore seem to be a protective one. It has been observed that organisms growing under favourable conditions with an ample supply of nutrients produce only small amounts of antibiotic compounds. Conversely, under unfavourable conditions, when the survival of the organisms might depend on the elimination of competitors, bio-synthesis of antibiotics increases.

Unfortunately, many lower plant antibiotics are toxic to animals and thus can-not be used as therapeutic agents. However, a number which do not cause serious side-effects are used in medicine today to treat bacterial infections. Such com-pounds include oxytetracycline, chlorotetracycline, erythromycin and rifamycin derivatives, while griseofulvin is used as a fungicide. The toxicity of natural com-pounds can sometimes be overcome by chemical modification, and synthetic and semisynthetic derivatives based on natural antibiotics play a useful role in chemo-therapy.

Some lower plant polyketide derivatives interfere with DNA or RNA biosyn-thesis and could, theoretically, be used against cells forming malignant tumours. Unfortunately, these compounds are usually too toxic to be used medicinally, but some experimental success has been achieved with the ansamycins and re-lated compounds.

Little is known of the function of higher plant polyketide derivatives, but they probably also protect from bacterial and fungal infection.

Polyketides in Chemosystematics

There has been little attempt to link the biosynthesis of polyketide derivatives with plant taxonomy, but some tentative conclusions can be drawn from the data at present available. The multiplicity of polyketide-derived compounds in lower plants and their relative rarity in higher plants indicates that the biosynthesis of such compounds is a primitive characteristic. In many cases, this pathway has been lost in the evolution of higher plants. Chlorination appears to be connected with polyketide biosynthesis, for although natural chloro compounds are rare amongst land plants, chlorinated polyketides are not unusual in the fungi. One of the few chlorinated compounds to be isolated from higher plants, 3-chloro-plumbagin (4.33), is derived from a polyketide.

Fungi biosynthesise many polyketide derivatives which vary in structure from the simple orsellinic acid (4.1) to the highly complex macrolides. Structural di-versity amongst the lichens, however, is much less, and these lower plants can be characterised by their usnic acid (4.24), depside and depsidone contents.

Anthraquinones such as emodin (4.34) and its derivatives are characteristic of the lichen genus *Caloplaca* which contains about 500 species. These species can be divided into groups according to their anthraquinone content. Methylene-*bis*-phloroglucinol derivatives are characteristic of ferns, although they also occur in higher plants.

There is no relationship between families of higher plants biosynthesising polyketides, and similar compounds are produced by both the relatively primitive Droseraceae and the advanced Plumbaginaceae. Thus, the loss of the polyketide biosynthetic pathway has little phylogenetic significance. However, it is interesting that the anthraquinones characteristic of the Rubiaceae, an advanced family, are not derived from polyketides, but are biosynthesised by the shikimic acid pathway (Chapter 7).

Quinones are biosynthesised in higher plants by several pathways, polyketide-derived naphthoquinones being characteristic of the Droseraceae, Ebenaceae and Plumbaginaceae families, while anthraquinones biosynthesised by the acetate-malonate pathway are found in the Polygonaceae, Leguminosae and Rhamnaceae families. Dianthrone derivatives, such as the sennosides (4.41), are characteristic of *Cassia* species (Leguminosae), and hypericin (4.44) is a good taxonomic marker for the *Hypericum* genus (Hypericaceae).

Other polyketides with taxonomic significance include the vesicant, long-side-chain phenols, such as urushiol (4.64), which are characteristic of the Anacardiaceae family.

Bibliography

Biosynthesis

'Biosynthesis of Phenols and Other Polyketides', in *Biosynthesis of Acetate-derived Compounds,* N. M. Packter, Wiley, 1973
'Biosynthesis of Phenolic Compounds', in *Biosynthesis*, vol. 1, T. A. Geissman (ed.), Specialist Periodical Reports, The Chemical Society, 1972
'Biosynthesis of Polyketides', in *Biosynthesis,* vols. 2, 4 and 5, *ibid.,* 1973, 1976 and 1977
'Secondary Metabolites Derived from Acetate: Fatty Acids and Polyketides', in *Secondary Metabolism*, J. Mann, Oxford University Press, 1978
'Compounds Formed by the Cyclisation of Polyketide Chains', in *Organic Chemistry of Secondary Plant Metabolism*, T. A. Geissman and D. H. G. Crout, Freeman, Cooper and Co., 1969
'Oxidative Coupling of Phenols', *ibid*.
'Miscellaneous Oxidative Processes', *ibid*.
'Biochemistry of Plant Phenolics', in *Recent Advances in Phytochemistry*, vol. 12, J. B. Harborne and C. F. Van Sumere (eds.), Plenum, 1979
'Carboaromatic and Related Compounds', in *Natural Products Chemistry*, vol. 2, K. Nakanishi *et al*. (eds.), Academic Press, 1975
'Secondary Metabolic Products', in *The Lichens*, V. Ahamadjian and M. E. Hale (eds.), Academic Press, 1973

Antibiotics

Fungal Metabolites, W. B. Turner, Academic Press, 1971
Biosynthesis of Antibiotics (continuing series, vol. 1, 1966), Academic Press
Biogenesis of Antibiotic Substances, Z. Vareh and Z. Hostalek (eds.), Academic Press, 1965
'Preservation of the Species: Toxins, Venoms and Means of Deception', section: Mycotoxins, Phytotoxins and Antibiotics, in *Introduction to Chemical Ecology*, M. Barbier, Longman, 1976

'Chemotherapy I: Bacteria, Fungi, Viruses', in *Gaddum's Pharmacology*, 8th edn,
A. S. V. Burgen and J. F. Mitchell, Oxford University Press, 1978

Aflatoxin

'Aflatoxin and Related Mycotoxins', in *Phytochemical Ecology*, J. B. Harborne
(ed.), Academic Press, 1972

Quinones

'Biosynthesis of Quinones', in *Biosynthesis*, vol. 3, *ibid*., 1975
'Quinones, Nature, Distribution and Biosynthesis', in *Chemistry and Biochemistry
of Plant Pigments*, vol. 1, T. W. Goodwin (ed.), Academic Press, 1976
'Naphthoquinones', in *Naturally Occurring Quinones*, 2nd edn, R. H. Thomson
(ed.), Academic Press, 1971
'Anthraquinones', *ibid*.
'Hypericin', *ibid*.
'The Cathartic Action of Anthraquinones', in *Pharmacology of Plant Phenolics*,
J. W. Fairbairn (ed.), Academic Press, 1959
'Photodynamic Compounds in Plants', *ibid*.
'Comparative Biosynthetic Pathways in Higher Plants', section: Different Bio-
synthetic Routes Leading to Identical or Similar Products, in *Chemistry in
Evolution and Systematics*, T. Swain (ed.), Butterworths, 1973

5 The Acetate-Mevalonate Pathway

Introduction

Numerous compounds classed as terpenoids or steroids are biosynthesised by the acetate–mevalonate pathway. The terms terpenoid and isoprenoid are interchangeable, isoprenoid referring to the five-carbon isoprene unit (5.1) from which all terpenoids are theoretically derived. This isoprene rule, which states that all terpenoids are multiples of the isoprene unit (i.e. C_{10}, C_{15}, C_{20}, etc.), is not strictly obeyed by natural products, although most compounds classed as terpenoids can be seen to be derived from such units (*Figure 5.1*).

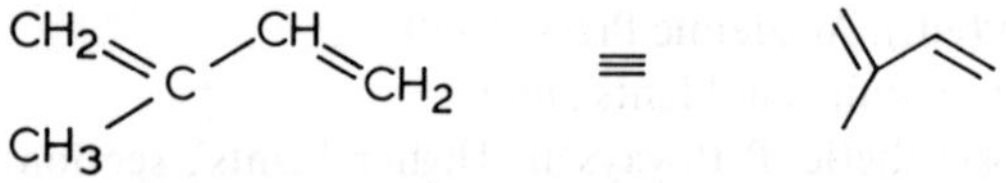

isoprene
(5.1)

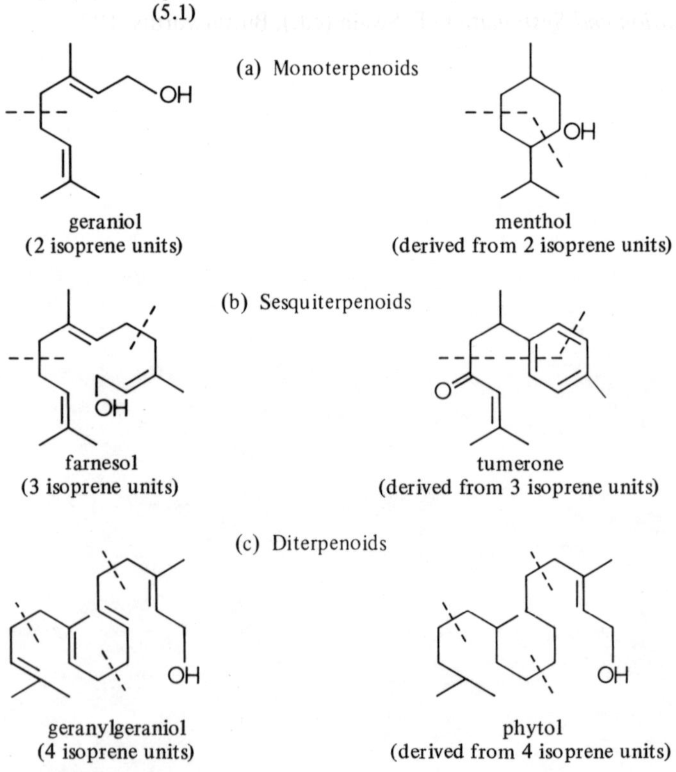

(a) Monoterpenoids

geraniol
(2 isoprene units)

menthol
(derived from 2 isoprene units)

(b) Sesquiterpenoids

farnesol
(3 isoprene units)

tumerone
(derived from 3 isoprene units)

(c) Diterpenoids

geranylgeraniol
(4 isoprene units)

phytol
(derived from 4 isoprene units)

Figure 5.1 Natural compounds following the isoprene rule.

Isoprene is not, however, the biogenic precursor of terpenoids and is only rarely found in nature.

It is probable that the number of terpenoids elaborated by plants is greater than that of any other group of natural products. Thus, these compounds are considered as groups, rather than individual compounds, according to the number of isoprene units they contain (i.e. hemiterpenoids, C_5; monoterpenoids, C_{10}; sesquiterpenoids, C_{15}; diterpenoids, C_{20}; triterpenoids, C_{30}; and carotenoids, C_{40}). Much work on stereochemical aspects of terpenoid and steroid biosynthesis has been carried out, but for reasons of space this has been omitted from our discussion. Details can be found in the books listed in the bibliography.

The acetate–mevalonate pathway is ubiquitous and all organisms can biosynthesise some terpenoids. The presence of this pathway is thus a primitive characteristic, but only angiosperms can elaborate all types of terpenoid and steroid. Animals are unable to biosynthesise carotenoids and insects are unable to biosynthesise steroids.

Mevalonic Acid

Mevalonic acid (5.2) is the primary precursor of all the terpenoids and steroids biosynthesised by plants (*Figure 5.2*). It is derived from acetyl CoA (5.3) through the intermediate formation of acetoacetyl CoA (5.4) and 3-hydroxy-3-methyl-glutaryl CoA (HMG CoA) (5.5), these reactions being catalysed by acetyl CoA acetyltransferase and HMG CoA synthase, respectively. Reduction of HMG CoA, catalysed by HMG CoA reductase, gives mevalonic acid (*Figure 5.3*).

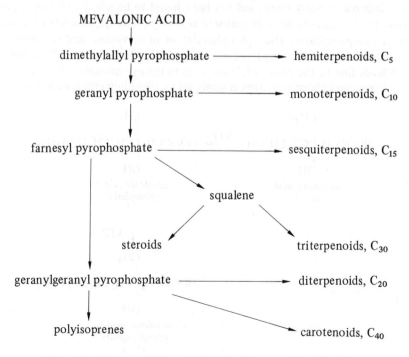

Figure 5.2 Compounds biosynthesised by the acetate–mevalonate pathway.

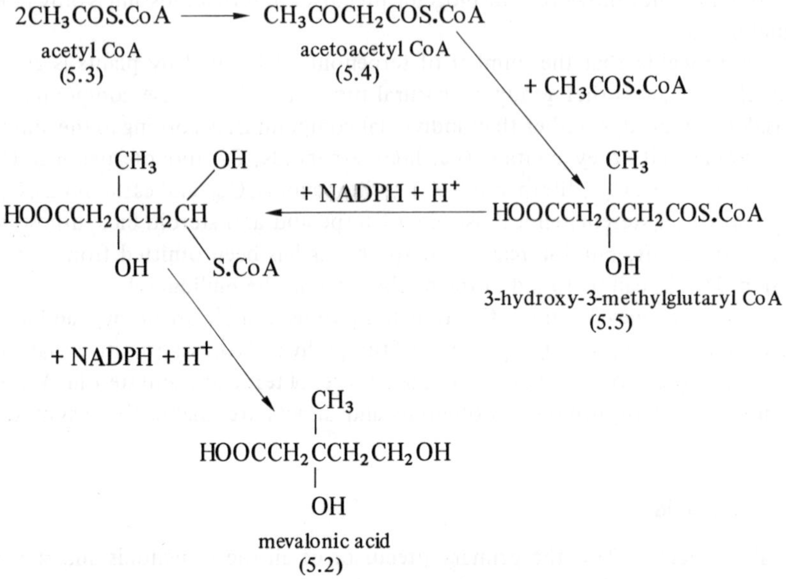

$2CH_3COS.CoA \longrightarrow CH_3COCH_2COS.CoA$

acetyl CoA   acetoacetyl CoA
(5.3)   (5.4)

$+ CH_3COS.CoA$

      CH$_3$ OH $+ NADPH + H^+$ CH$_3$

$HOOCCH_2CCH_2CH \longleftarrow HOOCCH_2CCH_2COS.CoA$

   OH   S.CoA   OH

3-hydroxy-3-methylglutaryl CoA
(5.5)

$+ NADPH + H^+$

                       CH$_3$

$HOOCCH_2CCH_2CH_2OH$

                       OH

mevalonic acid
(5.2)

Figure 5.3 The biosynthesis of mevalonic acid.

Before the pathway can continue, mevalonic acid must be catalytically phosphorylated by ATP and mevalonic acid kinase. Mevalonic acid kinase activity has been detected in many plants and has been found to be inhibited by such products of the acetate–mevalonate pathway as geranyl, farnesyl, geranylgeranyl and phytyl pyrophosphates. Thus, phosphorylation of mevalonic acid is a primary point at which control of terpenoid and steroid biosynthesis operates. Phosphorylation leads first to the mono- (5.6) and then to the pyrophosphate (5.7) (*Figure 5.4*). The second phosphorylation is catalysed by phosphomevalonate kinase.

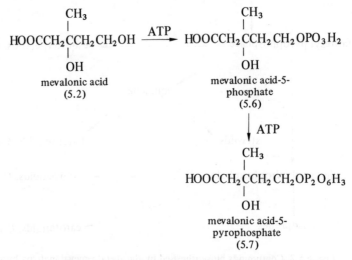

    CH$_3$ CH$_3$

$HOOCCH_2CCH_2CH_2OH \xrightarrow{\text{ATP}} HOOCCH_2CCH_2CH_2OPO_3H_2$

    OH   OH

mevalonic acid   mevalonic acid-5-
(5.2)   phosphate
                          (5.6)

$\downarrow$ ATP

                      CH$_3$

$HOOCCH_2CCH_2CH_2OP_2O_6H_3$

                      OH

mevalonic acid-5-
pyrophosphate
(5.7)

Figure 5.4 The phosphorylation of mevalonic acid.

Dimethylallyl pyrophosphate

A concerted decarboxylation–dehydration of mevalonic acid pyrophosphate gives isopentenyl pyrophosphate (5.8), the biogenic isoprene unit. This reaction, which is catalysed by pyrophosphomevalonate decarboxylase, is highly stereospecific and involves a *trans* elimination of the carboxyl and hydroxyl groups (*Figure 5.5*).

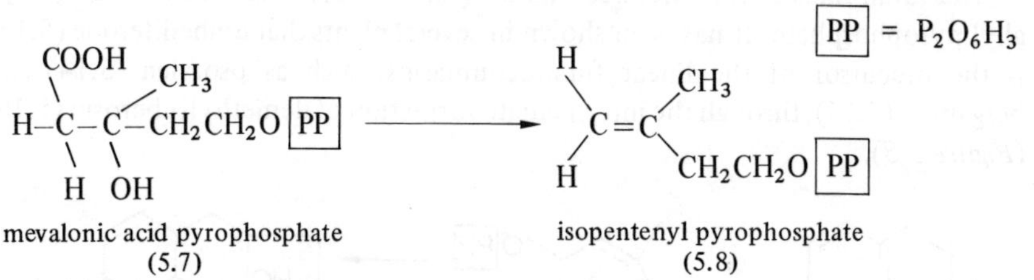

$$\boxed{PP} = P_2O_6H_3$$

mevalonic acid pyrophosphate
(5.7)

isopentenyl pyrophosphate
(5.8)

Figure 5.5 The derivation of isopentenyl pyrophosphate.

Under the influence of isopentenyl pyrophosphate isomerase, isopentenyl pyrophosphate is isomerised to the highly reactive dimethylallyl pyrophosphate (5.9), a reaction which involves an enzyme-bond intermediate (*Figure 5.6*).

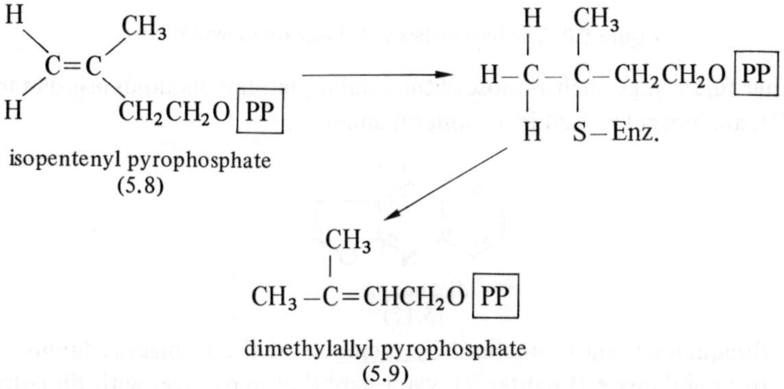

isopentenyl pyrophosphate
(5.8)

dimethylallyl pyrophosphate
(5.9)

Figure 5.6 The isomerisation of isopentenyl pyrophosphate to dimethylallyl pyrophosphate.

senecioic acid
(5.10)

leucine
(5.12)

isoamyl alcohol
(5.11)

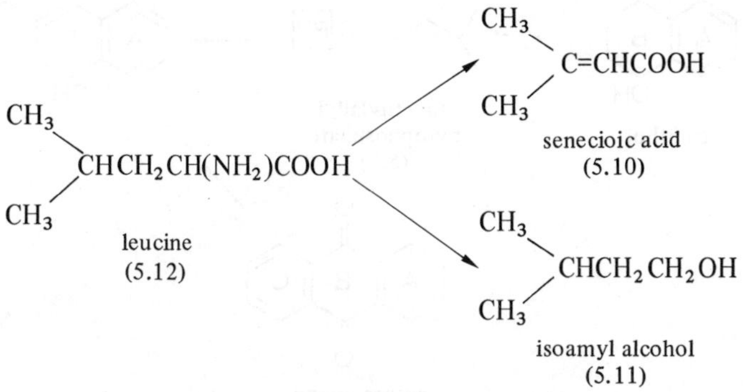

Figure 5.7 The derivation of simple hemiterpenoids.

Hemiterpenoids

Dimethylallyl pyrophosphate (5.9) is the precursor of the single isoprene units found in a number of plant compounds, but the simplest hemiterpenoids, such as senecioic acid (5.10) and isoamyl alcohol (5.11), are derived from the amino acid, leucine (5.12) (*Figure 5.7*).

The furan rings of the furanocoumarins (Chapter 6) are derived from dimethylallyl pyrophosphate. It has been shown in several plants that umbelliferone (5.13) is the precursor of the linear furanocoumarins, such as psoralen (5.14) and bergapten (5.15), through the intermediate formation of demethylsuberosin (5.16) (*Figure 5.8*).

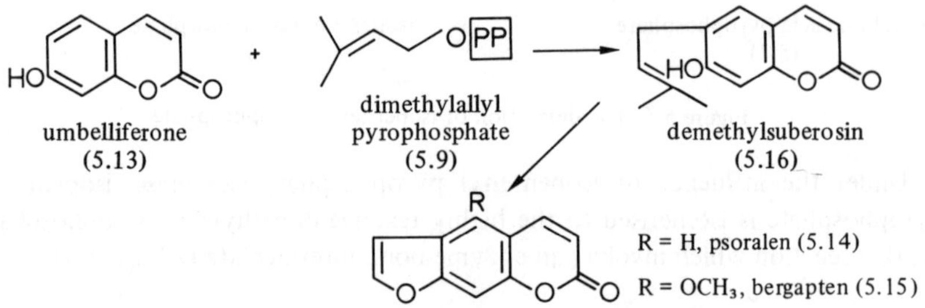

umbelliferone
(5.13)

dimethylallyl
pyrophosphate
(5.9)

demethylsuberosin
(5.16)

R = H, psoralen (5.14)
R = OCH₃, bergapten (5.15)

Figure 5.8 The biosynthesis of linear furanocoumarins.

Other furan rings, such as those of the furanoquinoline alkaloids (e.g. dictamnine (5.17)), are biosynthesised in a similar manner.

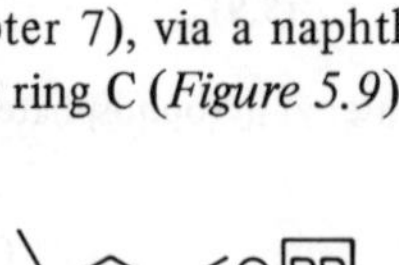

dictamnine
(5.17)

Anthraquinones biosynthesised by members of the Rubiaceae family are derived from shikimate (Chapter 7), via a naphthol derivative, with dimethylallyl pyrophosphate giving rise to ring C (*Figure 5.9*).

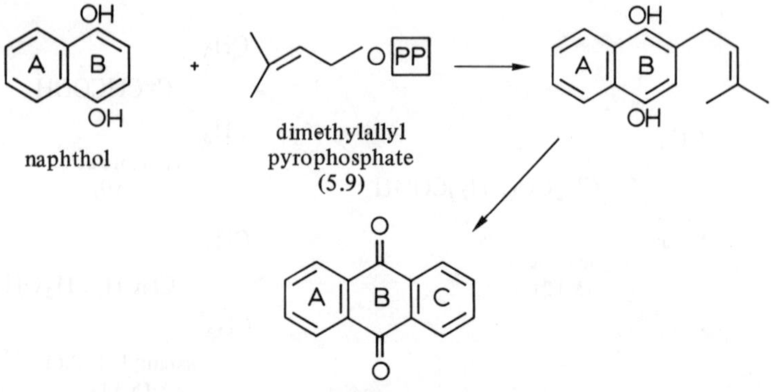

naphthol

dimethylallyl
pyrophosphate
(5.9)

anthraquinone

Figure 5.9 The biosynthesis of the Rubiaceae anthraquinones.

Geranyl Pyrophosphate and the Monoterpenoids

Head-to-tail condensation of dimethylallyl pyrophosphate with isopentenyl pyro-phosphate, catalysed by dimethylallyl transferase, gives the C_{10} monoterpenoid, geranyl pyrophosphate (5.18) (*Figure 5.10*).

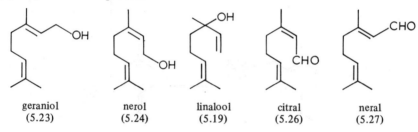

dimethylallyl   isopentenyl
pyrophosphate   pyrophosphate   geranyl pyrophosphate
(5.9)   (5.8)   (5.18)

Figure 5.10 The biosynthesis of geranyl pyrophosphate.

Monoterpenoids can be subdivided into acyclic, monocyclic and bicyclic com-pounds, as illustrated by the examples given in *Figure 5.11*.

(a) Acyclic monoterpenoids

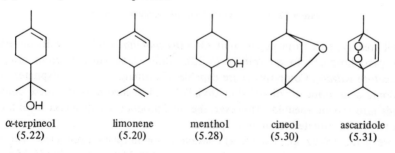

geraniol	nerol	linalool	citral	neral
(5.23)	(5.24)	(5.19)	(5.26)	(5.27)

(b) Monocyclic monoterpenoids

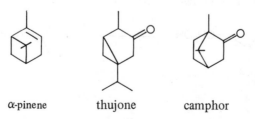

α-terpineol	limonene	menthol	cineol	ascaridole
(5.22)	(5.20)	(5.28)	(5.30)	(5.31)

(c) Bicyclic monoterpenoids

α-pinene   thujone   camphor

Figure 5.11 Some examples of monoterpenoids.

Orange-juice vesicles (*Citrus* sp., Rutaceae) contain an enzyme system which is capable of phosphorylating mevalonic acid and forming both isopentenyl pyrophosphate and dimethylallyl pyrophosphate, the final product being the monoterpene, linalool (5.19). Intact fruits and leaves convert linalool to limonene (5.20) and other cyclic monoterpenoids. Limonene can also be formed from geranyl pyrophosphate (5.18) and neryl pyrophosphate (5.21). It has been suggested that the *trans* double bond of geranyl pyrophosphate must be isomerised to the *cis* orientation of neryl pyrophosphate before cyclisation can occur (*Figure 5.12*).

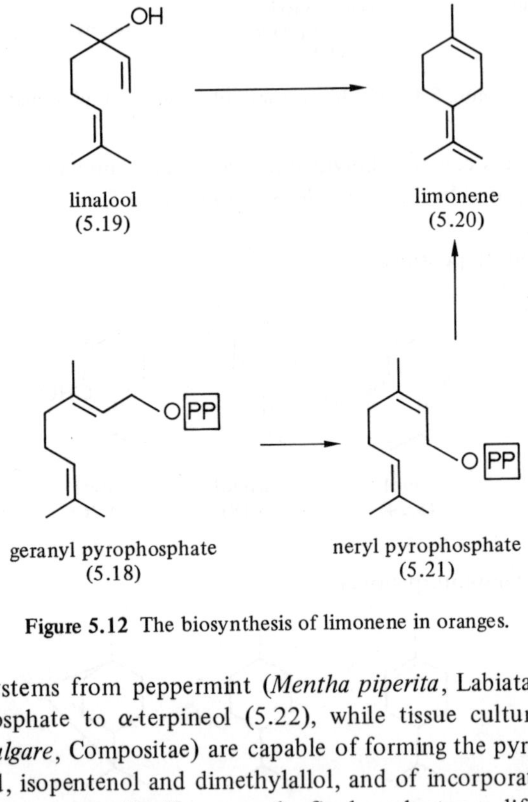

linalool
(5.19)

limonene
(5.20)

geranyl pyrophosphate
(5.18)

neryl pyrophosphate
(5.21)

Figure 5.12 The biosynthesis of limonene in oranges.

Cell free systems from peppermint (*Mentha piperita*, Labiatae) will convert neryl pyrophosphate to α-terpineol (5.22), while tissue cultures from tansy (*Tanacetum vulgare*, Compositae) are capable of forming the pyrophosphates of mevalonic acid, isopentenol and dimethylallol, and of incorporating these compounds into monoterpenoids. However, the final products are different from the monoterpenoids isolated from intact plants.

Rose petals (*Rosa* sp., Rosaceae) contain a monoterpenoid reductase which catalyses the reduction of geraniol (5.23) or nerol (5.24) to citronellol (5.25), a constituent of rose oil (*Figure 5.13*).

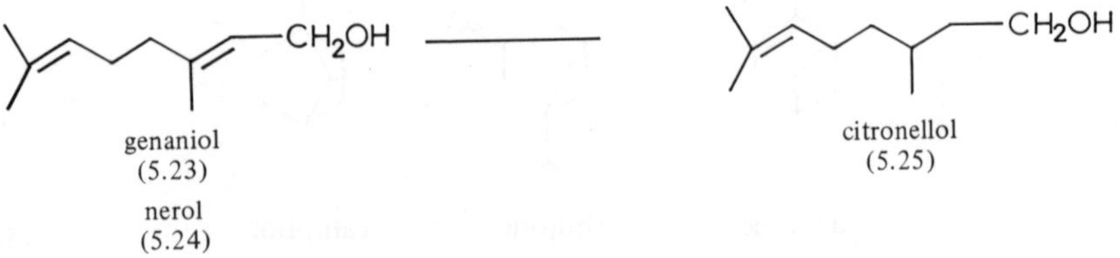

genaniol
(5.23)

nerol
(5.24)

citronellol
(5.25)

Figure 5.13 The formation of citronellol in rose petals.

Essential oils

Monoterpenoids are volatile and are responsible for the characteristic odours of many plants. They are constituents of the essential oils of plants, many of which are important to the perfume, flavouring and pharmaceutical industries. Natural perfumes and flavourings are expensive, as large quantities of plant material are required to produce a small amount of oil, 4000 kg of rose petals, for example, giving only around 1 kg of rose oil.

Perfumes and flavourings produced in quantity include rose oil, lemon grass oil from *Cymbopogon* species (Gramineae), which contains mainly citral (5.26) and neral (5.27), oil of neroli from orange blossom, which contains nerol (5.24) and limonene (5.20), and peppermint oil from *Mentha piperita* (Labiatae), which contains menthol (5.28). Citral is used industrially as a precursor of the ionones (5.29), β-carotene (5.132) and vitamin A_1 (5.135). The ionones have an odour resembling violets and are used in cheap perfumes.

β-ionone   α-ionone

(5.29)

Essential oils have useful carminative and expectorant properties and are used to treat digestive upsets and diseases of the respiratory tract, especially coughs. The constituents of peppermint oil are used for both purposes, while eucalyptus oil from *Eucalyptus* species (Myrtaceae), which contains cineole (5.30), is antiseptic and a treatment for respiratory diseases. Ascaridole (5.31) from *Chenopodium ambrosiodes* (Chenopodiaceae) is anthelmintic and has been used to expel intestinal worms.

Mixtures of structurally related monoterpenoids are usually present in essential oils and studies with peppermint have shown that the composition of the oil varies with the age of the plant and such external factors as temperature, etc. The production of monoterpenoids in the related water mint (*Mentha aquatica*) has been shown to be under genetic control.

Piperitenone (5.32) is the precursor of the peppermint monoterpenoids, which are biogenically related as shown in *Figure 5.14*. The menthones and menthols derived from piperitone (5.33) and pulegone (5.34) are stereoisomers.

Iridoids

The iridoids were first isolated from the Iridomyrex species of ant, from which they took their name. Many are insecticides and such compounds are widespread in plants, where they usually occur as glucosides, and are important as the precursors of several classes of alkaloid (see Chapter 9). Although not strictly terpenoids, the iridoids are derived from a monoterpenoid precursor.

The biosynthesis of loganin (5.35) has been studied in detail in the Madagascar periwinkle (*Catharanthus roseus* syn. *Vinca rosea*, Apocynaceae). Hydroxylation of geraniol (5.23), catalysed by geraniol hydratase, gives 10-hydroxygeraniol

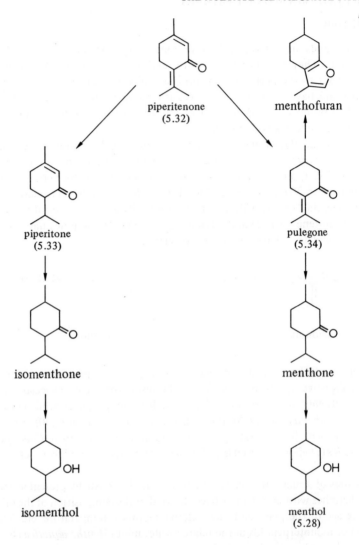

Figure 5.14 The derivation of some constituents of peppermint oil.

(5.36), which is isomerised to 10-hydroxynerol (5.37). The first glucosylated compound to be isolated, deoxyloganic acid (5.38), is converted to loganin (5.35) through the intermediate formation of either loganic acid (5.39) or deoxyloganin (5.40). Methylation, catalysed by methyltransferase, is by S-adenosylmethionine. A scheme for the biosynthesis of loganin based on available evidence is given in *Figure 5.15*.

Before loganin can be incorporated into the complex indole alkaloids (Chapter 9), it must be converted to secologanin (5.41). Other secoiridoids such as gentiopicroside (5.43) are also derived from loganin (*Figure 5.16*). Gentiopicroside, derived from sweroside (5.42), is the precursor of alkaloids such as gentianine (5.44) found in *Gentiana* species (Gentianaceae).

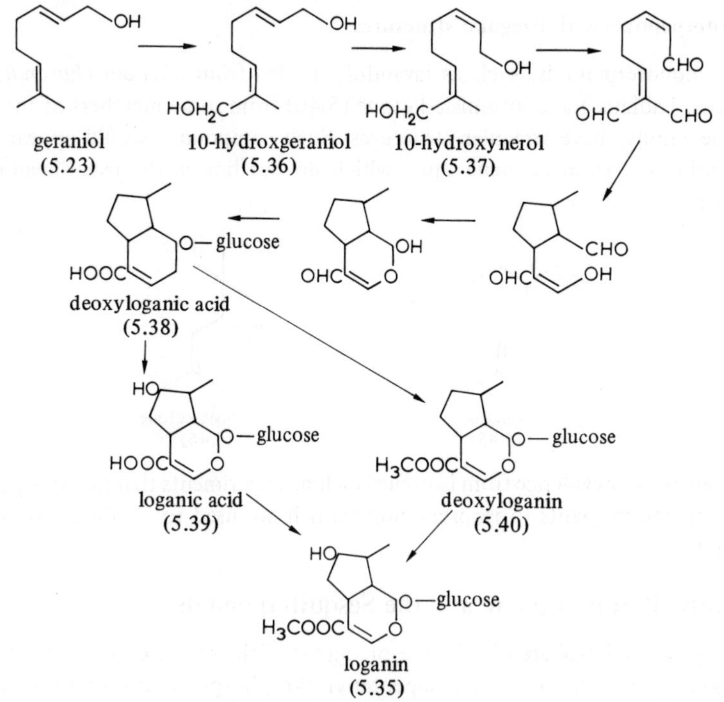

Figure 5.15 Some steps in the biosynthesis of loganin.

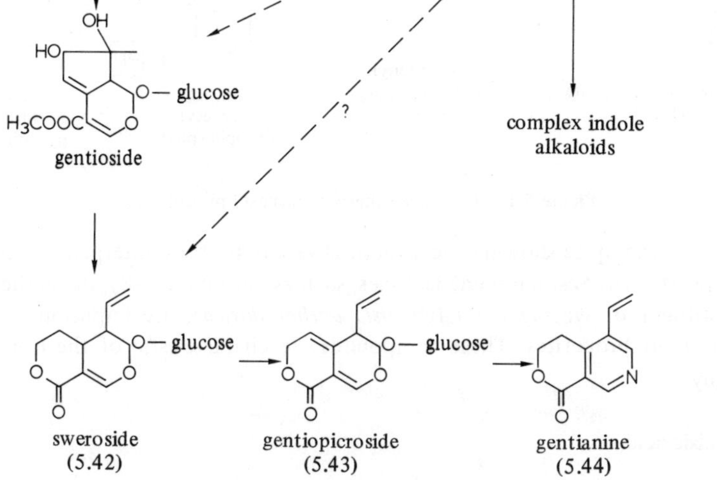

Figure 5.16 The derivation of some secoiridoids.

Monoterpenoids with irregular structures

Some monoterpenoids, such as lavandulol (5.45) from lavender (*Lavandula officinalis*, Labiatae) and artemisia ketone (5.46) from some members of the Compositae family, have irregular structures. Such compounds would appear to be biosynthesised from isoprene units which do not link in the usual head-to-tail manner.

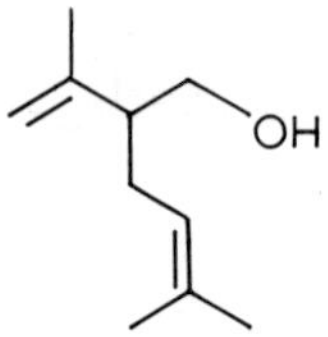

lavandulol
(5.45)

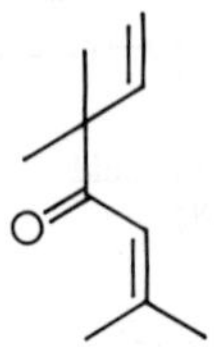

artemisia ketone
(5.46)

There is some evidence from isotopic labelling experiments that monoterpenoids are degraded in plants to isoprene units which are used to synthesise other terpenoids.

Farnesyl Pyrophosphate and the Sesquiterpenoids

Farnesyl pyrophosphate (5.47), the precursor of the sesquiterpenoids, is biosynthesised by the addition of an isopentenyl pyrophosphate unit (5.8) to geranyl pyrophosphate (5.18). The prenyl transferase which catalyses the formation of geranyl pyrophosphate appears also to catalyse the formation of the farnesyl derivative, although it is unable to lengthen the chain further. The product of the addition, 2-*trans*,6-*trans*-farnesyl pyrophosphate, is isomerised to the 2-*cis*,6-*trans* derivative before cyclisation occurs (*Figure 5.17*).

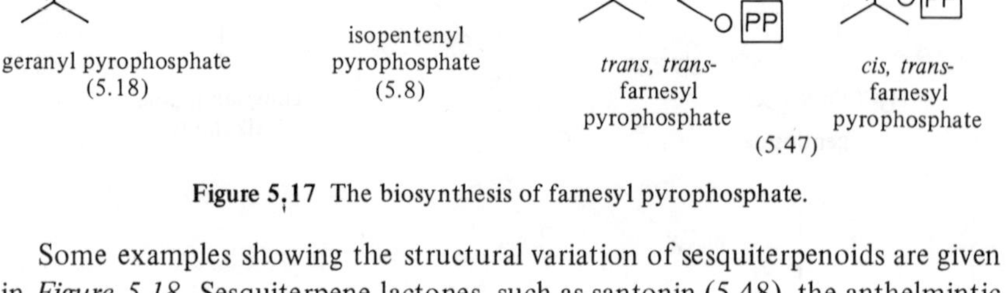

geranyl pyrophosphate
(5.18)

isopentenyl
pyrophosphate
(5.8)

trans, trans-
farnesyl
pyrophosphate
(5.47)

cis, trans-
farnesyl
pyrophosphate

Figure 5.17 The biosynthesis of farnesyl pyrophosphate.

Some examples showing the structural variation of sesquiterpenoids are given in *Figure 5.18*. Sesquiterpene lactones, such as santonin (5.48), the anthelmintic constituent of wormwood (*Artemisia anthelmintica*), are important for their antitumour properites. These compounds are characteristic of the Compositae family.

Abscisic acid

One of the most important of the higher plant sesquiterpenoids is the growth-inhibitory hormone, abscisic acid (5.49). This compound has been shown to be

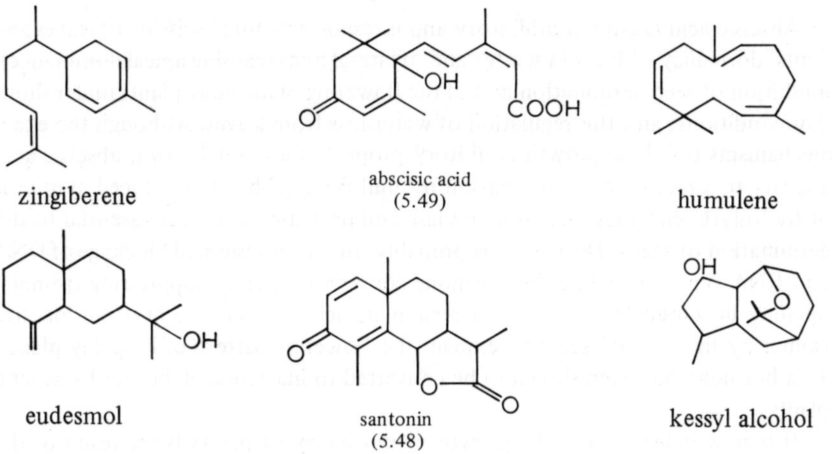

zingiberene

abscisic acid
(5.49)

humulene

eudesmol

santonin
(5.48)

kessyl alcohol

Figure 5.18 Some sesquiterpenoids.

derived from mevalonic acid (5.2) in several plants, and in grape (*Vitis* sp., Vitaceae) pericarp tissue cultures. Little is known of the biosynthetic pathway forming abscisic acid, although this hormone has been shown to be derived from both the epoxide (5.50) and xanthoxin (5.51). Xanthoxin is a degradation product of certain carotenoids, but both it and the epoxide (5.50) could be derived from farnesyl pyrophosphate (5.47) as shown in *Figure 5.19*.

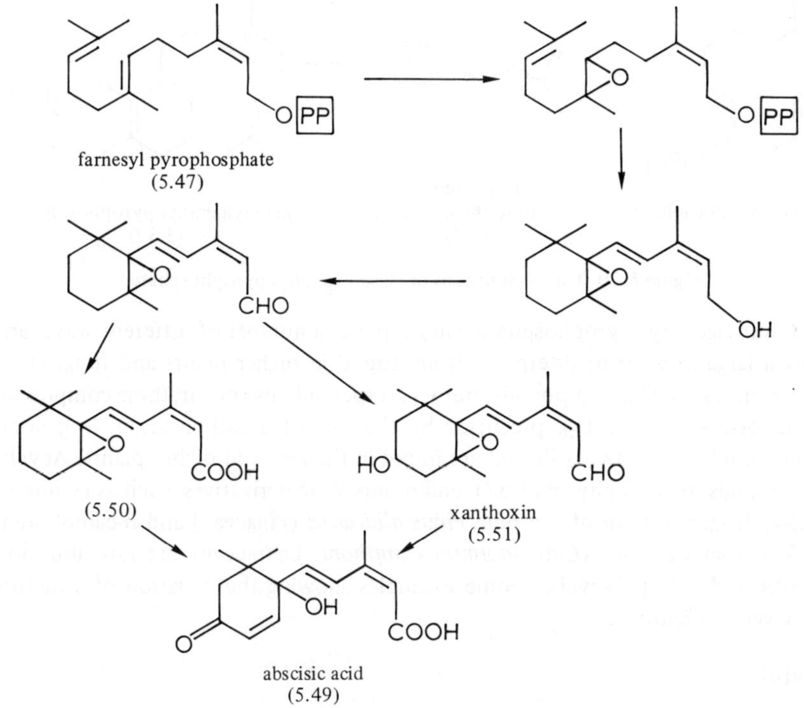

farnesyl pyrophosphate
(5.47)

(5.50)

xanthoxin
(5.51)

abscisic acid
(5.49)

Figure 5.19 Some steps in the possible biosynthesis of abscisic acid.

Abscisic acid is growth-inhibitory and is responsible for abscission of leaves and fruits; dormancy of buds in winter and of lateral buds (causing apical dominance); inhibition of seed germination and of the flowering of long-day plants under short-day conditions; and the regulation of water loss from leaves. Although the exact mechanisms for these growth-inhibitory properties are not known, abscisic acid appears to prevent seed germination by nullifying gibberellin-induced synthesis of hydrolytic enzymes such as α-amylase and protease, which are essential to the germination of seeds. Dormancy is probably due to abscisic acid blockage of DNA and RNA synthesis, while this hormone conserves water by suppressing stomatal opening in wilted leaves. The conversion of active abscisic acid to an inactive isomer by light would seem to explain the flowering pattern of long-day plants. This hormone has been shown to be converted to inactive metabolites by several plants.

It is now believed that the growth or dormancy of plants is the result of the relative concentration of hormones, rather than to the presence or absence of a specific compound, these concentrations being regulated by environmental factors.

Geranylgeranyl Pyrophosphate and the Diterpenoids

The addition of isopentenyl pyrophosphate (5.8) to farnesyl pyrophosphate (5.47), catalysed by geranylgeranyl pyrophosphate synthase, gives geranylgeranyl pyrophosphate (5.52) (*Figure 5.20*), the precursor of the diterpenoids.

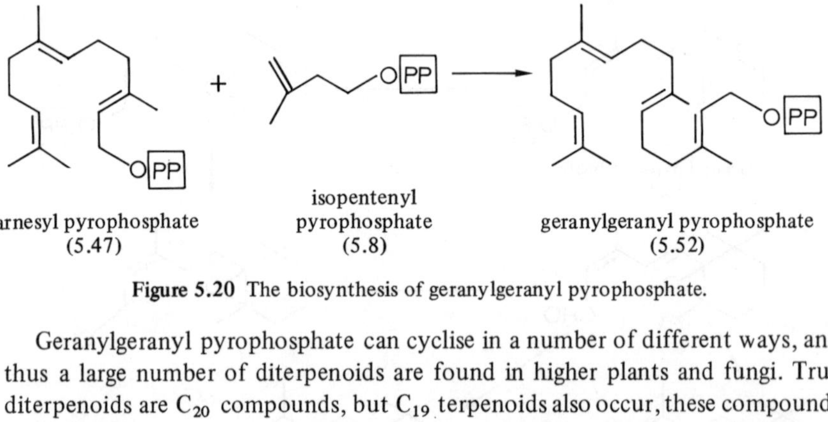

farnesyl pyrophosphate   isopentenyl   geranylgeranyl pyrophosphate
        (5.47)   pyrophosphate   (5.52)
                                    (5.8)

Figure 5.20 The biosynthesis of geranylgeranyl pyrophosphate.

Geranylgeranyl pyrophosphate can cyclise in a number of different ways, and thus a large number of diterpenoids are found in higher plants and fungi. True diterpenoids are C_{20} compounds, but C_{19} terpenoids also occur, these compounds being derived from a C_{20} precursor by the loss of a carbon atom. In general, diterpenoids are non-volatile and are found in the resins of higher plants. Acyclic compounds, such as phytol (5.53), and monocyclic derivatives, such as cembrene (5.54) from the resin of the pine *Pinus albicaulis* (Pinaceae) and α-camphorene (5.55) from camphor (*Cinnamomum camphora,* Lauraceae), are rare and most diterpenoids are polycyclic. Some examples showing the variation of structure are given in *Figure 5.21*.

Phytol

Phytol (5.53) is ubiquitous in higher plants as a component of chlorophyll in which it is esterified to chlorophyllide *a* (see Chapter 10). The esterification is

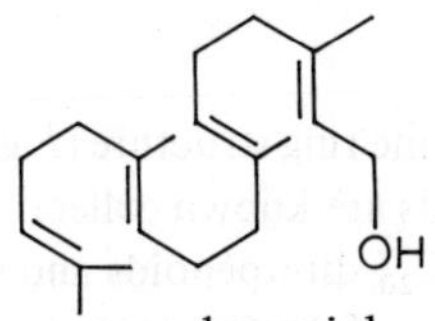

geranylgeraniol

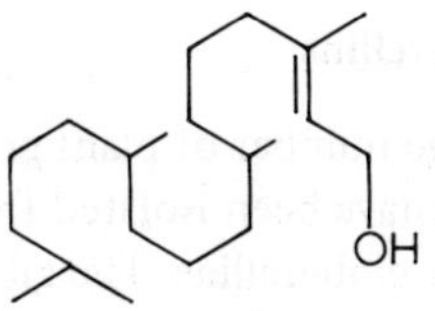

phytol
(5.53)

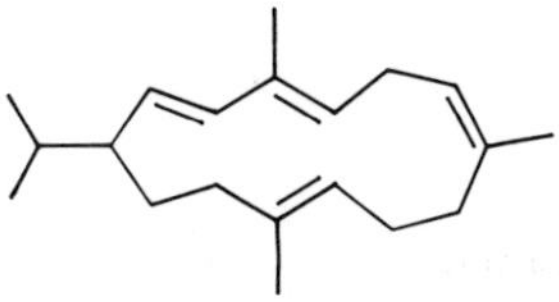

cembrene
(5.54)

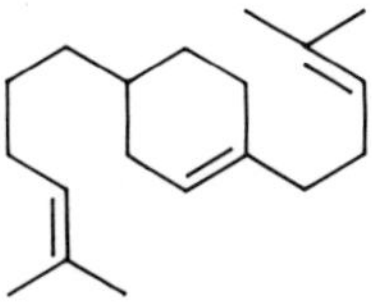

α-camphorene
(5.55)

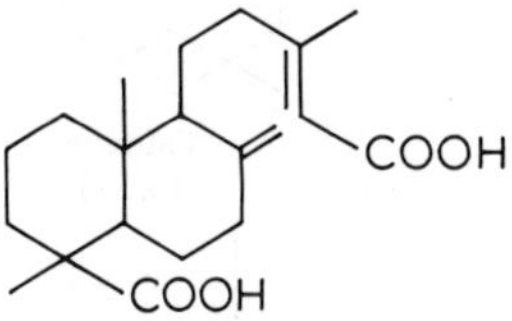

agathic acid

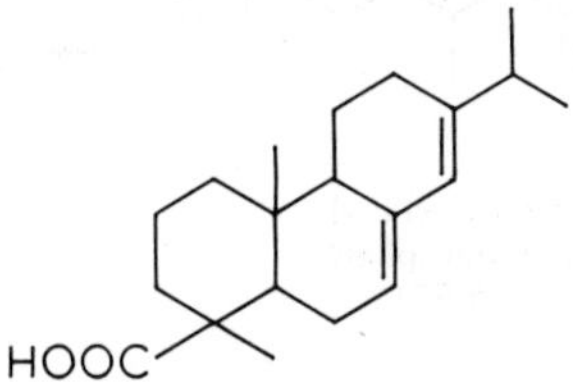

abietic acid

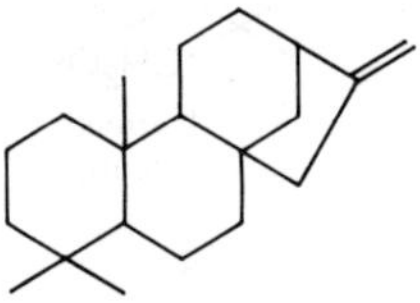

kaurene
(5.58)

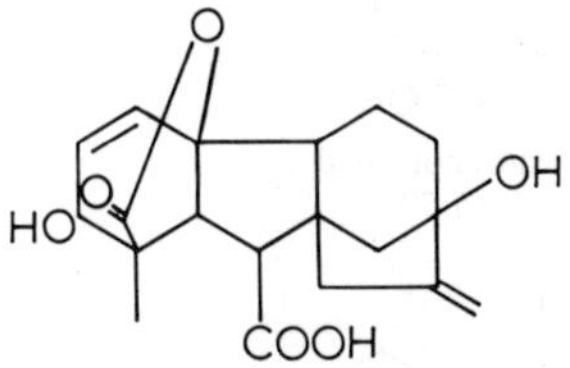

gibberellic acid
(C$_{19}$ derivative)
(5.56)

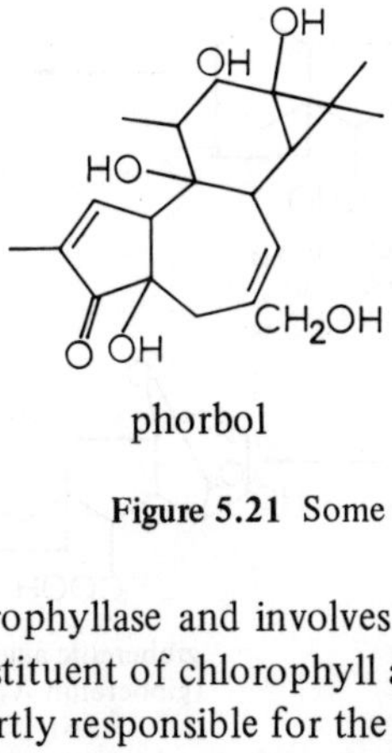

phorbol

Figure 5.21 Some diterpenoids.

catalysed by chlorophyllase and involves the unphosphorylated alcohol. Phytol is a necessary constituent of chlorophyll as it confers both lipid solubility on the molecule and is partly responsible for the biological activity of the compound.

Phytol is also a precursor of the tocopherols (vitamin E) and phylloquinone (vitamin K), which are described in Chapter 7.

Gibberellins

A large number of plant growth hormones with the gibbane ring structure (*Figure 5.22*) have been isolated from plants and such compounds are known collectively as the gibberellins. The gibberellins can be divided into C_{20} diterpenoids and C_{19} compounds derived from a C_{20} precursor. The best known of the C_{19} compounds, gibberellic acid (5.56) was first isolated from the fungus *Gibberella fujikuroi*

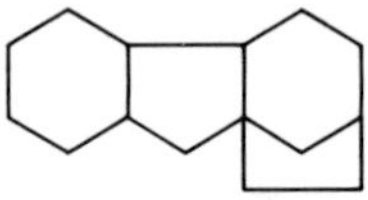

Figure 5.22 Gibbane ring structure.

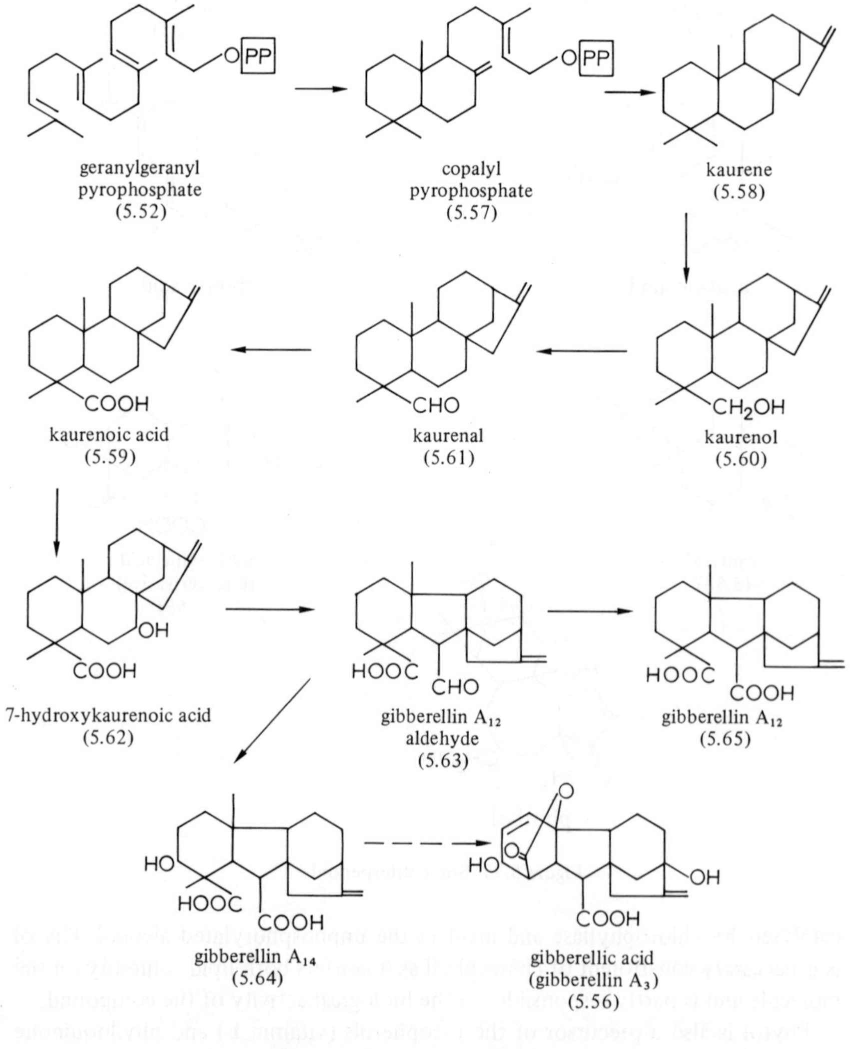

Figure 5.23 Some steps in the biosynthesis of the gibberellins.

(*Fasarium moniliforme*), which is pathogenic on rice (*Oryza sativa*, Gramineae) causing the bakanae disease of excessive growth.

Several investigations into the biosynthesis of the gibberellins have shown these compounds to be derived from mevalonic acid (5.2) through the formation of geranylgeranyl pyrophosphate (5.52). Cyclisation of geranylgeranyl pyrophosphate gives copalyl pyrophosphate (5.57), which is converted to kaurene (5.58) by a second cyclisation (*Figure 5.23*). Kaurene synthetase, the enzyme catalysing the cyclisations, has been isolated from *G. fujikuroi* and several higher plants. As the two cyclisations could not be separated, it seems that the enzyme exists as a complex and it is possible that copalyl pyrophosphate remains enzyme-bound throughout.

A series of reactions converts the C-19 methyl group of kaurene into the carboxyl group of kaurenoic acid (5.59), the intermediates being kaurenol (5.60) and kaurenal (5.61).

Hydroxylation of kaurenoic acid gives 7-hydroxykaurenoic acid (5.62), the immediate precursor of the gibbane ring system. In *G. fujikuroi* and some higher plants, the first gibberellin to be formed is that known as A_{12} aldehyde (5.63).

Gibberellin A_{12} aldehyde gives rise to two families of gibberellins in *G. fujikuroi* – those such as A_{14} (5.64) in which hydroxylation at C-3 has taken place, and those such as A_{12} (5.65) which do not contain a hydroxyl group at C-3. The C_{19} derivatives, such as gibberellic acid (5.56) itself (known as A_3), are formed by loss of C-20 and lactonisation. The biosynthetic pathway of the *G. fujikuroi* gibberellins is given in *Figure 5.23*.

In general, the gibberellins act to oppose the effects of abscisic acid (5.49). Thus, they stimulate the production of hydrolytic enzymes, such as α-amylase and protease in germinating seedlings, break dormancy in seeds and stimulate leaf and fruit growth and the flowering of certain long-day plants. Dwarf varieties of plants which are deficient in gibberellins have been bred and such plants can be made to grow normally by application of the growth hormones. Enhanced growth is due mainly to cell elongation and is associated with increased RNA synthesis.

Gibberellic acid is produced commercially from *G. fujikuroi* cultures and is used in the malting of barley, production of grapes and for various horticultural purposes.

Squalene and the Triterpenoids

Squalene (5.66) is important as the precursor of the cyclic triterpenoids and steroids. It is biosynthesised by the tail-to-tail linkage of two farnesyl pyrophosphate (5.47) molecules. The cyclopropane derivative, presqualene (5.67), is an intermediate and the reaction, which is catalysed by squalene synthase, requires NADPH for conversion of presqualene to squalene. Although the mechanism of squalene formation is not known in detail, there is some evidence that the cyclopropane ring expands to a cyclobutane ring before cleavage (*Figure 5.24*).

Before conversion to the cyclic triterpenoids or steroids, squalene is converted to its 2,3-epoxide (5.68). This reaction, which is catalysed by squalene epoxidase (squalene mono-oxygenase (2,3 epoxidising)), requires molecular oxygen and

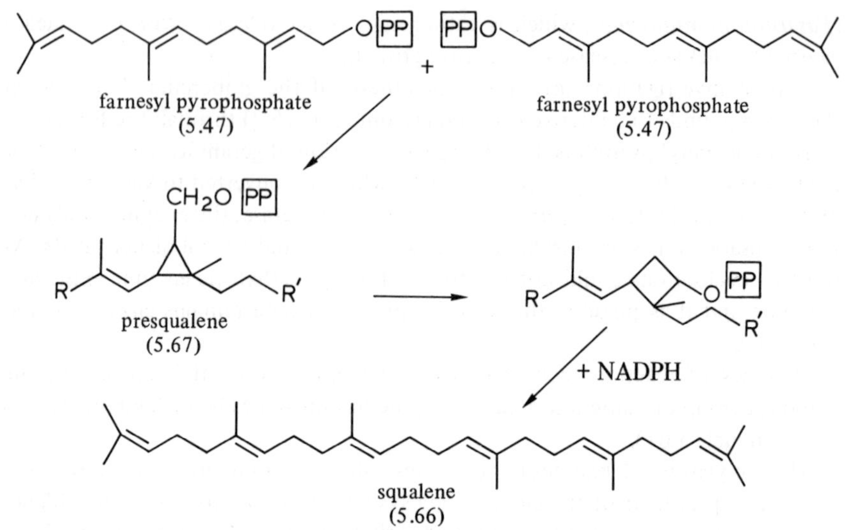

Figure 5.24 The biosynthesis of squalene.

NADPH (*Figure 5.25*). It has been shown that oxygen can add on at either end of the molecule, so that free squalene would seem to be the precursor, while the characteristic folding of the molecule prior to cyclisation must take place after epoxide formation, as shown in *Figure 5.25*.

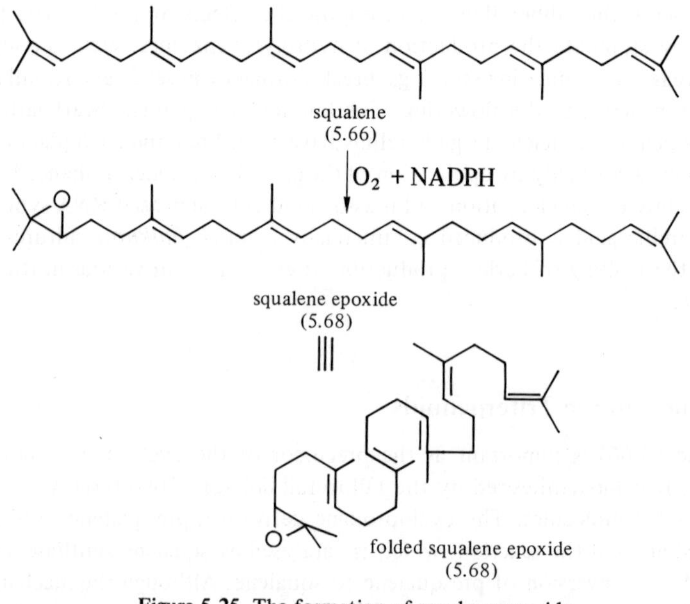

Figure 5.25 The formation of squalene epoxide.

Triterpenoids

The triterpenoids can be divided into two main classes — the tetracyclic compounds with four rings and the pentacyclic compounds with five rings.

Many triterpenoids do not obey the classical isoprene rule, whereby they can be seen to be composed of discrete isoprene units. However, all such compounds can be derived from a terpenoid precursor (squalene) by simple rearrangements and are therefore said to obey the *biogenic* isoprene rule. The cyclisation of squalene epoxide and its subsequent rearrangement to form the various families of triterpenoids are dependent on stereochemical concepts which are outside the scope of this book. (A good description is given in 'Higher terpenoids' in *Organic Chemistry of Secondary Plant Metabolism* listed in the bibliography.) A two-dimensional representation of the simplest cyclisation of squalene epoxide, that forming the dammarene derivative, dammarenediol (5.69), is shown in *Figure 5.26*. The pentacyclic triterpenes, such as lupeol (5.70) and β-amyrin (5.71) are biosynthesised from a tetracyclic precursor of the dammarene type by a rearrangement resulting in expansion of ring D and the formation of ring E (*Figure 5.27*). Such rearrangements have been shown to take place by examining the labelling patterns of pentacyclic triterpenoids formed from a squalene precursor of known isotopic labelling.

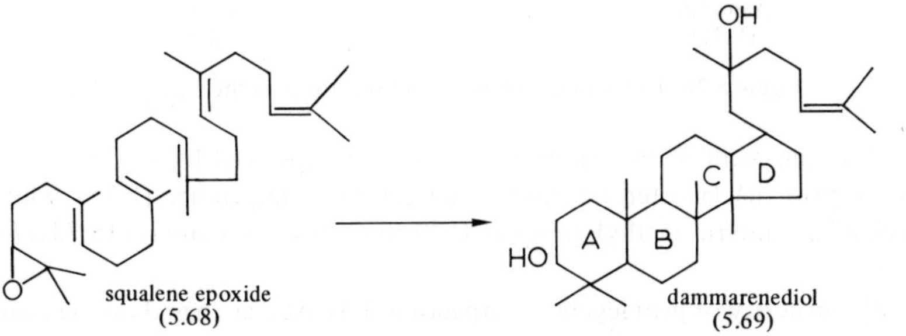

Figure 5.26 The cyclisation of squalene epoxide, forming tetracyclic triterpenoids.

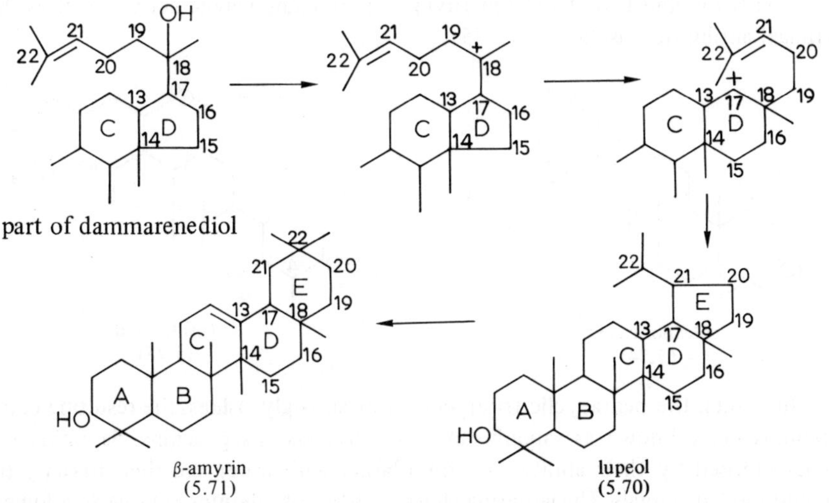

Figure 5.27 The derivation of rings D and E of the pentacyclic triterpenoids. (Note that the numbering shows the rearrangement of the precursor and is not the numbering of lupeol of β-amyrin.)

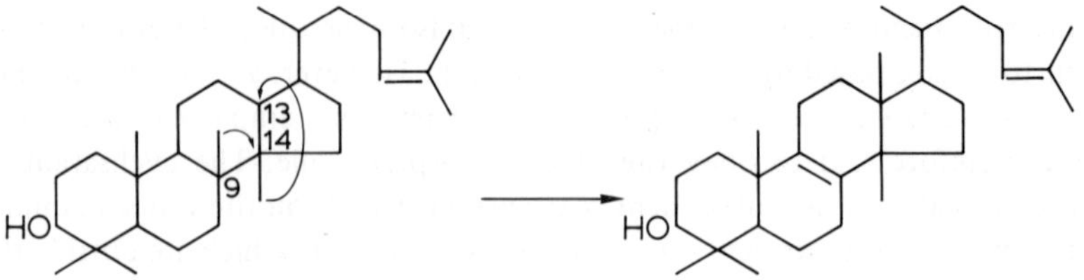

dammarene derivative

lanosterol, euphol, tirucallol
(stereoisomers)
(5.72)

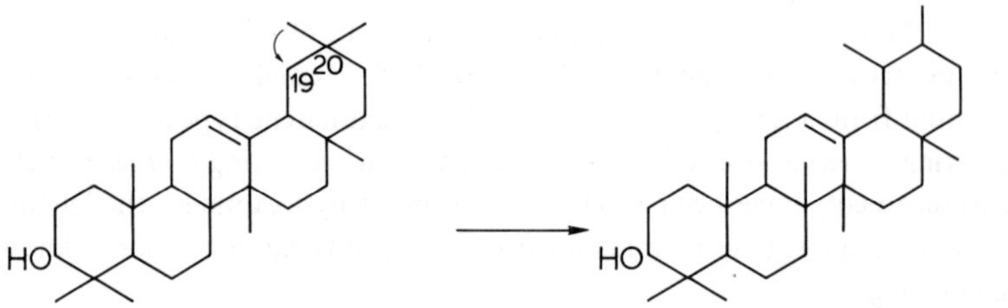

β-amyrin
(5.71)

α-amyrin
(5.73)

Figure 5.28 The rearrangement of methyl groups in the triterpenoids.

Rearrangement of the dammarene methyl groups (14-CH_3 → C-13, 9-CH_3 → C-14) gives the lanosterol family of tetracyclic triterpenoids (5.72), while the shift of a β-amyrin methyl group at C-20 to C-19 gives α-amyrin (5.73) *(Figure 5.28)*.

The majority of pentacyclic triterpenoids, however, are based on the β-amyrin structure, oxidation of the methyl group at C-17 to a carboxyl group being common, as in the widespread oleanolic acid (5.74).

Oxidation and loss of a C-4 methyl group of a dammarene derivative gives the fungal antibiotic, fusidic acid (5.75).

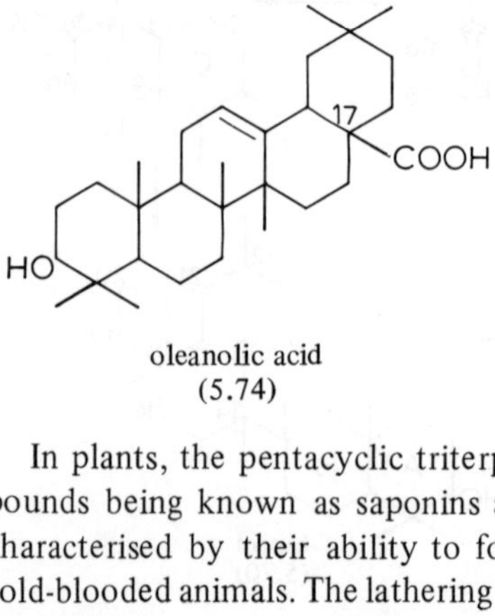

oleanolic acid
(5.74)

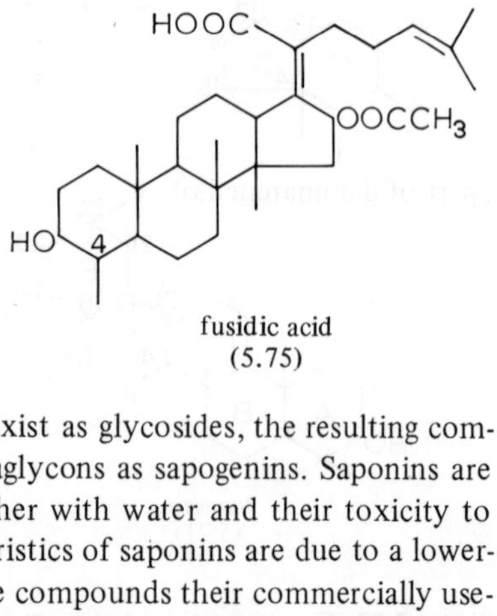

fusidic acid
(5.75)

In plants, the pentacyclic triterpenoids exist as glycosides, the resulting compounds being known as saponins and the aglycons as sapogenins. Saponins are characterised by their ability to form a lather with water and their toxicity to cold-blooded animals. The lathering characteristics of saponins are due to a lowering of surface tension, which also gives these compounds their commercially useful emulsifying properties. Toxicity to cold-blooded animals is due to an interference with their breathing mechanisms. Although not toxic to warm-blooded animals if ingested, saponins can be lethal if injected into the bloodstream, as they haemolyse the red blood corpuscles. This haemolytic property is variable and depends on the saponin. Large quantities of saponins are irritants which cause vomiting and diarrhoea. Saponins also have antilipemic properties and lower the serum cholesterol levels, while some are cytotoxic, antibiotic or fungicidal. In medicine, these compounds have been used for many purposes, particularly as anti-inflammatory agents and expectorants. Because of their toxicity to fish but not to man, saponins have been used as fish poisons in some parts of the world for thousands of years.

Triterpenoid saponins are widespread, if not universal, in plants, high concentrations being particularly characteristic of the Caryophyllaceae, Polygalaceae, Sapindaceae and Sapotaceae families. The steroidal saponins described below have a much more restricted occurrence.

The sugar portion of a saponin molecule can be complex, as for instance in gypsoside A (2.49) from *Gypsophila pacifica* (Caryophyllaceae) shown in Chapter 2.

Glycyrrhizin (5.76) from liquorice root (*Glycyrrhiza glabra*, Leguminosae) is anti-inflammatory and has been used to treat gastric ulcers and dermatitis. It also has a very sweet taste and is used commercially as a sweetening agent, while liquorice has mild laxative properties. Asiaticoside (5.77) from *Centella asiatica* (Umbelliferae) is a derivative of α-amyrin which is effective in the treatment of leprosy.

glycyrrhizin
(5.76)

asiaticoside
(5.77)

Although generally only toxic to cold-blooded animals, a few triterpenoid saponins are also poisonous to those with warm blood. The hederacosides from ivy (*Hedera helix*, Araliaceae), whose sapogenin is hederagenin (5.78a), are toxic to children and young animals, while the soyasapogenol saponins (soyasapogenol A (5.78b)) present in the leguminous fodder plants, lucerne (*Medicago sativa*) and white clover (*Trifolium repens*), cause bloat in animals eating large quantities

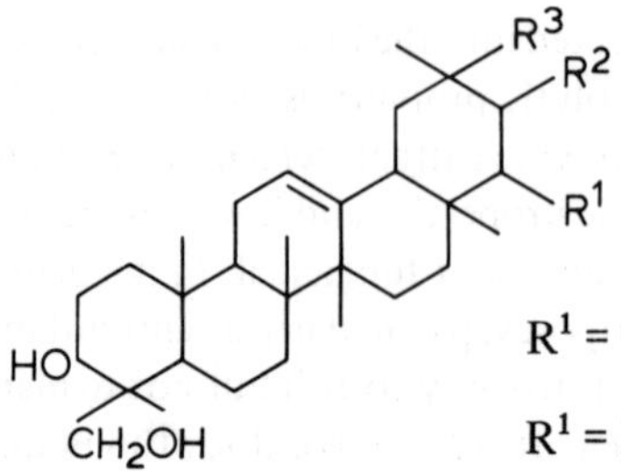

$R^1 = R^2 = H$, $R^3 = COOH$, hederagenin (5.78a)

$R^1 = R^2 = OH$, $R^3 = CH_3$, soyasapogenol A (5.78b)

of the fresh plants. The saponins present in corn cockle (*Agrostemma githago*, Caryophyllaceae) are responsible for githagism, a wasting disease caused by the accidental inclusion of corn cockle seeds with wheat when ground into flour.

Metabolism of Triterpenoids

The conversion of tetracyclic triterpenoids to the various classes of steroids is described in the following section. The pentacyclic triterpenoids appear to undergo

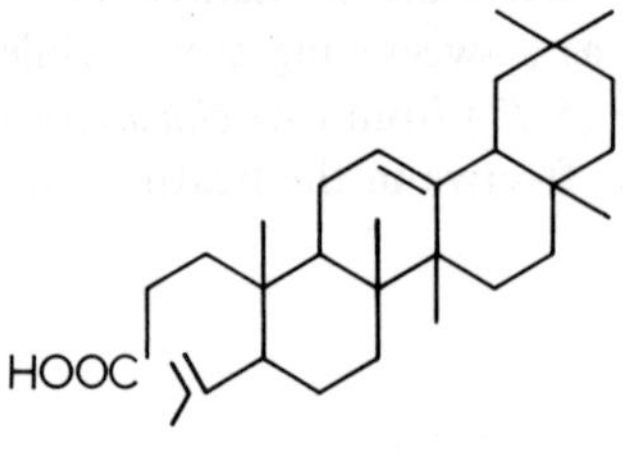

nyctanthic acid
(5.79)

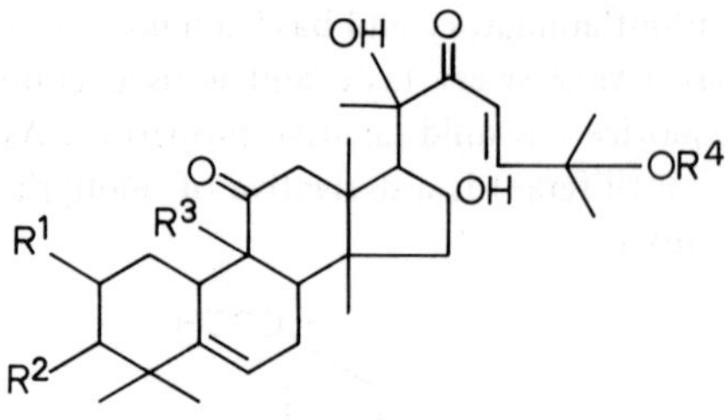

cucurbitacins
(5.80)

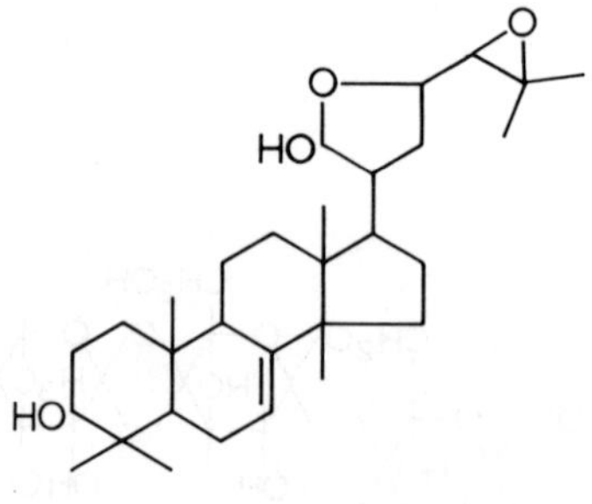

turraeanthin
(5.81)

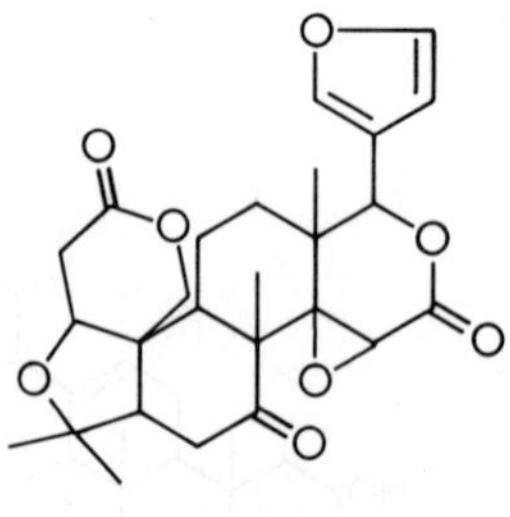

limonin
(5.82)

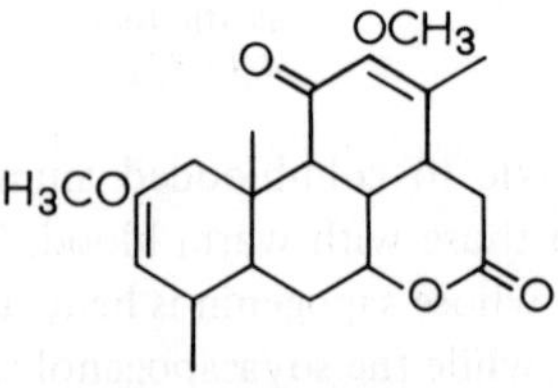

quassin
(5.83)

little metabolism, although nyctanthic acid (5.79) is presumably derived by oxidation of ring A of a pentacyclic precursor. In contrast, a large number of compounds isolated from plants are oxidation products of the tetracyclic triterpenoids. These include the cucurbitacins (5.80) which are responsible for the bitter flavour and toxicity of several members of the Cucurbitaceae family. Similar compounds known as meliacins (e.g. turraeanthin (5.81)), limonoids (e.g. limonin (5.82)) and quassins (e.g. quassin (5.83)) occur in the Meliaceae, Rutaceae and Simaroubaceae families, respectively. In general, such compounds have a bitter flavour and are insecticidal.

Cycloartenol and the Steroids

The steroids have the same ring structure as the tetracyclic triterpenoids, but are distinguished from these compounds by their lack of methyl groups at C-4 and C-14.

The fungal steroids are derived from lanosterol (5.72), the most common being ergosterol (5.84) which forms ergocalciferol (vitamin D_2 (5.85)) on irradiation with ultraviolet light (*Figure 5.29*). Although rare in higher plants, ergosterol has been detected in wheat. However, it probably makes little contribution to dietary vitamin D, which is mainly obtained from animal fats.

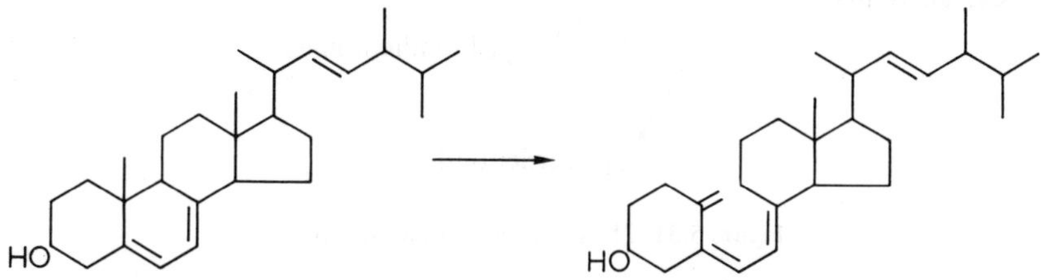

ergosterol
(5.84)

ergocalciferol
(5.85)

Figure 5.29 The formation of vitamin D_2.

The precursor of the higher plant steroids, the triterpenoid, cycloartenol (5.86), is formed by the cyclisation of squalene epoxide (5.68) (*Figure 5.30*).

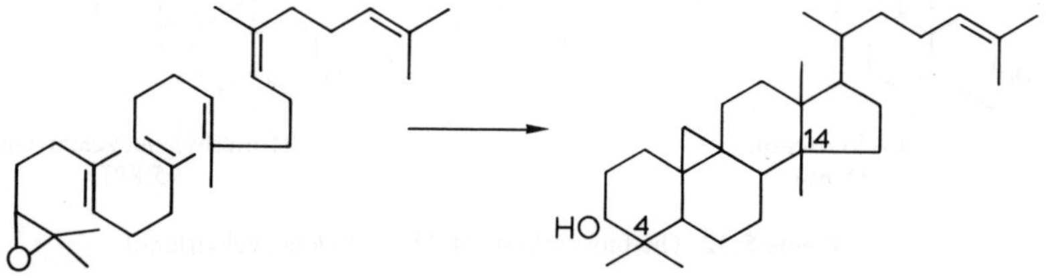

squalene epoxide
(5.68)

cycloartenol
(5.86)

Figure 5.30 The biosynthesis of cycloartenol.

Through its conversion to 24-methylenecycloartenol (5.88) or cholesterol (5.92), cycloartenol is the precursor of a host of compounds collectively known as the phytosteroids, which contain varying numbers of carbon atoms, as shown in *Figure 5.31*. In general, higher plant triterpenoids are converted to steroids by the loss of methyl groups in the order C-4, C-14, C-4. However, there are exceptions to this rule, and there appears to be no overall pathway of steroid biosynthesis applicable to all higher plants.

CYCLOARTENOL C_{27} insect moulting hormones

24-methylenecycloartenol   CHOLESTEROL

C_{29} steroids C_{28} steroids C_{27} sapogenins

                                                          C_{27} alkaloids

                    C_{21} steroids

C_{21} alkaloids C_{19} and C_{18} steroids

                                              C_{24} bufadienolides

                        C_{23} cardenolides

Figure 5.31 The derivation of phytosteroids.

Sitosterol (5.87), a C_{29} sterol, is the most common of the higher plant steroids. It is biosynthesised from cycloartenol (5.86) through the formation of 24-methylenecycloartenol (5.88). Methylation of cycloartenol by *S*-adenosylmethionine is followed by rearrangement of the double bond, as shown in *Figure 5.32*.

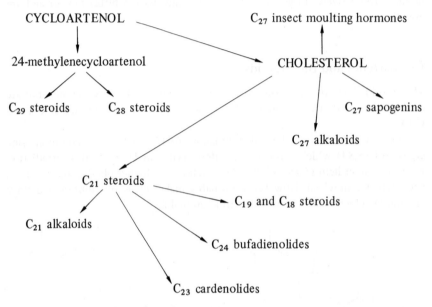

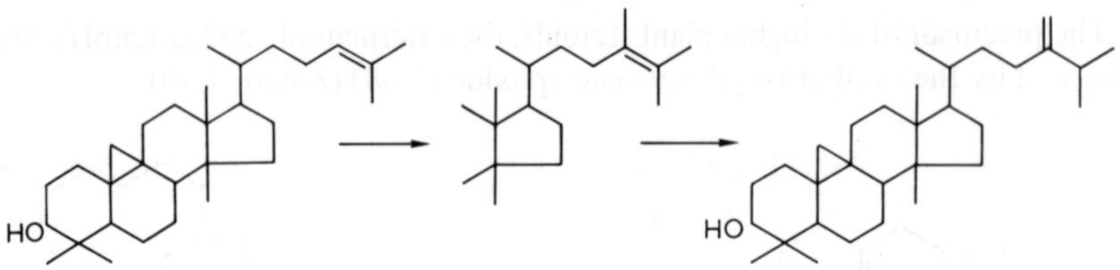

cycloartenol   24-methylenecycloartenol
(5.86)   (5.88)

Figure 5.32 The biosynthesis of 24-methylenecycloartenol.

The remaining steps in the biosynthetic pathway have not been established unequivocally, but the scheme shown in *Figure 5.33* seems the most plausible on the evidence at present available. However, many divergent pathways must occur

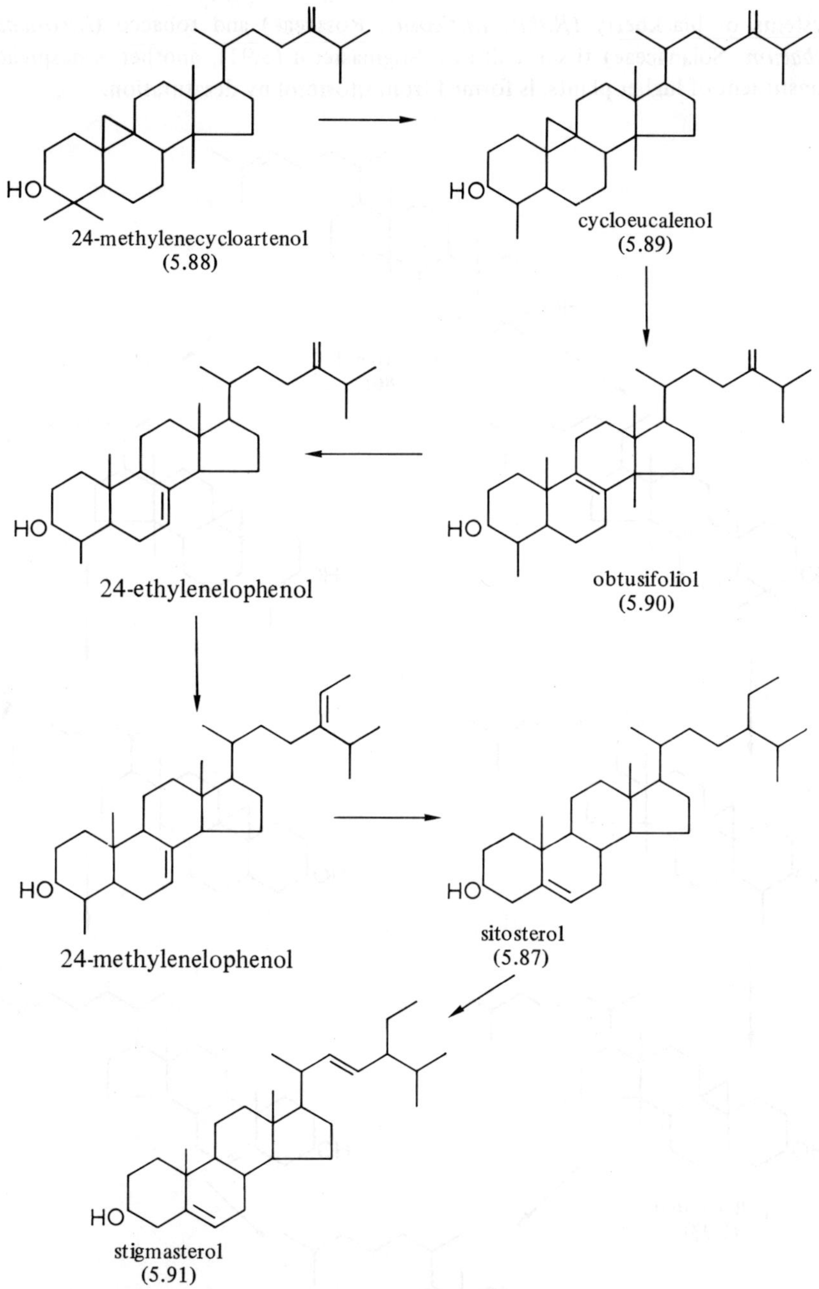

Figure 5.33 Some probable steps in the biosynthesis of the C_{29} steroids.

to account for the variety of C_{29} steroids found in plants, some of which contain a C-4 or C-14 methyl group.

The sequence converting cycloartenol (5.86) to obtusifoliol (5.90), through the intermediate formation of cycloeucalenol (5.89) has been established in cell free systems of blackberry (*Rubus fruticosus*, Rosaceae) and tobacco (*Nicotiana tabacum*, Solanaceae) tissue cultures. Stigmasterol (5.91), another widespread constituent of higher plants, is formed from sitosterol by desaturation.

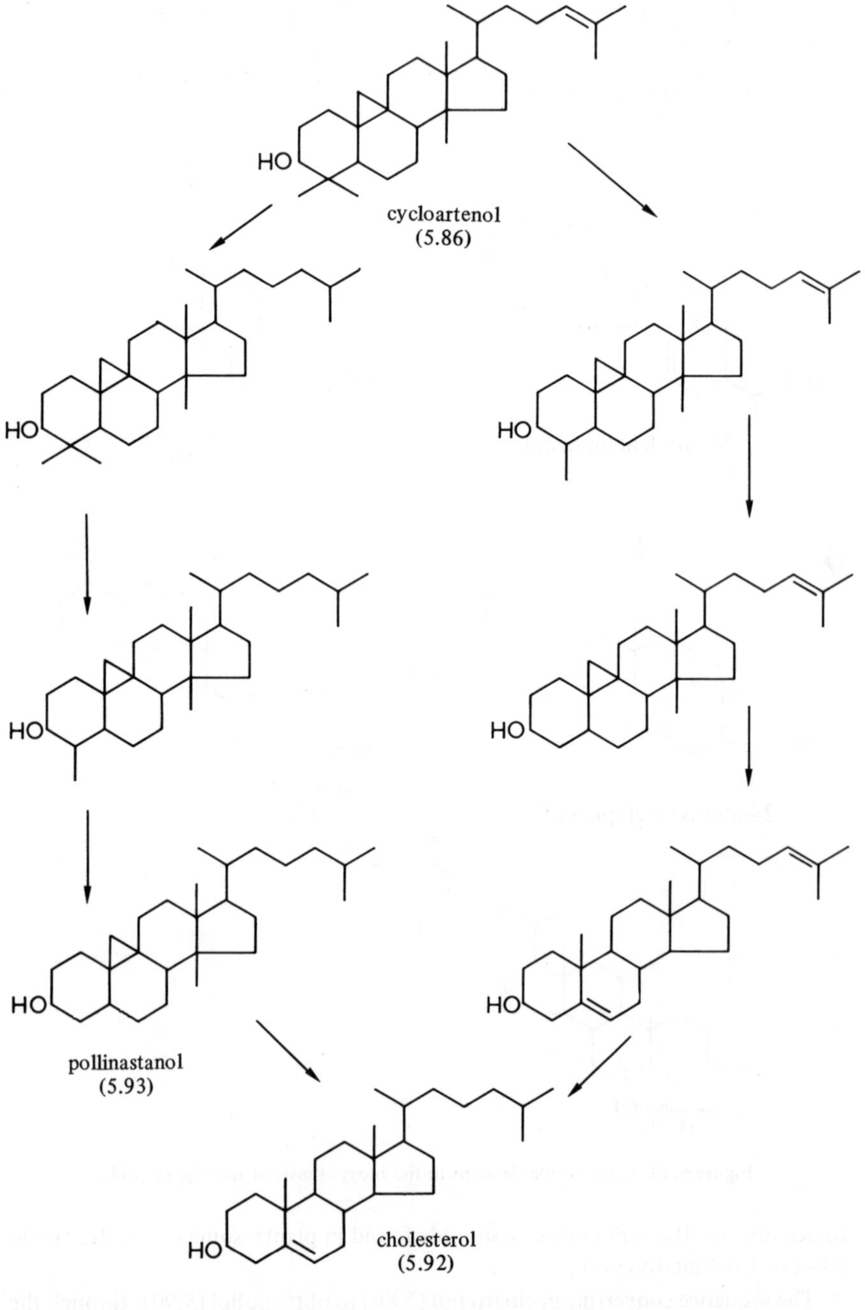

Figure 5.34 Possible pathways for the biosynthesis of cholesterol in plants.

Cholesterol (5.92), the precursor of a large number of plant steroids, is biosynthesised in higher plants from cycloartenol (5.86). It is unlikely that lanosterol (5.72) is an intermediate, as it is in the biosynthetic pathway of animals, as this compound is rare in higher plants. On an analogy with the biosynthesis of C_{29} steroids, demethylation at C-4 would be expected to precede ring opening, as shown in *Figure 5.34*. The intermediate pollinastanol (5.93) has been isolated from several plants.

Insect moulting hormones

A number of closely related steroids with the general formula shown in *Figure 5.35* are responsible for the shedding of insect larval skins. These phytoecdysones are the precursors of the moulting hormones, insects being unable to synthesise the steroid nucleus.

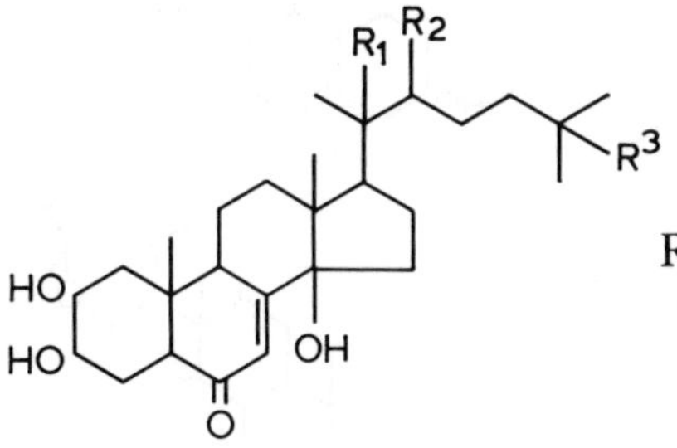

$$R^1, R^2, R^3 = H \text{ or } OH$$

Figure 5.35 General formula of the phytoecdysones.

The detailed biosynthesis of the phytoecdysones has not been established, but these compounds can be derived from cholesterol by oxidation reactions.

Sapogenins

Cholesterol is the precursor of the C_{27} sapogenins which have a spiroketal structure. Diosgenin (5.94) or its epimer, yamogenin, the first sapogenins to be biosynthesised in plants, are derived from 26-hydroxycholesterol (5.95). Other sapogenins, such as gitogenin (5.96), digitogenin (5.97) and hecogenin (5.98), are the result of oxidation reactions (*Figure 5.36*).

In plants, sapogenins occur combined at C-3 with sugars, the glycosides being known as saponins. In general, steroidal saponins have the same properties as triterpenoid saponins, but some are toxic to warm-blooded animals when ingested. These sapotoxins cause severe stomach upset, while their toxicity is due to the formation of compounds with lecithin, a major component of most membrane lipids in animal cells. This leads to paralysis of the nerve centres and the heart.

Steroidal saponins are less widespread than triterpenoid saponins but are common in the monocotyledonous Liliaceae, Agavaceae and Dioscoreaceae families. They are also characteristic of the Solanaceae family and the *Digitalis* genus (Scrophulariaceae). Diosgenin (5.94) from yams (*Dioscorea* sp., Dioscoreaceae)

and hecogenin (5.98) from sisal *(Agave sisalana,* Agavaceae) are important commercially as the starting point for the synthesis of synthetic hormones and steroid
drugs.

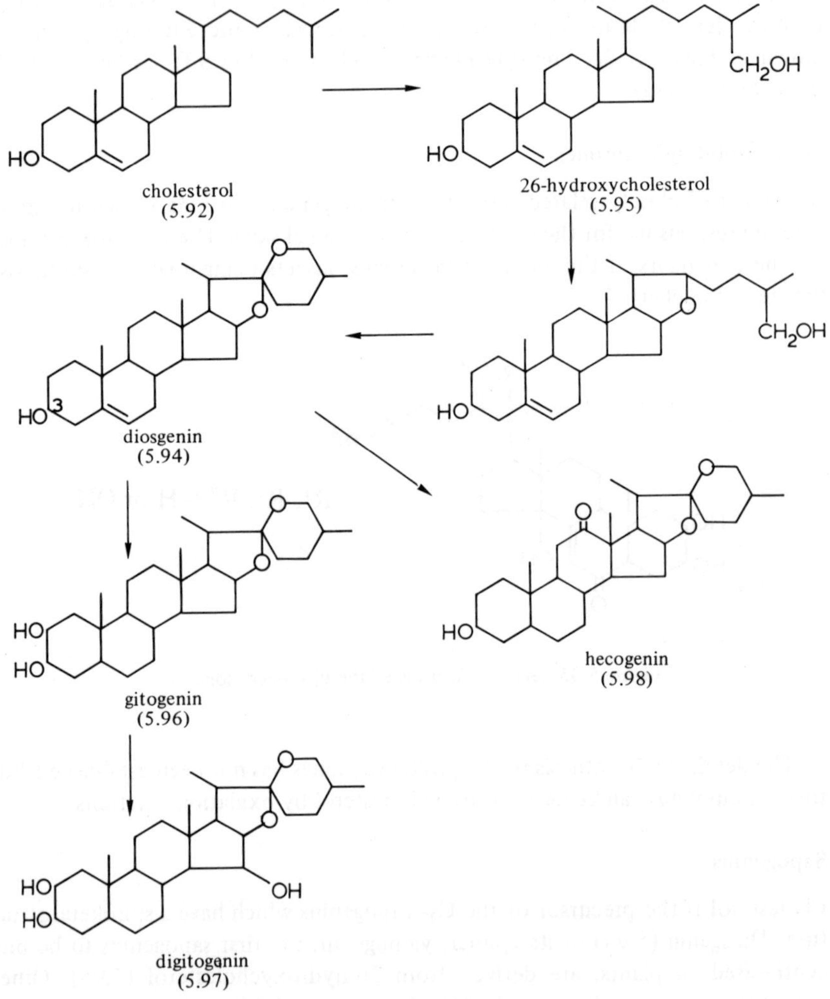

Figure 5.36 The biosynthesis of some steroidal sapogenins.

Steroidal alkaloids

Some C_{27} steroidal sapogenins occurring in the Solanaceae and Liliaceae families
contain nitrogen and are therefore classed as alkaloids. These compounds are biosynthesised from cholesterol (5.92), probably through the intermediate formation
of diosgenin (5.94), the nitrogen atom being introduced by a transamination reaction. The simpler Solanaceae alkaloids, such as solasodine (5.99), are then
formed by cyclisation of 26-aminodihydrodiosgenin (5.100) *(Figure 5.37)*. However, little is known of the final steps in the pathway biosynthesising the more

complex alkaloids, such as solanidine (5.101), although tomatillidine is a possible intermediate, as shown in *Figure 5.37*.

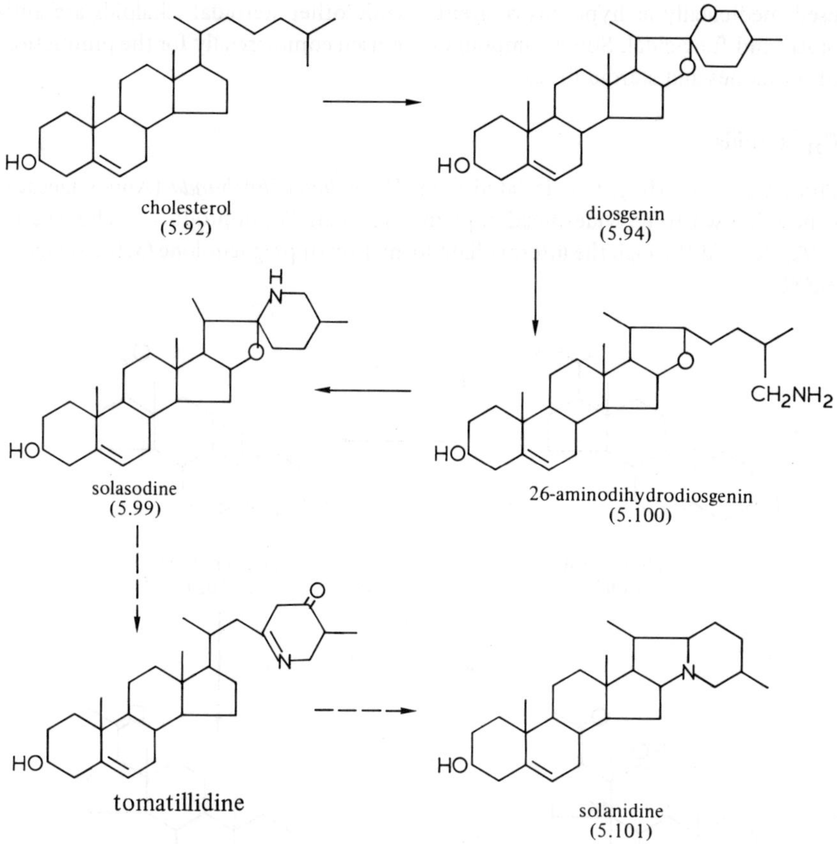

Figure 5.37 Some possible steps in the biosynthesis of Solanaceae steroidal alkaloids.

A number of alkaloids of the veramarine (5.102) type isolated from *Veratrum* and *Fritillaria* species (Liliaceae) have undergone contraction of ring C and expansion of ring D and are therefore no longer strictly steroids. However, steroidal alkaloids, such as veramine (5.103) also occur and these are probably the precursors of the more complex compounds.

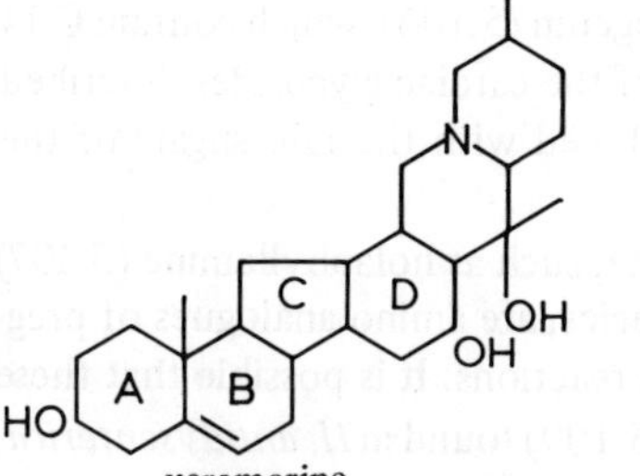

veramarine
(5.102)

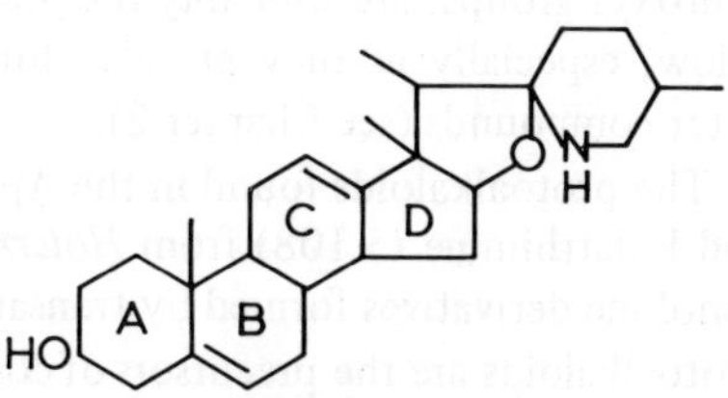

veramine
(5.103)

The C_{27} steroidal alkaloids occur in plants as glycosides and thus have all the properties of saponins. However, the presence of nitrogen increases their physiological action and these compounds are sapotoxins. The *Veratrum* alkaloids are used medicinally as hypotensive agents, while other steroidal alkaloids are antibiotic and fungicidal. Some compounds are used commercially for the production of hormones and steroid drugs.

C_{21} steroids

Progesterone (5.104), first isolated from *Holarrhena floribunda* (Apocynaceae), is now known to be widespread in plants as a degradation product of cholesterol (5.92) derived through the intermediate formation of pregnenolone (5.105) (*Figure 5.38*).

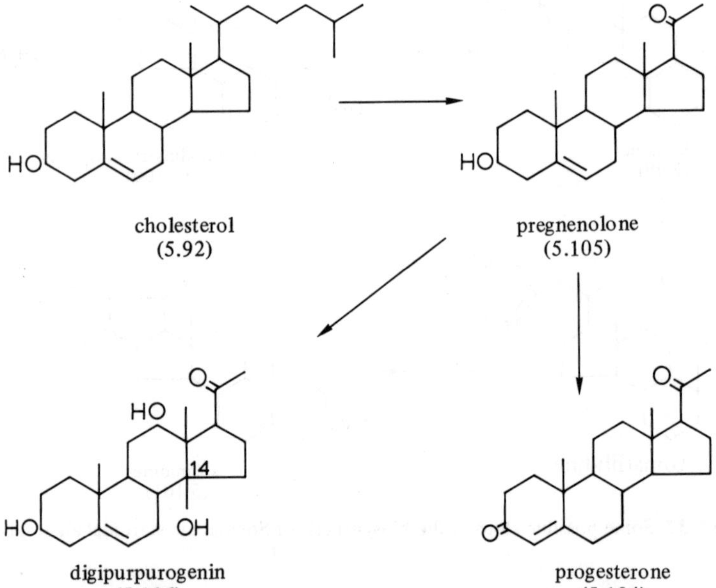

cholesterol
(5.92)

pregnenolone
(5.105)

digipurpurogenin
(5.106)

progesterone
(5.104)

Figure 5.38 The derivation of the C_{21} steroids.

Pregnenolone and progesterone are the precursors of the C_{21} steroids found in many plants. Metabolites such as digipurpurogenin (5.106), which contain C-14 hydroxyl groups, are probably the precursors of the cardiac glycosides described below, especially as they are also found combined with the rare sugars of the latter compounds (see Chapter 2).

The protoalkaloids found in the Apocynaceae, such as holaphyllamine (5.107) and holarrhimine (5.108) from *Holarrhena* species, are amino analogues of pregnenolone derivatives formed by transamination reactions. It is possible that these protoalkaloids are the precursors of conessine (5.109) found in *H. antidysenterica*, as shown in *Figure 5.39*.

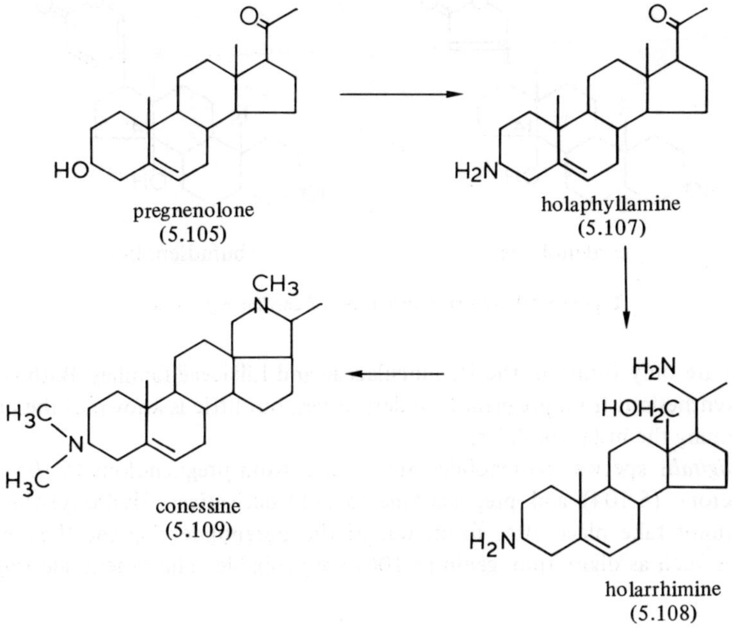

Figure 5.39 The probable biosynthesis of conessine.

Conessine has been used medicinally as an amoebicide in the treatment of dysentery. This alkaloid is also hypotensive and in overdoses causes death through respiratory paralysis.

Buxus species (Buxaceae) contain alkaloids such as cyclobuxamine H (5.110), which are obviously derived from cycloartenol (5.86) derivatives.

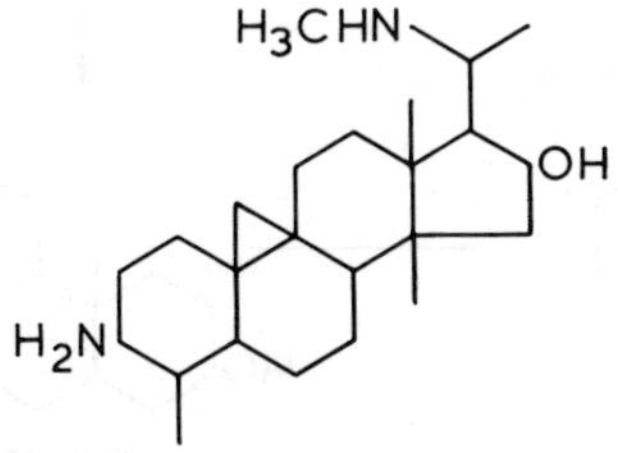

cyclobuxamine H
(5.110)

Cardiac glycosides

The genins of the cardiac glycosides are divided into the C_{23} cardenolides and C_{24} bufadienolides, which have the ring structures shown in *Figure 5.40*. Their cardio-activity is associated with the unsaturated lactone ring, the C-14 hydroxyl group and the stereochemistry of the molecule.

Cardenolides are found in several plant families, notably Apocynaceae and the *Digitalis* genus, while bufadienolides, which were first isolated from toad (*Bufa*)

cardenolides   bufadienolides

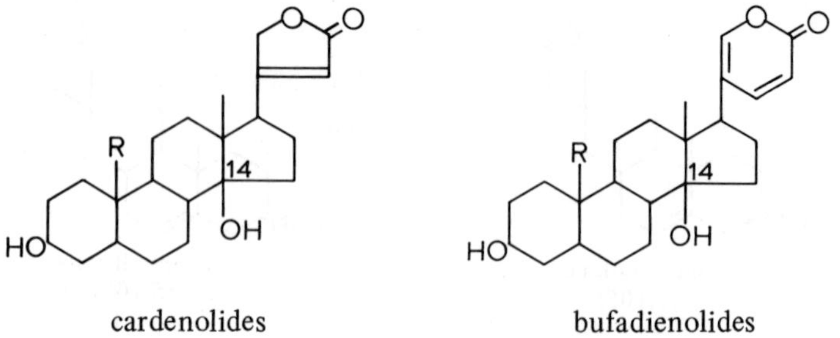

Figure 5.40 The ring structures of cardiac glycosides.

poisons, are only found in the Ranunculaceae and Liliaceae families. Both types
are biosynthesised from pregnenolone derivatives, but little is known of the pathway forming the bufadienolides.

In *Digitalis* species, cardenolides are formed from pregnenolone (5.105) via
progesterone (5.104) and pregnanolone (5.111) derivatives. Hydroxylation at
C-14 cannot take place after formation of the butenolide ring and thus intermediates such as digipurpurogenin (5.106) are probable. The butenolide ring is

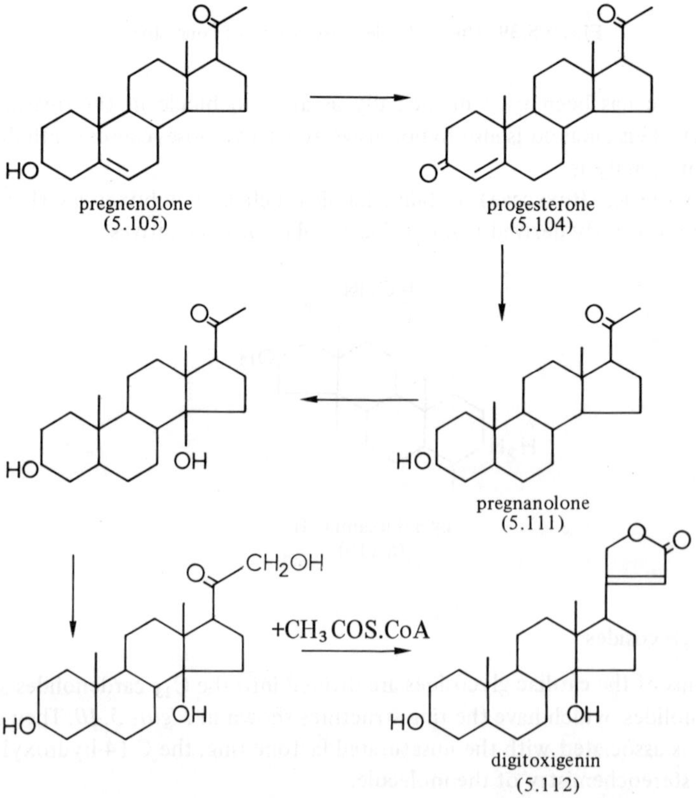

pregnenolone
(5.105)

progesterone
(5.104)

pregnanolone
(5.111)

+CH₃COS.CoA

digitoxigenin
(5.112)

Figure 5.41 Some steps in the biosynthesis of the cardenolides.

derived from acetate. A biosynthetic scheme for the formation of digitoxigenin (5.112) in *Digitalis* species is shown in *Figure 5.41*.

Digitoxigenin is the simplest cardenolide, but a wide variety of compounds have been isolated from plants, all of which can be theoretically derived from digitoxigenin by oxidation reactions. The series, periplogenin (5.113), strophanthidol (5.114), strophanthidin (5.115) and sarmentosigenin E (5.116), shows gradual increase in oxidation at C-19 (*Figure 5.42*).

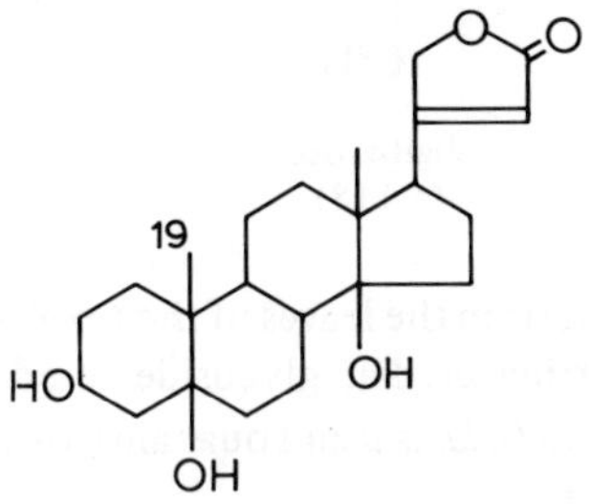

periplogenin
(5.113)

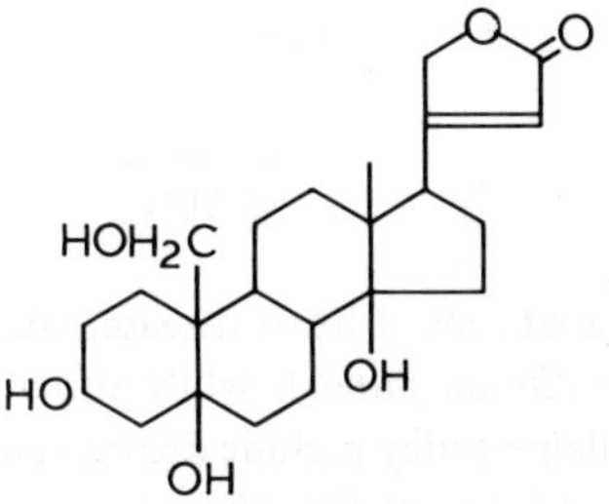

strophanthidol
(5.114)

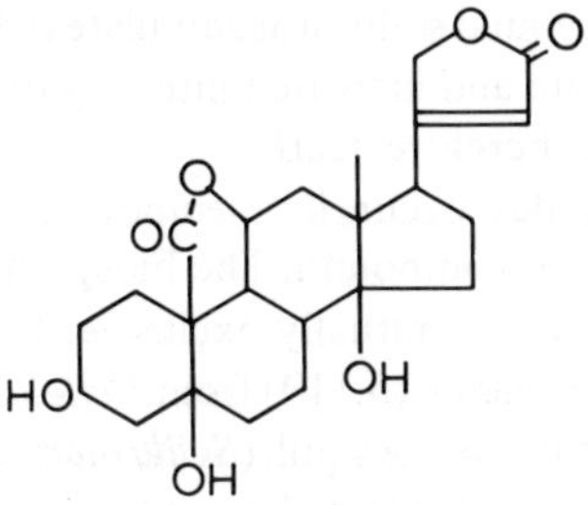

sarmentosigenin E
(5.116)

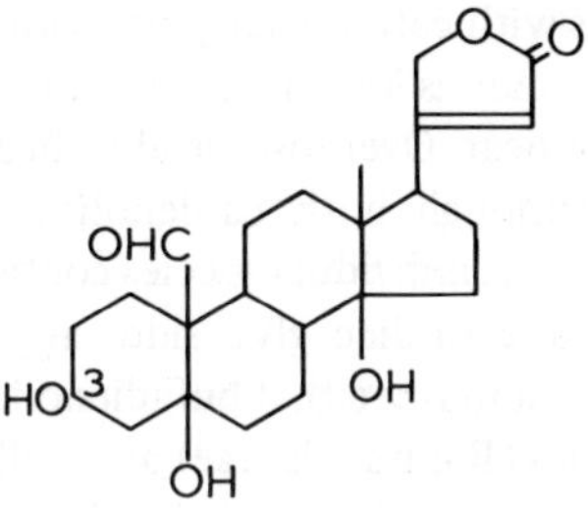

strophanthidin
(5.115)

Figure 5.42 Some cardenolides.

In plants, cardenolides occur linked at C-3, with sugars which are often rare (see Chapter 2, for examples) and not found combined with any other type of genin except pregnenolone derivatives. It is possible that the final stages of cardenolide biosynthesis take place on the glycosides, rather than the free genins. Convallatoxin, the rhamnoside of strophanthidin (5.115), is the most toxic of all the C_{23} cardiac glycosides. It occurs in the leaves and flowers of lily-of-the-valley (*Convallaria majalis*, Liliaceae). One of the most notorious of the cardiac glycosides, ouabain, is also a rhamnoside, the genin being ouabagenin (5.117). Ouabain, which occurs in *Acocanthera* and *Strophanthus* species (Apocynaceae), has been used as an arrow poison by African tribes for thousands of years and is still used today in the poaching of wild animals or for acts of homicide. Ouabain is only toxic if injected directly into the bloodstream, when it acts quickly, killing a large animal in minutes. The lack of toxicity of this and many other cardiac glycosides if ingested is probably due to the bulk of the molecule and its highly hydroxylated nature, both of which prevent absorption into the bloodstream.

Digitoxigenin (5.112) occurs combined with digitoxose (5.118) as digitoxin, the main constituent of digitalis, which has been used for many years to treat

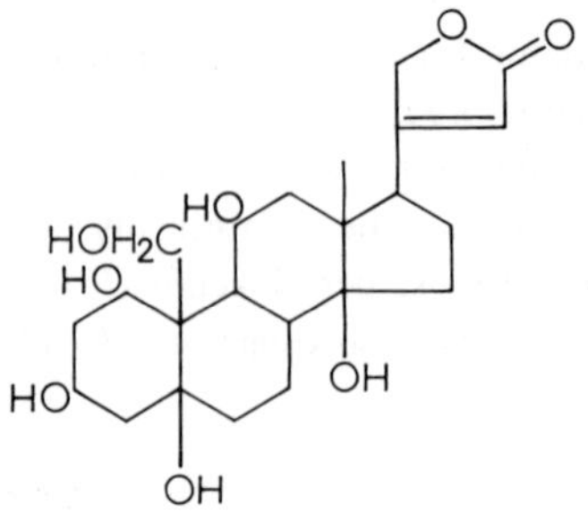

ouabagenin
(5.117)

CHO
|
CH$_2$
|
HCOH
|
HCOH
|
HCOH
|
CH$_3$

digitoxose
(5.118)

certain types of heart disease. Digitalis is extracted from the leaves of the foxglove (*Digitalis purpurea*), while other naturally occurring cardiac glycosides used in medicine today include digoxin and lantoside C from *D. lanata* and ouabain (strophanthin G) from *Strophanthus gratus* (Apocynaceae).

Cardiac glycosides act on cell membranes by inhibiting ATPase. This interferes with cation transport, with the result that cellular sodium accumulates while potassium is lost. This causes a decreased heart rate and increased intensity of the heart beat. Overdoses lead to heart stoppage and therefore death.

Although both cardenolides and bufadienolides occur in members of the Liliaceae, individual species contain only one type of compound. The biosynthesis of these cardiac glycosides appears, therefore, to be mutually exclusive. It has been established that bufadienolides, such as hellebrigenin (5.119) from *Helleborus* species (Ranunculaceae) and scillarenin (5.120) from white squill (*Scilla maritima*, Liliaceae) are biosynthesised from a C$_{21}$ steroid and a three-carbon unit, possibly propionyl CoA.

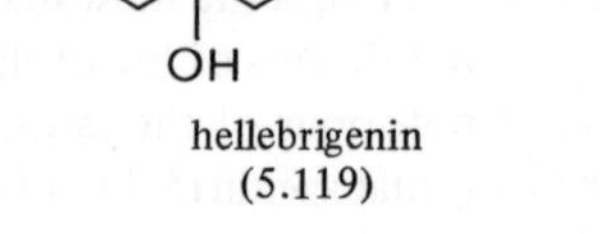

hellebrigenin
(5.119)

scillarenin
(5.120)

The squill glycosides have found some use in the treatment of heart disease but, in general, the bufadienolides are too toxic to be used medicinally.

Steroid degradation

As in animals, there is evidence that the side chain of plant steroids is completely degraded to the C$_{19}$ androgens, progesterone (5.104), for instance, resulting in the male sex hormone, testosterone (5.121). The loss of the C-10 methyl group

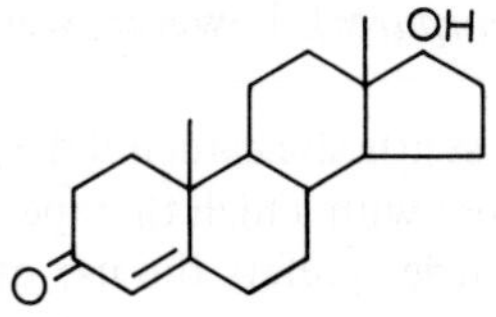

testosterone
(5.121)

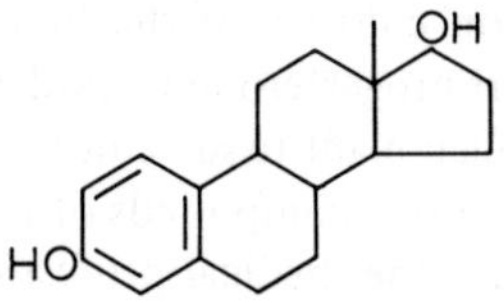

oestradiol
(5.122)

and aromatisation of ring A then leads to the C_{18} oestrogens, which include the female sex hormones such as oestradiol (5.122).

The amounts of these steroids in plants are both too small to be worth commercial extraction or to upset the hormonal balance of animals eating them. Although it has not been proved, it would seem likely that testosterone and the oestrogens are responsible for the male and female characteristics of flowers.

Steroids in tissue cultures

Cultures of microorganisms have been exploited commercially for the production of pharmaceutically useful steroids including corticosteroids, androgens and oestrogens. Precursors such as cholesterol, progesterone, diosgenin or the steroidal alkaloids are easily obtained from plant or animal extracts. Plant tissue cultures have not, as yet, been used for the same purposes, but are of potential interest. In general, transformations of the steroid molecule by plant tissue cultures involves oxidation of hydroxyl groups, reduction of keto groups and saturation of double bonds. Degradation of side chains has not been observed. It is interesting that glucoside formation or esterification with palmitic acid often takes place.

Callus tissue cultures from the hypocotyl of germinated *Holarrhena antidysenterica* seedlings biosynthesise cholesterol, 24-methylenecholesterol, 28-isofucosterol, sitosterol and stigmasterol. However, the alkaloids characteristic of this plant were not detected. Suspension cultures of *Dioscorea deltoidea* convert both cholesterol and sitosterol to diosgenin.

A number of plant tissue cultures will convert pregnenolone to progesterone or transform progesterone into pregnenolone or pregnanolone derivatives.

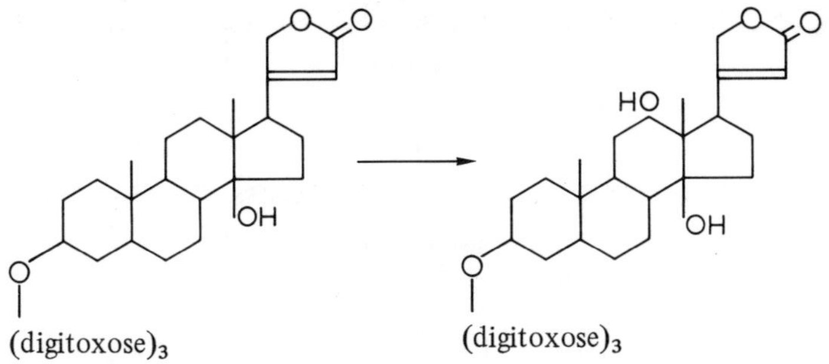

(digitoxose)$_3$

digitoxin
(5.124)

(digitoxose)$_3$

digoxin
(5.123)

Figure 5.43 The conversion of digitoxin to digoxin in *Digitalis lanata* tissue cultures.

Suspension cultures of the foxglove (*Digitalis purpurea*), however, were unable to convert progesterone to cardiac glycosides.

Although plant tissue cultures seem unable to synthesise cardenolides, they are able to convert compounds of low activity to those with a high therapeutic value. At present, the medical profession in some countries prefers the use of digoxin (5.123) to digitoxin (5.124) and the former compound can be formed in *D. lanata* tissue cultures through hydroxylation of the latter (*Figure 5.43*).

Carotenoids

The biosynthesis of the C_{40} terpenoids takes place in the chloroplasts of plants or chromatophores of bacteria. The precursor, geranylgeranyl pyrophosphate (5.52), undergoes a tail-to-tail condensation to give prephytoene (5.125), an ana-

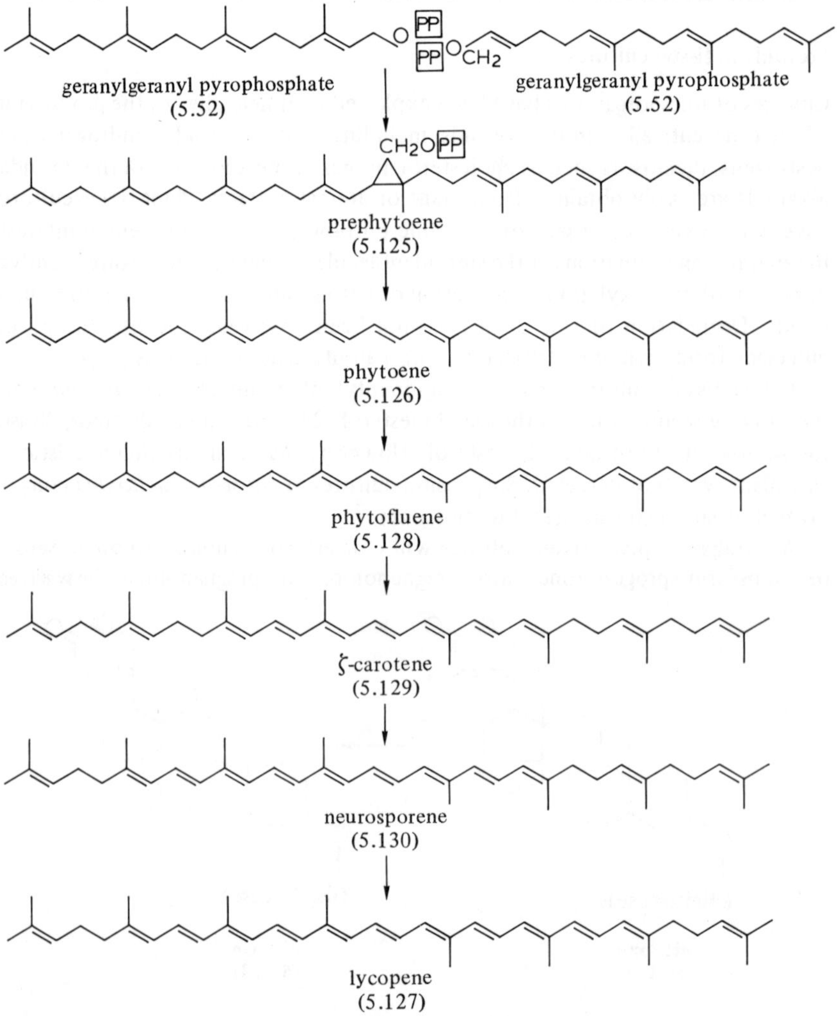

Figure 5.44 The biosynthesis of lycopene.

logue of presqualene (5.67). Conversion of prephytoene to phytoene (5.126) is followed by desaturation reactions which lead to lycopene (5.127), through the intermediate formation of phytofluene (5.128), ζ-carotene (5.129) and neurosporene (5.130) (*Figure 5.44*). The stereochemistry of the phytoene molecule is such that it cannot fold to form polycyclic compounds analogous to the squalene derivatives.

persicum esculentum, Solanaceae) and water melon (*Cucumis citrullus*, Cucurbitaceae). It is converted to α- (5.131) and β-carotene (5.132) by cyclisation through the formation of the monocyclic γ- and δ-carotenes (5.133, 5.134) (*Figure 5.45*).

β-Carotene is responsible for the orange colour of carrots (*Daucus carota*, Umbelliferae) and contributes to the yellow and orange colours of many flowers,

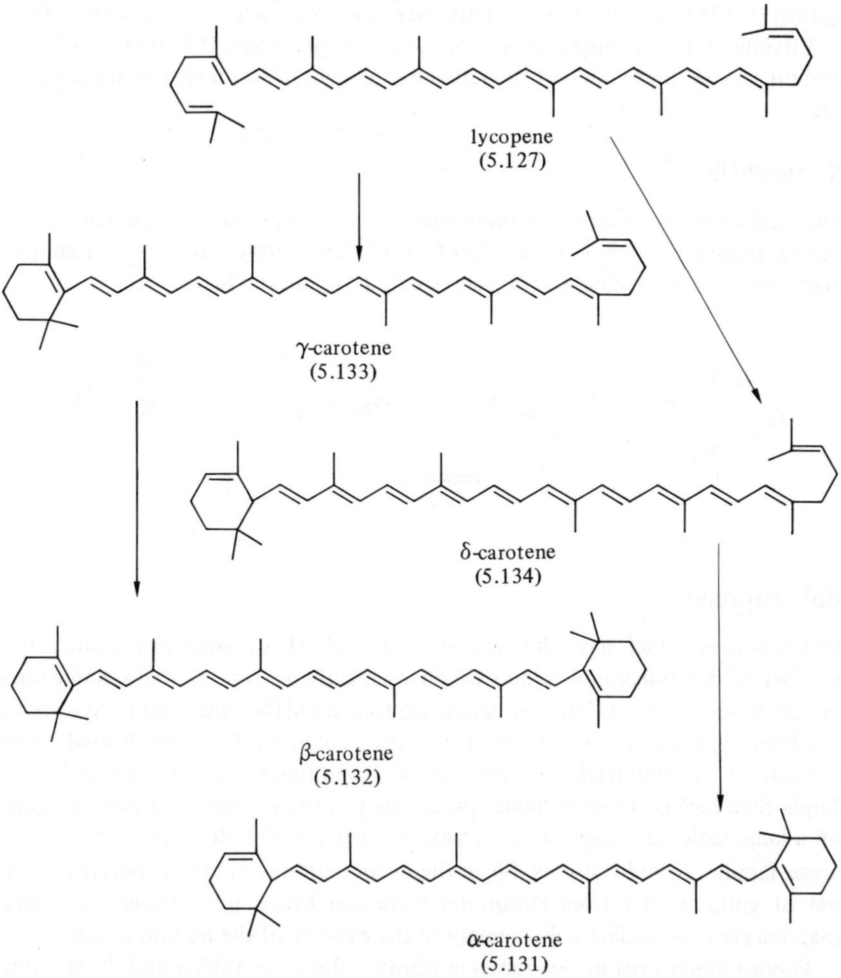

lycopene
(5.127)

γ-carotene
(5.133)

δ-carotene
(5.134)

β-carotene
(5.132)

α-carotene
(5.131)

Figure 5.45 The biosynthesis of α- and β-carotene.

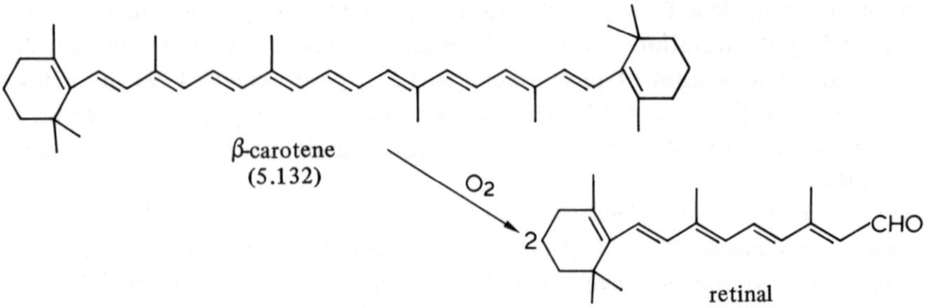

Figure 5.46 The formation of vitamin A aldehyde from β-carotene in man.

fruits and leaves. It is important to man as the precursor of vitamin A aldehyde (retinal (5.135)), obtained by oxidative degradation of β-carotene (*Figure 5.46*).

Vitamin A is concerned in several metabolic processes. It is particularly important to the functioning of the retina in dull light, thus preventing night blindness.

Xanthophylls

The oxidation of carotenes by molecular oxygen, catalysed by oxygenases, gives the xanthophylls, such as zeaxanthin (5.136). Xanthophylls occur in the chloroplasts and are responsible for the yellow colour of autumn leaves.

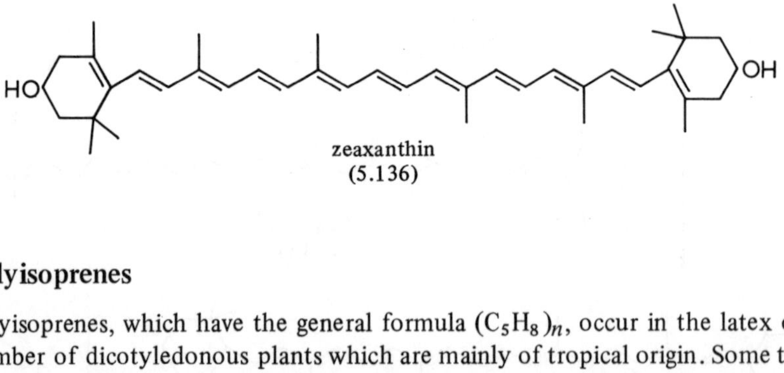

zeaxanthin
(5.136)

Polyisoprenes

Polyisoprenes, which have the general formula $(C_5H_8)_n$, occur in the latex of a number of dicotyledonous plants which are mainly of tropical origin. Some temperate members of the Compositae, such as golden rod (*Solidago* spp.) and Russian dandelion (*Taraxacum kok-saghyz*), however, contain rubber which used to be extracted for commercial use. Nowadays, the para rubber tree (*Hevea brasiliensis,* Euphorbiaceae) is the only rubber-producing plant to be exploited commercially on a large scale, although certain plants belonging to the Moraceae and Apocynaceae families have local uses. Since the introduction of synthetic polymers, the use of gutta percha from *Palaquium gutta* and balata from *Mimusops balata* (Sapotaceae) has declined, due mainly to the expense of the natural product.

Polyisoprenes exist in two forms in plants – the all-*cis* rubber and the all-*trans* gutta percha. Chicle from the South American *Achras sapota* (Sapotaceae) contains both the *cis* and *trans* isomers.

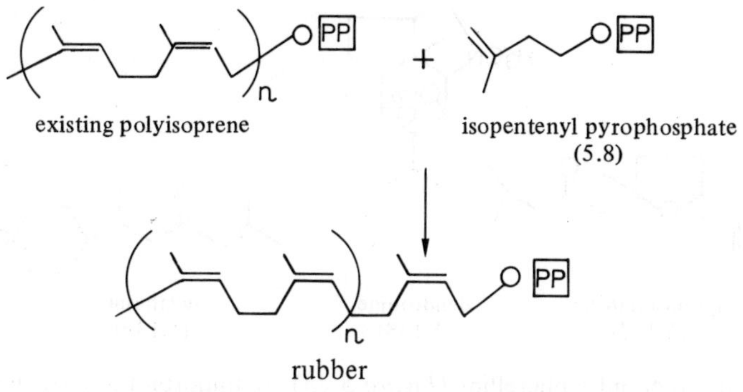

existing polyisoprene

isopentenyl pyrophosphate
(5.8)

rubber

Figure 5.47 The biosynthesis of rubber.

Rubber is biosynthesised by lengthening of molecules on the surface of existing rubber particles. The chain-lengthening unit, isopentenyl pyrophosphate (5.8), adds on to form a *cis* orientated molecule (*Figure 5.47*) in contrast to the *trans* orientation of geranyl pyrophosphate (5.18) or farnesyl pyrophosphate (5.47).

The biosynthesis of the existing polyisoprene molecules has not been established and little is known of the derivation of gutta percha. However, an all-*trans* compound could be derived by the continuing addition of isopentenyl pyrophosphate to a farnesyl pyrophosphate precursor.

Although rubber has many industrial applications, the bulk is used in the manufacture of tyres. Gutta percha has mostly been used in golf balls, while chicle is the basis of chewing gum.

The Function of Terpenoids and Steroids in Plants

Although once thought to be metabolic waste products, the functions of terpenoids and steroids in plants are now known to be as many and varied as their molecular structures. The main function of the volatile monoterpenoids is to attract insect pollinators and scavengers (predators, harmless to plants, which prey on plant-eating insects or fungal invaders) or to repel invaders. Plants containing large concentrations of monoterpenoids are unattractive to herbivores and birds, while some, such as peppermint, are good insect repellants. Monoterpenoids released into the environment exert allelopathic effects, preventing the germination of seeds of other plants. Particularly effective are cineole (5.30) and camphor (*Figure 5.11*).

Except for the growth inhibitory hormone, abscisic acid (5.49), the function of the sesquiterpenoids would seem to be protective, as they are strong bacterial, fungal and animal toxins. Phytoalexins, such as ipomeamarone (5.137) from the sweet potato (*Ipomoea batatas*, Convulvulaceae) are produced in response to fungal infection, while the picrotoxin alkaloids, such as dendrobine (5.138) from the orchid *Dendrobium nobile* (Orchidaceae), are toxic to animals. The insect juvenile hormone analogues such as juvabione (5.139) which occur mainly in the gymnosperms prevent metamorphosis of the larva to the adult form and thus prevent the breeding of harmful insects.

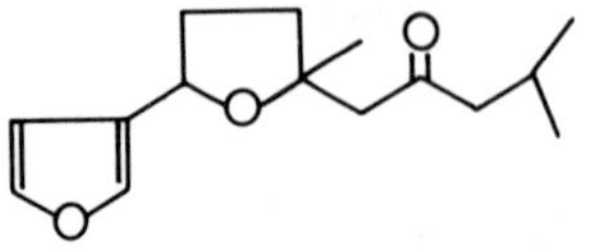

ipomeamarone   dendrobine   juvabione
(5.137)   (5.138)   (5.139)

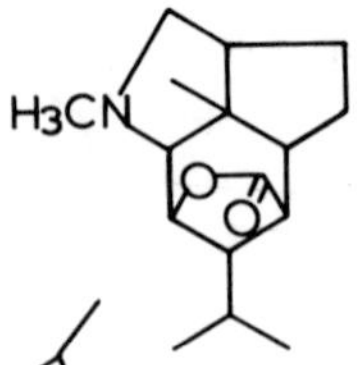

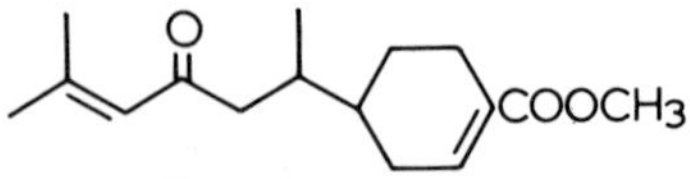

The diterpenoid gibberellins (*Figure 5.23*) are important as growth hormones in higher plants, while the trisporic acid (5.140) derivatives, modified diterpenoids, are reproductive agents in some fungi. The diterpenoid alkaloids, such as aconitine (5.141) from *Aconitum* species (Ranunculaceae), are highly toxic to animals.

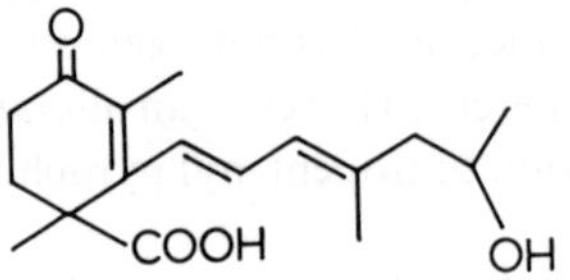

aconitine
(5.141)

trisporic acid
(5.140)

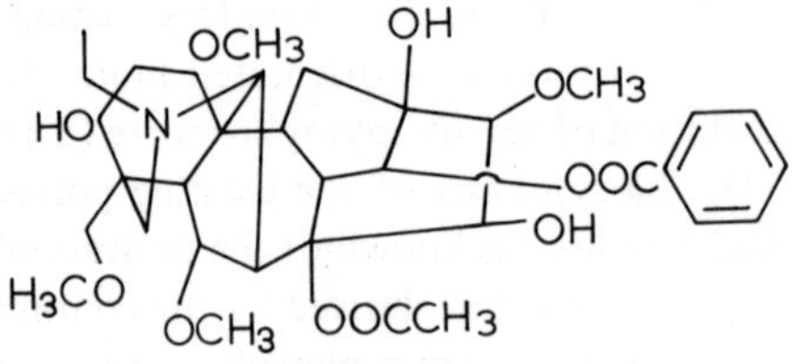

The triterpenoids and their derivatives, the steroids, probably have a wider variety of functions than any other class of plant compound. The triterpenoid and steroid saponins are toxic to insects, bacteria and fungi, while the alkaloids and cardiac glycosides are animal poisons. Other saponins promote seed germination and inhibit root growth. The compound withaferin A (5.142) from *Withania somniferum* (Solanaceae) is antimitotic, inhibiting the growth of plant and animal cells, including tumour cells.

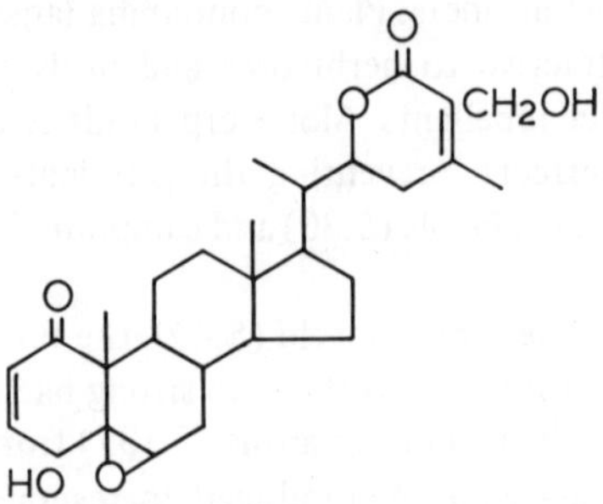

withaferin A
(5.142)

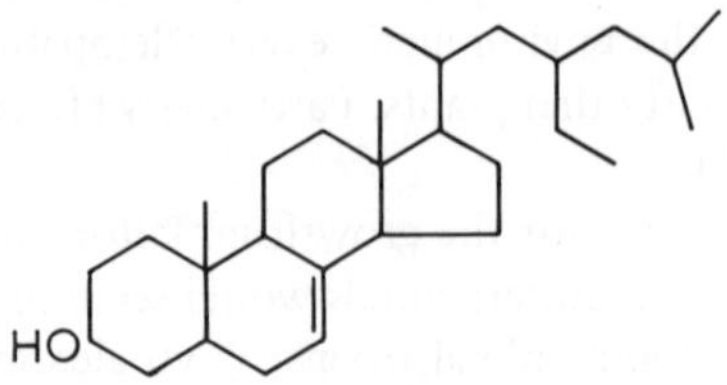

schottenol
(5.143)

Insects have turned the production of steroids by plants to their advantage, as such compounds are insect vitamins. As with the juvenile hormones, the production of insect moulting hormone analogues by plants is an attempt to interfere with the metabolism of phytophagous insects, preventing the larvae reaching the adult stage and breeding. Silkworms (*Bombax moi*), however, can detoxify such compounds, and this property is probably common amongst butterflies and moths (Lepidoptera). It is interesting that the steroid, schottenol (5.143), from senita cactus (*Lophocereus schottii*, Cactaceae), is a vitamin needed by its invader *Drosophila pachea* for larval development.

Steroids function as a source of energy for microorganisms and they are degraded to compounds with hormonal activity in animals. Thus, the widely occurring sitosterol (5.87) is oestrogenic and in concentration can interfere with the breeding of herbivores. Stigmasterol (5.91), however, is a vitamin (antistiffness factor) for the guinea pig.

Steroids are important constituents of cell membranes where they influence the permeability characteristics. There are also indications that steroids are necessary to the flowering process. Testosterone (5.121) and oestradiol (5.122) have sex hormone activity in yeast, while administration to cucumber (*Cucumis sativa*, Cucurbitaceae) flowers causes them to change sex.

The main purpose of carotenoids and xanthophylls in non-photosynthetic tissue is to attract birds and butterflies, which respond to their yellow, orange and red colours. Although these terpenoids appear to play no part in the photosynthetic process, their universal occurrence in chloroplasts and the chromatophores of photosynthetic bacteria is possibly to protect the cells against photo-oxidative destruction caused by the incidental absorption of visible light.

No specific function has yet been discovered for the polyisoprenes, rubber and gutta percha, but they probably act as feeding deterrents.

Terpenoids and Steroids in Chemosystematics

Terpenoids and steroids are some of the oldest compounds to be synthesised by living organisms, and thus it is not surprising that such a wide variety of structures should be generated by plants and animals today. The acetate–mevalonate pathway is one of the most primitive biosynthetic pathways, and some of the earliest biogenetically derived compounds to be discovered are the terpenoid derivatives, pristane (5.144) and phytane (5.145), detected in South African rocks which are over three billion years old. It has been suggested that phytane is the remains of the phytol residue of chlorophyll, but this has not been established unequivocally.

pristane
(5.144)

phytane
(5.145)

The spores of ferns, lycopods and liverworts are the oldest remains of land plants, while the pollen of gymnosperms has been found in Carboniferous coal deposits. The resistance to decay of spores and pollen is due to a substance, sporopollenin, found in the external walls. Sporopollenin is a polymerised carotenoid, again indicating the ancient nature of the acetate–mevalonate pathway.

All members of the plant kingdom contain carotenoids and terpenoid quinones such as the menoquinones and phylloquinones described in Chapter 7. The formation of squalene and the biosynthesis of sterols is also an ancient mechanism, as, although rare, steroids have been detected in blue-green algae (Cyanophyta), the most primitive of the algae. The formation of monoterpenoids, however, is evolutionarily a much more recent process; these compounds being rare in lower plants, but characteristic of seed-bearing plants. A scheme showing the main evolution of terpenoids and steroids is given in *Table 5.1*.

The wide variety of terpenoids and steroids in plants, together with their diverse occurrence, makes many of these compounds of little use in chemosystematics. The cardiac glycosides, for instance, occur sporadically in many families and even in the Apocynaceae, where these compounds are abundant, they have little taxonomic value. (For a discussion of the sugar components of cardiac glycosides, see Chapter 2.) The triterpenoid saponins are too widely distributed to have taxonomic value when considered alone. However, in combination with other chemical characteristics, such as phenols and flavonoids, these compounds have been used to show that the Pittosporaceae family has more affinity with the Araliaceae than the Saxifragaceae and should therefore be included in the order Araliales (Umbellales) rather than in the Rosales. The steroidal saponins have a much narrower distribution, being in general restricted to the monocotyledonous

Table 5.1 The evolution of terpenoids and steroids.

Evolution	Widely occurring terpenoid types
seed-bearing plants (Spermatophyta) ↑	carotenoids, quinones, steroids, triterpenoids, diterpenoids, sesquiterpenoids, monoterpenoids
ferns (Pteridophyta) ↑	carotenoids, quinones, steroids, triterpenoids, diterpenoids, sesquiterpenoids
liverworts (Bryophyta) ↑	carotenoids, quinones, steroids, triterpenoids, sesquiterpenoids
fungi	carotenoids, ubiquinones, steroids, triterpenoids, diterpenoids
brown algae (Phaeophyta) red algae (Rhodophyta) green algae (Chlorophyta)	carotenoids, quinones, steroids
blue-green algae (Cyanophyta) ↑	carotenoids, quinones
bacteria	carotenoids, menaquinones

families Liliaceae, Dioscoreaceae, Amaryllidaceae, Agavaceae and Bromeliaceae, although they also occur in the dicotyledonous Solanaceae and Scrophulariaceae. The Solanaceae is characterised by alkaloidal saponins which have not been found outside this family, while the related *Veratrum* alkaloids are confined to the Veratrieae tribe of the Liliaceae. Diterpenoid alkaloids of the aconitine type are characteristic of the related *Aconitum* and *Delphinium* genera of the Ranunculaceae family. They also occur in the Garryaceae, a family difficult to classify on morphological grounds (see also below).

Oxidised terpenoid derivatives are useful taxonomic markers, as the cucurbitacins are characteristic of the Cucurbitaceae family while the meliacins, limonoids and quassins are found in the closely related Meliaceae, Rutaceae and Simaroubaceae families.

The monoterpenoid composition of the *Pinus* genus (Pinaceae) has been useful in tracing hybridisation between species and the migration of the genus, while cultivars of the blue spruce (*Picea pungens,* Pinaceae) have been identified from their monoterpenoid composition.

The production of iridoids in plants has confirmed some recent changes in classification made on morphological grounds. Thus, aucubin (5.146) occurs in Cornaceae and Scrophulariaceae and also in *Buddleia* and Garryaceae. *Buddleia* was once included in the Loganiaceae but has now been moved to Buddleiaceae which is placed near Scrophulariaceae, while Garryaceae has been removed from the apetalous orders to the vicinity of Cornaceae.

aucubin
(5.146)

Although the sesquiterpene lactones have been detected in several plant families, it is only in the Compositae that they occur widely. These compounds can be divided into five structural types and their occurrence in the tribes of the Compositae is shown in *Figure 5.48*.

Sesquiterpene lactones have been isolated from all tribes except Astereae, Arctotideae and Mutisieae, the closely related Heliantheae and Helenieae being particularly rich in these compounds and the only tribes to synthesise a variety of ambrosanolides. These compounds are characteristic of *Ambrosia* and if this genus and its relatives *Iva, Franseria* and *Xanthium* were removed to a separate family, Ambrosiaceae, as suggested by some taxonomists on morphological grounds, the sesquiterpene lactone content of the Heliantheae would be similar to that of the Helenieae.

The Cynareae is characterised by a lack of tricyclic lactones, guaianolides occurring only rarely. Thus, it seems that most species belonging to this tribe have not evolved the enzymes necessary for the further cyclisation of germacranolides.

The Senecioneae is distinguished from all other tribes by its content of eremophilanolides, and the occurrence of such compounds in *Adenostyles* indicates

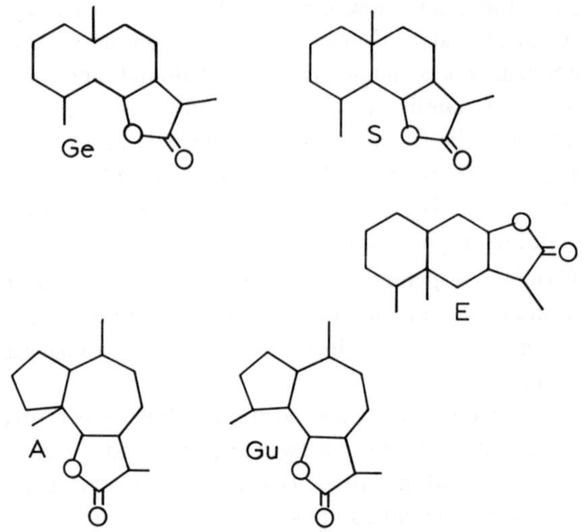

Ge = Germacranolides
S   = Santanolides
E   = Eremophilanolides
Gu = Guaianolides
A   = Ambrosanolides

Tribe	Ge	S	E	Gu	A
Vernonieae	+	+	−	−	−
Eupatorieae	+	−	−	+	+
Astereae	−	−	−	−	−
Inuleae	−	+	−	+	+
Heliantheae	+	+	−	+	+
Helenieae	+	+	−	+	+
Anthemideae	+	+	−	+	−
Senecioneae	−	−	+	−	−
Calenduleae	+	−	−	−	−
Arctotideae	−	−	−	−	−
Cynareae	+	−	−	+	−
Mutisieae	−	−	−	−	−
Cichorieae	−	+	−	+	−

Figure 5.48 Sesquiterpene lactones in the Compositae.

that this genus should be included in the Senecioneae and not the Eupatorieae, as in earlier classifications.

Bibliography

Isoprene Rule

'The Isoprene Rule and the Biogenesis of Terpenoid Compounds', L. Ruzicka in *Experientia*, vol. IX, 1953

Terpenoids – Biosynthesis

'The Biosynthesis of $C_5 - C_{20}$ Terpenoid Compounds', in *Biosynthesis*, Specialist Periodical Reports (continuing series, vol. 1, 1972), The Chemical Society 'Triterpenoids, Steroids and Carotenoids', *ibid*.

'Biosynthesis of Terpenoids and Steroids', in *Terpenoids and Steroids*, Specialist Periodical Reports (continuing series, vol. 1, 1971), The Chemical Society

'Terpenoid Compounds', in *Organic Chemistry of Secondary Plant Metabolism*, T. A. Geissman and D. H. G. Crout, Freeman, Cooper and Co., 1969

'Sesquiterpenes', *ibid*.

'Diterpenoid Compounds', *ibid*.

'Higher Terpenoids', *ibid*.

'Metabolic Degradation of Triterpenes', *ibid*.

'Alkaloids of Mixed Amino Acid–Mevalonate Origin', *ibid*.

'Metabolites Derived from Isoprenoids', in *Secondary Metabolism*, J. Mann, Oxford University Press, 1978

'Biosynthesis of Isoprenoid Compounds Derived from Geranylgeranyl Pyrophosphate', in *Biosynthesis of Acetate-derived Compounds*, N. M. Packter, Wiley, 1973

Natural Substances Formed Biologically from Mevalonic Acid, T. W. Goodwin (ed.), Academic Press, 1970

Aspects of Terpenoid Chemistry and Biochemistry, T. W. Goodwin (ed.), Academic Press, 1971

'Terpenoids: Structure, Biogenesis and Distribution', in *Recent Advances in Phytochemistry*, vol. 6, V. C. Runeckles and T. J. Mabry (eds.), Academic Press, 1973

'Biochemistry and Physiology of Lower Terpenoids', *ibid*.

'The Biosynthesis of Monoterpenoids', in *Progress in Phytochemistry*, vol. 5, L. Reinhold *et al*. (eds.), Pergamon Press, 1978

Terpenoids – Function

The Biochemical Functions of Terpenoids in Plants, T. W. Goodwin (ed.), The Royal Society, 1978

'Biochemical Interactions between Higher Plants', in *Introduction to Ecological Biochemistry*, J. B. Harborne, Academic Press, 1977

'Insect Feeding Preferences', *ibid*.

'Higher Plant–Lower Plant Interactions', *ibid*.

Terpenoid Plant Hormones

'Hormones', in *Plant Biochemistry*, 3rd edn, J. Bonner and J. E. Varner (eds.), Academic Press, 1976

'Recent Aspects of the Chemistry and Biosynthesis of the Gibberellins', in *Recent Advances in Phytochemistry*, vol. 7, V. C. Runeckles *et al*. (eds.), Academic Press, 1974

'Chemistry and Biochemistry of Abscisic Acid', *ibid*.

'Plant Growth Substances', in *Plant Biochemistry*, Biochemistry Series One, vol. 11, D. H. Northcote (ed.), Butterworths, 1974

Steroids – Biosynthesis

'Biosynthesis of Isoprenoid-derived Compounds Formed via Farnesyl Pyrophosphate: The Sterols', in *Biosynthesis of Acetate-derived Compounds, ibid*.

Steroid Biochemistry and Pharmacology, M. H. Briggs and J. Brotherton, Academic Press, 1970

Steroid Biochemistry, E. Heftmann, Academic Press, 1970

'Triterpenoids, Steroids and Carotenoids', in *Biosynthesis, ibid*.

'The Biosynthesis of Terpenoids and Steroids', in *Terpenoids and Steroids, ibid*.

'The Biosynthesis of Plant Sterols', in *Progress in Phytochemistry*, vol. 3, L. Reinhold and Y. Liwschitz (eds.), Interscience, 1972

'Sterol Formation and Transformation in *Digitalis*', in *Recent Advances in Phytochemistry*, vol. 3, C. Steelink and V. C. Runeckles (eds.), Meridith Corp., 1970

Steroids – Function

'Functions of Steroids in Plants', in *Progress in Phytochemistry*, vol. 4, L. Reinhold *et al*. (eds.), Pergamon Press, 1977

'Hormonal Interactions Between Plants and Animals', in *Introduction to Ecological Biochemistry, ibid*.

'The Ecological Importance of Sterols in Invertebrates', in *Introduction to Chemical Ecology*, M. Barbier, Longman, 1976

'The Plant–Insect Relationship', *ibid*.

Insect Juvenile Hormones, J. L. Mean and M. Beroz (eds.), Academic Press, 1972

Carotenoids

'Triterpenoids, Steroids and Carotenoids', in *Biosynthesis, ibid*.

'Biosynthesis of the Carotenoids', in *Chemistry and Biochemistry of Plant Pigments*, vol. 1, T. W. Goodwin (ed.), Academic Press, 1976

'Distribution of Carotenoids', *ibid*.

'Functions of Carotenoids other than in Photosynthesis', *ibid*.

Carotenoids, O. Isler (ed.), Burkhauser (Basel), 1971

Chemosystematics

'Chemical Patterns and Relationships of the Umbelliferae', in *The Biology and Chemistry of the Umbelliferae*, V. H. Heywood (ed.), Academic Press, 1971

'Taxonomic Evidence from Terpenoids and Steroids', in *The Chemotaxonomy of Plants*, P. M. Smith, Edward Arnold, 1976

'Chemical Evidence for the Classification of Some Plant Taxa', in *Perspectives in Phytochemistry*, J. B. Harborne and T. Swain (eds.), Academic Press, 1969

'Chemotaxonomy of the Sesquiterpenoids of the Compositae', *ibid*.

The Biology and Chemistry of the Compositae, vols. 1 and 2, V. H. Heywood *et al*. (eds.), Academic Press, 1977 (Terpenoids are discussed in several chapters)

'Some Aspects of the Distribution of Diterpenes in Plants', in *Chemistry in Botanical Classification*, G. Bendz and J. Santesson (eds.), Academic Press, 1974

'Chemistry of Geographical Races', sections V and VI, in *Chemistry in Evolution and Systematics*, T. Swain (ed.), Butterworths, 1973

6 Shikimic Acid Pathway Metabolites

Introduction

Benzenoid compounds in plants are biosynthesised by two main pathways: the shikimic acid pathway and the acetate-malonate pathway (Chapter 4). In higher plants, a large number of aromatic compounds are derived from phenylalanine, tyrosine and tryptophan, end-products of the shikimic acid pathway. Some of these compounds are discussed in this chapter, others will be found in Chapters 7 and 9.

The Shikimic Acid Pathway

Although the existence of the shikimic acid pathway (*Figure 6.1*) in plants has been known for almost 30 years, it is only recently that certain steps have been clarified and the enzymes involved isolated. Shikimic acid (6.1) was first isolated in the last century from the Japanese ashikimi (star-anise, *Illicium anisatum*, Illiciaceae) and is now known to be ubiquitous in plants, although not all accumulate the compound. The original sequence of reactions was deduced from studies with *Escherichia coli* mutants, but has since been shown to take place in higher plants.

The precursors of shikimic acid, erythrose-4-phosphate (6.2) and phosphoenol pyruvate (PEP) (6.3), are primary metabolites derived from glucose-6-phosphate by the oxidative pentose phosphate cycle (hexose monophosphate shunt) and the process of glycolysis. The precursors combine to form 3-deoxy-D-arabino-heptulosonic acid-7-phosphate (DAHP) (6.4), a reaction catalysed by phospho-2-oxo-3-deoxyheptonate aldolase. The enzyme, 3-dehydroquinate synthase, catalysing the cyclisation of DAHP to 3-dehydroquinic acid (6.5) requires cobalt(II) and nicotinamide adenine dinucleotide (NAD) as cofactors. It is probable therefore that the cyclisation is an oxidation/reduction reaction as shown in *Figure 6.2*.

The shikimic acid pathway contains several branch points, the first of these, dehydroquinic acid, can be converted either to 3-dehydroshikimic acid (6.6), which continues the pathway, or to quinic acid (6.7). The enzymes catalysing the dehydration of dehydroquinic acid are of two kinds. Form 1, associated with shikimate dehydrogenase, is independent of shikimate concentration, while form 2 is specifically activated by shikimate.

It has been suggested that the two forms provide a control in the utilisation of dehydroquinic acid (*Figure 6.3*), producing either shikimic acid or proto-catechuic acid. The 3-dehydroquinate dehydratase/shikimate dehydrogenase

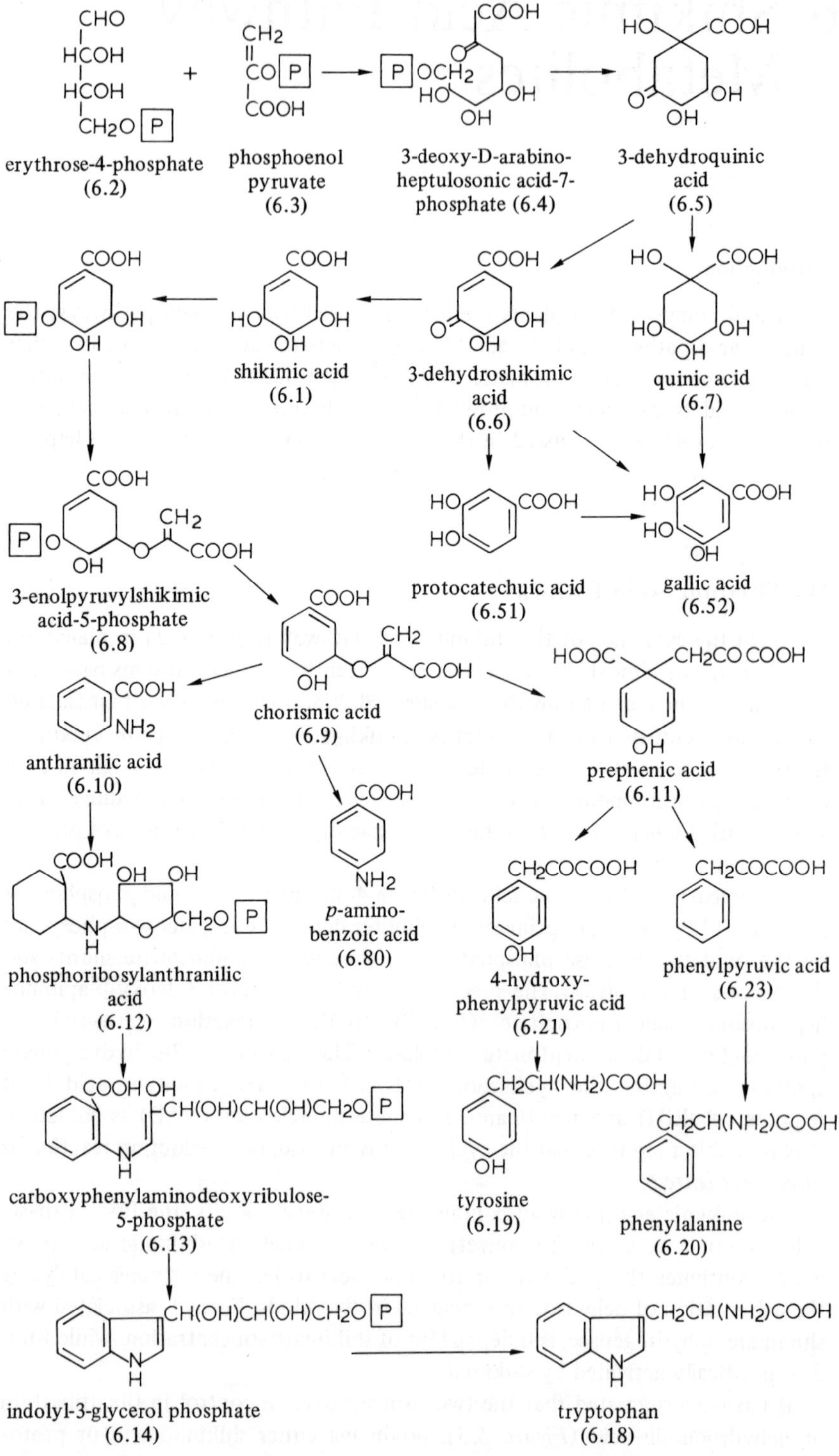

Figure 6.1 The shikimic acid pathway.

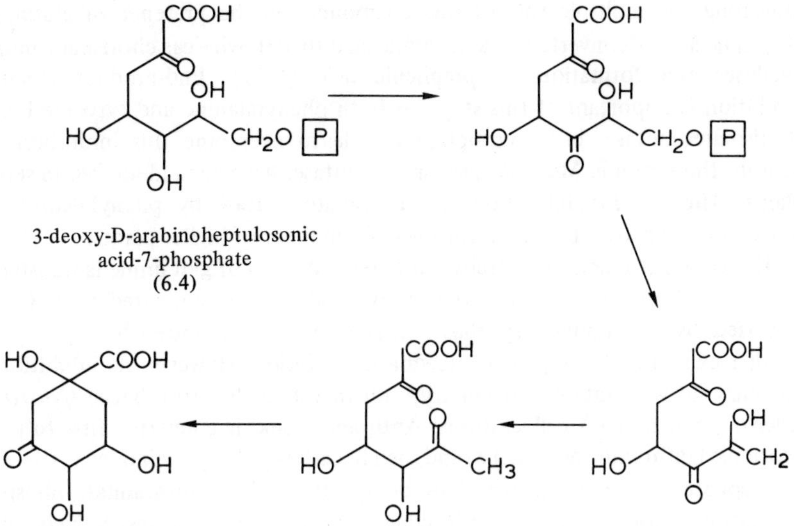

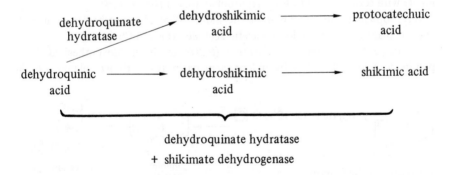

3-deoxy-D-arabinoheptulosonic
acid-7-phosphate
(6.4)

3-dehydroquinic acid
(6.5)

Figure 6.2 The cyclisation of 3-deoxy-D-arabinoheptulosonic acid-7-phosphate to 3-
dehydroquinic acid.

dehydroquinate   dehydroshikimic   protocatechuic
hydratase   acid   acid

dehydroquinic ⟶ dehydroshikimic ⟶ shikimic acid
acid   acid

dehydroquinate hydratase
+ shikimate dehydrogenase

Figure 6.3 The utilisation of dehydroquinic acid.

bifunctional enzyme has been isolated from several higher plants. In lower plants, these catalytic functions are performed by two separate enzymes.

After phosphorylation, catalysed by shikimate kinase, shikimic acid adds on enol pyruvate to form 3-enolpyruvylshikimic acid-5-phosphate (6.8). This reaction is catalysed by enolpyruvylshikimate phosphate synthase, while conversion to chorismic acid (6.9) is catalysed by chorismate synthase. The formation of chorismic acid is an important branch point in the shikimic acid pathway as this compound can undergo five different types of conversion (*Figure 6.1*). The name 'chorismic' is derived from a Greek word for separate, indicating the multiple role of this compound. In the presence of glutamine, chorismic acid is converted to anthranilic acid (6.10), whereas chorismate mutase catalyses the formation of prephenic acid (6.11). End-product feedback inhibition is important at this stage, as both phenylalanine and tyrosine inhibit chorismate mutase, but tryptophan is able to overcome this inhibition and activate the enzyme. Multiple chorismate mutases have been identified in several plants. The *in vivo* inhibition of chorismate mutase by phenylalanine and tyrosine is not shown by tissue cultures of a number of higher plants.

The complex reaction whereby the amide nitrogen of glutamine is transferred to C-2 of chorismic acid, and the pyruvyl side chain eliminated from C-3, is catalysed by anthranilate synthase. This enzyme is inhibited by tryptophan, another example of end-product feedback inhibition. However, no inhibition of this enzyme was obtained when tryptophan was added to tobacco (*Nicotiana tabacum*, Solanaceae) cell cultures. Anthranilic acid is converted first to phosphoribosylanthranilic acid (6.12) and then to carboxyphenylaminodeoxyribulose-5-phosphate (6.13), these reactions being catalysed by anthranilate phosphoribosyltransferase and phosphoribosylanthranilate isomerase, respectively. Ring closure to form indolyl-3-glycerol phosphate (6.14) is catalysed by indolylglycerol phosphate synthase. The enzyme catalysing the final reaction, tryptophan synthase, has been isolated from algae, tobacco and peas (*Pisum sativum*, Leguminosae). It consists of two components: component A catalyses the dissociation of indolylglycerol phosphate to indole (6.15) and glyceraldehyde-3-phosphate (6.16), while component B catalyses the direct condensation of indole with serine (6.17) to form tryptophan (6.18) (*Figure 6.4*).

Tryptophan is the precursor of the indole alkaloids described in Chapter 9 and the plant hormone, indolyl-3-acetic acid, described in Chapter 8.

Tyrosine (6.19) and phenylalanine (6.20) are both biosynthesised from prephenic acid (6.11), but by independent pathways. Although plants contain

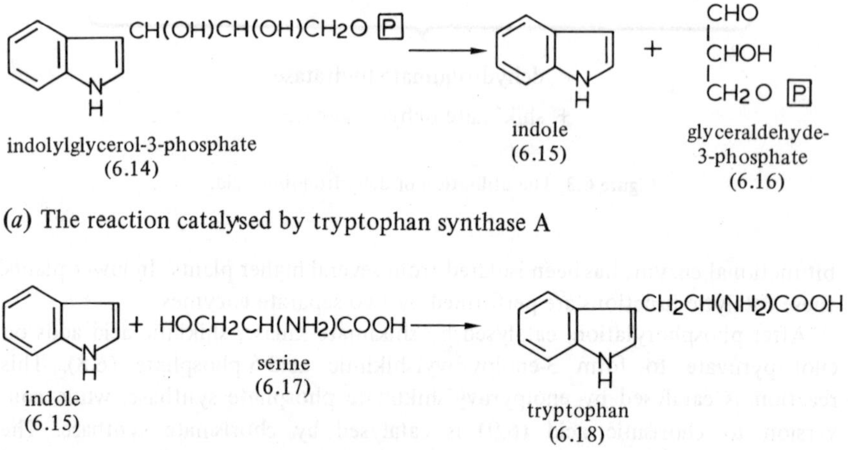

indolylglycerol-3-phosphate
(6.14)

indole
(6.15)

glyceraldehyde-
3-phosphate
(6.16)

(*a*) The reaction catalysed by tryptophan synthase A

indole
(6.15)

serine
(6.17)

tryptophan
(6.18)

(*b*) The reaction catalysed by tryptophan synthase B

Figure 6.4 Final stages in the biosynthesis of tryptophan.

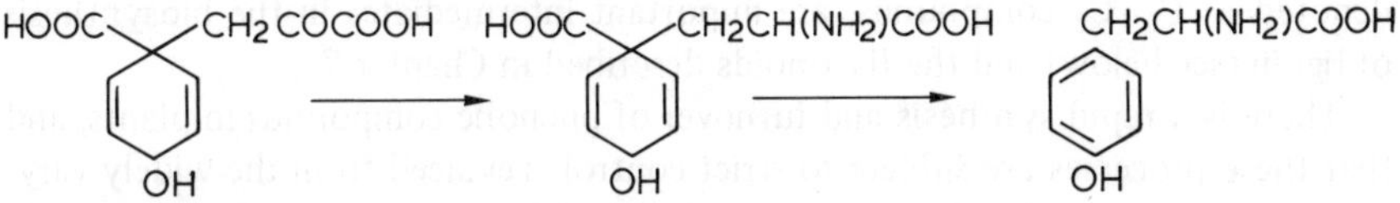

prephenic acid
(6.11)

pretyrosine
(6.22)

tyrosine
(6.19)

Figure 6.5 A novel route to tyrosine.

enzymes which will catalyse the hydroxylation of phenylalanine to tyrosine, this is not the main biosynthetic pathway, as it is in animals. In the formation of tyrosine (*Figure 6.1*), prephenic acid is first aromatised to 4-hydroxyphenyl-pyruvic acid (6.21), a reaction catalysed by prephenate dehydrogenase. Trans-amination, catalysed by tyrosine aminotransferase, then gives tyrosine. Recently, a second pathway from prephenic acid to tyrosine has been detected in blue-green algae. In this pathway, the reactions are reversed, prephenic acid being first transaminated to pretyrosine (6.22) and then aromatised (*Figure 6.5*). There are indications that this pathway is also operative in *Phaseolus* (Leguminosae). It has been suggested that the pretyrosine pathway is the more primitive, as it does not have the refined regulation of the hydroxyphenyl-pyruvate pathway.

The biosynthesis of phenylalanine (6.20) (*Figure 6.1*) involves first the aromatisation of prephenic acid to phenylpyruvic acid (6.23), a reaction catalysed by prephenate dehydratase, and then transamination catalysed by phenylalanine aminotransferase.

Cinnamic Acid Derivatives

Higher plants accumulate a wide range of phenolic compounds which are derivatives or metabolites of cinnamic acid. The phenylpropanes, generally

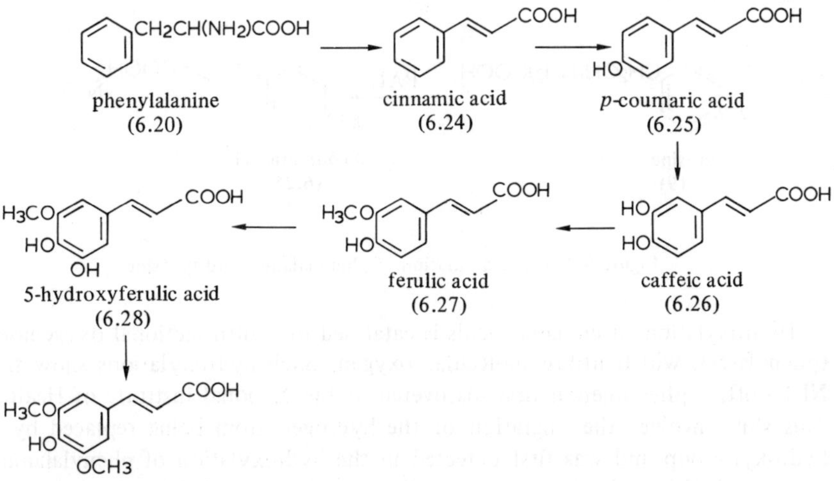

phenylalanine
(6.20)

cinnamic acid
(6.24)

p-coumaric acid
(6.25)

5-hydroxyferulic acid
(6.28)

ferulic acid
(6.27)

caffeic acid
(6.26)

sinapic acid
(6.29)

Figure 6.6 The biosynthesis of cinnamic acid derivatives.

denoted as C_6-C_3 compounds, are important intermediates in the biosynthesis of lignin (see below) and the flavonoids described in Chapter 7.

There is a rapid synthesis and turnover of phenolic compounds in plants, and that these processes are subject to strict control is evinced from the widely varying concentrations, which can be correlated with various internal and external factors, such as age of tissue, time of year, etc.

The common phenylpropane derivatives, cinnamic acid (6.24), p-coumaric acid (6.25), caffeic acid (6.26), ferulic acid (6.27), 5-hydroxyferulic acid (6.28) and sinapic acid (6.29), are interrelated and biosynthesised from phenylalanine, as shown in *Figure 6.6*. It is generally found that, except for members of the Gramineae and Compositae families, tyrosine is a poor precursor of the hydroxyphenylpropanes. The enzymes catalysing the conversions shown in *Figure 6.6* have been characterised, and it has been suggested that they exist as a multienzyme complex.

Phenylalanine ammonia-lyase (PAL) catalyses the deamination of phenylalanine and also of tyrosine in the Gramineae family members, wheat (*Aestivum sativum*) and maize (*Zea mays*). In both cases, a *trans*-cinnamic acid results as shown in *Figure 6.7*. PAL has been found to exist as several isozymes in some plants, each isozyme being inhibited by a specific phenol. PAL will not catalyse the deamination of the unnatural D-amino acids; it is, however, unaffected by substituents on the aromatic ring. In dark tissue, PAL is inactive, and there appears to be a strict control over the activation of this enzyme, and therefore the conversion of the amino acids to cinnamic acid derivatives.

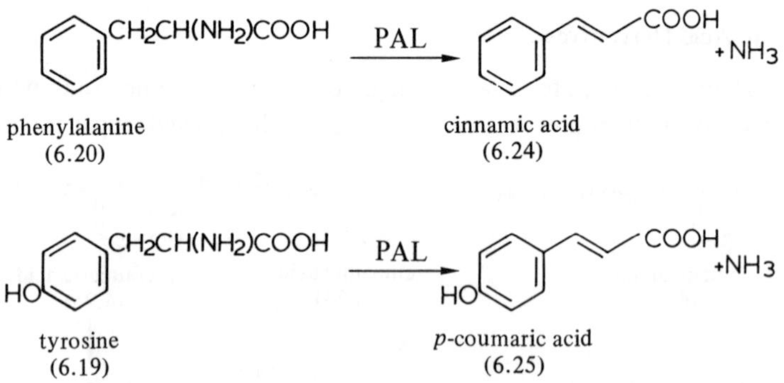

	phenylalanine	cinnamic acid
	(6.20)	(6.24)

	tyrosine	p-coumaric acid
	(6.19)	(6.25)

Figure 6.7 The deamination of phenylalanine and tyrosine.

Hydroxylation of cinnamic acids is catalysed by multifunctional oxygenases (phenolases), which utilise molecular oxygen. Such hydroxylations show the NIH shift, a phenomenon first discovered at the National Institute of Health. This shift involves the migration of the hydrogen atom being replaced by a hydroxyl group and was first detected in the hydroxylation of phenylalanine which took place in bacteria or mammalian liver. Tritium labelling the 4 position of phenylalanine was found to be retained in tyrosine but to have migrated to the 3 position (*Figure 6.8*).

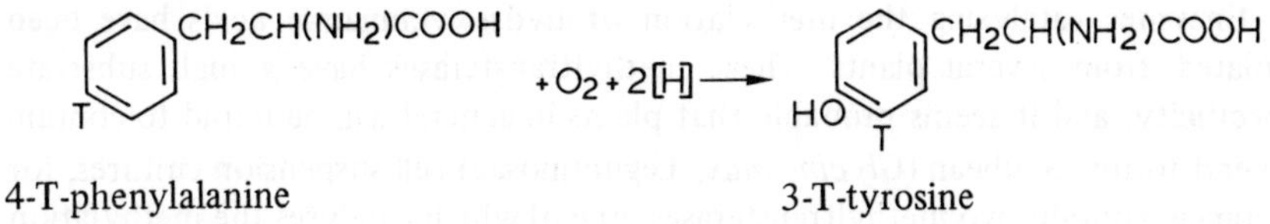

4-T-phenylalanine   3-T-tyrosine

Figure 6.8 The NIH shift.

Similar results have since been achieved in higher plants with 4-tritium-labelled cinnamic acid, which gives *p*-coumaric acid (6.25) and 4-hydroxybenzoic acid (6.30) with migration of tritium to the 3 position (*Figure 6.9*).

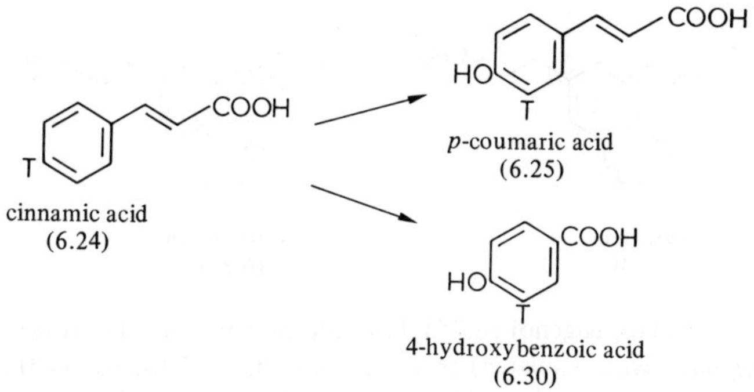

p-coumaric acid
(6.25)

cinnamic acid
(6.24)

4-hydroxybenzoic acid
(6.30)

Figure 6.9 NIH shift during 4-hydroxylation.

The NIH shift also occurs during 2-hydroxylation, the hydrogen atom migrating to the 3 position. Thus, feeding 2,6-tritium-labelled cinnamic acid to sweet clover (*Melilotus alba*, Leguminosae) resulted in *o*-coumaric acid (6.31) in which 92% of the tritium was retained (*Figure 6.10*).

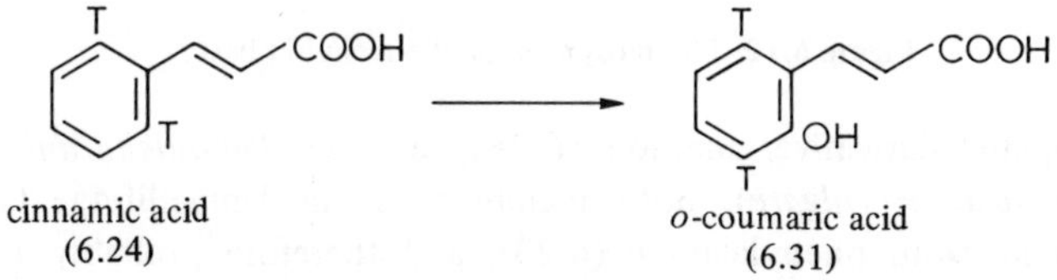

cinnamic acid
(6.24)

o-coumaric acid
(6.31)

Figure 6.10 NIH shift during 2-hydroxylation.

It is interesting that in wintergreen (*Gaultheria procumbens*, Ericaceae), a plant which accumulates salicylic acid (6.32), hydroxylation takes place after degradation of the cinnamic acid side chain, but the NIH shift is still observed (*Figure 6.11*).

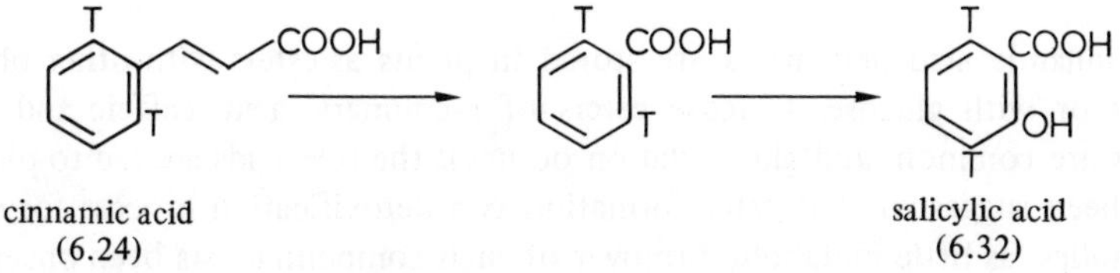

cinnamic acid
(6.24)

salicylic acid
(6.32)

Figure 6.11 The biosynthesis of salicylic acid in wintergreen.

Enzymes catalysing the methylation of hydroxycinnamic acids have been isolated from several plants. These methyltransferases have a high substrate specificity, and it seems probable that plants in general will be found to contain several forms. Soybean (*Glycine max,* Leguminosae) cell suspension cultures, for instance, contain two methyltransferases, one of which catalyses the methylation of caffeic (6.26) and 5-hydroxyferulic acids (6.28), and the other that of the flavonoids, luteolin and quercetin.

The essential oils which are allyl- or propenylphenol derivatives often occur together in the same plant, e.g. myristicin (6.33) and isomyristicin (6.34) in nutmeg (*Myristica fragrans,* Myristicaceae), indicating a common biosynthetic pathway.

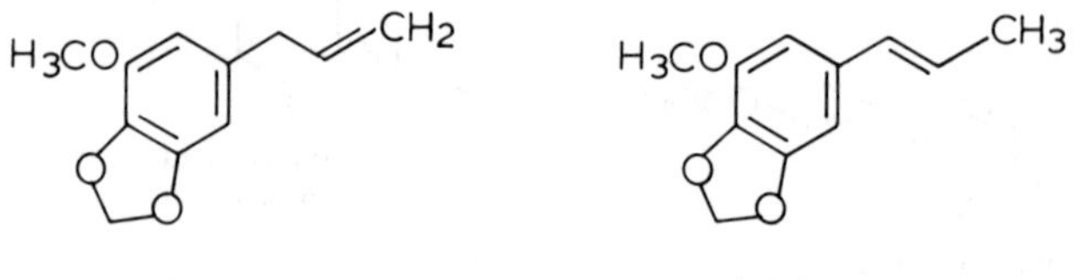

myristicin
(6.33)

isomyristicin
(6.34)

The allylphenol, eugenol (6.35), has been shown to be biosynthesised in basil (*Ocimum basilicum,* Labiatea) from glucoferulic acid (6.36) via the propenyl alcohol glucoside, coniferin (6.37) (*Figure 6.12*).

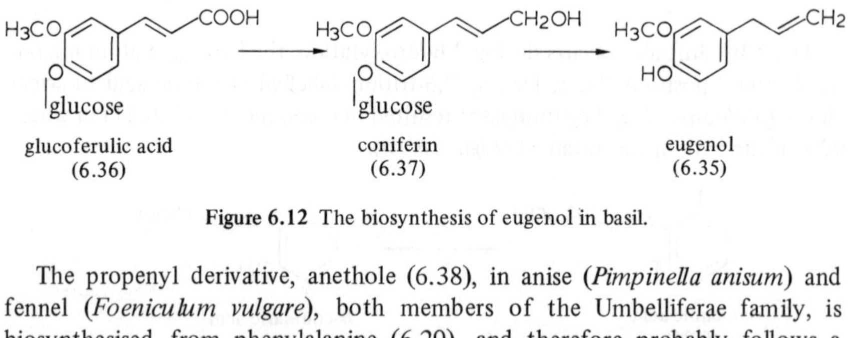

glucoferulic acid
(6.36)

coniferin
(6.37)

eugenol
(6.35)

Figure 6.12 The biosynthesis of eugenol in basil.

The propenyl derivative, anethole (6.38), in anise (*Pimpinella anisum*) and fennel (*Foeniculum vulgare*), both members of the Umbelliferae family, is biosynthesised from phenylalanine (6.20), and therefore probably follows a pathway similar to that of eugenol.

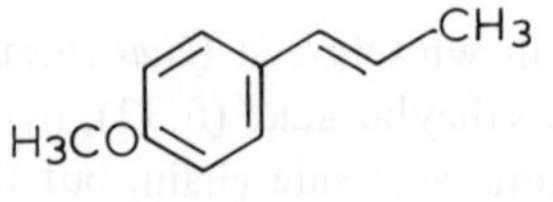

anethole
(6.38)

Cinnamic acid derivatives are stored in plants as esters with other phenolic acids or with glucose. Glucose esters of *p*-coumaric acid, caffeic and ferulic acids are common, and glucosylation occurs if the free acids are fed to plants. It has been suggested that ester formation is a detoxification process for storing phenolics, as little metabolic turnover of such compounds has been observed in living cells.

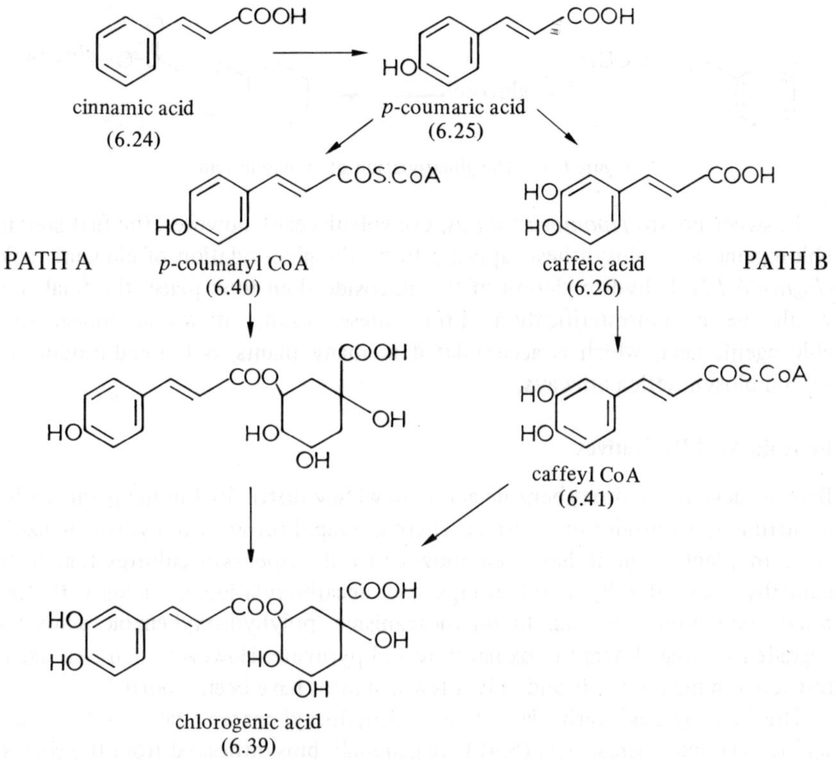

Figure 6.13 The biosynthesis of chlorogenic acid in the Solanaceae.

The widespread chlorogenic acid, 3-caffeylquinic acid (6.39), is an ester of quinic acid (6.7), a product of the shikimic acid pathway (*Figure 6.1*), and caffeic acid (6.26). Chlorogenic acid is particularly abundant in coffee beans (*Coffea* spp., Rubiaceae). The soluble content of coffee can contain as much as 13% by weight of chlorogenic acid. The enzymes catalysing the formation of chlorogenic acid in tomato fruits *(Lycopersicum esculentum*, Solanaceae) have been isolated and identified, while the kinetics of the synthesis of this acid have been studied in *Cestrum peoppigii*, another member of the Solanaceae family. The results of these studies indicate that two pathways are operative (*Figure 6.13*), path A being the more important, as quinate transferase shows a greater affinity for *p*-coumaryl CoA (6.40) than for caffeyl CoA (6.41).

There is evidence that, in the potato (*Solanum tuberosum*, Solanaceae), ester formation precedes hydroxylation, as 3-cinnamylquinic acid (6.42) is an active intermediate in chlorogenic acid biosynthesis.

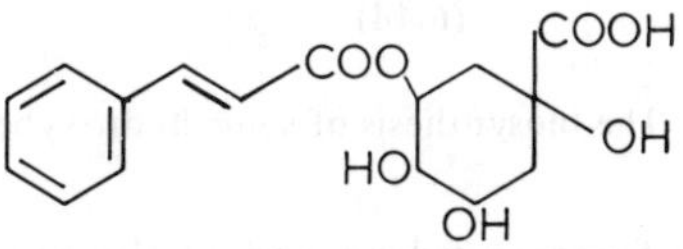

3-cinnamylquinic acid
(6.42)

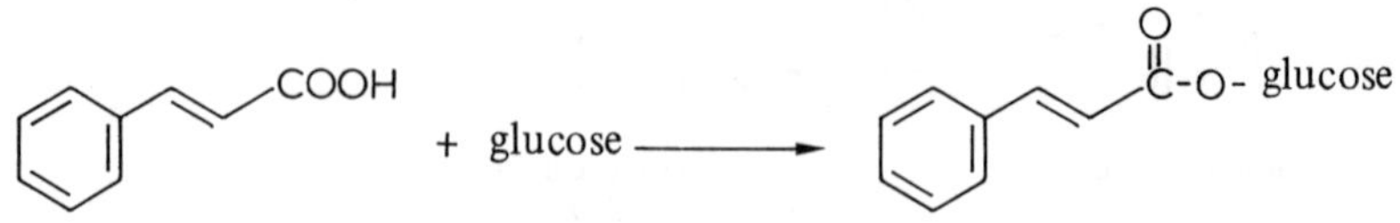

Figure 6.14 The glucosylation of cinnamic acid.

In sweet potato *(Ipomoea batatas,* Convolvulaceae), however, the first step in chlorogenic acid biosynthesis appears to be the glucosylation of cinnamic acid *(Figure 6.14)*. If hydroxylation of the glucoside then took place, the final step would be a transesterification. From these results, it would appear that chlorogenic acid, which is accumulated in many plants, is the end-product of several biosynthetic pathways.

Benzoic Acid Derivatives

Benzoic acid and hydroxybenzoic acids are widely distributed in the plant world, occurring as glycosides or as esters. There is a rapid turnover of hydroxybenzoic acids in plants, and it has been shown in cell suspension cultures that both demethylation of *p*-hydroxyl groups and decarboxylation of acids with free *p*-hydroxyl groups occurs. In microorganisms, polyhydroxybenzoic acids are degraded by ring cleavage to oxalacetate and pyruvate. However, such cleavage is restricted in higher plants and only a few instances have been reported.

The benzoic acid derivatives such as 4-hydroxybenzoic acid (6.43), vanillic acid (6.44) and syringic acid (6.45) are generally biosynthesised from the corresponding cinnamic acid derivatives *(Figure 6.15)* by β-oxidation of the propenyl side chain after ester formation with coenzyme A. It has also been shown in wintergreen that hydroxylation of the benzene ring can take place.

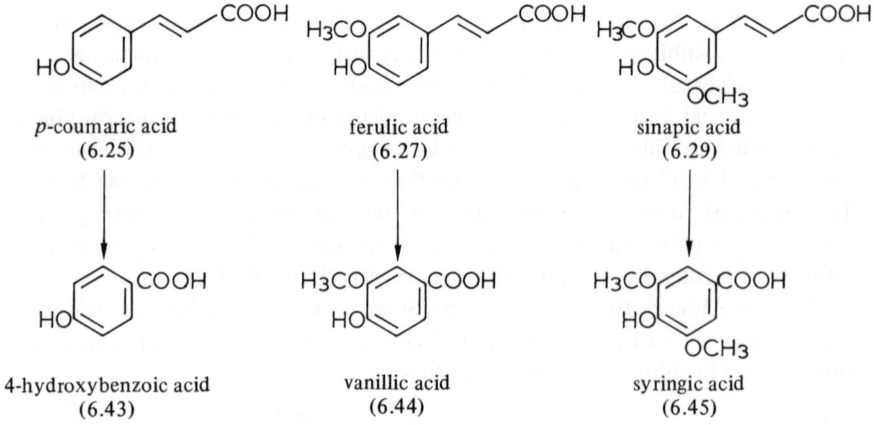

p-coumaric acid (6.25)

ferulic acid (6.27)

sinapic acid (6.29)

4-hydroxybenzoic acid (6.43)

vanillic acid (6.44)

syringic acid (6.45)

Figure 6.15 The biosynthesis of some hydroxybenzoic acids.

In cell free extracts of potato tubers and of the fungus, *Polyporus hispidus,* 4-hydroxybenzoic acid was shown to be derived from *p*-coumaric acid, not by β-oxidation, but by the scheme shown in *Figure 6.16*. 4-Hydroxybenzoic acid is important as a precursor of the ubiquinones and plastoquinones described in Chapter 7.

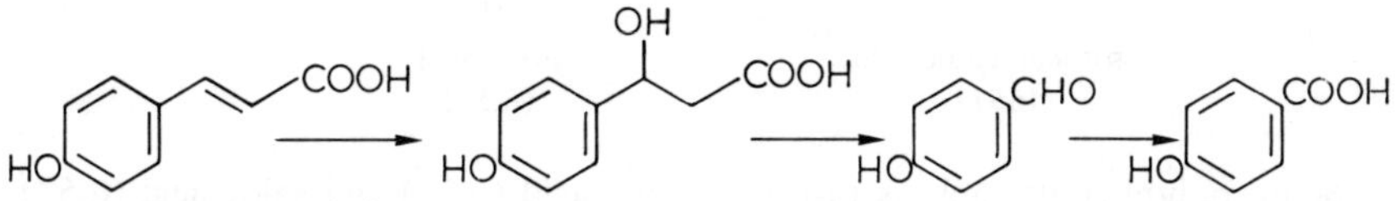

p-coumaric acid
(6.25)

4-hydroxybenzoic acid
(6.43)

Figure 6.16 A biosynthetic route to 4-hydroxybenzoic acid.

Although salicylic acid (6.32) is derived from phenylalanine, 6-methylsalicylic acid (6.46) is biosynthesised from acetate in both microorganisms and higher plants (Chapter 4).

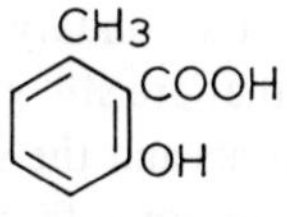

6-methylsalicylic acid
(6.46)

In willow (*Salix purpurea*, Salicaceae), the glucoside salicin (6.47) is derived from benzoic acid (6.48) or *o*-coumaric acid (6.31) and not from salicylic acid or salicyl alcohol (saligenin). *o*-Coumaric acid is the better precursor and the pathways are similar, involving salicylaldehyde (6.49) and helicin (6.50) as intermediates (*Figure 6.17*).

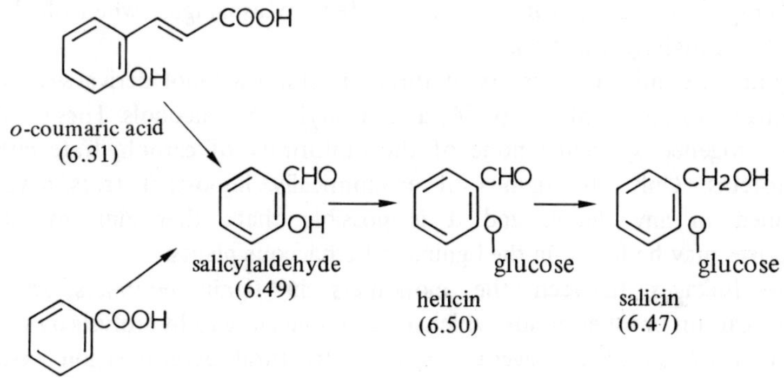

o-coumaric acid
(6.31)

salicylaldehyde
(6.49)

helicin
(6.50)

salicin
(6.47)

benzoic acid
(6.48)

Figure 6.17 The biosynthesis of salicin.

Salicin was the forerunner of the synthetic acetylsalicylic acid or aspirin, with its useful analgesic and antipyretic properties. Salicin is too toxic to be taken internally, but this glucoside is a useful external analgesic, especially for rheumatic pain.

HO⬡COOH
HO

protocatechuic acid
(651)

HO⬡COOH
HO
OH

gallic acid
(6.52)

Some polyphenols such as protocatechuic acid (6.51) and gallic acid (6.52) are biosynthesised by direct aromatisation of 3-dehydroshikimic acid (6.6) (*Figure 6.1*), although they can also be formed by hydroxylation of cinnamic and benzoic acid derivatives.

Lignin

The term 'lignin' is derived from the Latin *lignum* for wood, and woody plants or parts of plants contain large quantities of this substance. Lignin is an important skeletal component of secondary cell walls and is thus not found in young plants or parts of plants that are still growing. It is lignin that provides the hardness and rigidity of tree trunks and the stems of perennial plants. In the cell wall, lignin is bound to other components, notably hemicelluloses (Chapter 2). The lignin-cellulose-hemicellulose complex has economic importance in the production of paper from wood. The strong bonds linking cellulosic material to lignin render the former inaccessible to enzyme hydrolysis, and thus lignin has a direct influence on the digestibility, and hence the nutritive value, of herbage plants to grazing animals. In herbage plants, lignin is also bonded to cell wall protein.

Lignification of plants affects the palatability of fruits and vegetables. Thus, the unpleasant 'stone cells' in pears are due to lignification, and many root and stem crops, such as carrots, beetroot, celery or asparagus, when old become inedible through lignification.

Lignins are mixed polymers of three cinnamyl alcohol derivatives, namely *p*-coumaryl (6.53), coniferyl (6.54) and sinapyl (6.55) alcohols. These polymers are heterogeneous, having none of the uniformity of carbohydrate polymers or proteins. Only the lignins of economically important trees have been examined in any detail, and it is possible that other cinnamyl alcohol derivatives may be found in the lignins of herbaceous plants.

The linkages between the monomers in lignin polymers are either carbon–carbon or ether bonds, which are not amenable to hydrolysis by enzymes or the usual chemical reagents, making structural determination extremely difficult.

The random polymerisation of the alcohols is non-enzymatic (see below), so that in all probability no two lignin polymer molecules are identical. However, certain structural properties have been elucidated, mainly from model systems in which the cinnamyl alcohols are made to polymerise under certain conditions.

Softwood lignins differ from hardwood polymers in the proportion of the three alcohols present. Softwood lignin contains about 80% coniferyl alcohol and only around 14% *p*-coumaryl and 6% sinapyl alcohols, while hardwood lignin contains about 56% coniferyl alcohol, 40% sinapyl alcohol and only around

4% *p*-coumaryl alcohol. The lignins of herbaceous plants have received little attention, due to their lower concentrations and the difficulty of freeing the polymer from complexes with protein, etc. However, it has been found that lignin of members of the Gramineae has a high content of *p*-coumaryl alcohol.

The biosynthetic pathway to the formation of lignins is basically similar in all plants, the three alcohols being derived from the corresponding cinnamic acid derivatives (*Figure 6.18*). The *O*-methyltransferases (OMT) catalysing the methylation of the hydroxycinnamic acids are specific for lignin synthesis. It has been shown in bamboo (*Bambusa* sp., Gramineae) that the OMT of this plant is bifunctional, catalysing both the methylation of caffeic acid (6.26) to ferulic acid (6.27) and the methylation of 5-hydroxyferulic acid (6.28) to sinapic acid (6.29) (*Figure 6.6*). As the substrates are in competition with one another and 5-hydroxyferulic acid shows the greater affinity for the enzyme, a feedback inhibition operates. In has been suggested that this pathway is applicable to all angiosperms, but that in gymnosperms the OMT is mono-functional, catalysing only the methylation of caffeic acid to ferulic acid. This could be the reason for the lack of sinapyl units in the lignins of softwoods.

The conversion of the cinnamic acid derivatives to the alcohols (*Figure 6.18*) is catalysed by specific dehydrogenases, cinnamyl alcohol oxidoreductase being the first dehydrogenase found in plants to have absolute specificity towards cinnamyl derivatives. The first step in the reduction is the formation of the coenzyme A ester and cinnamyl aldehyde is an intermediate.

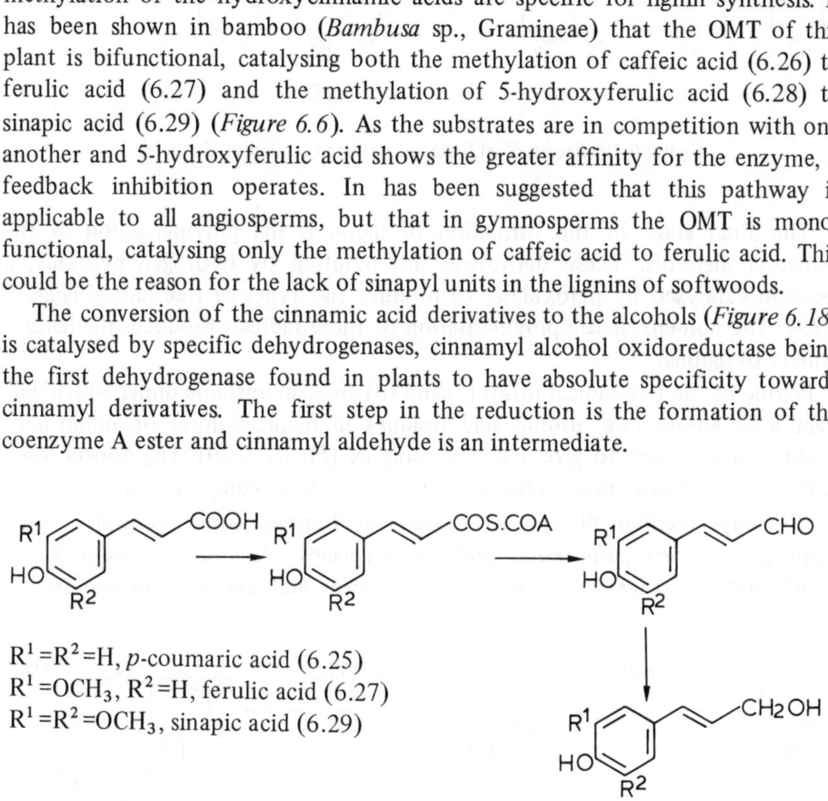

$R^1 = R^2 = H$, *p*-coumaric acid (6.25)
$R^1 = OCH_3$, $R^2 = H$, ferulic acid (6.27)
$R^1 = R^2 = OCH_3$, sinapic acid (6.29)

$R^1 = R^2 = H$, *p*-coumaryl alcohol (6.53)
$R^1 = OCH_3$, $R^2 = H$, coniferyl alcohol (6.54)
$R^1 = R^2 = OCH_3$, sinapyl alcohol (6.55)

Figure 6.18 The biosynthesis of the cinnamyl alcohols.

Coniferyl alchohol dehydrogenase activity has been detected in many plants, and multiple forms of the enzyme have been found. The level of activity of this enzyme can be correlated with the degree of lignification.

Cell suspension cultures of soybean (*Glycine max*, Leguminosae) produce cinnamyl alcohol from cinnamic acid, and the lignin produced by such cultures contains cinnamyl units. However, it is not known if the lignin of intact plants contains these units.

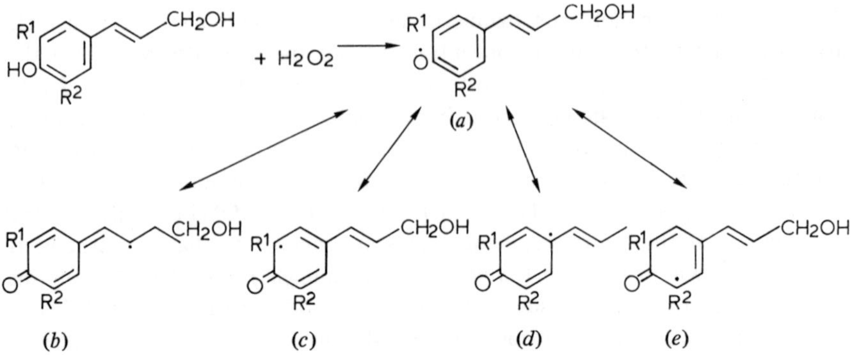

Figure 6.19 Free-radical forms of oxidised cinnamyl alcohols.

The final stage in the formation of lignin is the polymerisation of the cinnamyl alcohols. These derivatives are oxidised by hydrogen peroxide, a reaction catalysed by peroxidase, to produce five types of free radical (*Figure 6.19*). The non-enzymatic polymerisation of these radicals produces the heterogeneity of lignins.

Pairing of the free-radical forms (*a*) and (*b*) gives an unstable dimer which will react with anions (e.g. uronic acid residues in hemicelluloses or amino acid residues in proteins) to give a mixed complex (*Figure 6.20*). The dimers react further with the various free radicals eventually to form complex polymers.

Cinnamyl alcohol dimers are known as lignans, some examples being pinoresinol (6.56) from *Pinus* and *Picea* species (*Pinaceae*), sesamin (6.57) from benniseed (*Sesamum* species, Pedaliaceae) and podophyllotoxin (6.58)

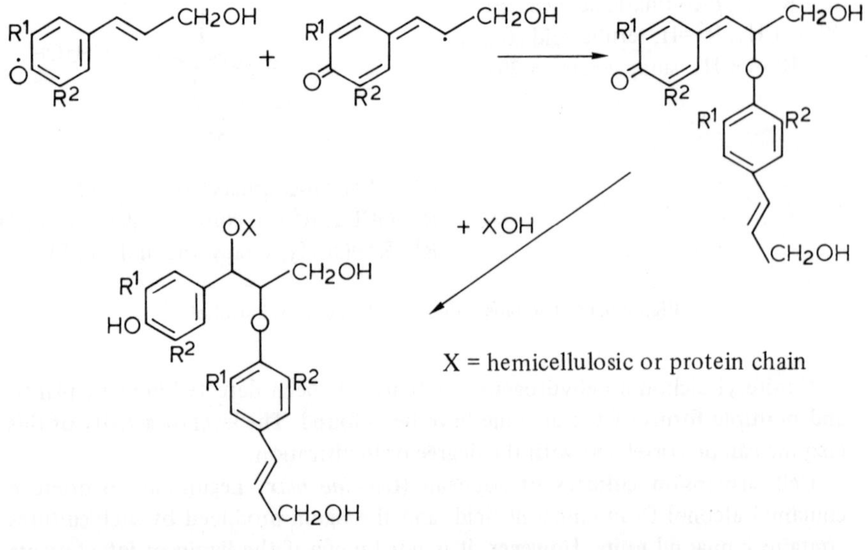

X = hemicellulosic or protein chain

Figure 6.20 The binding of lignin to other cell wall components.

from *Podophyllum* species (Podophyllaceae). Some lignans are used as anti-oxidants in food, and sesamin has found importance as a synergist in pyrethrum insecticides. Podophyllotoxin is a cytotoxin that has shown promise in the treatment of certain types of malignant tumours.

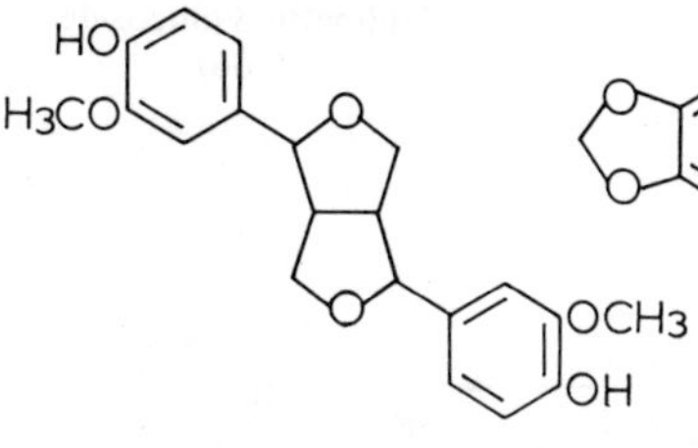

pinoresinol
(6.56)

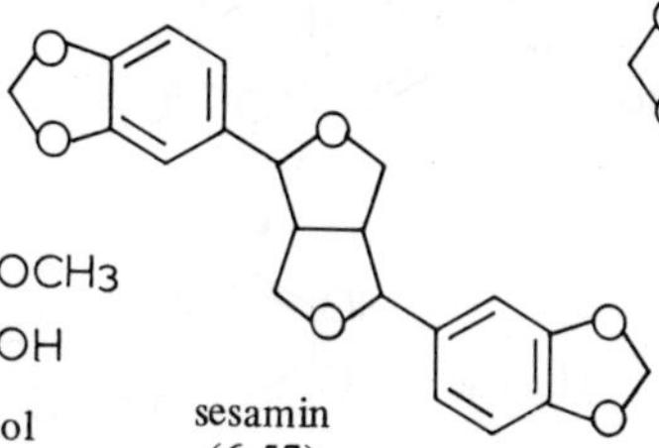

sesamin
(6.57)

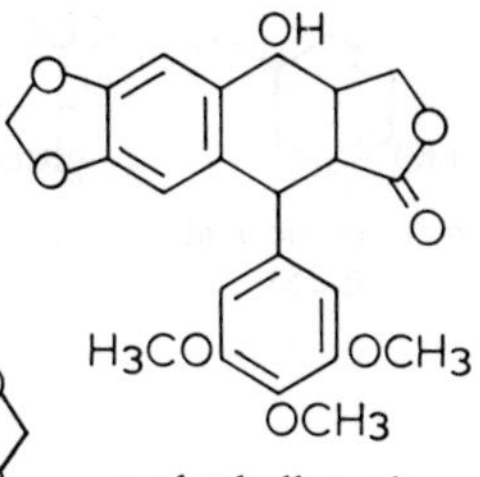

podophyllotoxin
(6.58)

Coumarins

The key step in the biosynthesis of the coumarins is the introduction of an *o*-hydroxyl group to a cinnamic acid derivative. The phenolase catalysing this reaction has been isolated from the chloroplasts of sweet clover (*Melilotus alba*, Leguminosae). The *o*-hydroxyl group subsequently undergoes glucosylation which facilitates the isomerisation of *trans*-cinnamic acid derivatives to the *cis* form. Hydrolysis of the sugar then leads to spontaneous cyclisation (*Figure 6.21*)

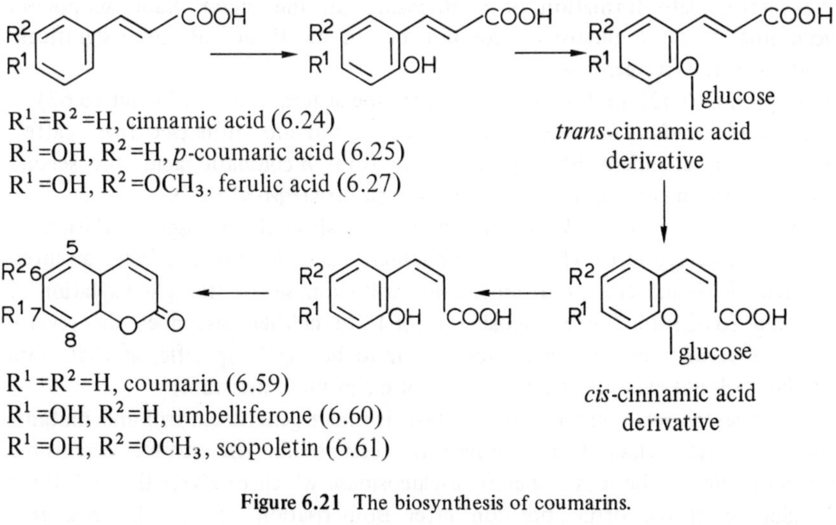

$R^1 = R^2 = H$, cinnamic acid (6.24)
$R^1 = OH$, $R^2 = H$, *p*-coumaric acid (6.25)
$R^1 = OH$, $R^2 = OCH_3$, ferulic acid (6.27)

trans-cinnamic acid
derivative

cis-cinnamic acid
derivative

$R^1 = R^2 = H$, coumarin (6.59)
$R^1 = OH$, $R^2 = H$, umbelliferone (6.60)
$R^1 = OH$, $R^2 = OCH_3$, scopoletin (6.61)

Figure 6.21 The biosynthesis of coumarins.

Most coumarins contain an oxygen atom in the 7 position of coumarin (6.59) and this is introduced by *p*-hydroxylation of cinnamic acid prior to *o*-hydroxyl-ation. Umbelliferone (6.60) is the parent compound of this group of coumarins. Other hydroxyl groups can be introduced and methylated before isomerisation and ring closure, ferulic acid (6.27), for instance, giving scopoletin (6.61). How-ever, in tobacco tissue cultures, ferulic acid is converted to scopoletin without

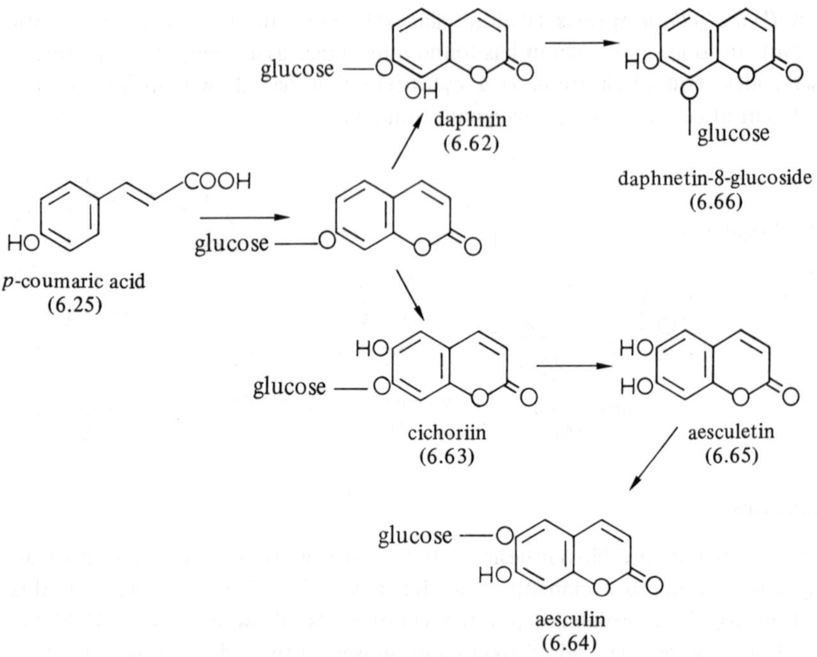

Figure 6.22 The biosynthesis of daphnin and cichoriin.

the intermediate formation of a glucoside. In the intact plant, scopoletin accumulates as the glucoside, scopolin, but in the tissue cultures a significant amount of the free genin was detected.

Daphnin (6.62) in *Daphne odora* (Thymelaeaceae) and cichoriin (6.63) in chicory (*Cichorium intybus,* Compositae) are not biosynthesised from caffeic acid (6.26), as might be expected, but from *p*-coumaric acid, the second hydroxyl group being introduced after ring closure (*Figure 6.22*).

An enzyme isolated from chicory will catalyse the transglucosylation of cichoriin to aesculin (6.64), the free coumarin, aesculetin (6.65), being an intermediate. A similar enzyme from *Daphne* will catalyse the transglucosylation of daphnin (6.62) to the 8-glucoside (6.66). In neither case does the reverse reaction occur, and these enzymes appear to be highly specific, as that from *Daphne* will not catalyse the conversion of cichoriin to aesculin.

In sweet clover, coumarin production is under genetic control and mutants have been bred which do not contain this compound. The enzyme responsible for coumarin synthesis is a specific β-glucosidase which catalyses the hydrolysis of glucose in the ortho position after isomerisation of the cinnamic acid derivative, thus allowing spontaneous ring closure. 'Bound' coumarin occurs in sweet clover as melilotic acid β-glucoside (6.67). In mature plants, this glucoside is biosynthesised by reduction of coumarin (6.59), ring cleavage and glucosylation (*Figure 6.23*). In young shoots, however, melilotic acid glucoside can be formed directly from *o*-coumaric acid glucoside (6.68) (*Figure 6.23*), a process which bypasses the genetically controlled β-glucosidase.

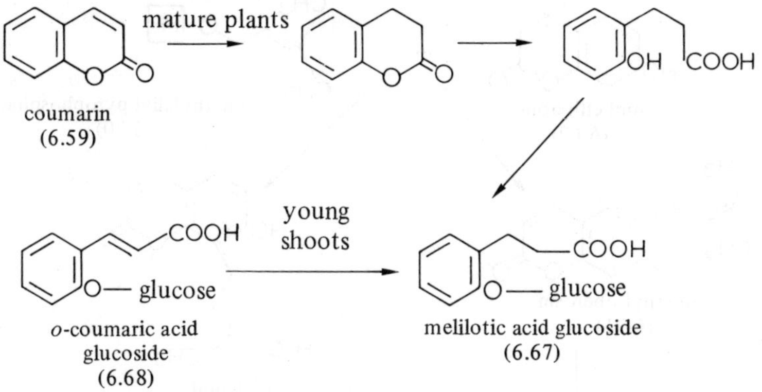

Figure 6.23 The biosynthesis of melilotic acid glucoside in sweet clover.

The presence of coumarin and melilotic acid glucoside in clovers and grasses is of concern to farmers, as fermentation of these plants in hay can result in the production of dicoumarol (3,3'-methylene-*bis*-4-hydroxycoumarin (6.69)), a blood anticoagulant which causes fatal internal haemorrhages in cattle. A mould, *Aspergillus fumigatus*, has been isolated from spoilt hay which will convert melilotic acid glucoside to dicoumarol as shown in *Figure 6.24*

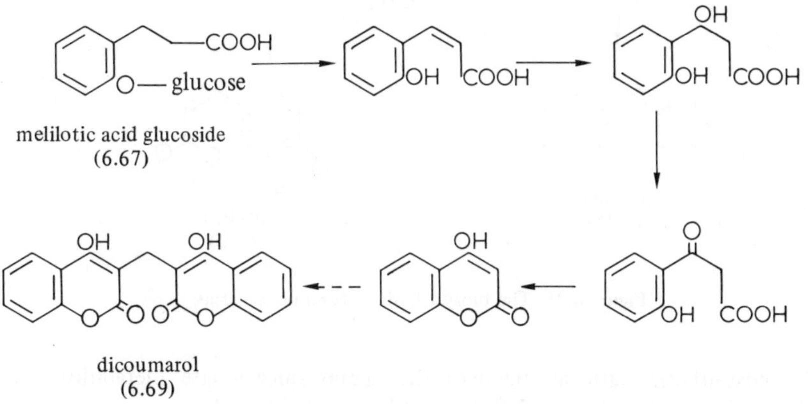

Figure 6.24 The production of dicoumarol in spoilt hay.

Dicoumarol and synthetic analogues are used therapeutically as anticoagulants. They act by inhibiting the synthesis of prothrombin and some of the thrombo-plastin components, proteins concerned in the blood-clotting mechanisms. Vitamin K (Chapter 7) promotes the clotting of blood and it seems that dicoumarol, an analogue of the vitamin, acts as a competitive inhibitor of the enzymes concerned in the synthesis of the blood-clotting proteins.

Furanocoumarins

As the furanocoumarins are closely related to the coumarins, they are con-sidered here, although they more properly belong to the following chapter, as

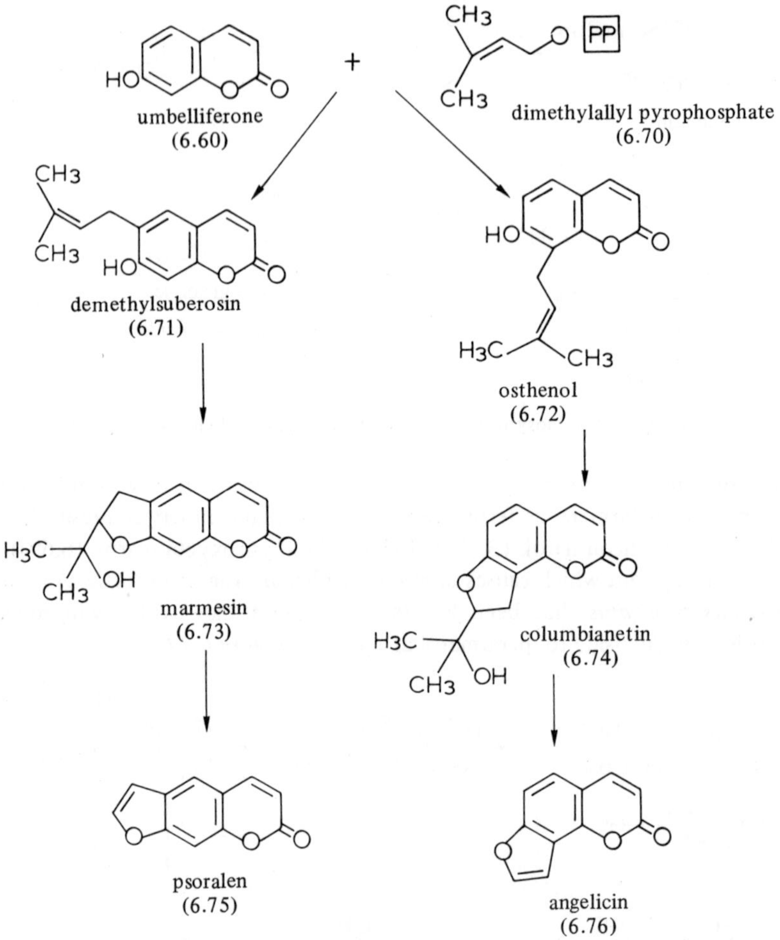

Figure 6.25 The biosynthesis of furanocoumarins.

their biosynthetic pathway involves the acetate-mevalonate metabolite, di-methylallyl pyrophosphate (6.70). Umbelliferone (6.60) is the precursor of most furanocoumarins, and an enzyme has been isolated from rue (*Ruta graveolens*, Rutaceae) which catalyses the reaction between this coumarin and dimethylallyl pyrophosphate giving demethylsuberosin (6.71), the precursor of the linear furanocoumarins (*Figure 6.25*). Osthenol (6.72), the precursor of the angular furanocoumarins, is probably biosynthesised by a similar reaction. Ring closure of the dimethylallyl derivatives gives marmesin (6.73) and columbianetin (6.74), respectively, subsequent degradation of the side chain resulting in psoralen (6.75) from *Psoralea corylifolia* (Leguminosae) and angelicin (6.76) from angelica (*Angelica archangelica*, Umbelliferae) (*Figure 6.25*).

Furanocoumarins are the active ingredients of some fish poisons, while the dimethylallylcoumarin derivative, mammein (6.77) from mammy apple (*Mammea americana*, Guttiferae), is insecticidal. Psoralen and its derivatives show photodynamic activity, and they have been taken orally to promote suntanning of the skin.

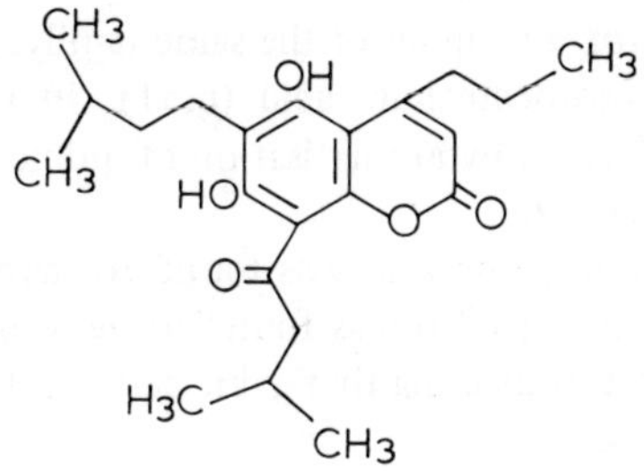

mammein
(6.77)

The Hydrolysable Tannins

Tannins are compounds with an astringent taste which have the ability to tan leather. During the tanning process, proteins in leather are bound to tannin in such a way that they are no longer susceptible to fungal or bacterial attack. The astringent taste of tannins is due to lower oligomers reacting with proteins in the saliva, thus reducing its lubricating properties.

Tannins are usually divided into two categories, namely the hydrolysable and the condensed tannins. The hydrolysable tannins can be hydrolysed with hot, dilute acid, whereas the condensed tannins cannot. The condensed tannins are flavonoid derivatives and they are discussed in Chapter 7.

The most common of the hydrolysable tannins are esters of gallic acid (6.52) (*Figure 6.1*) and ellagic acid (6.78) with sugars. It is usual to find all the hydroxyl groups of the sugar esterified with a phenolic acid and extra acid molecules linked together by depside linkages. Chinese gallotannin from the galls of the sumac plant (*Rhus semialata*, Anacardiaceae) contains only gallic acid and glucose, and a probable structure is shown in *Figure 6.26*. Commercial tannic acid is a mixture of gallic acid with various galloyl esters of glucose.

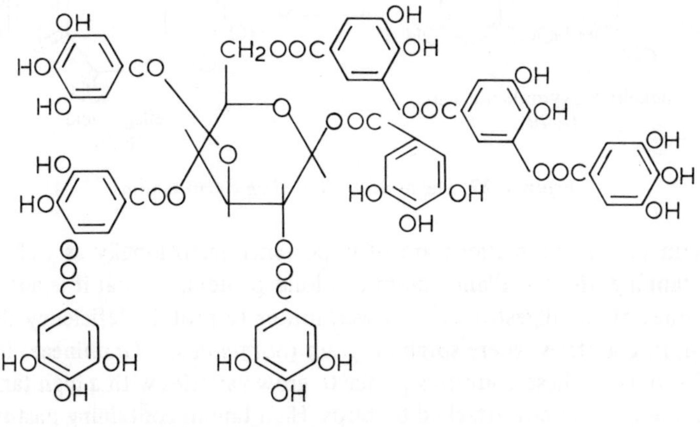

Figure 6.26 A probable structure for Chinese gallotannin.

Gallic acid can be biosynthesised in plants by two distinct pathways involving shikimic acid metabolites. In *Geranium pyrenaicum* (Geraniaceae), this phenolic acid was shown to be derived directly from 3-dehydroshikimic acid (6.6), while in *Pelargonium hortorum,* a member of the same family, gallic acid was obtained by hydroxylation of protocatechuic acid (6.51). In various microorganisms, however, gallic acid is formed by aromatisation of quinic acid (6.7). These results are summarised in *Figure 6.1*.

The second pathway to gallic acid was found to take place in *Rhus typhina.* In this plant, phenylalanine (6.20) was found to be a much better precursor of the acid than was glucose, indicating that gallic acid must be biosynthesised from cinnamic acid derivatives.

Other important tannins include the ellagic acid (6.78) derivatives, one of the simplest being corilagen (*Figure 6.27*) from divi-divi (*Caesalpinia coriaria,* Leguminosae). In this ellagitannin, glucose is esterified with gallic acid and hexahydroxydiphenic acid (6.79). On hydrolysis of the tannin, the diphenic acid lactonises to ellagic acid. It is probable that hexahydroxydiphenic acid is derived from gallic acid and that dimerisation takes place after tannin formation.

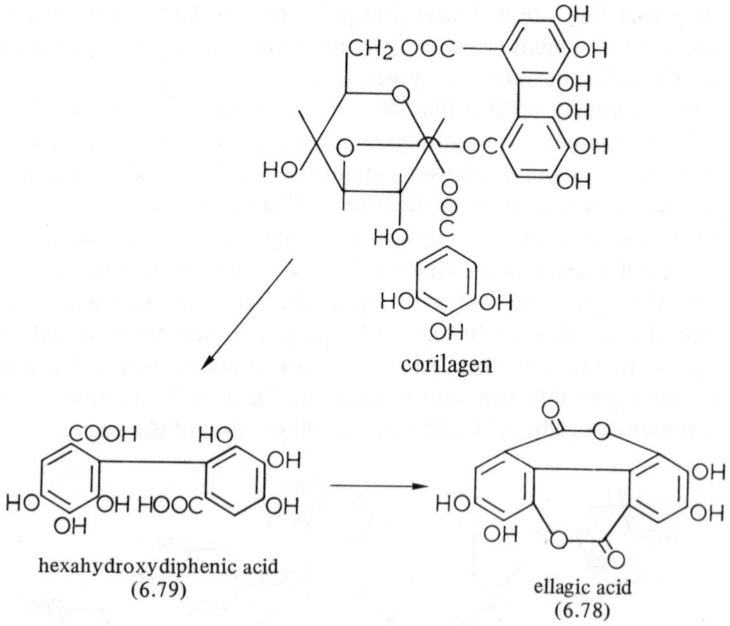

corilagen

hexahydroxydiphenic acid
(6.79)

ellagic acid
(6.78)

Figure 6.27 The hydrolysis of ellagic acid tannins.

Protein–tannin interactions are of importance nutritionally as well as in the leather-tanning process. Plant tannin can bind protein, so that it is not attacked by enzymes of the digestive tract. Kwashiorkor (a protein-deficiency disease) is common in countries where sorghum grain (*Sorghum* sp., Gramineae) is a staple food. Farmers in these countries prefer to grow varieties with a high tannin content, as the grain is not attacked by birds. High tannin-containing pasture plants are also of concern in animal nutrition.

p-Aminobenzoic Acid

p-Aminobenzoic acid (6.80) is an important constituent of tetrahydrofolate or coenzyme F, a complex compound shown in *Figure 6.28*.

Folic acid was first isolated from green leaves (foliage) and was shown to be essential to the nutrition of monkeys (vitamin M) and chickens (vitamin Bc). Coenzyme F, now known to occur in all living organisms, is involved in the transfer of one-carbon groups such as formyl, which are attached to N^5 and N^{10} (see also Chapter 10).

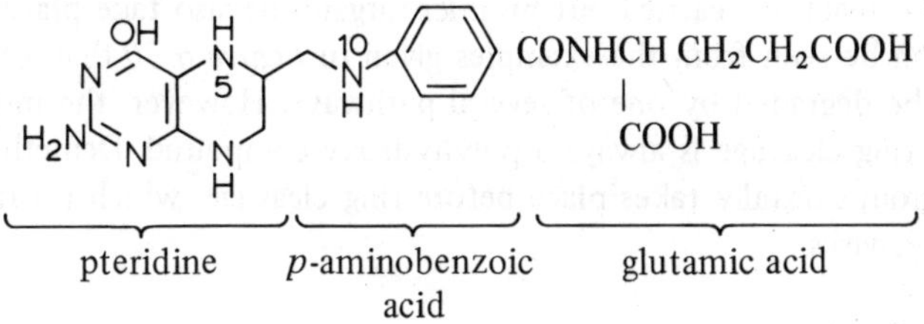

Figure 6.28 Coenzyme F.

It has recently been shown that p-aminobenzoic acid is biosynthesised directly from chorismic acid (6.9) (*Figure 6.29*). Nitrogen is donated by glutamine but no intermediates have been isolated.

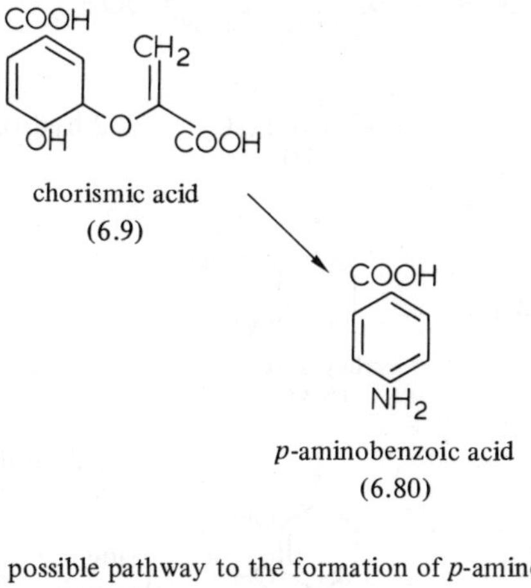

chorismic acid
(6.9)

p-aminobenzoic acid
(6.80)

Figure 6.29 A possible pathway to the formation of p-aminobenzoic acid.

p-Aminobenzoic acid is a growth factor for certain bacteria. The success of the sulphonamide drugs in treating infections such as those caused by the streptococci bacteria is due to competitive inhibition of this growth factor.

The Catabolism of Shikimic Acid Pathway Metabolites

It is well known that microorganisms, particularly soil fungi and bacteria, can utilise phenolic compounds as a source of energy and carbon. In these organisms, shikimic acid pathway metabolites are first oxidised to polyhydroxyphenols and then degraded to aliphatic acids by ring cleavage. Such catabolism of higher plant and animal phenolics by lower organisms is an important part of the carbon cycle.

It is only recently that similar degradations have been shown to take place in higher plants. Previously, despite the varying concentrations and rapid turnover rates of phenols in plants, evidence of catabolism was lacking. This had led to the conclusion that metabolically active phenols were converted either to lignins and tannins, or were detoxified by glucosylation and the glucosides stored in the vacuoles. However, sufficient evidence has now been accumulated to show that the catabolic reactions carried out by microorganisms also take place in higher plants. It can be seen from the examples given in *Figure 6.30* that a given compound can be degraded by one of several pathways. However, the intermediate undergoing ring cleavage is always a polyhydroxy compound. Demethylation of methoxyl groups usually takes place before ring cleavage, which is catalysed by various oxygenases.

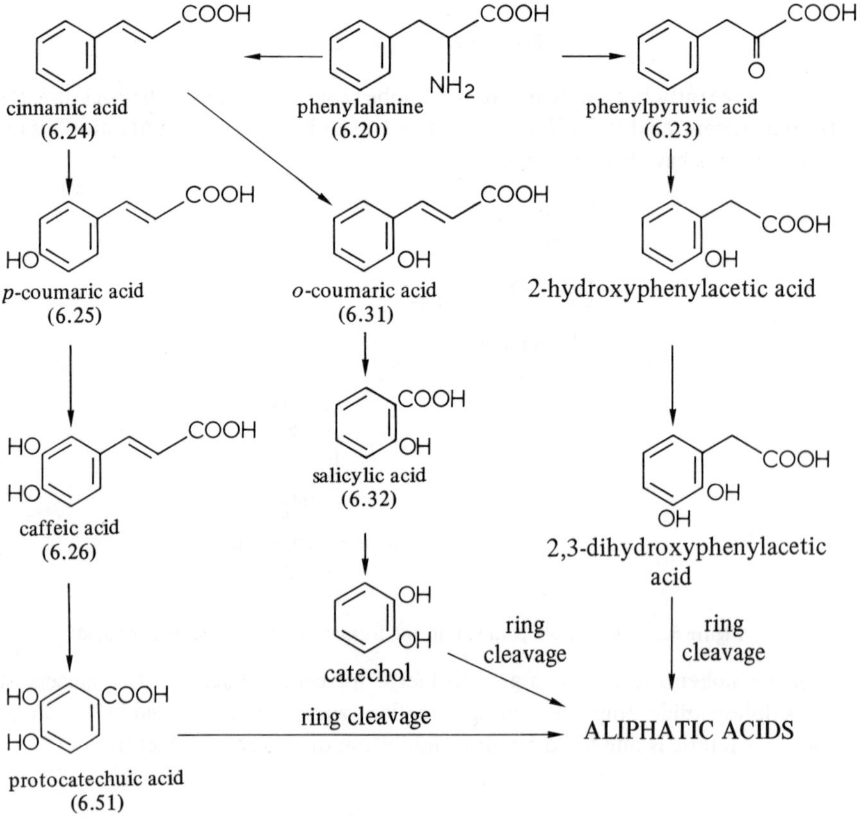

Figure 6.30 The catabolism of some shikimic acid pathway metabolites.

The Function of Phenols in Higher Plants

Phenolic compounds produced by higher plants have a definite effect on the growth both of the plants themselves and of invaders such as fungi or viruses. A number of such compounds including 4-hydroxybenzoic acid, salicylic acid, p-coumaric acid, gallic acid, coumarin and scopoletin are plant growth inhibitors at high concentrations. Such compounds are synthesised in the green tissue and accumulate in woody tissue during dormancy. They depress seed germination, root formation and bud opening and the sprouting of potatoes. It has been found that the concentration of phenolic compounds is much reduced in the relevant tissues immediately before seed germination or bud opening. These growth inhibitory effects appear to be due to interference with the plant growth hormones, particularly indolylacetic acid.

The inhibition of growth by phenolic compounds is utilised by higher plants as a protection against invasion by fungi or viruses. It has been suggested that phenolic compounds normally produced by a plant confer natural immunity against certain fungal and virus diseases. Such compounds accumulate in cells adjacent to those infected by disease.

There are many examples of phenolic compounds inhibiting the growth or spore germination of particular fungi. The lignan, hydroxymatairesinol, inhibits the growth of the fungus, *Fomes annosus,* which infects many trees, while 3,4-dihydroxybenzaldehyde is produced in the outer skin of green Cavendish bananas (*Musa,* Musaceae) as a defence against *Gloeosporium musarum,* the fungus causing ripe fruit rot. Coumarin and its derivatives inhibit the growth of many fungi, while tannin protects against both fungal and virus attack. In the latter case, the virus is prevented from multiplying by protein–tannin interactions.

Some plants produce phytoalexins in response to invasion. These compounds are not normally found in the plant but are synthesised as a direct result of infection. Examples are 6-methoxymellein (6.81) produced by carrots, which is biosynthesised from acetate, and orchinol (6.82) produced by orchids (Orchidaceae). Other phytoalexins are flavonoid derivatives and will be discussed in the following chapter.

6-methoxymellein
(6.81)

orchinol
(6.82)

Various phenolic compounds produced by plants, such as p-coumaric acid, ferulic acid, salicylic acid and 4-hydroxybenzoic acid are allelopaths, which prevent the growth of other plants. These chemicals are not toxic to the plants producing them as they are generally in the form of inactive glucosides. These are washed into the soil by rain where they are hydrolysed and oxidised by soil bacteria. Quinones are cell toxins which act by combining with protein.

The phenolic compounds discussed in this chapter have little effect on higher animals. They are mostly excreted as conjugates with sulphuric or glucuronic acids.

Simple Phenols, Phenolic Acids and Coumarins in Chemosystematics

The simple phenols are uncommon as plant constituents and their occurrence appears to have little systematic relevance. Only in the Rutaceae and Liliaceae families do these compounds occur with any frequency. Hydroquinone, or its glucoside, arbutin, is the most common of the simple phenols, occurring particularly in the Ericaceae, Rosaceae, Saxifragaceae, Proteaceae and Compositae families. It is a useful taxonomic marker in the Rosaceae as it occurs only in the closely related pear (*Pyrus*) and *Docynia* genera.

The hydroxybenzoic acids are widely distributed in the angiosperms, but members of the Primulaceae and Liliaceae families contain compounds with unusual hydroxylation patterns. 2-Hydroxy-4-methoxy-, 2-hydroxy-5-methoxy- and 3,5-dihydroxybenzoic acid have been isolated from the cowslip (*Primula veris*, Primulaceae), while 2-hydroxy-6-methoxybenzoic acid occurs in the Turk's Cap lily (*Gloriosa superba*) and the autumn crocus (*Colchicum autumnale*), both members of the Liliaceae family. Salicylic acid and the related *o*-pyrocatechuic acid (2,3-dihydroxybenzoic acid) are characteristic of the Ericaceae family, while salicyl alcohol appears to be restricted to willows (*Salix* spp., Salicaceae) and some members of the Rosaceae, where it occurs as the glucoside, salicin.

Gallic acid is not widespread despite its occurrence in the hydrolysable tannins. It has been suggested that the biosynthesis of gallic acid (from 3-dehydroshikimic acid or quinic acid) is a primitive feature which has been irreversibly lost by many of the evolutionarily advanced plants. However, gallic acid and the gallotannins are present in most families belonging to the order Rosales. Ellagic acid, the dimer of gallic acid, is infrequently found in the Dicotyledones, but is absent from the Monocotyledones, Gymnospermae and ferns. It is a useful taxonomic marker in the subfamily Rosoideae of the Rosaceae. Rosoideae contains seven tribes and of these only Kerrieae does not contain ellagic acid. As none of the remaining four subfamilies of Rosaceae contain this compound, it would appear on chemosystematic grounds that Kerrieae should be removed from Rosoideae.

4-Hydroxybenzoic and 4-hydroxyphenylacetic acids have been detected in eight genera of the Saxifragaceae family, but in *Astilbe* these compounds are replaced by 2-hydroxyphenylacetic acid. This acid is thus a useful taxonomic marker for the *Astilbe* genus.

The four cinnamic acid derivatives, *p*-coumaric, caffeic, ferulic and sinapic acids, are widespread, although their concentrations vary considerably. Sinapyl alcohol is a major constituent of hardwood lignin, but is absent from, or only a minor constituent of, softwood lignin. Derivatives of *p*-coumaric acid are abundant in the Monocotyledones, *p*-coumaryl alcohol occurring in the lignin of members of the Gramineae, and *p*-methoxycinnamic acid occurring in some members of the Liliaceae family.

Coumarin is a common volatile constituent contributing to the scent of many plants. The hydroxylated derivatives, however, have a much more restricted distribution. Umbelliferone occurs widely in the Umbelliferae family, while in the Compositae it is characteristic of the hawkweed genus (*Hieracium*), where it has proved a useful taxonomic marker. Species which have recently been separated from *Hieracium* and grouped as the *Pilosella* genus on morphological and genetic grounds do not contain umbelliferone.

Daphnetin has been detected in five *Daphne* species (Thymelaeaceae), but also occurs widely in the Saxifragaceae family. The Saxifragaceae is one of the few families whose members contain isocoumarin derivatives. The furanocoumarins are particularly characteristic of the Rutaceae family.

Bibliography

The Shikimic Acid Pathway

The Shikimate Pathway, E. Haslam, Butterworths, 1974
'From Glucose to Aromatics', B. Ganem in *Tetrahedron*, Vol. 34, 1978

Biosynthesis of Phenolic Compounds

'Biosynthesis of Phenolic Compounds Derived from Shikimate', in *Biosynthesis*, Specialist Periodical Reports (continuing series, vol. 1, 1972), The Chemical Society
Biochemistry of Phenolic Compounds, J. B. Harborne (ed.), Academic Press, 1964
'Compounds Derived from Shikimic Acid', in *Organic Chemistry of Secondary Plant Metabolism*, T. A. Geissman and D. H. G. Crout, Freeman, Cooper and Co., 1969
'Extension of the Phenylpropanoid Unit', *ibid.*
'Metabolites Derived from Shikimic Acid', in *Secondary Metabolism*, J. Mann, Oxford University Press, 1978
'Biochemistry of Plant Phenolics', in *Recent Advances in Phytochemistry*, vol. 12, J. B. Harborne and C. F. Van Sumere (eds.), Plenum, 1979
'Phenylalanine Ammonia-Lyase and Phenolic Metabolism', in *Recent Advances in Phytochemistry*, vol. 8, V. C. Runeckles and E. E. Conn (eds.), Academic Press, 1974
'Possible Multienzyme Complexes Regulating the Formation of C_6-C_3 Phenolic Compounds and Lignin in Higher Plants, *ibid.*

Lignin

'Lignin', in *Phytochemistry*, vol. 3, L. P. Miller (ed.), Van Nostrand Reinhold Co., 1973
'Lignin', in *Chemistry and Biochemistry of Plant Herbage*, vol. 1, G. W. Butler and R. W. Bailey (eds.), Academic Press, 1973
'Oxidative Coupling of Phenols', section: Lignans and Lignins, in *Organic Chemistry of Secondary Plant Metabolism*, *ibid.*

Ecology of Phenolic Compounds

'The Toxicity of Plant Phenolics', in *Pharmacology of Plant Phenolics*, J. Fairbairn (ed.), Academic Press, 1959

'Biochemical Interactions Between Higher Plants', in *Introduction to Ecological Biochemistry*, J. B. Harborne, Academic Press, 1977

'Phytoalexins', in *Phytochemical Ecology*, J. B. Harborne (ed.), Academic Press, 1972

Chemosystematics

'Coumarin Patterns in the Umbelliferae', in *The Biology and Chemistry of the Umbelliferae*, V. H. Heywood (ed.), Academic Press, 1971

'Flavonoid and Phenylpropanoid Patterns in the Umbelliferae', *ibid.*

'Anthemideae — Chemical Review', section: Coumarins, in *The Biology and Chemistry of the Compositae*, vol. 2, V. H. Heywood *et al.* (eds.), Academic Press, 1977

'Chemistry of Geographical Races', section IV, in *Chemistry in Evolution and Systematics*, T. Swain (ed.), Butterworths, 1973

7 Compounds with a Mixed Biogenesis

Introduction

Aromatic compounds in plants are generally biosynthesised by one of three main routes — the acetate–malonate, acetate–mevalonate or shikimic acid pathways. These pathways have been described in Chapters 4, 5 and 6, respectively. A number of aromatic compounds, however, have a mixed biogenesis, such that they are derived from products of two or more of the main pathways. Such compounds include the flavonoids, xanthones and stilbenes, whose biosynthesis involves both the shikimic acid and acetate–malonate pathways, while many plant quinones are products of the shikimic acid and acetate–mevalonate pathways.

Flavonoids

The flavonoid and isoflavonoid ring structures are of mixed biosynthetic origin (*Figure 7.1*), ring A being derived from three acetate units condensed head to tail, while ring B and the three carbons of the central ring are derived from cinnamic acid. As the acetate units are first converted to malonyl CoA (7.1), both the acetate–malonate (Chapter 3) and the shikimic acid (Chapter 6) pathways contribute to flavonoid biosynthesis.

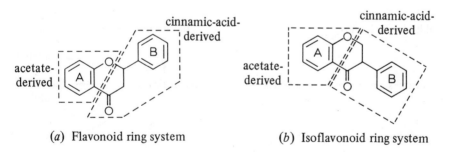

(*a*) Flavonoid ring system   (*b*) Isoflavonoid ring system

Figure 7.1 The derivation of the flavonoid and isoflavonoid ring systems.

The addition of malonyl CoA units to cinnamic acid derivatives is catalysed by the enzyme, flavanone synthase. It has recently been shown in parsley (*Petroselinum crispum*, Umbelliferae) that this enzyme is highly specific and will only utilise p-coumaric acid (7.2) as substrate (*Figure 7.2*). Previously, it had been concluded from the results of feeding isotopically labelled ferulic and sinapic acids to *Petunia hybrida* (Solanaceae) that the hydroxylation pattern of ring B was determined by the cinnamic acid derivative incorporated. However,

Hess's cinnamic acid starter hypothesis cannot apply to parsley, where hydroxyl-ation of the various flavonoids must occur after the flavanone synthase stage.

Chalcones

The first product of the addition reaction is a chalcone derivative (*Figure 7.2*), and it has been suggested that during the addition of all three malonyl units the substrate remains enzyme-bound.

Chalcones can be considered as the precursors of all other classes of flavonoid (*Figure 7.3*). They have been isolated from many plants, particularly members of the Oxalidaceae, Compositae, Scrophulariaceae, Gesneriaceae, Acanthaceae and Liliaceae families, where their bright-yellow colour contributes to flower pigmentation.

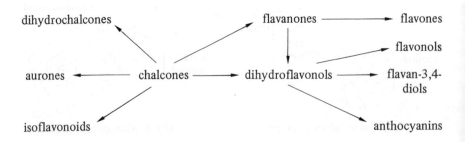

Figure 7.2 The biosynthesis of chalcones.

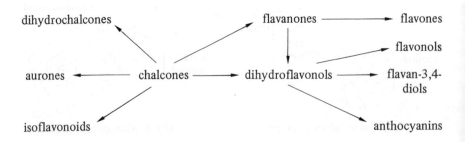

Figure 7.3 The biogenetic relationship of the flavonoids.

Recently a number of colourless α-hydroxychalcones have been isolated from various unrelated plants. It has been suggested that such compounds could be biosynthesised from 4-hydroxyphenylpyruvic acid (7.3) as shown in *Figure 7.4*.

Chalcone, itself, has not been detected in plants, and all naturally occurring compounds are hydroxylated. The hydroxyl groups are often methylated or

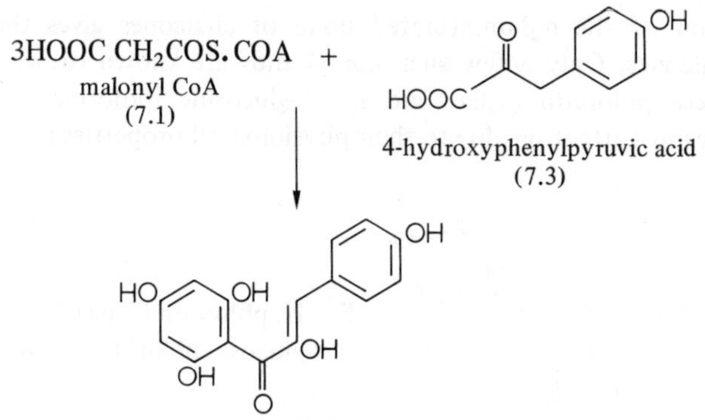

3HOOC CH$_2$COS.COA   +
malonyl CoA
(7.1)

4-hydroxyphenylpyruvic acid
(7.3)

α-hydroxychalcone

Figure 7.4 The possible biosynthesis of α-hydroxychalcones.

glucosylated. One of the most commonly occurring chalcones is butein (7.4a), which is found in the wood or bark of several trees belonging to the Leguminosae family. The 4'-glucoside, coreopsin (7.4b), has been isolated from *Butea monosperma* (Leguminosae) and several members of the Compositae family, including *Coreopsis* species.

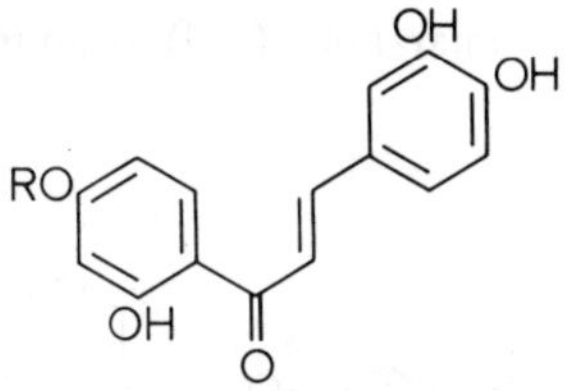

R = H, butein (7.4a)
R = glucose, coreopsin (7.4b)

The usual hydroxylation patterns of chalcones are resorcinol (2',4'-dihydroxy-) or phloroglucinol (2',4',6'-trihydroxy-) substitution on ring A and 4-hydroxy- or 3,4-dihydroxy- substitution on ring B. Recently, the abnormal hydroxylation pattern encountered in echinatin (7.5), isolated from liquorice (*Glycyrrhiza echinata,* Leguminosae), has been shown to occur through transposition of the A and B rings. The normal 4,4',6'-trihydroxychalcone is first formed and this is rearranged to the retrochalcone (*Figure 7.5*).

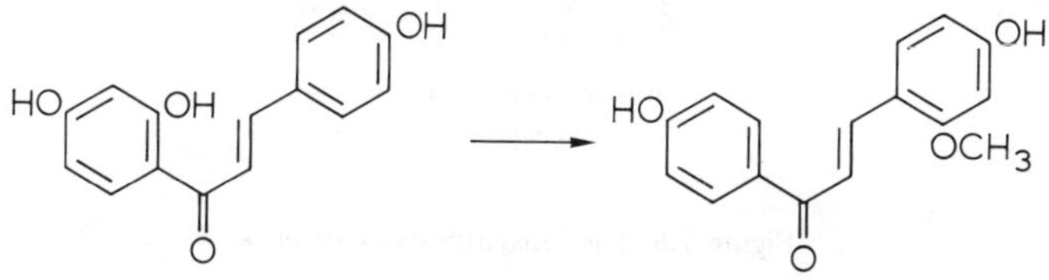

normal chalcone

retrochalcone (echinatin)
(7.5)

Figure 7.5 The biosynthesis of echinatin.

Reduction of the α,β-unsaturated bond of chalcones gives the colourless dihydrochalcones. Only a few such compounds are known to occur naturally, and of these phloretin (7.6a) and its 2'-glucoside, phloridzin (7.6b), have received the most attention, due to their physiological properties (see Chapter 2).

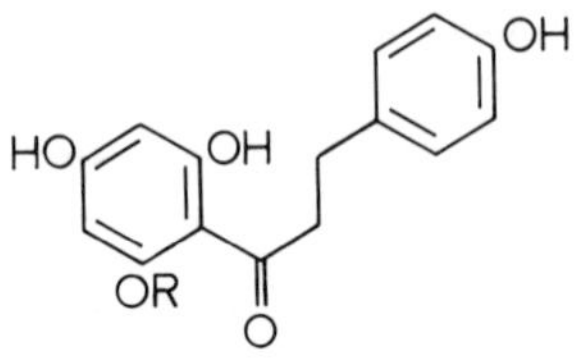

R = H, phloretin (7.6a)
R = glucose, phloridzin (7.6b)

Aurones

The aurones are bright-yellow compounds which are particularly common in members of the Compositae family, where they contribute to the flower colour of many species. Although mainly found in the flowers, aurones have also been detected in the bark, wood or leaves of some plants.

It has been suggested that the aurone ring system (*Figure 7.6*) is biosynthesised from a chalcone precursor through a coumaranone intermediate. Cell free extracts of soybeans (*Glycine max,* Leguminosae) will convert 2',4',4-trihydroxy-chalcone (7.7) to hispidol (7.8) with the coumaranone (7.9) as an intermediate (*Figure 7.6*).

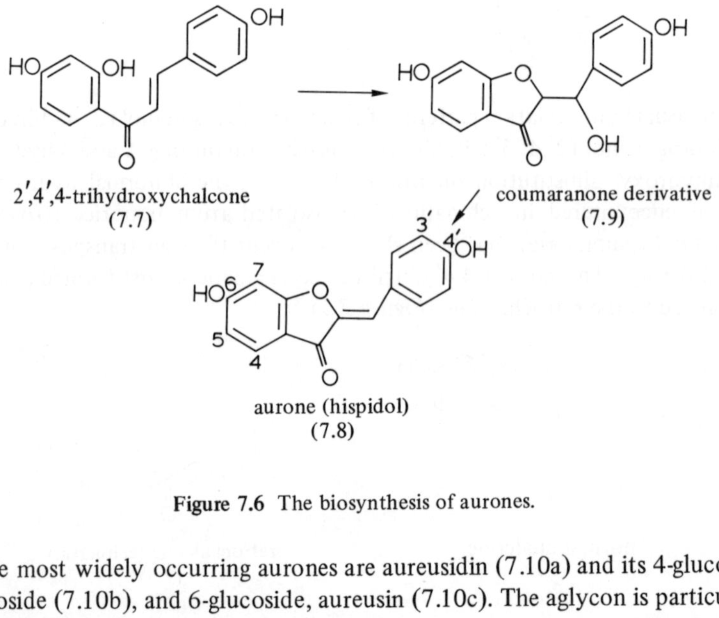

2',4',4-trihydroxychalcone
(7.7)

coumaranone derivative
(7.9)

aurone (hispidol)
(7.8)

Figure 7.6 The biosynthesis of aurones.

The most widely occurring aurones are aureusidin (7.10a) and its 4-glucoside, cernuoside (7.10b), and 6-glucoside, aureusin (7.10c). The aglycon is particularly characteristic of the Cyperaceae family, while the glucosides have been detected in plants belonging to several different families.

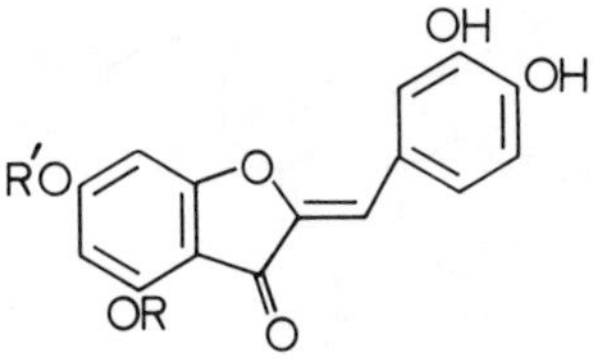

R=R'=H, aureusidin (7.10a)
R'=H, R=glucose, cernuoside (7.10b)
R=H, R'=glucose, aureusin (7.10c)

Flavanones

The isomerisation of chalcones to flavanones (*Figure 7.7*) is catalysed by chalcone–flavanone isomerase, an enzyme which has been isolated from several plants. This enzyme can be separated into varying numbers of isoenzymes which are specific for particular A ring hydroxylation patterns. The chalcone–flavanone isomerase from parsley, for instance, will only catalyse the isomerisation of 2′,4′,6′,4-tetrahydroxychalcone (7.11), the precursor of the flavonoids occurring in this plant. The product of the isomerisation is the flavanone, naringenin (7.12), which has been detected in plants belonging to several different families. Chalcone–flavanone isomerase will only utilise free chalcones as substrates and has no action on the glucosides.

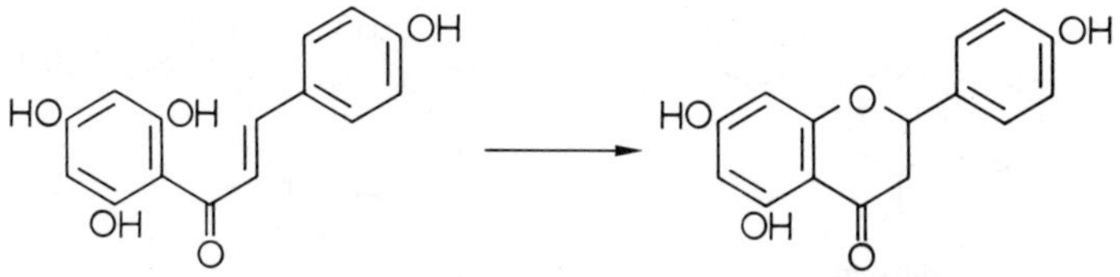

2′,4′,6′, 4-tetrahydroxychalcone
(7.11)

naringenin
(7.12)

Figure 7.7 The reaction catalysed by chalcone–flavanone isomerase.

The flavanones are widespread in higher plants and *O*-glycosides can contain such complex sugars as rutinose and neohesperidose. Naringenin, for instance, occurs in *Citrus* species (Rutaceae) as the 7-neohesperidoside (naringin), 7-rutinoside, 7-neohesperidoside-4′-glucoside and 7-rutinoside-4′-glucoside. The bitterness of certain *Citrus* species, especially grapefruit (*C. paradisi*), is due to flavanone glycosides.

Flavones

The 2,3-dehydrogenation of flavanones giving the flavone ring system (*Figure 7.8*) is catalysed by flavanone oxidase, an enzyme that utilises molecular oxygen. In parsley, this enzyme catalyses the oxidation of naringenin to apigenin (7.13) (*Figure 7.8*) and is specific for flavanones, no oxidation of the corresponding chalcone being observed. As it has recently been shown that naringenin can be produced directly from *p*-coumaric acid and malonyl CoA in parsley, there is some doubt as to the necessity of the chalcone stage in the biosynthesis of flavones.

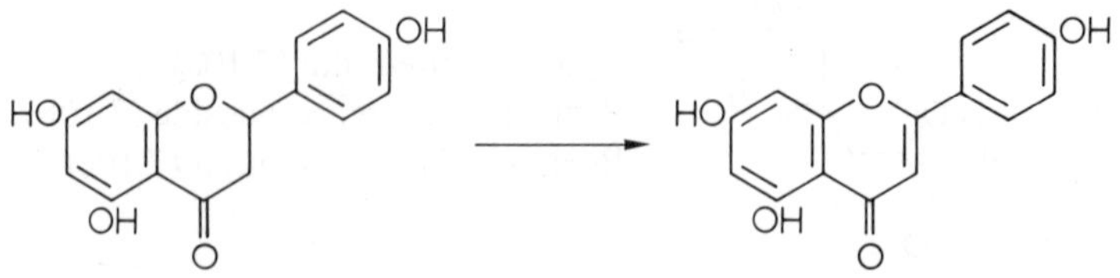

naringenin
(7.12)

apigenin
(7.13)

Figure 7.8 The biosynthesis of flavones.

Substitution patterns of flavones range from flavone (7.14), itself, which has been found in the farina of *Primula* and *Dionysia* species (Primulaceae), to flavones with seven hydroxyl groups, some of which are methylated. Examples of the latter are gardenin A (7.15a), C (7.15b) and E (7.15c) from the gum exudate of *Gardenia lucida* (Rubiaceae). Flavones with a hydroxyl group at position 3 are known as flavonols and are considered separately below.

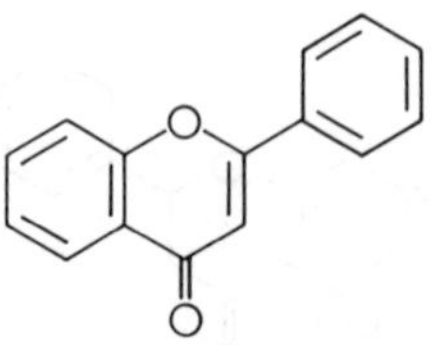

flavone
(7.14)

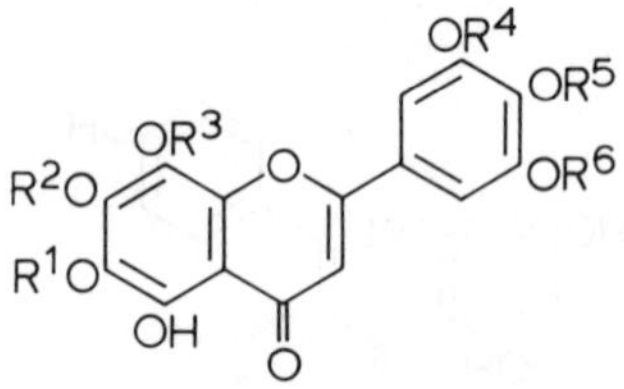

$R^1=R^2=R^3=R^4=R^5=R^6=CH_3$, gardenin A (7.15a)
$R^1=R^2=R^3=R^5=R^6=CH_3$, $R^4=H$, gardenin C (7.15b)
$R^1=R^2=R^3=R^5=CH_3$, $R^4=R^6=H$, gardenin E (7.15c)

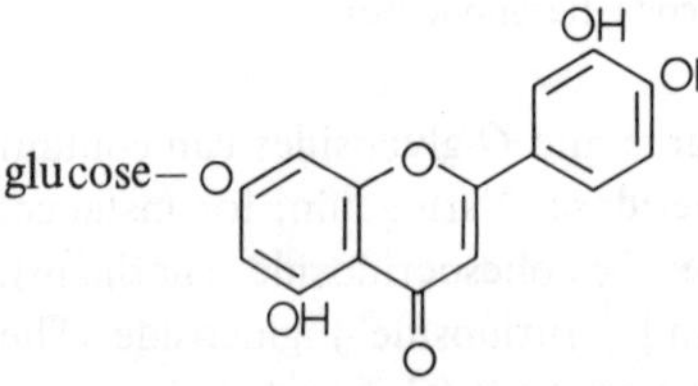

luteolin-7-glucoside
(7.16)

Flavones with free hydroxyl groups usually exist as *O*-glycosides, although over 100 free flavones have been detected in plants. The most commonly occurring glycoside is luteolin-7-glucoside (7.16), but around 20 different glycosides of luteolin from plants have been characterised. The mono-saccharides, D-glucose, D-galactose, D-glucuronic acid, D-xylose, L-rhamnose, L-arabinose and D-apiose, are commonly found linked to flavones (see Chapter 2 for sugar structures), while both disaccharides and trisaccharides occur. Rutinose is the most common disaccharide and flavone-7-neohesperidosides have been isolated from several plants.

A number of flavone *C*-glycosides have also been isolated from plants. These are usually apigenin or luteolin derivatives and the best known is vitexin (7.17), first isolated from the wood of the New Zealand puriri tree *(Vitex lucens,* Verbenaceae) as long ago as 1898. The results of tracer experiments to date indicate that *C*-glycosylation occurs at the chalcone/flavone stage of biosynthesis, but the exact mechanism is not known.

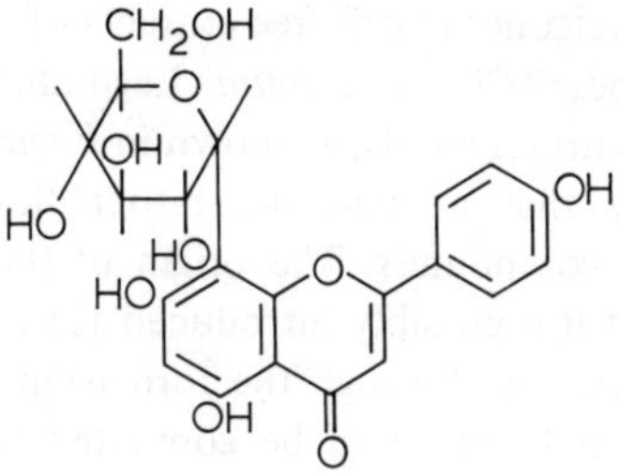

vitexin
(7.17)

The flavones of *Artocarpus* and *Morus* species (Moraceae) are characterised by their isoprenoid substituents. Artocarpin (7.18) from jack fruit (*A. hetero-phyllus*) and mulberrin (7.19) from mulberry (*M. alba*) are probably biosynthesised from the chalcone derivative (7.20) and dimethylallyl pyrophosphate (7.21) as shown in *Figure 7.9*. The postulated intermediate, artocarpesin (7.22), has been isolated from jack fruit.

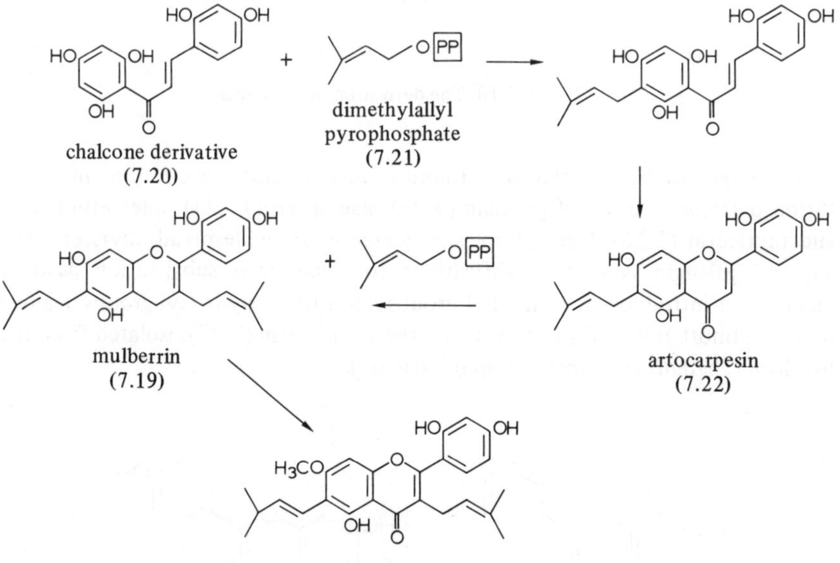

chalcone derivative
(7.20)

dimethylallyl
pyrophosphate
(7.21)

mulberrin
(7.19)

artocarpesin
(7.22)

artocarpin
(7.18)

Figure 7.9 The biosynthesis of mulberrin and artocarpin.

Flavonols

Flavonols are restricted almost totally to higher plants, where they occur in the flowers or are associated with the lignin of wood and leaves. Unchelated flavonols give body to white flowers which would otherwise be almost transparent. Their absorption in the near ultraviolet is probably important, as many insects can 'see' in this range of the spectrum. The flavonoid derivatives have been shown to be biosynthesised from chalcones in cell free extracts of soybeans (*Glycine max*, Leguminosae) and chick peas (*Cicer arietinum*, Leguminosae) through the intermediate formation of dihydroflavonols, as shown in *Figure 7.10*. However, it has been proved conclusively that a chalcone, rather than a flavanone, is the precursor of the dihydro compounds. The origin of the 3-hydroxyl group has yet to be determined, and it is possibly introduced at the chalcone stage or even derived from phenylpyruvic acid through the formation of α-hydroxychalcones (see above). Flavones do not appear to be converted to flavonols, despite the similarity of their structures.

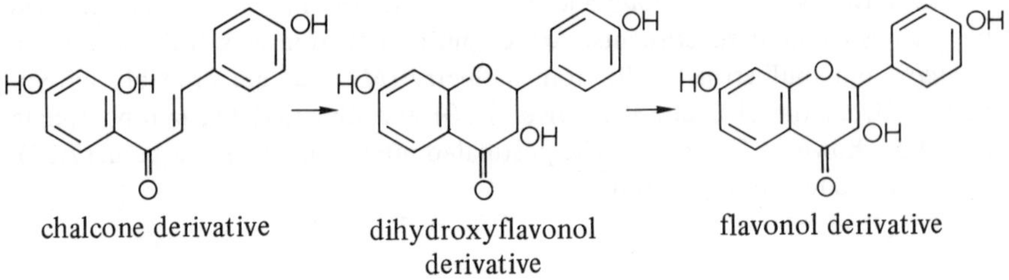

chalcone derivative   dihydroxyflavonol   flavonol derivative
                                     derivative

Figure 7.10 The derivation of flavonols.

The large numbers of flavonols found in higher plants have four main substitution patterns — those of galangin (7.23), kaempferol (7.24), quercetin (7.25) and myricetin (7.26). Kaempferol and quercetin are widespread, myricetin less so, and galangin is rare. Modifications to these basic substitution patterns include the introduction of methyl groups and further hydroxyl groups, leading to such almost fully substituted structures as digicitrin (7.27), isolated from the foxglove (*Digitalis purpurea*, Scrophulariaceae).

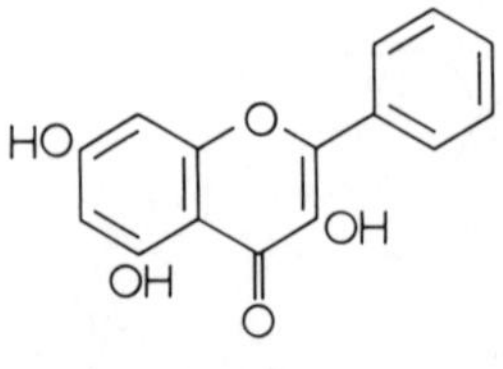

galangin
(7.23)

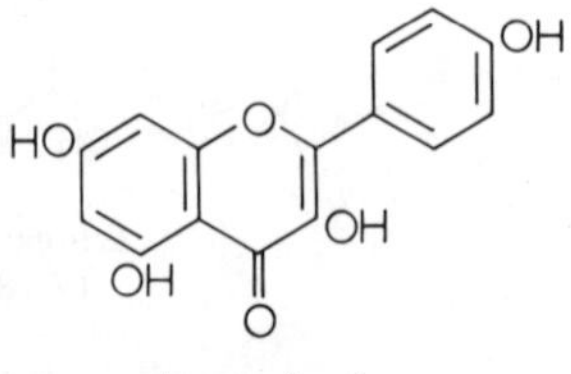

kaempferol
(7.24)

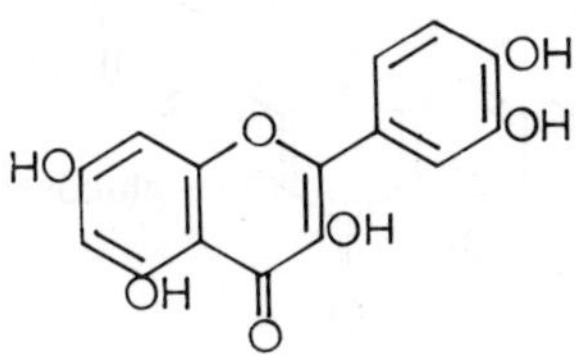

quercetin
(7.25)

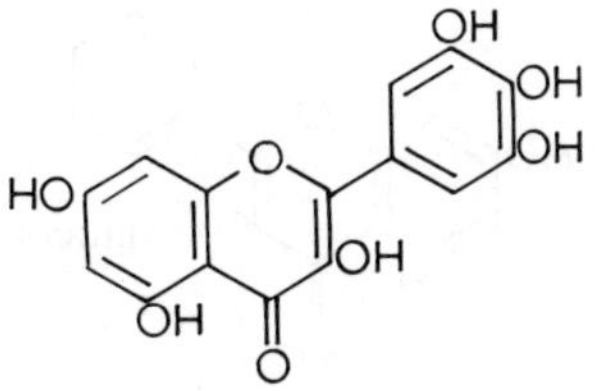

myricetin
(7.26)

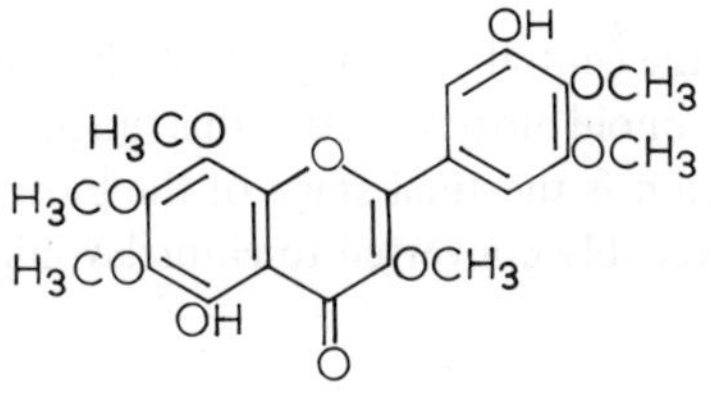

digicitrin
(7.27)

Flavonols with an extra hydroxyl group at C-6 or C-8, such as gossypetin (7.28) from cotton (*Gossypium* spp., Malvaceae) and birdsfoot trefoil (*Lotus corniculatus*, Leguminosae) and quercetagetin (7.29) from some species of Compositae, are bright yellow and contribute to the colour of some flowers.

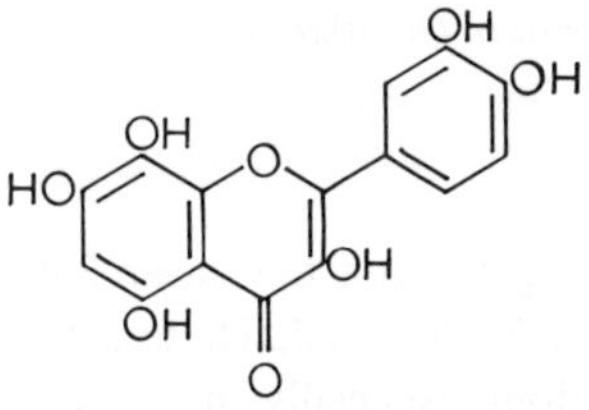

gossypetin
(7.28)

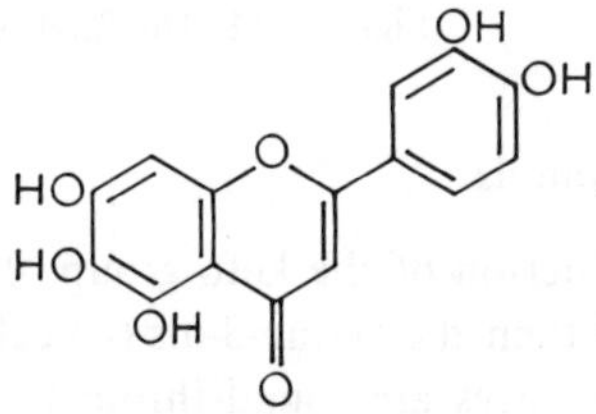

quercetagetin
(7.29)

Hydroxyflavonols are usually found in plants as *O*-glycosides, the most common being the 3-glycosides. The sugars encountered are similar to those found in flavone glycosides. About 60 glycosides of kaempferol and 70 glycosides of quercetin have been characterised, the 3-glucosides, astragalin (7.30) and isoquercitrin (7.31), being widespread. Sugars are rarely attached to the 5- or 4'-hydroxyl groups of kaempferol, but although 3-glycosides of quercetin are the most common, 7-, 4'-, 5- and 3'-glycosides are also encountered.

Only about 12 glycosides of myricetin are known, while the 3-rutinoside from *Datisca cannabina* (Datiscaceae) is the only glycoside of galangin which has been characterised.

One of the few flavonoids found in lower plants, chlorflavonin (7.32) from the fungus, *Aspergillus candidus*, has an unusual 2'-hydroxylation pattern. Chlorflavonin is unique in also containing a chlorine atom. Feeding experiments

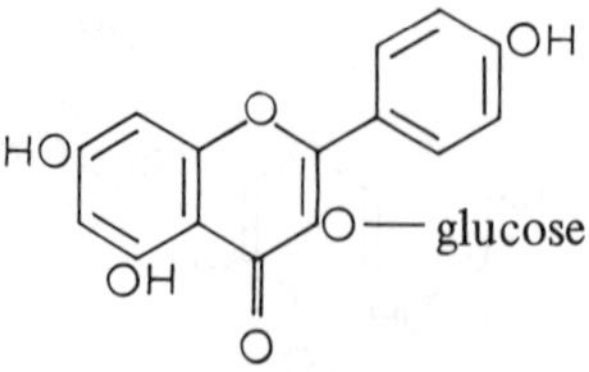

astragalin
(7.30)

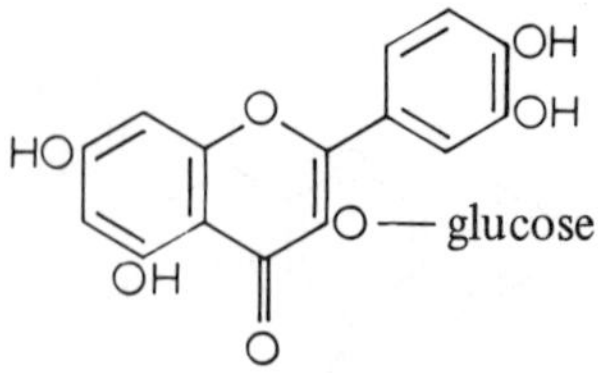

isoquercitrin
(7.31)

have shown ring B and the central C_3 unit to be derived from phenylalanine, and thus the pathway of flavonoid biosynthesis in this fungus appears to follow that of higher plants. Chlorination is the final stage of the biosynthesis as the dechloro-flavonol (7.33) was irreversibly converted to chlorflavonin (*Figure 7.11*).

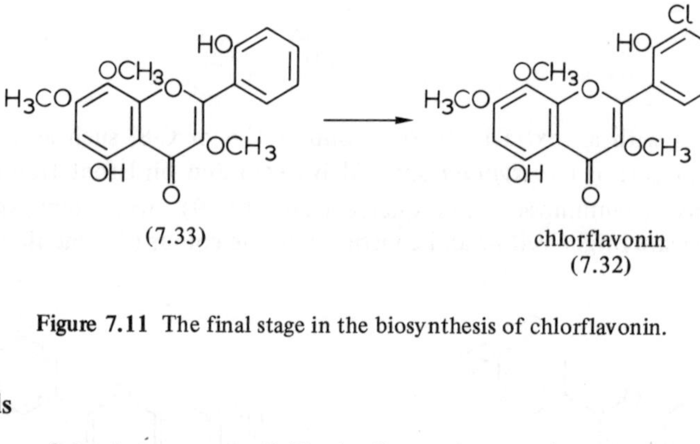

(7.33)

chlorflavonin
(7.32)

Figure 7.11 The final stage in the biosynthesis of chlorflavonin.

Flavanols

Reduction of the keto group of dihydroflavonols gives first the flavan-3,4-diols and then the flavan-3-ols (catechins), as shown in *Figure 7.12*. These colourless substances are found throughout the plant kingdom, especially in the wood of higher plants.

The flavan-3-ols exist as two series of stereoisomers—those compounds in which the 2,3-hydrogen atoms are *trans* being known as catechins, while in the *epi*-catechins these hydrogen atoms are *cis*. (The chiral C-2 and C-3 give rise to (+) and (–) optical isomers, those occurring naturally being (+)catechin and (–)*epi*-catechin).

Flavan-3,4-diols are also known as leucoanthocyanidins, as on treatment with acid they give the corresponding anthocyanidin, although in plants the latter compounds are not biosynthesised from their leuco counterparts.

Dihydrokaempferol (7.34) is incorporated into the flavan-3-ols of both tea (*Camellia sinensis*, Theaceae) and raspberry (*Rubus* sp., Rosaceae), while in raspberry flavan-3,4-diols are also formed (*Figure 7.12*).

Proanthocyanidins

Condensation of catechin or *epi*-catechin with flavan-3,4-diols gives the dimers known as proanthocyanidins (flavolans). Two types of dimer are known, those

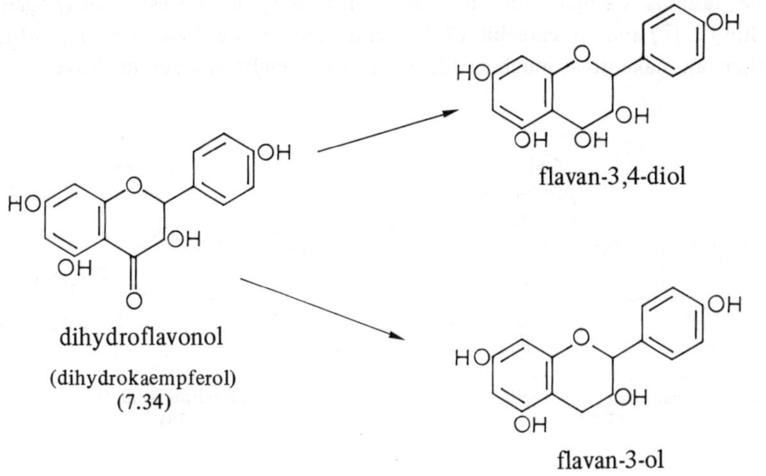

Figure 7.12 The formation of flavan-3-ols in tea and raspberry and flavan-3,4-diols in raspberry.

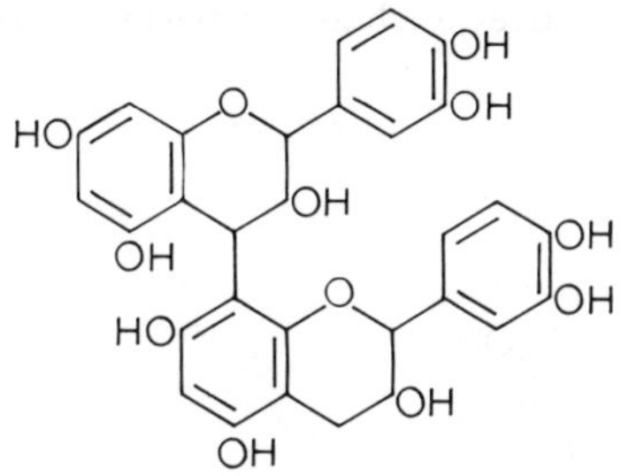

procyanidin B
(7.35)

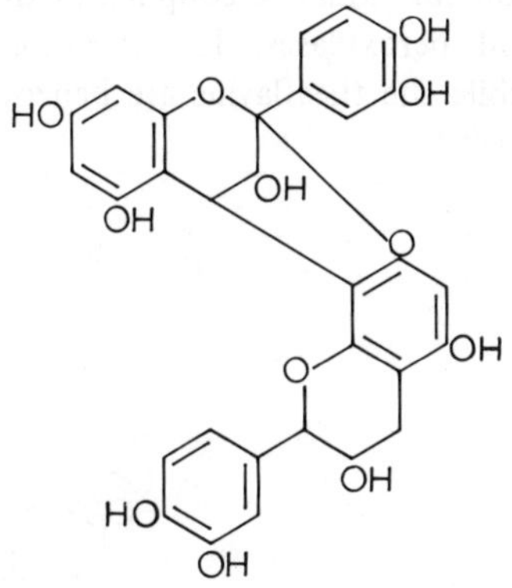

proanthocyanidin A2
(7.36)

such as procyanidin B (7.35), with one linkage between the monomers, and those such as proanthocyanidin A2 (7.36), with two such linkages.

Further polymerisation eventually gives the condensed tannins. The mechanism of polymerisation is not known, but it appears to be non-enzymatic and somewhat similar to that of lignin formation. Like the hydrolysable tannins (Chapter 6), the condensed tannins are astringent and tan leather. It is the condensed tannins which are responsible for the astringency of unripe fruits, as they react with proteins in the saliva.

The Flavan-3-ols and flavan-3,4-diols do not normally form glycosides, but the formation of flavan-3-ol gallate esters is unique amongst the flavonoids. Such esters are characteristic of tea and *Bergenia* spp. (Saxifragaceae), where they appear to replace the more common proanthocyanidins.

The 'tannins' of black tea are derived from the 3-gallate esters of *epi*-gallo-catechin (7.37) and *epi*-catechin (7.38), and their unesterified derivatives which together can make up as much as 30% of the dry weight of green tea leaves.

epi-gallocatechin-3-gallate
(7.37)

epi-catechin-3-gallate
(7.38)

The fermentation of green tea converts the four monomers into thearubigins and theaflavins, the colouring matter of black tea. Fermentation involves the random oxidative coupling of the monomers, a process catalysed by phenolases and peroxidases. The thearubigins are partly polymeric proanthocyanidins, while the theaflavins are benzotropolone derivatives, such as theaflavin (7.39), itself.

theaflavin
(7.39)

Anthocyanins

Of all the flavonoids, the anthocyanins are probably the best known, as they are responsible for the colour of blue, mauve and red flowers, fruits and leaves. These pigments are glycosides of the unstable anthocyanidins, compounds based on the flavylium ion (*Figure 7.13*). In general, anthocyanidins are charac-terised by their blue colour in alkaline solution, which is due to the formation of an anhydro base (*Figure 7.13*).

Although over 100 anthocyanins are known, the number of aglycons is small, and the most common anthocyanins are glycosides of pelargonidin (7.40), cyanidin (7.41), peonidin (7.42), delphinidin (7.43), petunidin (7.44) and malvidin (7.45).

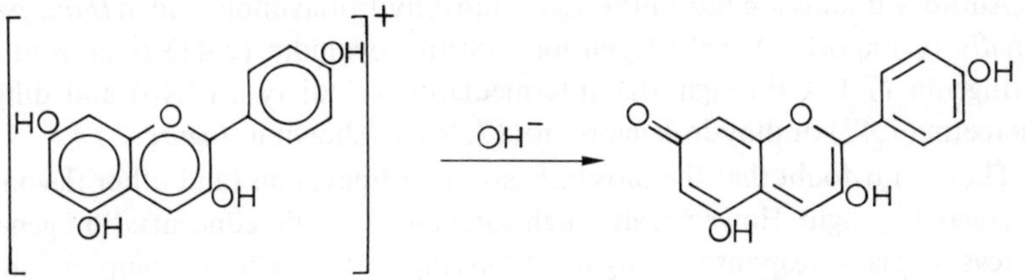

flavylium ion   anhydro base
(red)   (blue)

Figure 7.13 The formation of anthocyanidin anhydro bases.

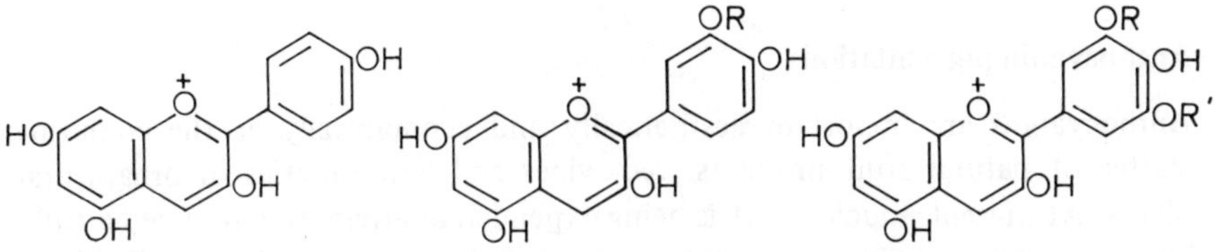

pelargonidin
(7.40)

R = H, cyanidin (7.41)   R = R' = H, delphinidin (7.43)
R = CH$_3$, peonidin   R = H, R' = CH$_3$, petunidin (7.44)
(7.42)   R = R' = CH$_3$, malvidin (7.45)

Glucose is the most common sugar to occur in anthocyanidin glycosides, although almost 40 different types of glycoside have been isolated. Acylation of sugars with p-coumaric, caffeic, ferulic or sinapic acids also occurs.

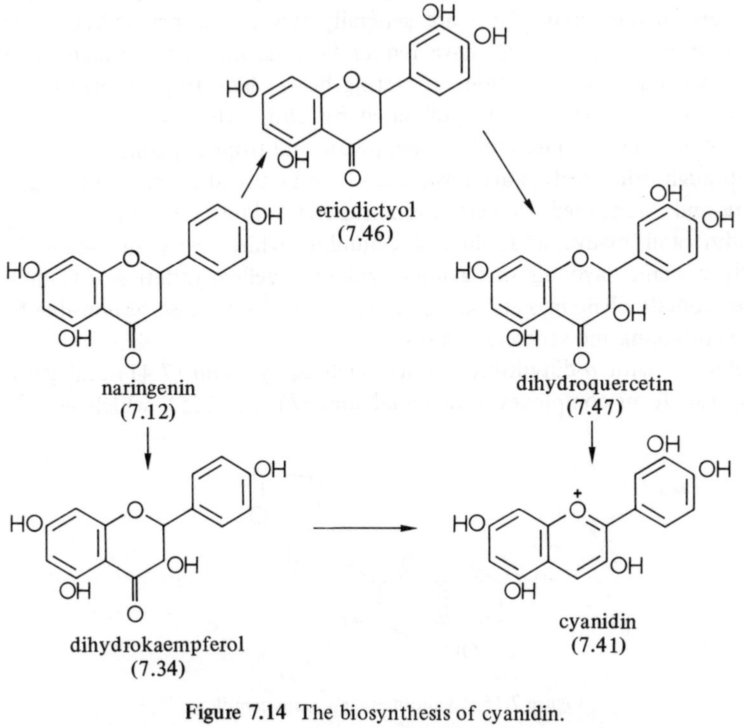

eriodictyol
(7.46)

naringenin
(7.12)

dihydroquercetin
(7.47)

dihydrokaempferol
(7.34)

cyanidin
(7.41)

Figure 7.14 The biosynthesis of cyanidin.

Anthocyanidins are biosynthesised from dihydroflavonols, and in *Haplopappus gracilis* (Compositae) cell suspension cultures cyanidin (7.41) is derived from naringenin (7.12) through the intermediates, eriodictyol (7.46) and dihydroquercetin (7.47) or dihydrokaempferol (7.34), as shown in *Figure 7.14*.

There is no doubt that the biosynthesis of anthocyanins (and other flavonoids) is affected by light. However, although total anthocyanin concentration generally increases, plant response to light of varying wavelengths is complex, and no overall conclusions can be drawn.

Plants with white flowers often contain a gene block which prevents the formation of anthocyanins. The point at which such a block occurs can be deduced by feeding appropriate precursors.

Anthocyanin pigmentation

Anthocyanins are important aesthetically and economically as the colouring matter of natural fruit products, red wines and new varieties of ornamental plants. At present, much effort is being expended in attempting to breed a truly blue rose. Anthocyanin pigmentation of plants is complex and controlled by a number of factors, some of which are discussed below.

The nature of the aglycon of an anthocyanin affects its colour, increasing hydroxylation leading to bluer colours, while methylation increases redness. Thus delphinidin (7.43), with six hydroxyl groups, is the bluest of all the anthocyanidins. Unfortunately, delphinidin is absent from the genus *Rosa*, a fact which considerably reduces the chances of breeding a blue rose. Most flower colour variation is due to mutations in anthocyanidin hydroxylation. Such mutations in temperate plants are generally towards decreased hydroxylation, such that new varieties will have redder flowers. Blue predominates amongst wild flowers as it is the colour attracting bees. Many tropical plants have red flowers, however, as they are pollinated by birds, while white flowers, which attract moths, are common in both temperate and tropical plants.

Although other factors are involved, it can be broadly assumed that all pink, scarlet and orange-red flowers contain pelargonidin; crimson and magenta, cyanidin; and mauve and blue, delphinidin. White, ivory or cream flowers usually contain flavonols or flavones, while the yellow varieties of Compositae species contain chalcones and aurones. Anthocyanins are also responsible for the colours of young and senescent leaves.

Aglycons with *o*-dihydroxy groups such as cyanidin (7.41) and petunidin (7.44) can form complexes with metal ions (*Figure 7.15*), which leads to an

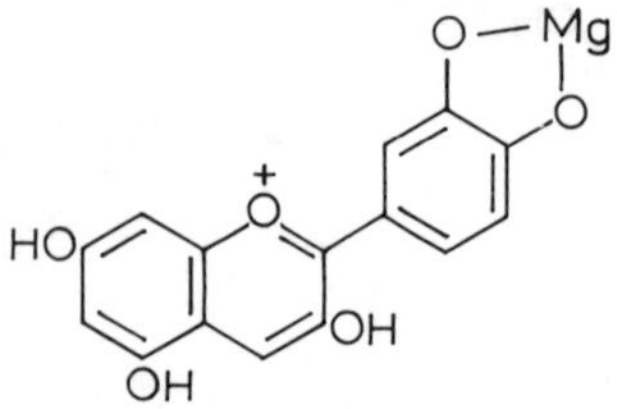

Figure 7.15 Cyanidin–magnesium complex.

increase in blueness. The petals of blue cornflowers (*Centaurea cyanus*, Compositae) contain cyanin complexes with iron and aluminium or magnesium, but mutant varieties containing only pelargonidin (7.40) cannot form such complexes and their petals are red or pink. Cyanidin is present in many wild species of *Rosa*, and thus the only hope of breeding a blue rose is by inducing complex formation or by increasing copigmentation (see below).

It is well known that *Hydrangea* (Hydrangeaceae) flowers can be turned from pink to blue by the application of aluminium salts or iron chelates to the soil. This blueing is due to magnesium, iron or aluminium complexes of delphinidin.

Flowers containing the same anthocyanins are often of different colours, while those containing different anthocyanins can have similar colours. Petal colour, therefore, depends on factors other than the nature of the aglycon, and the most important of these is copigmentation. Copigments, which alter the colour of anthocyanins, include pectin, soluble sugars, flavone *C*-glycosides, flavonol glycosides, the aurone, aureusidin (7.10a), and the xanthone, mangiferin (7.64). The complexing of an anthocyanin with pectin, for example, increases the blueness of petals, while the colour of red wines is due to complexes with phenols and tannins.

Isoflavonoids

Although mainly restricted to the Leguminosae family, the isoflavonoids are important because of their physiological effects on plants and animals. Isoflavonoids differ from flavonoids in that ring B is attached to the α-carbon atom (chalcone numbering) rather than the β (*Figure 7.16*).

Like the flavonoids, the isoflavonoids are biosynthesised from chalcone precursors and ring migration has been postulated as an oxidation of a flavanone intermediate leading to an isoflavone (*Figure 7.16*).

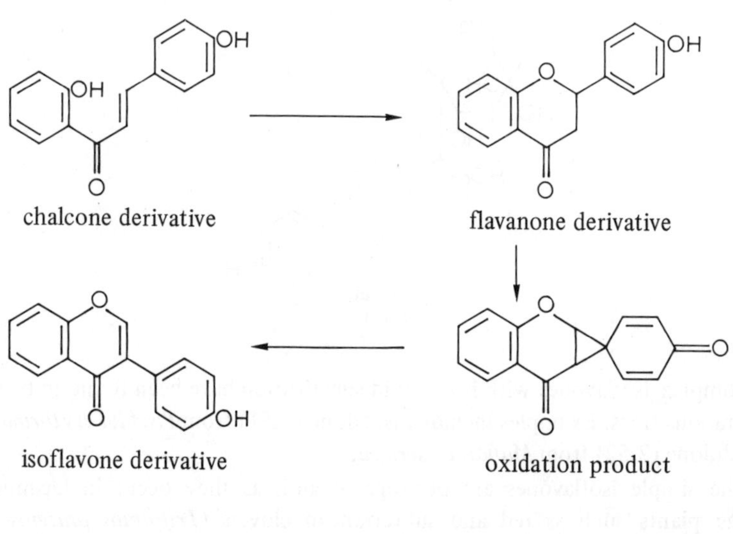

chalcone derivative   flavanone derivative

isoflavone derivative   oxidation product

Figure 7.16 A postulated mechanism for the biosynthesis of isoflavones.

It has been shown in mung bean (*Phaseolus aureus*, Leguminosae) seedlings that the isoflavanone, dihydrodaidzein (7.48), is biosynthesised from 2',4',4-trihydroxychalcone (7.7) with the isoflavone, daidzein (7.49), as an intermediate (*Figure 7.17*). Thus, the isoflavonoid biosynthetic pathway does not run parallel to that of the flavonoids.

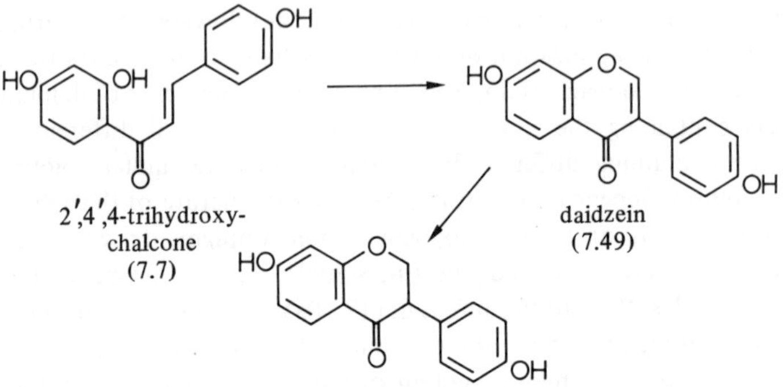

2',4',4-trihydroxy-chalcone
(7.7)

daidzein
(7.49)

dihydrodaidzein
(7.48)

Figure 7.17 The biosynthesis of dihydrodaidzein in mung beans.

The isoflavones are the largest class of isoflavonoids, and about 60 aglycons are known. Fewer glycosides have been characterised than for most classes of flavonoid, but it is probable that this will be remedied in the future. O-glycosylation occurs mainly at the C-7 hydroxyl group, and 7-glucosides and 7-rhamnoglucosides are the most common. A few C-glycosides have been characterised, such as puerarin (daidzein-8-C-glucoside) (7.50), isolated from *Pueraria thunbergiana* (Leguminosae).

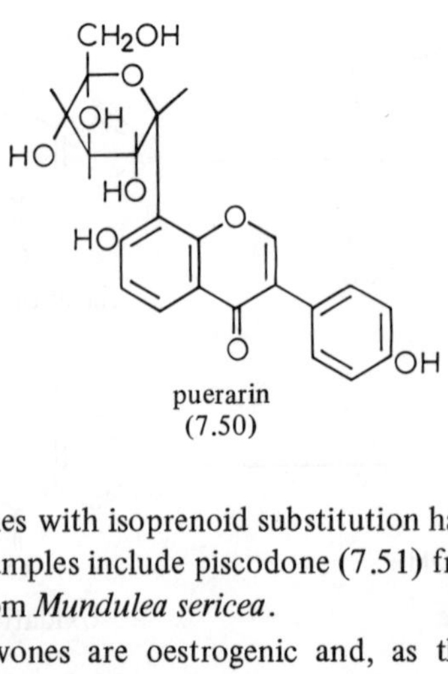

puerarin
(7.50)

Complex isoflavones with isoprenoid substitution have been found in tropical leguminous trees. Examples include piscodone (7.51) from *Piscidia erythrina* and mundulone (7.52) from *Mundulea sericea*.

The simple isoflavones are oestrogenic and, as they occur in leguminous forage plants such as red and subterranean clovers (*Trifolium pratense* and *T. subterraneum*), they are of economic importance. Permanent infertility

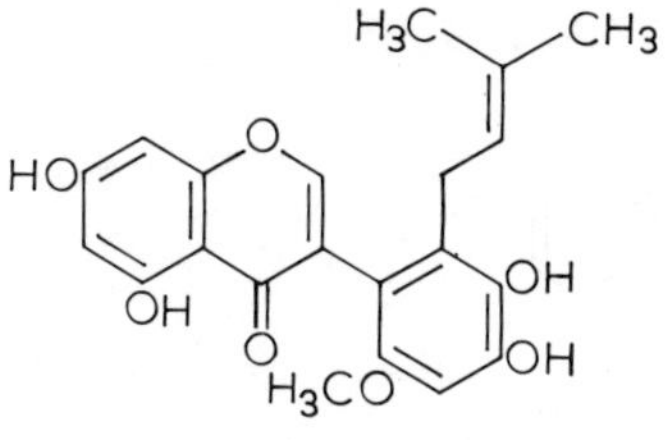

piscodone
(7.51)

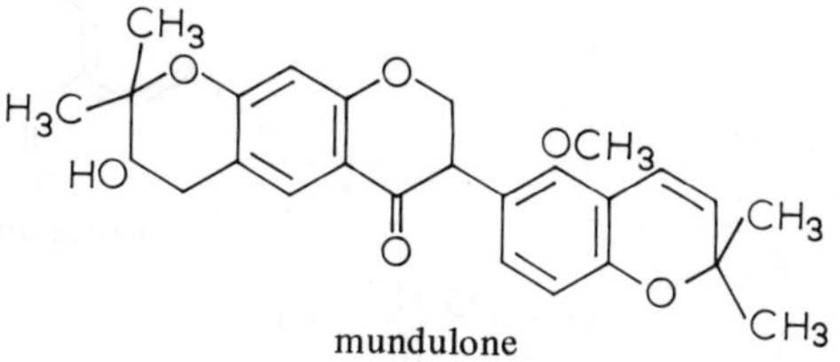

mundulone
(7.52)

occurs if sheep are allowed to graze such plants for longer than six months, while temporary infertility can occur if the animals graze oestrogenic plants during the mating season. The oestrogenic properties of clover appear to be due to formononetin (7.53), which is degraded to the oestrogen, equol (7.54), in the rumen of sheep (*Figure 7.18*).

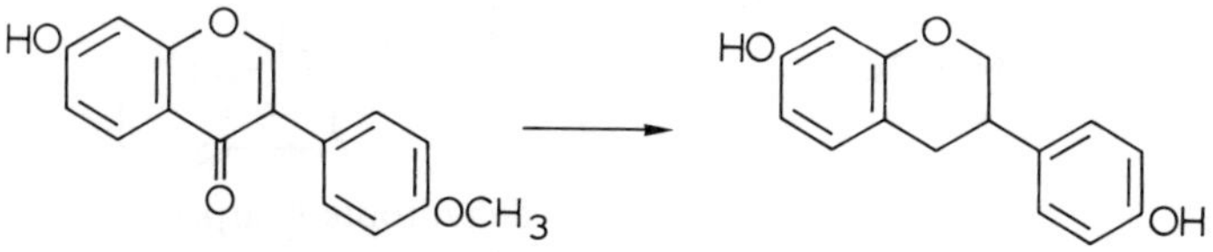

formononetin
(7.53)

equol
(7.54)

Figure 7.18 The degradation of isoflavones in the rumen of sheep.

It has been suggested that compounds such as equol and the coumestans (see below) are oestrogenic because their molecular structure bears some similarity to that of such potent oestrogens as diethylstilboestrol (7.55).

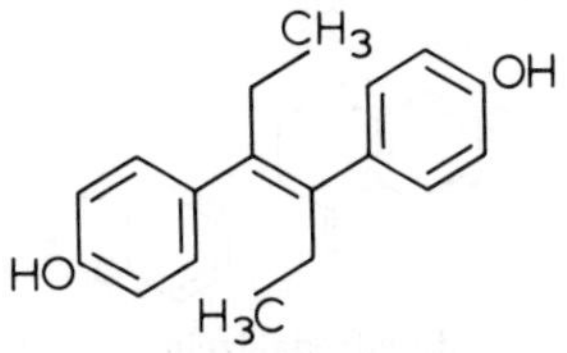

diethylstilboestrol
(7.55)

Isoflavones are probably the precursors of the other classes of isoflavonoids interrelated as shown in *Figure 7.19*. Although evidence for this scheme is sparse, coumestrol (7.56) has been shown to be derived from daidzein (7.49) in mung bean seedlings, as shown in *Figure 7.20*. Coumestrol, which occurs in the leguminous forage plants, white clover (*T. repens*) and alfalfa (*Medicago sativa*), is a much stronger oestrogen than the isoflavones.

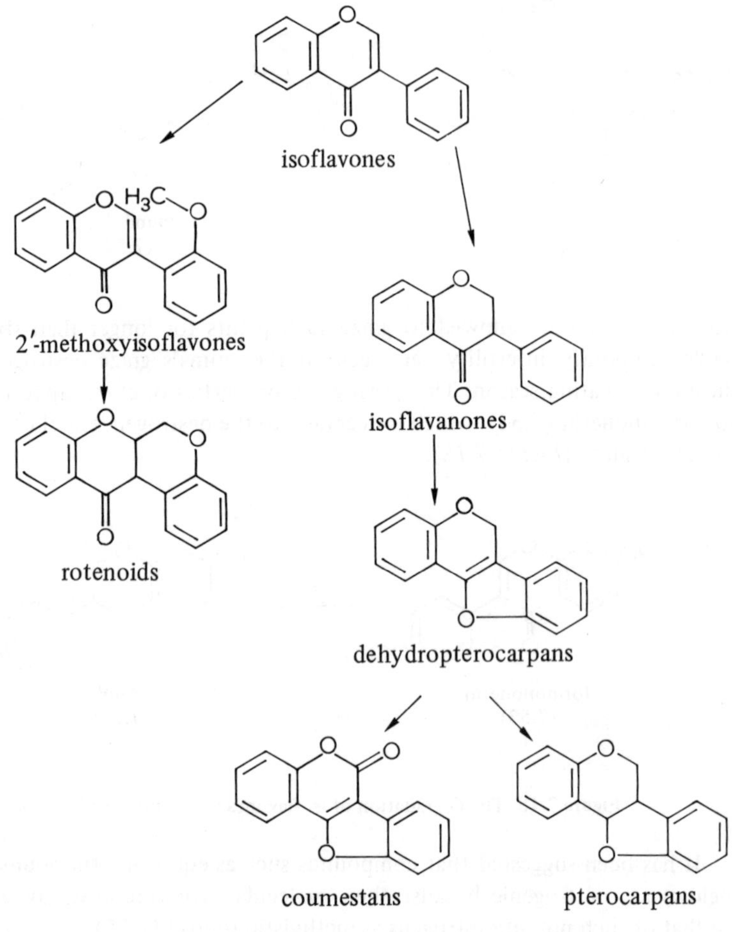

Figure 7.19 The biosynthetic relationship of the isoflavonoids.

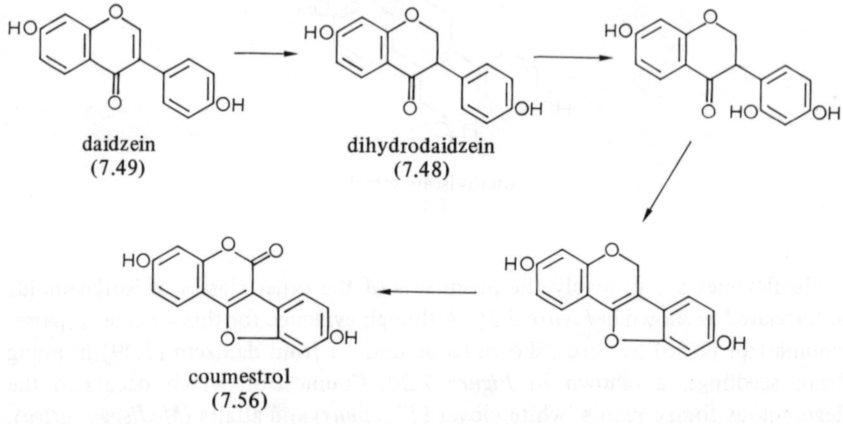

Figure 7.20 The biosynthesis of coumestrol.

The pterocarpans occur in the heartwood of leguminous trees and a number are phytoalexins, compounds which are synthesised as a result of fungal infection (see Chapter 6). Pisatin (7.57) and phaseollin (7.58) are phytoalexins produced by peas (*Pisum sativum*) and French beans (*Phaseolus vulgaris*), respectively.

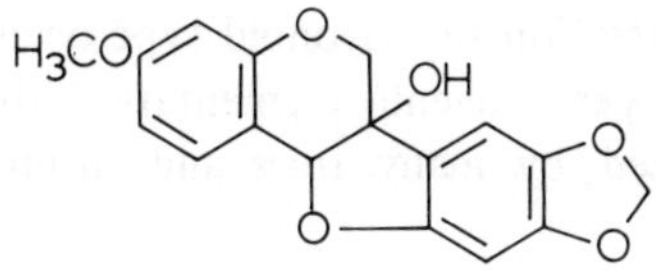

pisatin
(7.57)

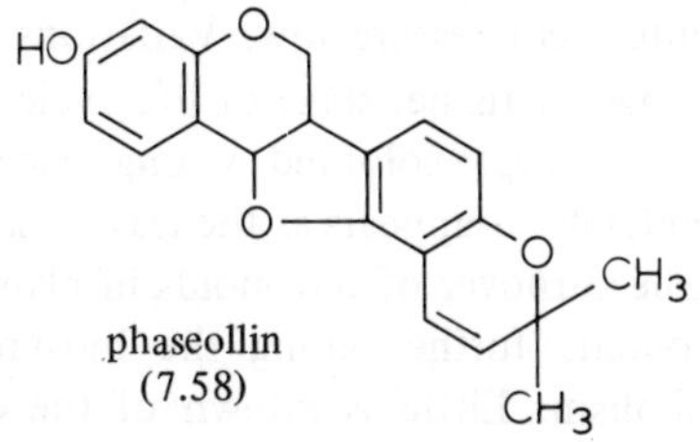

phaseollin
(7.58)

The rotenoids are highly complex isoflavonoids which are important as insecticides. Rotenone (7.59) is the active principle of derris powder obtained from the roots of *Derris elliptica* (Leguminosae).

Rotenoids are biosynthesised from 2′-methoxyisoflavones, the extra carbon atom being derived from methionine. A complete pathway has been established for the synthesis of amorphogenin (7.60) in *Amorpha fruticosa* (Leguminosae), as shown in *Figure 7.21*.

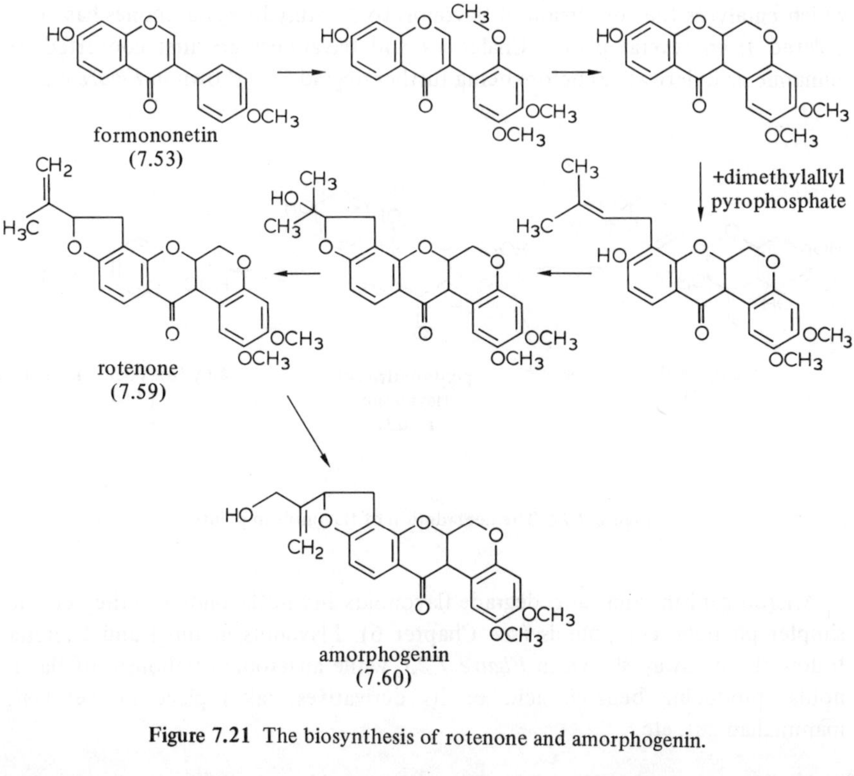

formononetin
(7.53)

rotenone
(7.59)

+dimethylallyl
pyrophosphate

amorphogenin
(7.60)

Figure 7.21 The biosynthesis of rotenone and amorphogenin.

Although the rotenoids have a low toxicity to mammals, they are highly toxic to insects and fish. This action is due to inhibition of mitochondrial oxidation.

Catabolism of Flavonoids

Flavonoids in plants have rapid turnover rates and the stationary concentrations of intermediates are low. Variations in concentration or flavonoid type occur with age of tissue, stage of life cycle or time of year. Juvenile pigmentation, the red- or purple-coloured young leaves produced by many trees and shrubs, completely disappears as the leaves mature.

The turnover of flavonoids in plants is due to both anabolism (conversion to flavonoids further along the biosynthetic pathway or to polymers) and to catabolism. Little is known of the catabolism of polymers, such as the condensed tannins, and polymer formation appears to be the preferred route for isoflavonoid detoxification.

The first step in the catabolism of flavonoids is the removal of the glycosidic sugars, and enzymes catalysing such reactions have been isolated from many plants. Studies on cell suspension cultures of a number of plants have shown that the A ring of flavonoids is completely degraded to carbon dioxide, while the B ring is converted to benzoic acid derivatives. Intermediates have only been established for a few compounds. Kaempferol (7.24), for instance, is converted to carbon dioxide and 4-hydroxybenzoic acid (7.61) through the intermediate formation of 2,3,5,7,4'-pentahydroxyflavanone (7.62) (*Figure 7.22*). An enzyme which catalyses the conversion of flavonols to 2,3-dihydroxyflavanones has been isolated from several plants. Chalcones and flavanones are first converted to cinnamic acid derivatives before being further degraded as shown in *Figure 6.30*.

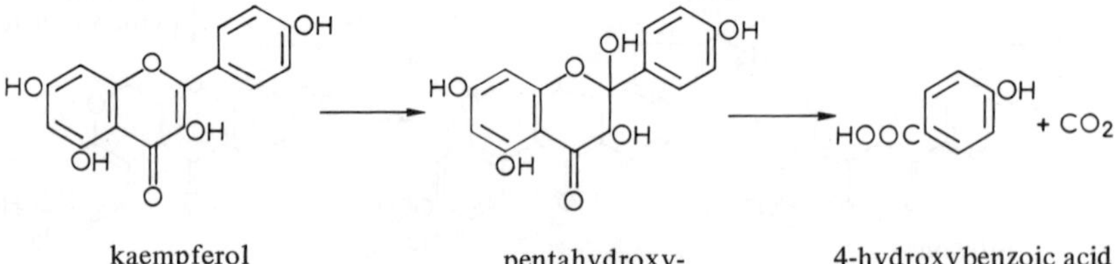

kaempferol	pentahydroxy-	4-hydroxybenzoic acid
(7.24)	flavanone	(7.61)
	(7.62)	

Figure 7.22 The degradation of flavonols in plants.

Microorganisms and fungi degrade flavonoids and isoflavonoids as they do the simpler phenolic compounds (*see* Chapter 6). Flavonols in fungi and bacteria follow the pathway shown in *Figure 7.23*, while anaerobic catabolism of flavonoids, producing benzoic acid or its derivatives, takes place in wet soil, mammalian gut, etc.

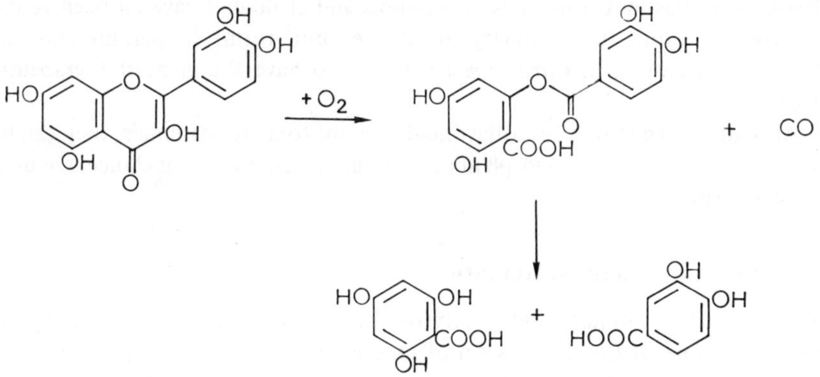

Figure 7.23 The degradation of flavonols by fungi and bacteria.

The Function and Physiological Effects of Plant Flavonoids

Flavonoids in plants have many functions, some of which are still not fully understood. For instance, flavonoids exert profound effects on growth hormones *in vitro*, but results from *in vivo* experiments are conflicting. Undoubtedly, flavonoids interfere in growth mechanisms, but whether they activate or inhibit these mechanisms depends on many factors. It has also been suggested that certain flavonoids are themselves growth regulators. There is some evidence that naringenin increases dormancy, but it has not been established that this compound induces natural dormancy. Experiments with callus tissue cultures have shown that a number of flavonoids increase the growth of cells.

Flavonoids are well known enzyme inhibitors, which, in general, act by complexing with protein. It has been found that the glycosides are much weaker inhibitors than the aglycons, lending support to the theory that glycoside formation is a detoxifying process.

The anthocyanins are obviously biosynthesised for their colours, which attract insects, birds and animals, the agents of pollination and seed dispersal. It is interesting that most members of the Gramineae do not synthesise anthocyanins, presumably because their flowers are wind pollinated. The purpose, if any, of anthocyanin production in leaves has not been clarified, although it has been suggested that they, and other flavonoids, act as ultraviolet light filters.

Probably the most important function of the isoflavonoids is to protect against invaders. The pterocarpans in the heartwood of leguminous trees probably protect against fungal attack, as do the phytoalexins produced by some plants in response to fungal infection. The rotenoids are insecticides, and the coumestan and isoflavone oestrogenic properties can be considered a defence against grazing animals.

The majority of flavonoids and isoflavonoids are non-toxic to mammals, and some are effective antibiotics against certain virus diseases. Certain naturally occurring flavonoids, generally gathered under the general heading of bioflavonoids or vitamin P, have been found to have beneficial effects in over fifty diseases associated with capillary fragility or abnormal capillary permeability. Flavanones, flavonols, isoflavones, catechins, flavandiols and chalcones have all been found effective in promoting capillary resistance, but rutin, hesperidin and the flavanones occurring in *Citrus* species appear to have the greatest therapeutic effect.

Although flavonoids and isoflavonoids are not toxic to mammals, they can be highly poisonous to fish, and plants containing rotenoids, for instance, are used as fish poisons.

Flavonoids in Chemosystematics

Because of their ubiquity and variety of chemical structures, the flavonoids are potentially one of the most useful of taxonomic markers. Using paper or thin-layer, one- or two-dimensional, chromatography, it is quick and easy to survey large numbers of plants for flavonoids. The spots are generally detected visually or with a UV spectrophotometer. In recent years, a number of flavonoid surveys have given useful results, some of which are described below, and it is to be expected that many more will follow.

Although present in all higher plants, flavonoids are absent from bacteria and the majority of algae. The most primitive plants to contain these compounds are the stoneworts (Charophyceae), a type of advanced green algae. Flavonoids such as chlorflavonin (7.32) have been found in fungi, but such compounds are rare. Flavonoids isolated from the primitive Bryophyta include simple flavone-*C*-glycosides and 3-deoxyanthocyanins, while complex *O*-glycosides, aurones and isoflavones have been found in the most advanced angiosperms.

Woody plants, whether occurring in the Dicotyledones or Monocotyledones, are considered more primitive than their herbaceous neighbours. Except in some isolated cases, it is found that woody plants contain leucoanthocyanins (colour-less flavonoids which give anthocyanins on treatment with hot acid), whereas herbaceous plants do not. Thus, nearly all the woody dicotyledon families from Casuarinaceae to Ebenaceae contain leucoanthocyanins, but in the herbaceous families, Aristolochiaceae to Compositae, these compounds are generally absent. In the Monocotyledones, all species of the woody Palmae family contain leucoanthocyanins, but such compounds are extremely rare in the herbaceous Gramineae and Orchidaceae. As might be expected for such an advanced family, the phenolic constituents of the Orchidaceae are unusual and many have not yet been identified.

The red anthocyanin and betacyanin pigments are mutually exclusive, as are the yellow betaxanthins and the carotenoids or yellow flavonoids. A survey of the Centrospermae has shown that all families except the Caryophyllaceae and Molluginaceae contain betacyanins and/or betaxanthins, the Caryophyllaceae and Molluginaceae containing anthocyanins, in common with non-Centrospermae plants. Thus, from a chemosystematic viewpoint, these families should be re-moved to a separate order (see also Chapter 9).

Yellow pigments have shown interesting correlations with systematics in other families, notably Primulaceae and Gesneriaceae. In the former, species belonging to two sections of the *Primula* genus, Vernales and Sikkimenses, con-tain the flavonols, gossypetin (7.28) or herbacetin (8-hydroxykaempferol), as flower pigments, while species belonging to the remaining sections contain carotenoids. In Gesneriaceae, several species of the subfamily, Cyrtandroideae, have flowers coloured with chalcones and aurones, but those of the subfamily, Gesnerioideae, contain carotenoids. Chalcones and aurones are also characteristic of the subtribe Coreopsidineae of the Compositae.

Plant flavonoid patterns have given useful correlations with established morphological taxonomy in a number of instances. Thus, the two tribes, Plumbagineae and Staticeae, making up the Plumbaginaceae family, show characteristic differences in leaf and flower flavonols and anthocyanins. It has been suggested that the two tribes should be designated as separate families on the grounds of both their morphological and their chemical differences.

In the *Pinus* genus, the presence or absence of flavones in the heartwood differentiates between the subgenera, Haploxylon, which contains such compounds, and Diploxylon, which does not.

The variable vegetative habit of members of the Ericaceae makes it difficult to classify such plants, and thus their content of unusual flavonoids is a useful taxonomic marker. The subfamily, Rhododendroideae, is distinguished from other subfamilies by the frequent presence of gossypetin, caryatin (quercetin-3, 5-dimethylether), 5-O-methylflavonols and dihydroflavonols. Gossypetin is particularly characteristic of the tribes, Rhodoreae and Phyllodoceae. The subgenus, Hymenanthes, of the *Rhododendron* genus, is characterised by the presence of caryatin, while the absence of gossypetin characterises the subgenus, Pentanthera.

In some plant genera, it has been found that the flavonoid pattern of each species is distinct and can be used as a 'fingerprint'. The *Baptisia* genus (Leguminosae), for instance, contains 62 flavonoids and each of 17 species has a characteristic pattern. However, in the closely related *Thermopsis* genus with 31 flavonoids, 13 species have almost identical patterns.

Plant populations within a species need not be chemically identical, and by examining the flavonoid patterns of different populations it has been found that distinct chemical races exist in, for instance, wild carrot (*Daucus carota*) and fennel (*Foeniculum vulgare*), both members of the Umbelliferae.

The geographical origin of a plant can often be correlated with its chemical characteristics. Thus, the species of *Geranium* (Geraniaceae) contain a wide range of flavonoids. Primitive compounds, such as the leucoanthocyanidins, predominate in Central Eurasian species, but progressively disappear in species found to the east and west.

The parents of natural hybrids can often be deduced if flavonoid biosynthesis is assumed to be additive. For example, the *Baptisia* hybrids, *B. leucantha* x *B. sphaerocarpa* and *B. alba* x *B. tinctoria*, contain quercetin-3-rutinoside-7-glucoside, a glycoside not present in any pure species. However, it was found that one set of parents biosynthesised quercetin-3-rutinoside and the other quercetin-7-glucoside. Thus, glycoside biosynthesis had become additive in the offspring.

Flavonoid patterns are also useful in determining the parents of cultivars in crop or ornamental plants. Such methods have been used to trace the origin of *Citrus* varieties.

Xanthones and Stilbenes

Xanthones are fairly common plant constituents, but stilbenes are much rarer. About 80 xanthones have been isolated from higher plants, the best-known being gentisin (7.63), the yellow pigment of *Gentiana lutea* (Gentianiaceae) roots, and mangiferin (7.64), a xanthone-*C*-glucoside first isolated from the roots of the mango tree (*Mangifera indica*, Anacardiaceae).

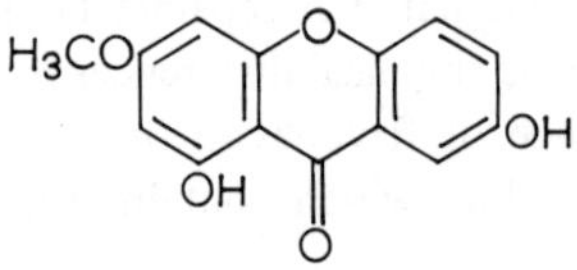

gentisin
(7.63)

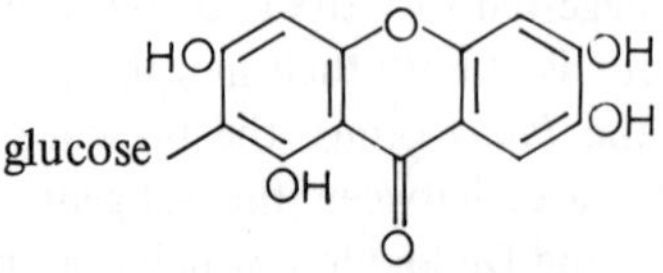

mangiferin
(7.64)

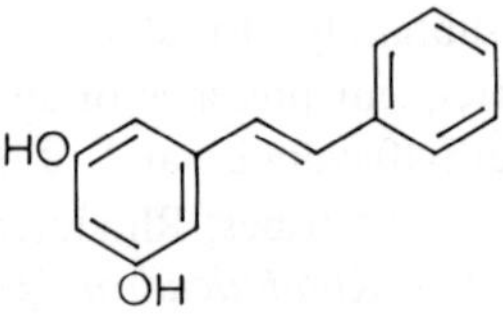

pinosylvin
(7.65)

Stilbenes have been found in a few unrelated plants, but except for the *Pinus* (Pinaceae) genus, where pinosylvin (7.65) is common, these compounds are rare plant constituents. Probably the most interesting is the dihydrostilbene, lunularic acid (7.66) (*Figure 7.25*), which acts as a growth inhibitor in liverworts. This compound also occurs in algae and is probably the most primitive of all compounds formed by the mixed shikimic acid/acetate–malonate pathways.

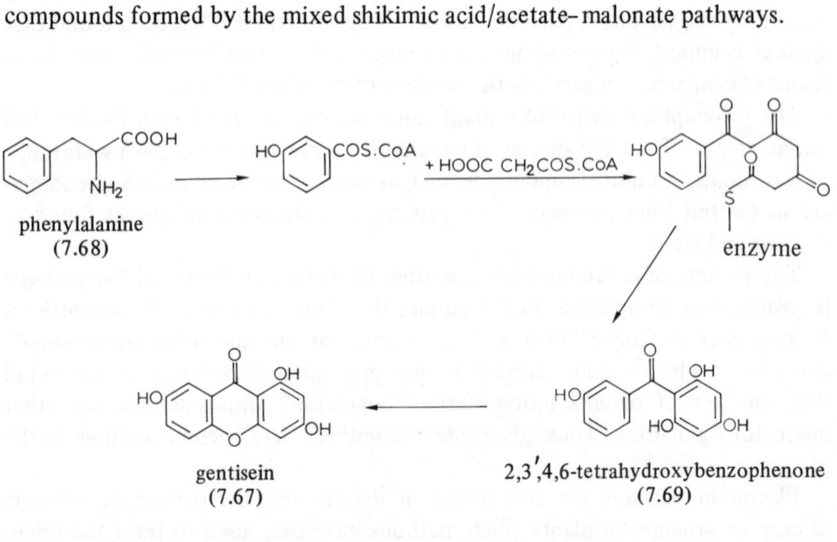

phenylalanine
(7.68)

enzyme

gentisein
(7.67)

2,3',4,6-tetrahydroxybenzophenone
(7.69)

Figure 7.24 The biosynthesis of gentisein.

Xanthones are biosynthesised from shikimate-derived benzoic acid derivatives and malonate. In *G. lutea* it has been shown that gentisein (7.67) is derived from phenylalanine (7.68), through the intermediate formation of 2,3',4,6-tetrahydroxybenzophenone (7.69) (*Figure 7.24*), a compound which also occurs in this plant.

Stilbenes are biosynthesised from cinnamate and malonate, probably by the pathway shown in *Figure 7.25*. Both phenylalanine (7.68) and acetate were incorporated into lunularic acid (7.66) in *Lunularia cruciata*, while the incorporation of hydrangenol (7.70) suggests that this compound is an intermediate (*Figure 7.25*). Lunularic acid was found to be metabolised to lunularin (7.71) in *L. cruciata*.

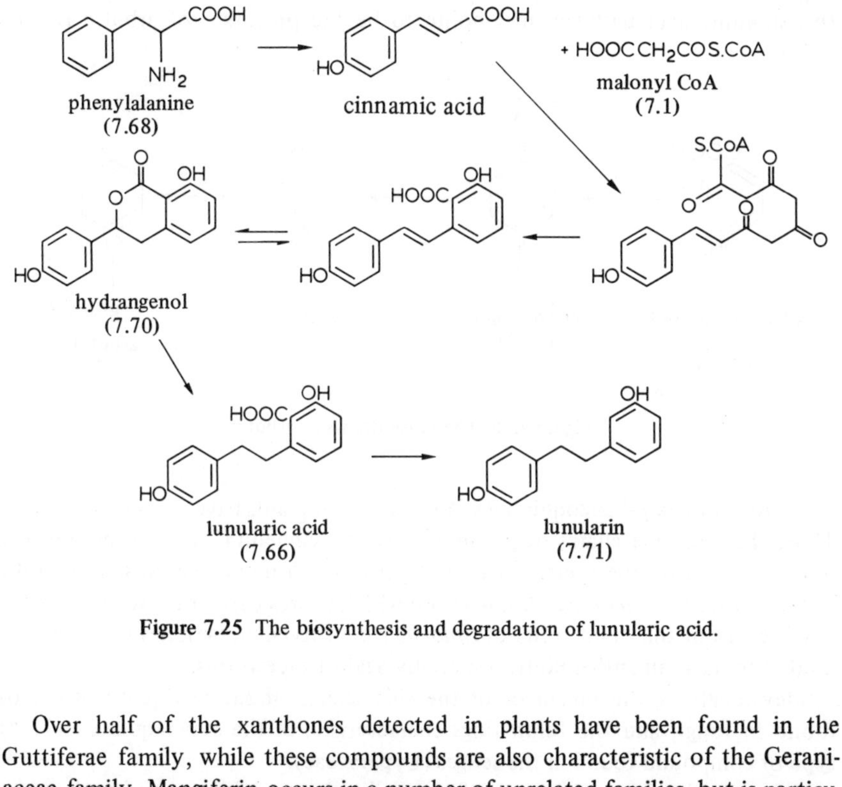

Figure 7.25 The biosynthesis and degradation of lunularic acid.

Over half of the xanthones detected in plants have been found in the Guttiferae family, while these compounds are also characteristic of the Geraniaceae family. Mangiferin occurs in a number of unrelated families, but is particularly characteristic of the Iridaceae. Stilbenes appear to be generally species-specific except in the *Pinus* genus where pinosylvin occurs in the heartwood of at least 50 species.

Quinones

Quinones are widely distributed in the plant world and have been found in most phyla except the ferns and mosses. The more common quinones range in colour from pale yellow through red to black; blue and green quinones being rare.

Quinones contribute little to plant pigmentation, however, as they generally occur in the roots, wood or bark.

The ubiquitous ubiquinones, plastoquinones, tocopherols, phylloquinones and bacterial menaquinones are concerned in redox systems, but the functions of other quinones in plants are not known. Many naturally occurring quinones have been used as dyes and medicinally, and some are important drugs in present-day medicine.

Benzoquinones

The simplest quinone, p-benzoquinone (7.72), does not occur naturally, but the quinol glucoside, arbutin (7.73), has been isolated from many plants. Feeding experiments with isotopically labelled 4-hydroxybenzoic acid (7.61) have shown this shikimic acid pathway metabolite to be the precursor of arbutin (*Figure 7.26*).

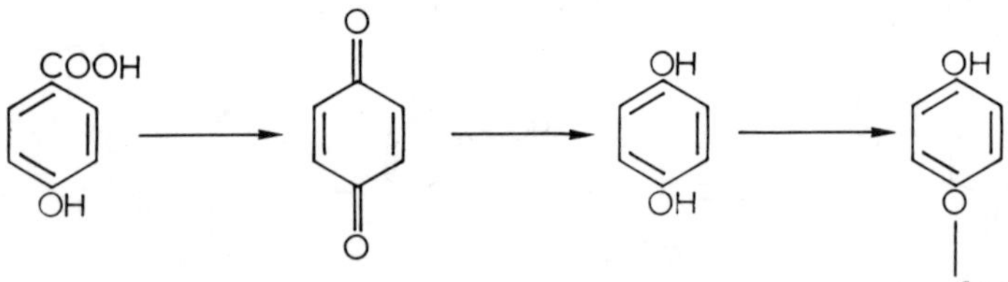

Figure 7.26 The biosynthesis of arbutin.

More complex p-benzoquinones found in higher plants have a mixed biogenesis, Thus, the long side chains of primin (7.74) and embelin (7.75) are presumably biosynthesised by the acetate–malonate pathway. Primin is the substance on the leaves of *Primula obconica* (Primulaceae) which causes dermatitis, while embelin, obtained from the dried fruits of *Embelia ribes* (Myrsinaceae), has for long been used in India as an anthelmintic, especially against tapeworms.

Mevalonate is the precursor of the side chains of the boviquinones (7.76) found in fungi, and the furan rings characteristic of the cyperaquinone (7.77) type of compound found in the sedges (Cyperaceae).

Substituted p-benzoquinones of the dalbergione (7.78) type, found in the heartwood of *Dalbergia* and *Macherium* species (Leguminosae), are usually included with the neoflavonoids (4-arylcoumarins and 4-arylchromenes). Little is known of the biosynthesis of these compounds due to the difficulty of studying them *in vivo*.

Ubiquinones, Plastoquinones and Tocopherols

The most important of the naturally occurring p-benzoquinones are the iso-prenoid derivatives which include the ubiquinones (coenzyme Q), plastoquinones

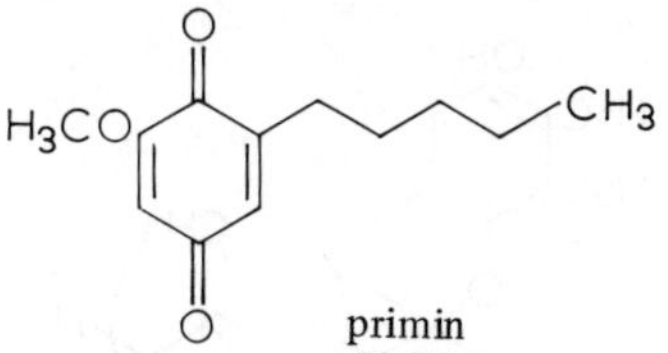

primin
(7.74)

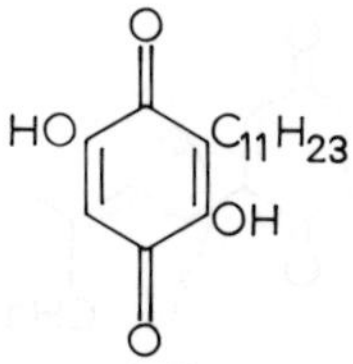

embelin
(7.75)

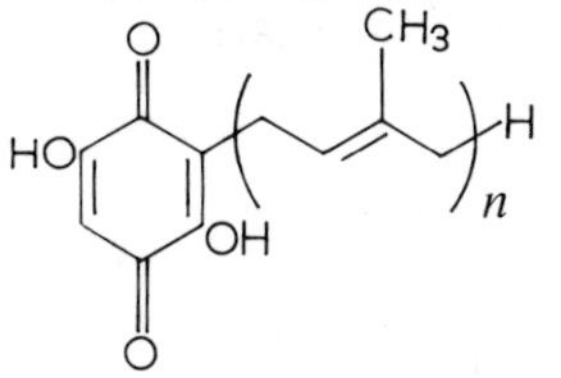

boviquinones
(7.76)

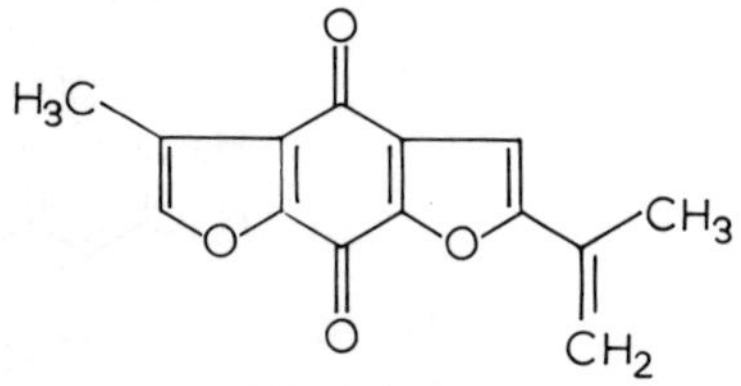

cyperaquinone
(7.77)

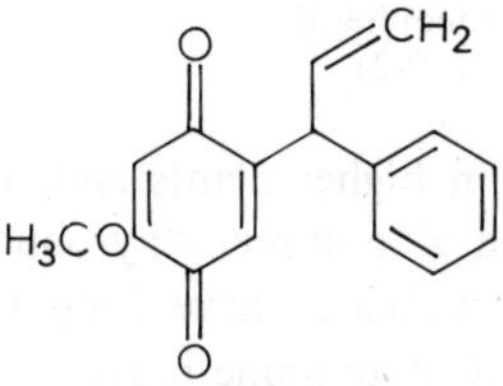

dalbergione
(7.78)

and tocopherols (vitamin E). In these compounds, the isoprenoid side chains are derived from mevalonate. The ubiquinones are widely distributed in nature, and only the gram positive bacteria and blue-green algae do not contain these derivatives of 5,6-dimethoxy-3-methyl-2-polyprenyl-1,4-benzoquinone (7.79, $n=1$ to 12, but $n=8, 9$ or 10 are the most common).

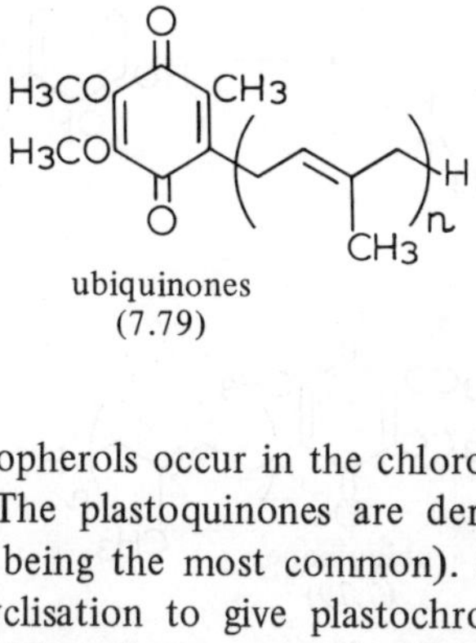

ubiquinones
(7.79)

Plastoquinones and tocopherols occur in the chlorophyll-containing tissue of higher plants and algae. The plastoquinones are derivatives of 2,3-dimethyl-benzoquinone (7.80, $n=9$ being the most common). Side chain hydroxylation (plastoquinone C) and cyclisation to give plastochromanol-8 (7.81) are also common. α-Tocopherol (7.82) is the most common form of vitamin E.

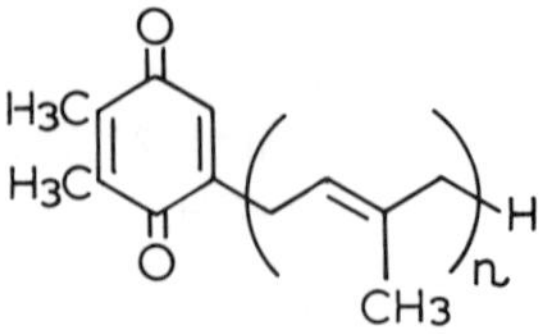

plastoquinones
(7.80)

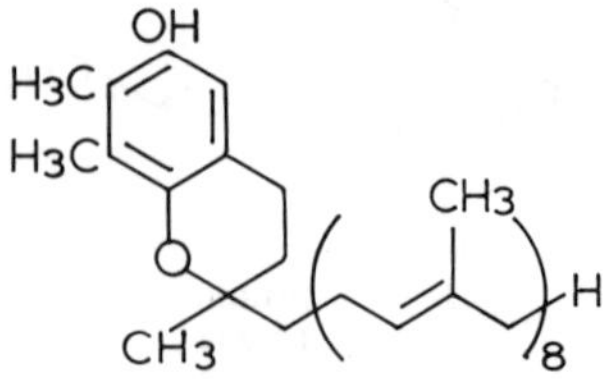

plastochromanol-8
(7.81)

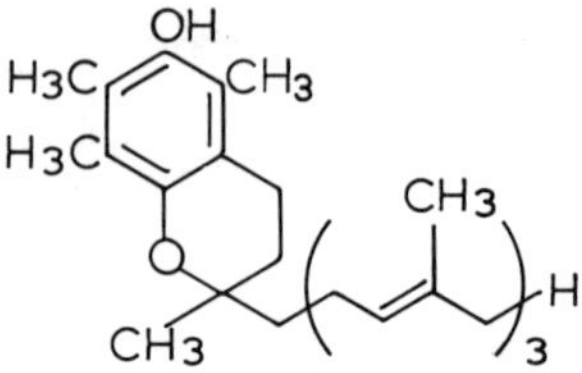

α-tocopherol
(7.82)

Ubiquinones are biosynthesised in higher plants and other organisms from 4-hydroxybenzoic acid (7.61) and a polyprenylpyrophosphate of requisite length. (Polyprenylpyrophosphate synthases have been isolated from several living systems.) As the pathways of ubiquinone biosynthesis in rats and lower organisms are similar, they are probably also applicable to higher plants (*Figure 7.27*), especially as 4-hydroxybenzoatepolyprenyltransferase activity has been detected in cell free homogenates of broad bean (*Vicia faba*, Leguminosae) seeds.

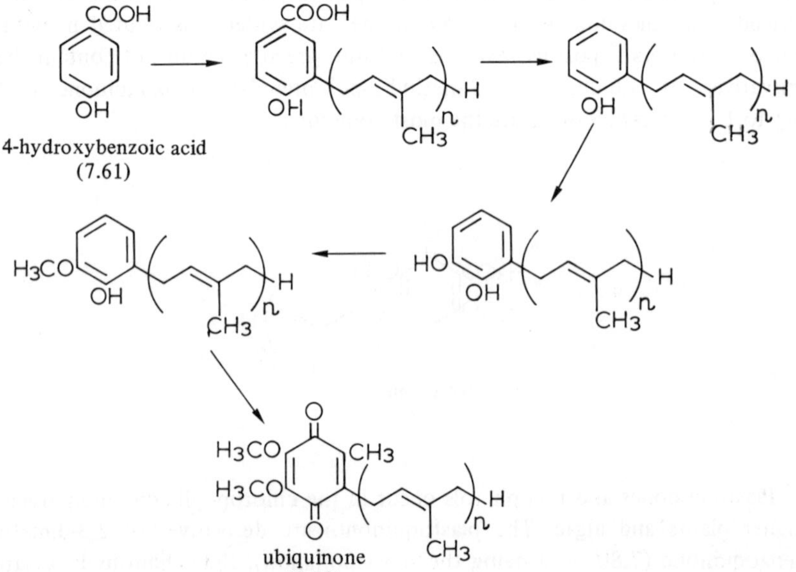

4-hydroxybenzoic acid
(7.61)

ubiquinone
(7.79)

Figure 7.27 Some steps in the biosynthesis of the ubiquinones.

The aromatic ring and one methyl group of the plastoquinones and tocopherols are derived from phenylalanine (7.68) or tyrosine (7.83) by the pathway shown in *Figure 7.28*. The remaining methyl groups are donated by methionine.

Although homogentisic acid (7.84) has been shown to be a good precursor of the plastoquinones and tocopherols, the intervening steps in the biosynthesis of these quinones are not known.

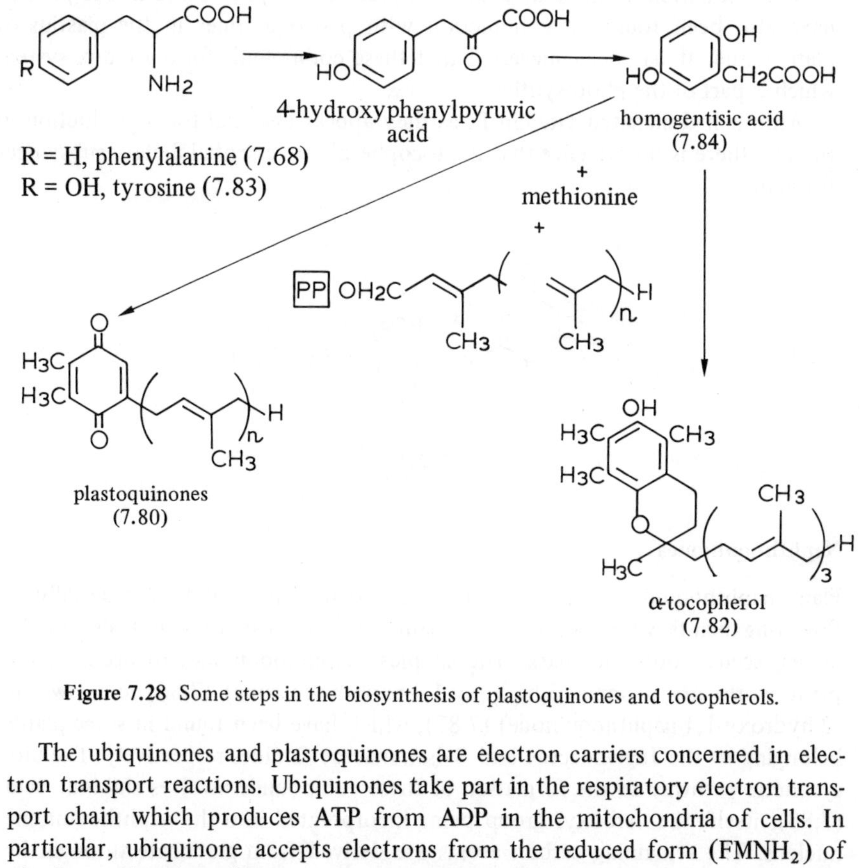

Figure 7.28 Some steps in the biosynthesis of plastoquinones and tocopherols.

The ubiquinones and plastoquinones are electron carriers concerned in electron transport reactions. Ubiquinones take part in the respiratory electron transport chain which produces ATP from ADP in the mitochondria of cells. In particular, ubiquinone accepts electrons from the reduced form ($FMNH_2$) of riboflavin phosphate (flavin mononucleotide) oxidising this compound to FMN (see also Chapter 10) and being itself reduced to the quinol form (*Figure 7.29*). Ubiquinol is reoxidised to the quinone when electrons are passed to cytochrome b.

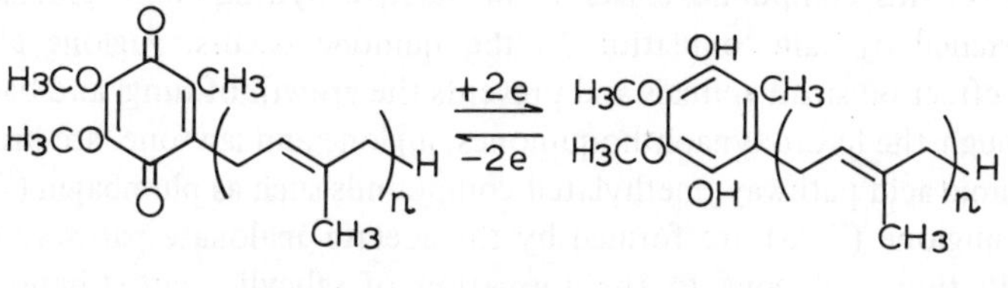

Figure 7.29 The ubiquinone redox system.

Plastoquinones are concerned in the electron transport reactions and photo-phosphorylations which take place in the chloroplasts. These quinones act in an analogous way to the ubiquinones, accepting electrons from the unknown compound generally represented as Q and passing them to cytochrome b.

Much less is known of the function of the tocopherols, which appear to act as antioxidants, being themselves easily oxidised to the quinone form (7.85). The high concentrations of vitamin E found in seed oils are probably necessary to prevent oxidation of these unsaturated lipids. α-Tocopherol and α-tocoquinone have also been found in conjunction with plastoquinones in the plastids of plant tissue. It has been suggested that these compounds form a redox system which is part of the photosynthetic process.

Although designated vitamin E, as they appear essential for reproduction in the rat, there is no evidence that the tocopherols are essential dietary substances for man.

tocoquinone
(7.85)

Naphthoquinones

Plant naphthoquinones are more numerous than benzoquinones, especially in flowering plants where they occur mainly in the heartwood but also in the leaves, seeds, roots and bark. The simplest naphthoquinones to occur in the plant world are juglone (5-hydroxy-1,4-naphthoquinone) (7.86) and lawsone (2-hydroxy-1,4-naphthoquinone) (7.87), which have been found in some plants belonging to the Juglandaceae and Balsaminaceae families respectively. The biosynthetic pathway to the formation of these compounds involves shikimic acid (7.88), itself, the carboxyl group being incorporated into the naphthoquinone nucleus. The remaining carbon atoms are derived from glutamate through the formation of 2-oxoglutarate (7.89) and o-succinoylbenzoic acid (7.90) is an intermediate (*Figure 7.30*). Naphthoquinone (7.91) was shown to be an intermediate in the formation of juglone. Juglone is an allelopath, responsible for the lack of growth of grasses, etc., under walnut trees (*Juglans* sp., Juglandaceae). In the tree, this compound exists as an inactive hydroquinone glucoside, but, when leached by rain, oxidation to the quinone occurs. Juglone also has a sedative effect on small animals and prevents the growth of fungi and bacteria.

Although the hydroxynaphthoquinones, juglone and lawsone, are products of the shikimic acid pathway, methylated compounds such as plumbagin (7.92) and 7-methyljuglone (7.93) are formed by the acetate–malonate pathway (Chapter 4), a situation analogous to the formation of salicylic acid (Chapter 6) and 6-methylsalicylic acid (Chapter 4).

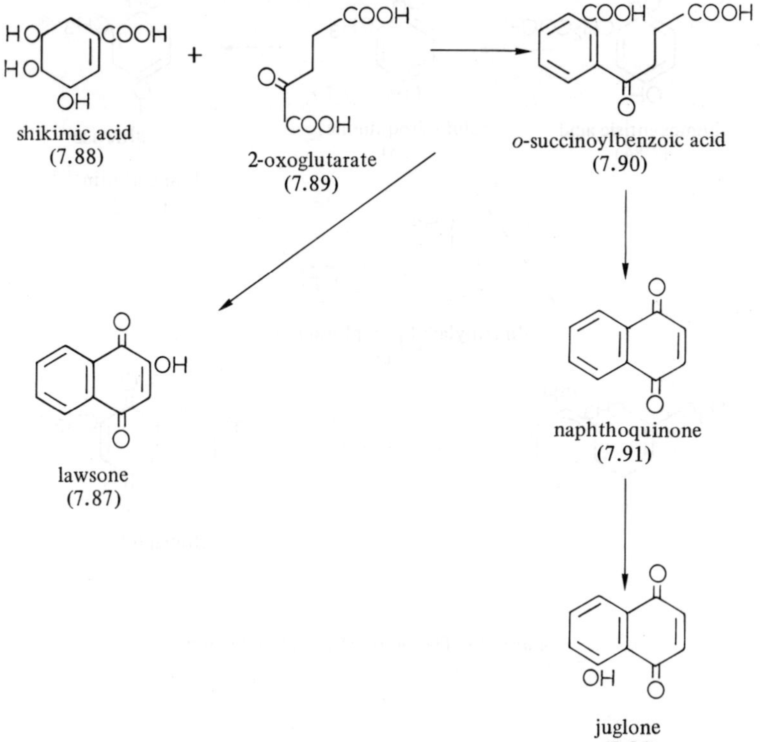

Figure 7.30 The biosynthesis of juglone and lawsone.

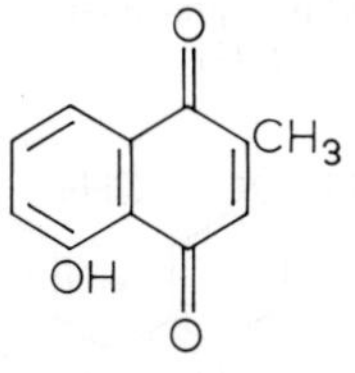

plumbagin
(7.92)

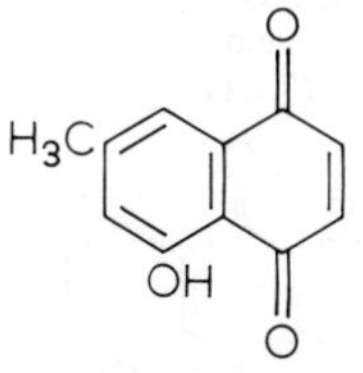

7-methyljuglone
(7.93)

In *Chimaphila umbellata* (Pyrolaceae), the naphthoquinone, chimaphilin (7.94), is a product of the addition of mevalonate (in the form of dimethylallyl pyrophosphate (7.21)) to toluhydroquinone (7.95), derived from the shikimic acid pathway metabolite, homogentisic acid (7.84) (see *Figure 7.28*), as shown in *Figure 7.31*. Homoarbutin (7.96) has also been detected in this plant.

In plants belonging to the Boraginaceae family, the shikimic acid metabolite 4-hydroxybenzoic acid (7.61), adds on two molecules of mevalonate (in the form of geranyl pyrophosphate (7.97)) to give quinones of the alkannin (7.98) type (*Figure 7.32*).

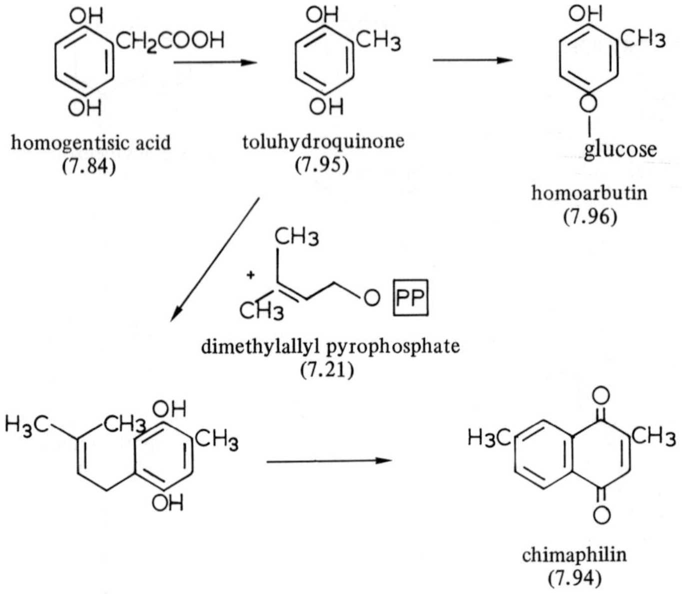

Figure 7.31 The biosynthesis of chimaphilin.

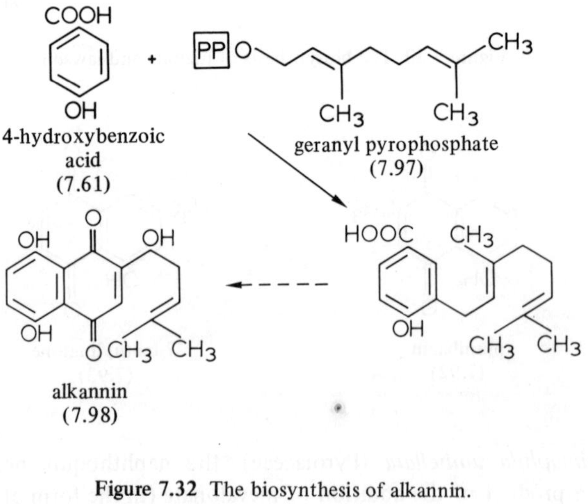

Figure 7.32 The biosynthesis of alkannin.

A number of naphthoquinones have been used as natural dyes, including lawsone (7.87), the active ingredient of henna, and alkannin (7.98) from dyer's bugloss (*Alkanna tinctoria*, Boraginaceae). Others, such as plumbagin (7.92) and chimaphilin (7.94) are antibiotics, while lapachol (7.99) has antitumour properties.

lapachol
(7.99)

Phylloquinones and Menaquinones

The most important of the naphthoquinones are the phylloquinones (7.100, $n=3$ being the most common) occurring in higher plants and algae, and the bacterial menaquinones (7.101, $n=1$ to 13), which also occur in some fungi. These compounds are known as vitamins K_1 and K_2 respectively.

phylloquinones
(7.100)

menaquinones
(7.101)

The biosynthesis of the naphthoquinone moiety of these compounds follows the pathway given in *Figure 7.30*, while the side chains are derived from mevalonate and the ring methyl groups from methionine.

Phylloquinone occurs in the plastids of plants and is probably concerned in the photosynthetic electron transport reactions. The menaquinones take the place of the ubiquinones in the respiratory chains of bacteria. These quinones are important to animals, as they are concerned in the blood-clotting mechanisms, and vitamin K is sometimes known as the antihaemorrhagic vitamin. Deficiency diseases in man are rare, however, as the intestinal flora are able to synthesise vitamin K.

Anthraquinones

The anthraquinones are the largest group of naturally occurring quinones and about 200 compounds have been identified. They are particularly abundant in the Rubiaceae, Scrophulariaceae, Leguminosae, Rhamnaceae, Polygonaceae and Liliaceae families, and also occur in fungi, lichens and bacteria. Unlike the benzoquinones (except arbutin) and naphthoquinones, the anthraquinones in higher plants occur mainly as glycosides. They are found in all parts of the plant except the flowers, and often several different compounds occur in the same plant. For example, *Digitalis* (Scrophulariaceae) leaves and madder *(Rubia tinctorum,* Rubiaceae) roots each contain over 20 anthraquinones. Anthraquinone, itself, has only been found in the cuticular wax of the perennial rye grass *(Lolium perenne,* Gramineae). Anthraquinones substituted with chlorine, such as fragilin (7.102) from *Sphaerophorus fragilis,* are characteristic of the lichens.

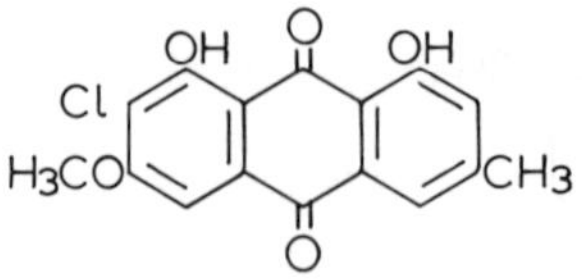

fragilin
(7.102)

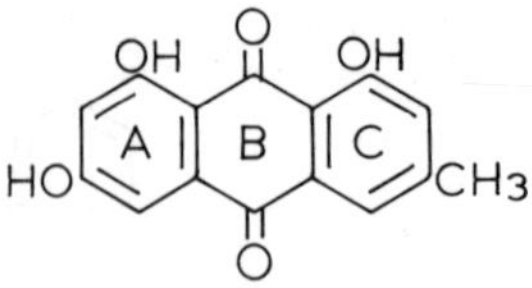

emodin
(7.103)

Anthraquinones of the emodin (7.103) type with substituents in both rings A and C are biosynthesised in both higher and lower plants by the acetate–malonate pathway (see Chapter 4). Those of the alizarin (7.104) type with substituents in ring C only, however, are derived from the shikimic acid–*o*-succinoylbenzoic acid pathway (*Figure 7.30*) and mevalonate, as shown in *Figure 7.33*. The steps forming the quinone ring and the point of entry of mevalonate are not known, but naphthoquinone does not appear to be a natural intermediate.

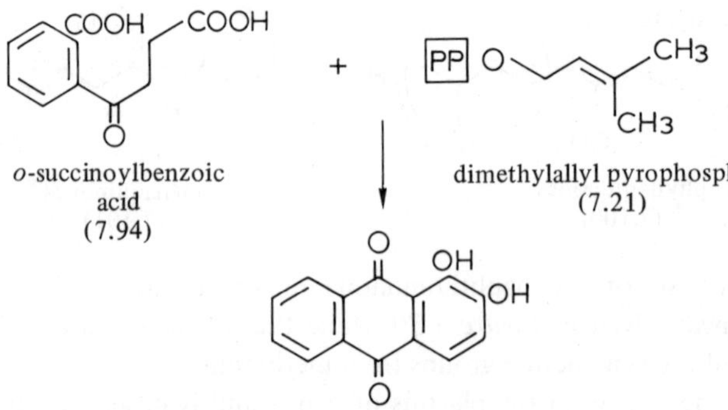

o-succinoylbenzoic
acid
(7.94)

dimethylallyl pyrophosphate
(7.21)

alizarin
(7.104)

Figure 7.33 The biosynthesis of alizarin.

Although morindone (7.105), which occurs with alizarin in *Morinda citrifolia* (Rubiaceae), is substituted in both rings A and C, it is biosynthesised from *o*-succinoylbenzoic acid (7.90), hydroxylation of ring A taking place after formation of the anthraquinone ring system.

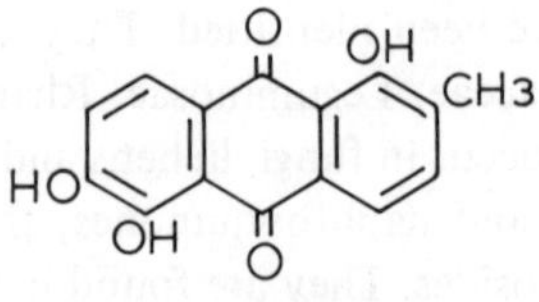

morindone
(7.105)

Quinones in chemosystematics

The ubiquinones, plastoquinones and tocopherols are widespread and have little taxonomic significance. Although the phylloquinones are confined to algae and higher plants, and the menaquinones to bacteria and some fungi, their distribution is too great to be of use in chemosystematics. A number of benzoquinones, naphthoquinones and anthraquinones, however, have a restricted distribution, which makes them potentially useful compounds in plant classification. Unlike the flavonoids, there has been little attempt to screen large numbers of species for quinones, and thus the phylogenetic significance of such compounds is unknown in most cases. However, a few tentative conclusions, from the evidence at present available, are presented below.

In the Dicotyledones, the polyprenylated benzoquinones, such as embelin (7.75), are good taxonomic markers for the Myrsinaceae family, while primin (7.74), with a shorter side chain, occurs in the related Primulaceae family. In the Monocotyledones, the sedges (Cyperaceae) also contain polyprenylated benzoquinones, but in the majority the side chains are cyclised to form furan rings, as in cyperaquinone (7.77). Such compounds are characteristic of the Cyperaceae family and have been used within the family to help in classification, which is difficult on morphological grounds.

Although naphthoquinones are biosynthesised by four pathways, there is little obvious correlation between these and plant evolution. The acetate-malonate pathway by which most lower plant quinones are biosynthesised would seem to be more primitive than the pathways involving shikimic acid. However, plumbagin (7.92) occurs in the Droseraceae and Plumbaginaceae and 7-methyljuglone (7.93) in the Droseraceae and Ebenaceae families, while the shikimic acid derived juglone (7.86) is characteristic of the less highly evolved Juglandaceae. Naphthoquinones of the alkannin (7.98) type, biosynthesised from 4-hydroxybenzoic acid and geranyl pyrophosphate, have only been found in the advanced Tubiflorales, however, alkannin being characteristic of the Boraginaceae, while lapachol (7.99) is widespread in the Verbenaceae and Bignoniaceae families.

The distribution of anthraquinones in the Dicotyledones has more phylogenetic relevance, as those compounds such as emodin (7.103) biosynthesised by the acetate–malonate pathway are found mainly in the Leguminosae, Polygonaceae and Rhamnaceae families, while anthraquinone derivatives derived from shikimic acid are found in the more advanced Rubiaceae, Verbenaceae and Bignoniaceae families.

Bibliography

Flavonoids – General

The Flavonoids, J. B. Harborne *et al.* (eds.), Chapman and Hall, 1975
'Nature and Properties of Flavonoids', in *Chemistry and Biochemistry of Plant Pigments,* vol. 1, T. Goodwin (ed.), Academic Press, 1976

Flavonoids – Biosynthesis

'Biosynthesis of Phenolic Compounds Derived from Shikimate', in *Biosynthesis,* Specialist Periodical Reports (continuing series, vol. 1, 1972), The Chemical Society

'Metabolites of the Shikimate Pathway', in *The Shikimate Pathway*, E. Haslam, Butterworths, 1974

'The Shikimate Pathway in Higher Plants', *ibid.*

'Biosynthesis of Flavonoids', in *Chemistry and Biochemistry of Plant Pigments, ibid.*

'Flavonoids', in *Organic Chemistry of Secondary Plant Metabolism*, T. A. Geissman and D. H. G. Crout, Freeman, Cooper and Co., 1969

Ecology of Flavonoids

'Functions of Flavonoids in Plants', in *Chemistry and Biochemistry of Plant Pigments, ibid.*

'Biochemical Interactions Between Higher Plants', in *Introduction to Ecological Biochemistry*, J. B. Harborne, Academic Press, 1977

'Higher Plant–Lower Plant Interactions', *ibid.*

'Biochemistry of Plant Pollination', *ibid.*

'Plant Phenolics Possessing Oestrogenic Activity', in *Pharmacology of Plant Phenolics*, J. W. Fairbairn (ed.), Academic Press, 1959

'Capillary Structure and the Action of Flavonoids', *ibid.*

Chemosystematics of the Flavonoids

Comparative Biochemistry of the Flavonoids, J. B. Harborne, Academic Press, 1967

'Evolution of Flavonoid Pigments', in *Comparative Phytochemistry*, T. Swain (ed.), Academic Press, 1966

'Dihydrochalcones', *ibid.*

'Flavonoid C-glycosides', *ibid.*

'Flavonoid and Phenylpropanoid Patterns in the Umbelliferae', in *The Biology and Chemistry of the Umbelliferae*, V. H. Heywood (ed.), Academic Press, 1971

'Flavonoids as Systematic Markers in the Angiosperms', in *Chemistry in Botanical Classification*, G. Bendz and J. Santesson (eds.), Academic Press, 1974

'Chemistry of Geographical Races', section IV, in *Chemistry in Evolution and Systematics*, T. Swain (ed.), Butterworths, 1973

'Comparative Biosynthetic Pathways in Higher Plants', *ibid.*

The Biology and Chemistry of the Compositae, vols. 1 and 2, V. H. Heywood et al. (eds.), Academic Press, 1977 (flavonoids are discussed in many chapters)

'Distribution of Flavonoids in the Leguminosae', in *Chemotaxonomy of the Leguminosae*, J. B. Harborne et al. (eds.), Academic Press, 1971

Quinones

'Quinones, Nature, Distribution and Biosynthesis', in *Chemistry and Biochemistry of Plant Pigments, ibid.*

'Biosynthesis of Quinones', in *Biosynthesis*, vol. 3, T. A. Geissman (ed.), Specialist Periodical Reports, The Chemical Society, 1975

'Benzoquinones', in *Naturally Occurring Quinones,* 2nd edn, R. H. Thomson, Academic Press, 1971

'Naphthoquinones', *ibid.*

'Anthraquinones', *ibid.*

'Comparative Biochemistry of Hydroxyquinones', in *Comparative Phytochemistry,* T. Swain (ed.), Academic Press, 1966

Xanthones and Stilbenes

'Extension of the Phenylpropanoid Unit', in *Organic Chemistry of Secondary Plant Metabolism, ibid.*

'Biosynthesis of Phenolic Compounds', section: Xanthones and Stilbenes, in *Biosynthesis, ibid.,* vol. 1, 1972

8 Compounds Derived from Amino Acids

Introduction

The amino acids are an important group of compounds concerned in the biosynthesis of a variety of nitrogen-containing plant metabolites, including proteins, enzymes, nucleic acids, hormones, chlorophyll, amines, alkaloids, cyanogenic glycosides and glucosinolates. Metabolic deamination of amino acids gives various products, including oxoacids, the terpene-like senecic acids found in the pyrrolizidine alkaloids (Chapter 9), and ethylene, an important growth hormone, formed from methionine.

Proteins are considered primary metabolites and are not, therefore, considered in this book, while the many alkaloids synthesised by plants are described in Chapter 9. The porphyrins and nucleic acids and their derivatives are discussed in Chapter 10.

All amino acids are derived directly or indirectly from glutamate (8.1) or glutamine (8.2), which experimental work has shown to be the only primary products of ammonia assimilation. In plants, there appear to be two pathways to the formation of glutamic acid. It has been clearly shown in yeast, *Candida utilis*, that glutamate (8.1) is formed from 2-oxoglutarate (8.3) and ammonia, the reaction being catalysed by glutamate dehydrogenase (*Figure 8.1*).

$$^-OOCCH_2CH_2COCOO^- + NH_3 \rightarrow {}^-OOCCH_2CH_2CH(NH_2)COO^-$$

2-oxoglutarate   glutamate

(8.3)   (8.1)

Figure 8.1 The biosynthesis of glutamate in yeast.

Amination of 2-oxoglutarate takes place in two steps with enzyme-bound γ-glutaryl phosphate as an intermediate. Recent work on tissue cultures of some higher plants has shown the existence of a second pathway involving the enzymes, glutamine synthase and glutamate synthase (*Figure 8.2*). Glutamate (8.1) is first aminated to glutamine (8.2), which then reacts with 2-oxoglutarate (8.3) to give two molecules of glutamate.

All other amino acids are formed by transamination, a reaction catalysed by aminotransferases (transaminases). A general reaction for the formation of an amino acid from an oxoacid is given in *Figure 8.3*. The reaction takes place in two stages — first a Schiff base is formed between enzyme-bound pyridoxal phosphate and the amino acid; this complex then reacts with the oxoacid to form a new amino acid, regenerating pyridoxal phosphate.

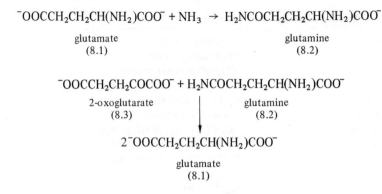

$$^-OOCCH_2CH_2CH(NH_2)COO^- + NH_3 \rightarrow H_2NCOCH_2CH_2CH(NH_2)COO^-$$

glutamate   glutamine
(8.1)   (8.2)

$$^-OOCCH_2CH_2COCOO^- + H_2NCOCH_2CH_2CH(NH_2)COO^-$$

2-oxoglutarate   glutamine
(8.3)   (8.2)

$$2^-OOCCH_2CH_2CH(NH_2)COO^-$$

glutamate
(8.1)

Figure 8.2 The biosynthesis of glutamate in higher plants.

$$R_1CH(NH_2)COOH + R_2COCOOH \rightarrow R_1COCOOH + R_2CH(NH_2)COOH$$

Figure 8.3 Transamination.

    The oxoacids from which many amino acids are formed are derived from the primary processes of carbon fixation, or glycolysis, or are products of the tricarboxylic acid cycle.

Non-protein amino acids

About 300 amino acids have been isolated from plants, of which only 22 normally occur in plant proteins (*Table 8.1*). The majority of the non-protein amino acids arise as a result of minor deviations in the biosynthetic pathways of the protein amino acids. These deviations are not due only to low specificity of enzymes catalysing protein amino acid synthesis, as specific enzymes catalysing the synthesis of non-protein amino acids have been established in several cases.

    The function of protein amino acids is clear — they are the fundamental units of proteins, and proteins are essential constituents of all living matter. The function of non-protein amino acids is much less clear, however, but as they are so numerous and widely distributed in the plant kingdom they are obviously necessary plant constituents and not just metabolic waste products or 'mistakes' by enzymes synthesising protein amino acids.

    Investigations to date indicate that the non-protein amino acids have several functions. The diamines, which accumulate in the seeds of some plants, rapidly disappear on germination and are, therefore, a medium for storing the nitrogen needed by the young plants. Many non-protein amino acids are analogues of the protein acids, and these act as metabolic inhibitors, especially to microorganisms. In some cases, they are also toxic to insects, animals, including man, or to other plants. Thus, such compounds are an important defence mechanism for the plant producing them. Several non-protein amino acids such as ornithine (8.4), citrulline (8.5) and *O*-acetylserine (8.6) are important metabolic intermediates, while the

Table 8.1 Amino acids found in plant proteins.

glycine	$CH_2(NH_2)COOH$
alanine	$CH_3CH(NH_2)COOH$
valine	$(CH_3)_2CHCH(NH_2)COOH$
leucine	$(CH_3)_2CHCH_2CH(NH_2)COOH$
isoleucine	$C_2H_5(CH_3)CHCH(NH_2)COOH$
aspartic acid	$HOOCCH_2CH(NH_2)COOH$
glutamic acid	$HOOCCH_2CH_2CH(NH_2)COOH$
asparagine	$H_2NOCCH_2CH(NH_2)COOH$
glutamine	$H_2NOCCH_2CH_2CH(NH_2)COOH$
proline	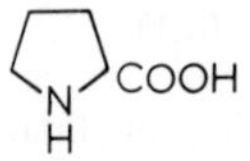
serine	$HOCH_2CH(NH_2)COOH$
threonine	$CH_3CH(OH)CH(NH_2)COOH$
lysine	$H_2NCH_2CH_2CH_2CH_2CH(NH_2)COOH$
arginine	$H_2N(=NH)CNHCH_2CH_2CH_2CH(NH_2)COOH$
histidine	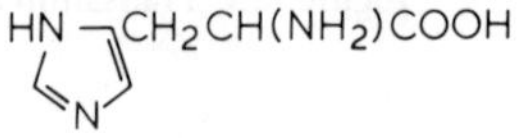
cysteine	$HSCH_2CH(NH_2)COOH$
methionine	$CH_3SCH_2CH_2CH(NH_2)COOH$
phenylalanine	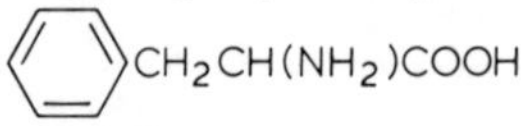
tyrosine	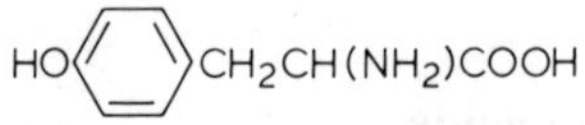
tryptophan	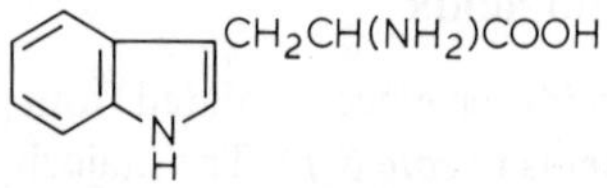
cystine	$HOOCCH(NH_2)CH_2SSCH_2CH(NH_2)COOH$
4-hydroxyproline	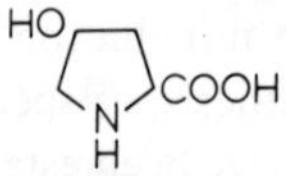

prevention of the germination of foreign pollen by non-protein amino acids in stylar tissue ensures the purity of the species and prevents hybridisation.

$H_2NCH_2CH_2CH_2CH(NH_2)COOH$ $H_2NOCNHCH_2CH_2CH_2CH(NH_2)COOH$

ornithine   citrulline
(8.4)   (8.5)

$H_3CCOOCH_2CH(NH_2)COOH$

O-acetylserine
(8.6)

In general, a non-protein amino acid can be said to act as an analogue of a protein amino acid if its antimetabolic activity is reversed by the protein amino acid. The non-protein amino acids which are analogues of the protein amino acids are toxic to microorganisms, other plants and, in sufficient concentration, to animals. The most potent have a shape similar to the protein amino acid, show similar ionisation behaviour and are, in general, slightly smaller, rather than larger, than the protein amino acid. Thus, the smaller azetidine-2-carboxylic acid (8.7) is a potent analogue of the protein amino acid, proline (8.8), whereas the larger pipecolic acid (8.9) has little analogue activity. The toxicity of some non-protein amino acids to animals is not always due to analogue activity but may have other causes.

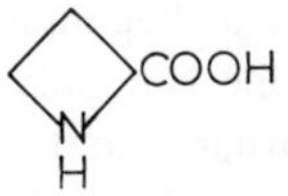

azetidine-2-
carboxylic acid
(8.7)

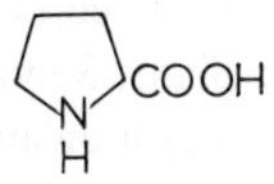

proline
(8.8)

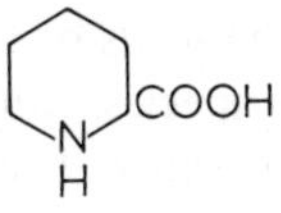

pipecolic acid
(8.9)

Protein amino acid analogues can interfere with protein synthesis in a number of ways. They may act as substrates for aminoacyl tRNA synthetases, which are involved in the synthesis of normal proteins, and thus become incorporated into a protein molecule. Replication of this protein with its 'foreign' amino acid will then take place, and if this amino acid replaces a normal protein amino acid at an active site in an enzyme, that enzyme will probably no longer be able to carry out its catalytic functions. The non-protein amino acid may also inhibit the formation of the protein amino acid or prevent its uptake into the cell, thus causing a deficiency of the essential protein amino acid. Such interference with normal protein synthesis will eventually cause the cell to die.

Plants producing protein amino acid analogues have developed ways of overcoming the antimetabolic properties of these substances. Thus, in some plants, non-protein amino acids are stored in the vacuoles away from the active sites of metabolism. In other plants, a mutation of enzymes has increased their specificity towards the protein amino acid, and the non-protein analogue can no longer act as a substrate. Such increased specificity occurs in some members of the Liliaceae synthesising azetidine-2-carboxylic acid (8.7), whose prolyl tRNA synthetase will accept proline or analogues slightly larger but not the smaller azetidine-2-carboxylic acid.

In view of the large number of non-protein amino acids, we are only going to describe in detail those which are particularly toxic to man, animals or other plants.

3-Cyanoalanine

O-acetylserine (8.6) is the precursor of a number of amino acids, both protein and non-protein. Its reaction with sulphide (*Figure 8.4*) to give the protein amino acid, cysteine (8.10), is the main entrance of sulphur into plant metabolites.

$$H_3COOCH_2CH(NH_2)COOH + H_2S \rightarrow HSCH_2CH(NH_2)COOH + H_3CCOOH$$

O-acetylserine
(8.6)

cysteine
(8.10)

Figure 8.4 The biosynthesis of cysteine.

$$HSCH_2CH(NH_2)COOH + CN^- \rightarrow N\equiv CCH_2CH(NH_2)COOH + HS^-$$

cysteine   3-cyanoalanine
(8.10)   (8.11)

Figure 8.5 The biosynthesis of 3-cyanoalanine.

Cysteine is the precursor of the toxic amino acid, 3-cyanoalanine (8.11), which occurs in the seeds of the common vetch (*Vicia sativa*, Leguminosae) and some other *Vicia* species. In the living plant, 3-cyanoalanine is mainly bound as its γ-glutamyl derivative. The enzyme, β-cyanoalanine synthase, catalysing the reaction between cysteine and cyanide (*Figure 8.5*) has been isolated from seedlings of the blue lupin (*Lupinus angustifolius*, Leguminosae).

3-Cyanoalanine is a neurotoxin, causing convulsions and muscular rigidity in rats and chicks. In rats, the compound also causes cystathioninuria, the lethal dose being 20 mg/100 g body weight. This metabolic defect can be prevented by injection of pyridoxal phosphate. Thus, β-cyanoalanine interferes with pyridoxal metabolism and inhibits the enzyme γ-cystathionase.

In non-accumulating plants, 3-cyanoalanine is enzymatically hydrolysed to the protein amino acid, asparagine, by 3-cyanoalanine hydrolase. Thus, this non-protein amino acid could be a ubiquitous constituent of plants, but may only accumulate in those species which do not contain 3-cyanoalanine hydrolase.

Mimosine

Mimosine (8.12) occurs in *Mimosa* and *Leucaena* species (Leguminosae), where it is found in both the seeds and the leaves, occurring partly as the glucoside.

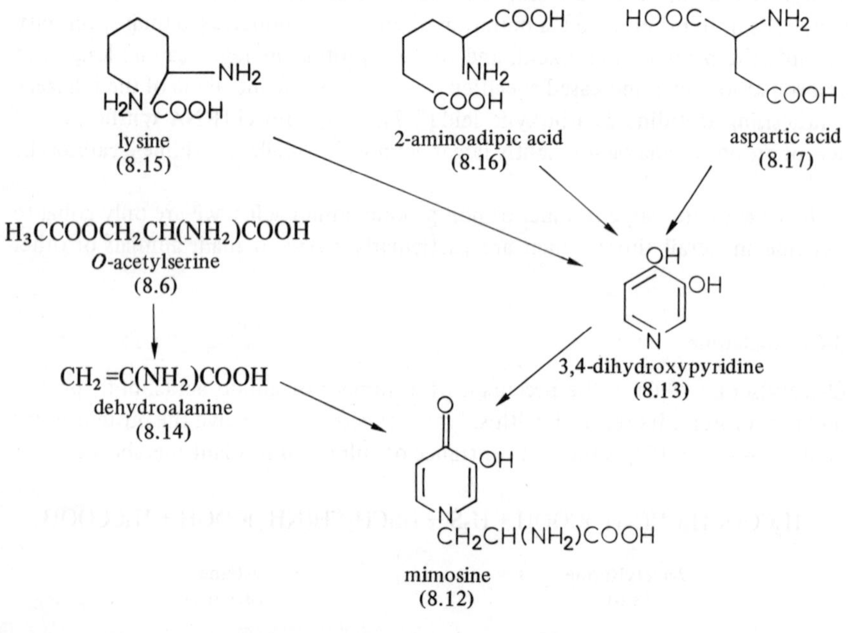

lysine
(8.15)

2-aminoadipic acid
(8.16)

aspartic acid
(8.17)

$H_3CCOOCH_2CH(NH_2)COOH$
O-acetylserine
(8.6)

$CH_2=C(NH_2)COOH$
dehydroalanine
(8.14)

3,4-dihydroxypyridine
(8.13)

$CH_2CH(NH_2)COOH$

mimosine
(8.12)

Figure 8.6 The biosynthesis of mimosine.

Mimosine is biosynthesised from 3,4-dihydroxypyridine (8.13) and O-acetylserine (8.6) with dehydroalanine (8.14) as a possible intermediate (*Figure 8.6*). The pyridine ring can be derived from lysine (8.15), 2-aminoadipic acid (8.16) or aspartic acid (8.17).

Mimosine is responsible for loss of hair in horses, donkeys, mules and pigs. In ruminants, it is degraded by stomach bacteria to dihydroxypyridine, and is thus not toxic to cattle. Loss of wool in sheep, however, has occurred in Australia, and can be so serious as to cause the death of the animals. Other symptoms of mimosine poisoning are loss of hoof, eye cataract formation, loss of fertility and haemorrhagic enteritis.

Mimosine can affect enzymes in three ways — chelating with metals in metal-containing enzymes, especially with iron; inhibiting pyridoxal phosphate; and competing with tyrosine. These enzyme inhibitions can be counteracted by supplying animals with excess iron, pyridoxal phosphate and tyrosine or phenylalanine.

Both O-acetylserine and mimosine are degraded by plant extracts, O-acetylserine giving pyruvic acid, ammonia and acetate through the intermediate formation of dehydroalanine (8.14), while mimosine is degraded to 3,4-dihydroxypyridine (8.13), pyruvate and ammonia.

N^3-oxalyldiaminopropionic Acid

N^3-oxalyldiaminopropionic acid (8.18) is the toxin present in the Indian pea (*Lathyrus sativus*, Leguminosae) responsible for neurolathyrism, a disease prevalent amongst the poor of India in times of famine, when pea meal is the only food available. This non-protein amino acid is also found in some other *Lathyrus* species and in two species of *Crotalaria* (Leguminosae). Neurolathyrism causes muscular rigidity, paralysis of the legs, convulsions and, in extreme cases, death. Experimental work has shown that these effects are not due to hydrolysis of N^3-oxalyldiaminopropionic acid, as neither oxalic acid (8.19) nor diaminopropionic acid (8.20) produce similar effects.

N^3-oxalyldiaminopropionic acid is biosynthesised from diaminopropionic acid and oxalyl CoA (*Figure 8.7*). The enzymes catalysing the synthesis of oxalyl CoA

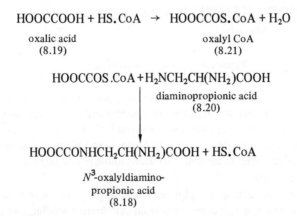

Figure 8.7 The biosynthesis of N^3-oxalyldiaminopropionic acid.

from oxalate, ATP and coenzyme A and the conversion of the diamino acid to its oxalyl derivative have been extracted from *L. sativus*.

Strains of *L. sativus* with a reduced neurotoxin content have now been bred as a result of repeated irradiation and selection.

Canavanine

Canavanine (8.22) (2-aminoguanidoxybutyric acid) has been detected in the seeds of at least 300 Papilionoideae species, and is the major non-protein amino acid in jack bean seeds (*Canavalia ensiformis*). It seems probable that this amino acid is biosynthesised by a pathway analogous to that by which ornithine (8.4) is converted to the protein amino acid, arginine. Thus, an enzyme has been obtained from jack bean seeds which catalyses both the reaction of canaline (8.23) with carbamyl phosphate (8.24) to form *O*-ureidohomoserine (8.25), and that between ornithine and carbamyl phosphate to form citrulline (8.5). Transamination of citrulline affords arginine, and it is possible that the same enzyme catalyses the transamination of *O*-ureidohomoserine to canavanine. Canaline can be derived from homoserine (8.26) and the probable biosynthesis of canavanine is given in *Figure 8.8*.

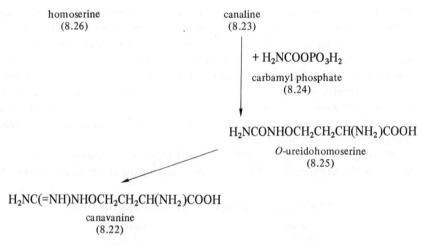

$$HOCH_2CH_2CH(NH_2)COOH \rightarrow H_2NOCH_2CH_2CH(NH_2)COOH$$

homoserine   canaline
(8.26)   (8.23)

$$+ H_2NCOOPO_3H_2$$

carbamyl phosphate
(8.24)

$$H_2NCONHOCH_2CH_2CH(NH_2)COOH$$

O-ureidohomoserine
(8.25)

$$H_2NC(=NH)NHOCH_2CH_2CH(NH_2)COOH$$

canavanine
(8.22)

Figure 8.8 The biosynthesis of canavanine.

The similarity of canavanine to arginine is sufficient to permit arginase to catalyse the hydrolysis of canavanine to canaline and urea. Canavanine competes with arginine in many metabolic reactions, and is thus a potent arginine analogue exerting antimetabolic effects on many bacteria and plants not synthesising this non-protein amino acid. In the concentrations found in jack bean seeds, canavanine is also toxic to animals.

That canavanine acts as a readily available supply of nitrogen is shown by the rapid disappearance of this compound from germinating seedlings and the appearance of newly formed homoserine.

Hypoglycin A

Hypoglycin A (3-(methylenecyclopropyl)alanine) (8.27) and its γ-glutamyl derivative were first isolated from the unripe fruits and seeds of akee (*Blighia sapida*,
Sapindaceae), while non-protein amino acids of related structure have been subsequently found in several species belonging to Sapindaceae and the closely related
Hippocastanaceae family. In animals, including man, the hypoglycins cause convulsions and, in extreme cases, death, due to low blood-sugar levels (hypoglycaemia),
a condition known as akee-akee in Jamaica, where akee fruits are eaten with salt
cod as a national dish.

When administered to animals, the hypoglycins cause the disappearance of
glycogen from the liver and the accumulation of fat particles. These effects are
due to interference with long-chain fatty acid oxidation, although the short-chain
acids are not affected. It appears that in the liver the hypoglycins are degraded to
methylenecyclopropylacetic acid (8.28), a compound which forms stable derivatives with carnitine and thus prevents the latter compound from forming the long-
chain fatty acyl carnitines essential to fatty acid oxidation.

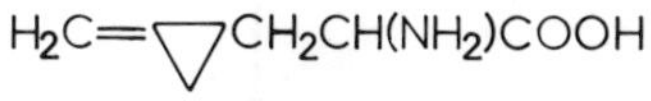

$$H_2C\!=\!\triangledown CH_2CH(NH_2)COOH \qquad\qquad H_2C\!=\!\triangledown CH_2COOH$$

hypoglycin A   methylenecyclopropylacetic acid
(8.27)   (8.28)

Little is known of the biosynthetic pathway forming hypoglycin A, but isoleucine (8.29) appears to be a precursor, as this compound was incorporated into
hypoglycin A in the seeds of akee. A possible pathway is given in *Figure 8.9*; the
isoleucine homologue (8.30) has been isolated from *Aesculus californica* (Hippocastanaceae). Chain lengthening of amino acids by acetate is a well established
process in plants .

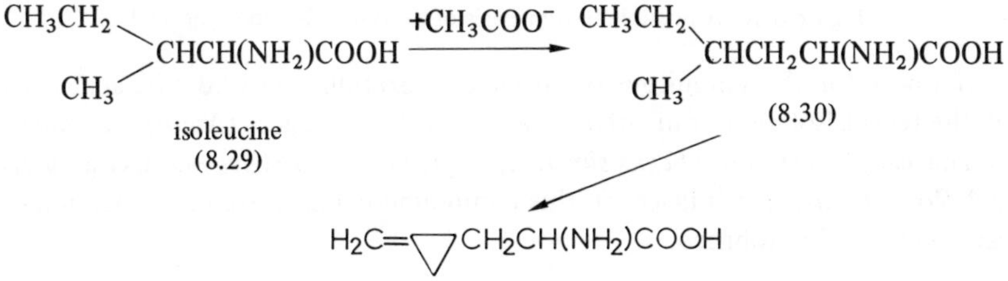

$$CH_3CH_2 \diagdown$$
$$\qquad\qquad CHCH(NH_2)COOH \xrightarrow{+CH_3COO^-} \begin{array}{c} CH_3CH_2\diagdown \\ \qquad CHCH_2CH(NH_2)COOH \\ CH_3\diagup \end{array}$$
$$CH_3\diagup$$

isoleucine   (8.30)
(8.29)

$$H_2C\!=\!\triangledown CH_2CH(NH_2)COOH$$

hypoglycin A
(8.27)

Figure 8.9 The probable biosynthesis of hypoglycin A.

Azetidine-2-carboxylic Acid

Azetidine-2-carboxylic acid (8.7) was once thought to be characteristic of the
Liliaceae, but it is now known to occur in the Amaryllidaceae, Agavaceae and
Leguminosae families, and as a very minor component of sugar beet (*Beta vulgaris*
subsp. *esculenta* var. *altissima*, Chenopodiaceae). Possibly this non-protein amino

acid is a ubiquitous constituent of plants but in quantities too small to be detected by conventional experimental methods.

Azetidine-2-carboxylic acid is a powerful analogue of the protein amino acid, proline (8.8), and has been shown to inhibit the formation of the latter compound. It also acts as a substrate for prolyl tRNA synthetase in many plants, animals and microbes, where it is incorporated into protein, including collagen, causing inhibition of growth. However, the prolyl tRNA synthetase of plants producing azetidine-2-carboxylic acid will not accept this substance as a substrate, but will accept non-protein amino acids which are slightly larger analogues of proline.

Methionine (8.31) appears to be the precursor of azetidine-2-carboxylic acid, as all four of its carbon atoms (but not the S-methyl group) are incorporated into this non-protein amino acid. Possible intermediates are homoserine (8.26), 2,4-diaminobutyric acid (8.32) and 4-amino-2-oxobutyric acid (8.33), as shown in *Figure 8.10*.

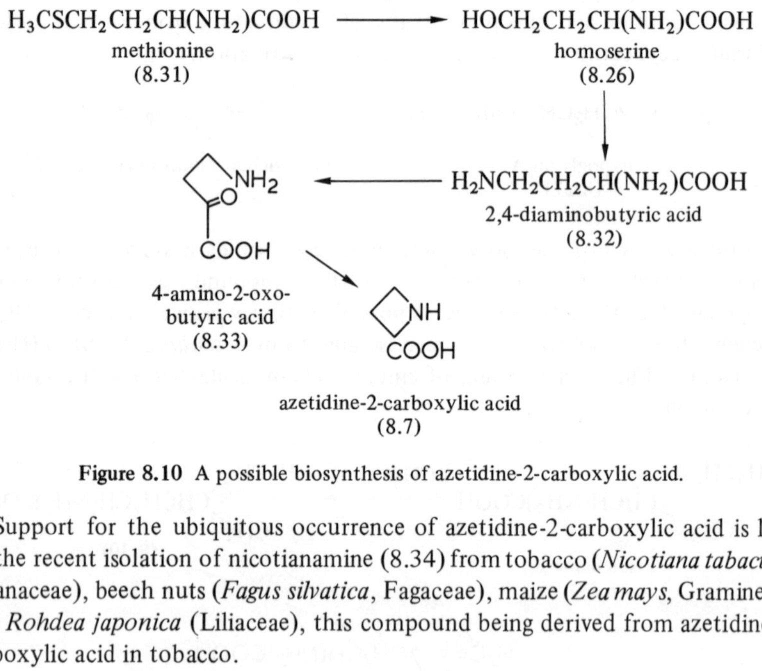

H₃CSCH₂CH₂CH(NH₂)COOH $\longrightarrow$ HOCH₂CH₂CH(NH₂)COOH

methionine   homoserine
(8.31)   (8.26)

H₂NCH₂CH₂CH(NH₂)COOH

2,4-diaminobutyric acid
(8.32)

4-amino-2-oxo-
butyric acid
(8.33)

azetidine-2-carboxylic acid
(8.7)

Figure 8.10 A possible biosynthesis of azetidine-2-carboxylic acid.

Support for the ubiquitous occurrence of azetidine-2-carboxylic acid is lent by the recent isolation of nicotianamine (8.34) from tobacco (*Nicotiana tabacum*, Solanaceae), beech nuts (*Fagus silvatica*, Fagaceae), maize (*Zea mays*, Gramineae) and *Rohdea japonica* (Liliaceae), this compound being derived from azetidine-2-carboxylic acid in tobacco.

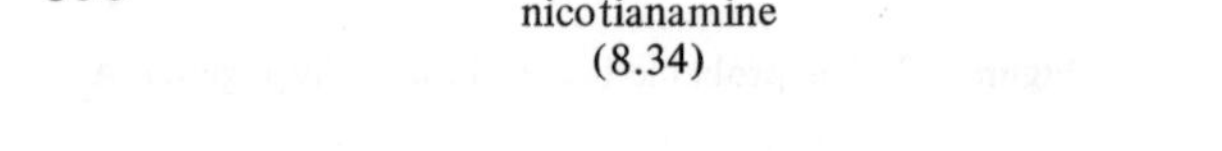

N⁺HCH₂CH₂CH(COOH)NHCH₂CH₂CH(NH₂)COOH
COO⁻
nicotianamine
(8.34)

Selenoamino Acids

Some plants are natural selenium accumulators, while others absorb this element if grown on seleniferous soils. At least 500 species of *Astragalus* (Leguminosae) grow in North America, and of these about 25 accumulate selenium in concentrations of up to 5000 ppm. In plant metabolism, selenium replaces sulphur, and the major non-protein amino acid in selenium-accumulating plants is Se-methyl-selenocysteine (8.35), while in non-accumulating plants it is canavanine (8.22). However, if grown on seleniferous soils, the non-accumulating plants synthesise Se-methylselenomethionine (8.36), which is incorporated into proteins. Accumulating plants do not incorporate selenium-containing amino acids into protein, and thus the aminoacyl tRNA synthetases of these plants are able to distinguish between sulphur and selenium, while those of non-accumulating plants cannot.

Other selenium-accumulating plants include *Haplopappus fremontii* (Compositae) and *Stanleya pinnata* (Cruciferae, which synthesise Se-methylselenocysteine, while the Australian *Neptunia amplexicaulis* (Leguminosae) and *Morinda reticulata* (Rubiaceae), together with *Lecythis ollaria* (Lecythidaceae), synthesise selenocystathionine (8.37). The nuts of *L. ollaria* contain sufficient of this compound to be toxic to man.

Non-accumulating forage plants which absorb selenium when grown on seleniferous soils include the clovers, *Trifolium repens* and *T. pratense* (Leguminosae), ryegrass (*Lolium perenne*, Gramineae) and wheat (*Aestivum sativum*, Gramineae), and these plants are responsible for alkali disease which affects animals grazing on plants growing on seleniferous soils. The selenium-accumulating plants can absorb this element from soils which contain only a few parts per million and they, especially *Astragalus* species, are responsible for the disease called 'blind staggers' which affects animals grazing them. Both types of disease can be fatal and cause much stock loss.

Work with *Astragalus bisulcatus* has shown that a common enzyme system is involved in the synthesis of the sulphur-containing amino acid, S-methylcysteine (8.38), and the selenium-containing Se-methylselenocysteine (8.35), and that these syntheses are competing. Similar results were obtained when the seeds of lima beans (*Phaseolus lunatus*, Leguminosae) were fed sulphate or selenate. Se-methylselenocysteine is formed from serine (8.39) through the intermediate, selenocysteine (8.40), and this pathway parallels that forming S-methylcysteine (*Figure 8.11*).

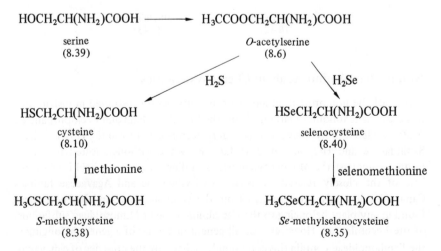

Figure 8.11 The biosynthesis of S-methylcysteine and Se-methylselenocysteine.

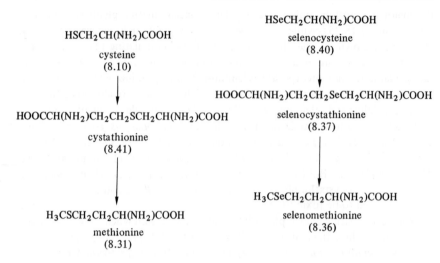

Figure 8.12 The biosynthesis of methionine and selenomethionine.

As cystathionine (8.41) is an intermediate in the conversion of cysteine (8.10) to methionine (8.31), it seems probable that selenocystathionine (8.37) is similarly an intermediate between selenocysteine (8.40) and selenomethionine (8.36) (*Figure 8.12*). If this is so, then selenium-accumulating plants are able to stop the synthesis of selenomethionine at either the selenocysteine stage or the seleno-cystathionine stage, but in non-accumulating plants the synthesis continues.

The unpleasant odours of many selenium-accumulating plants are due to volatile selenium compounds, such as dimethylselenide (8.42) and dimethyldiselenide (8.43), which are formed by enzymatic cleavage of selenium-containing amino acids.

$$H_3CSeCH_3 \qquad\qquad H_3CSeSeCH$$

dimethylselenide   dimethyldiselenide
(8.42)   (8.43)

Non-protein Amino Acids in Chemosystematics

Several of the non-protein amino acids are sufficiently restricted in distribution to be useful taxonomic markers. Thus, the branched-chain and cyclopropane derivatives related to hypoglycin A have only been isolated from the closely related Sapindaceae and Hippocastanaceae families, while, although found in various legumes, amongst the Monocotyledones, azetidine-2-carboxylic acid is characteristic of the closely related Liliaceae, Amaryllidaceae and Agavaceae families. Canavanine has only been isolated from the Papilionoideae, and has never been found in any plant belonging to the Caesalpinioideae or Mimosoideae subfamilies of the Leguminosae. However, not all genera or species of a genera belonging to the Papilionoideae contain this compound, so that only the presence of canavanine is significant. A recent classification divides the Phaseoleae tribe of Papilionoideae into seven subtribes. Of these, canavanine is characteristic of Diocleinae and Kennediinae, absent from Cajaninae and Ophrestiinae, and occurs infrequently in the remainder.

Non-protein amino acids have been found to be systematically significant within some families or genera. Thus, the tribes of the Cucurbitaceae family can be differentiated according to their non-protein amino acid content, and the presence or absence of substituted asparagines, m-carboxyphenylalanine or 3-pyrazol-l-yl-alanine has helped to place several genera or species whose morphological relationships were unclear.

The genera, *Lathyrus* and *Vicia*, of the Papilionoideae are so closely related that it is not always possible to identify isolated seeds on purely morphological grounds. However, the presence or absence of certain characteristic non-protein amino acids (*Table 8.2*) makes identification certain.

Probably the most consistent occurrence of a non-protein amino acid is that of N-acetyldjenkolic acid in the Gummiferae section of the *Acacia* genus (Mimosoideae). Every species assigned to this section on morphological grounds contains N-acetyldjenkolic acid as the major non-protein amino acid.

Table 8.2 Non-protein amino acids characteristic of *Lathyrus* and *Vicia*.

Lathyrus	Vicia
homoarginine	canavanine
2,4-diaminobutyric acid	3-cyanoalanine
N^3-oxalyldiaminopropionic acid	
lathyrine	

Amines

Amines occurring in plants can be generally divided into the volatile, aliphatic compounds, the di- and polyamines and the aromatic compounds which, because of their physiological properties, were once included with the alkaloids but are now classified as protoalkaloids. The volatile amines have unpleasant, fishy odours and are produced by some flowers, especially the aroids, to attract insects. Although

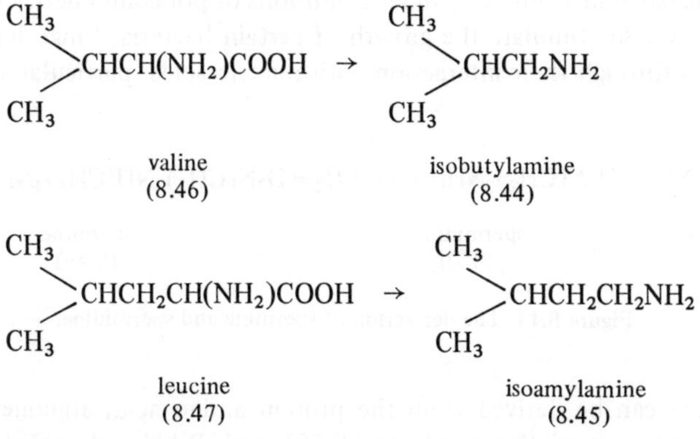

$$\begin{array}{c} CH_3 \\ {>} CHCH(NH_2)COOH \\ CH_3 \end{array} \rightarrow \begin{array}{c} CH_3 \\ {>} CHCH_2NH_2 \\ CH_3 \end{array}$$

valine (8.46) → isobutylamine (8.44)

$$\begin{array}{c} CH_3 \\ {>} CHCH_2CH(NH_2)COOH \\ CH_3 \end{array} \rightarrow \begin{array}{c} CH_3 \\ {>} CHCH_2CH_2NH_2 \\ CH_3 \end{array}$$

leucine (8.47) → isoamylamine (8.45)

Figure 8.13 The biosynthesis of isobutylamine and isoamylamine.

many of the lower aliphatic amines appear to be biosynthesised by transamination of aldehydes, rather than decarboxylation of the corresponding amino acids, iso-butylamine (8.44) and isoamylamine (8.45) are derived from the protein amino acids, valine (8.46) and leucine (8.47), respectively, in the flowers of *Sorbus aucuparia* and *Crataegus monogyna* (Rosaceae) (*Figure 8.13*).

3-Aminopropionitrile

3-Aminopropionitrile (8.48) is the compound present in some *Lathyrus* species (Leguminosae), especially sweet-pea seeds (*L. odoratus*), causing osteolathyrism in man and animals. Although 3-aminopropionitrile is generally present in the fresh plant as the γ-glutamyl derivative, only the free amine is an effective lathyrogen.

Osteolathyrogens cause skeletal deformations due to interference with the cross-linking between polypeptide chains in the connective tissue components, elastin and collagen. This interference is partly due to inhibition of the synthesis of desmosine and isodesmosine, amino acids which effect the cross-linking in elastin. Cross-linking interference of collagen can be prevented in chicks by feeding excess calcium.

$$N{\equiv}CCH_2CH_2NH_2$$

3-aminopropionitrile
(8.48)

It is possible that 3-aminopropionitrile is a degradation product of 3-cyano-alanine (8.11), although in many plants the latter compound is converted to asparagine and it has not been detected in *Lathyrus* species.

Polyamines

The di- and polyamines are important as growth regulators in plants, properties which are related to their basic nature and ability to bind with acidic groups, especially those of the nucleic acids. Spermine (8.49) and spermidine (8.50) are widespread in higher plants, and together with putrescine (8.51), from which they are derived (*Figure 8.14*), these polyamines are responsible for maintaining the acid/base balance in plant cells under conditions of potassium deficiency or high acidity. They also stimulate the growth of certain bacteria, fungi, higher plants and animals through their interaction with nucleic acids, particularly ribosomal RNA.

$$H_2N(CH_2)_4NH_2 \rightarrow H_2N(CH_2)_3NH(CH_2)_4NH_2 \rightarrow H_2N(CH_2)_3NH(CH_2)_4NH(CH_2)_3NH_2$$

putrescine   spermidine   spermine
(8.51)   (8.50)   (8.49)

Figure 8.14 The derivation of spermine and spermidine.

Putrescine can be derived from the protein amino acid, arginine (8.52), in higher plants by two pathways (*Figure 8.15*), path B being the more important, although decarboxylation of ornithine (8.4) has been observed in pea seedlings

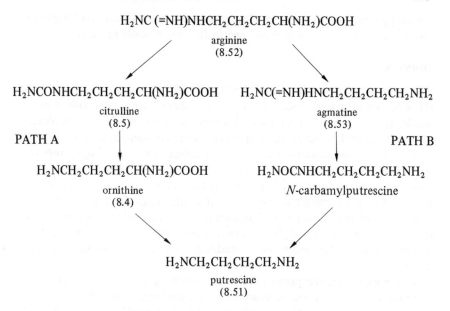

$$H_2NC(=NH)NHCH_2CH_2CH_2CH(NH_2)COOH$$

arginine
(8.52)

$$H_2NCONHCH_2CH_2CH_2CH(NH_2)COOH \qquad H_2NC(=NH)HNCH_2CH_2CH_2CH_2NH_2$$

citrulline   agmatine
(8.5)   (8.53)

PATH A   PATH B

$$H_2NCH_2CH_2CH_2CH(NH_2)COOH \qquad H_2NOCNHCH_2CH_2CH_2CH_2NH_2$$

ornithine   *N*-carbamylputrescine
(8.4)

$$H_2NCH_2CH_2CH_2CH_2NH_2$$

putrescine
(8.51)

Figure 8.15 The biosynthesis of putrescine.

and tobacco plants. Both putrescine and agmatine (8.53) accumulate in plants suffering from potassium deficiency, and the enzymes, arginase decarboxylase and *N*-carbamylputrescine amidohydrolase, are more active under such conditions.

Many plants contain amine oxidases which catalyse the oxidation of amines to aldehydes. After oxidation, putrescine (8.51) and cadaverine (8.54) spontaneously cyclise to form 1-pyrroline (8.55) and 1-piperideine (8.56), respectively (*Figure 8.16*), the basic units of some alkaloids (see Chapter 9). In non-alkaloid-

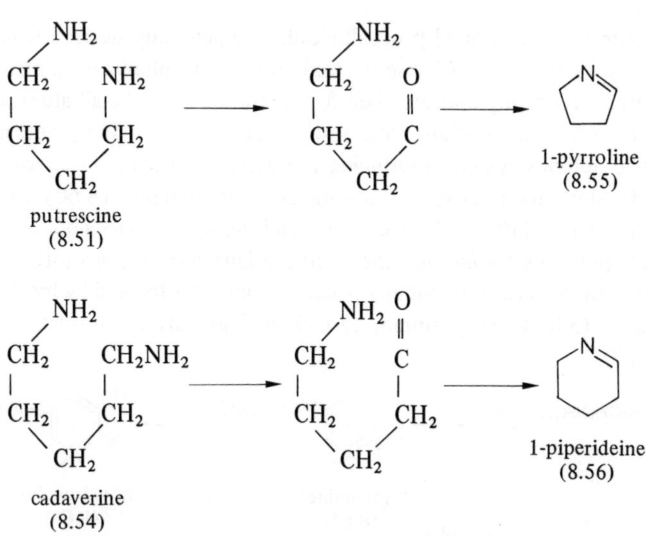

Figure 8.16 The oxidation of putrescine and cadaverine.

forming plants, these oxidation products may polymerise. *Helianthus* (Compositae) tissue is able to degrade spermine to spermidine and possibly putrescine.

Histamine

The heterocyclic amine, histamine (8.57), is the plant constituent responsible for the wheals and itching associated with many stinging hairs, especially those of nettles (*Urtica* species, Urticaceae). This compound only produces its unpleasant effects if injected, as when ingested it is converted to non-physiologically active compounds in the digestive tract. The physiological effects of histamine vary with the animal species; in man an injection of this amine causes a fall in arterial blood pressure, blushing and headache. At the site of injection, oedema occurs, forming the white bumps characteristic of nettle stings. Itching is a result of stimulation of the sensory nerves. In large doses, or particularly sensitive people, histamine can produce anaphylaxis, a hyposensitive state such that the introduction of further doses of the amine causes breathlessness, pallor, collapse and sometimes death.

Histamine is formed in plants and animals from the protein amino acid, histidine (8.58) (*Figure 8.17*), whose biosynthesis is described in Chapter 10.

Histamine has an erratic occurrence in the plant world, but is particularly common in the Chenopodiaceae family.

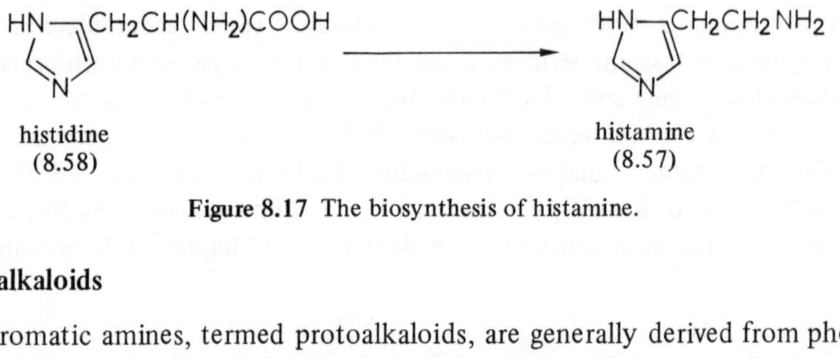

histidine
(8.58)

histamine
(8.57)

Figure 8.17 The biosynthesis of histamine.

Protoalkaloids

The aromatic amines, termed protoalkaloids, are generally derived from phenylalanine or tyrosine (8.59). (Those derived from tryptophan have a nitrogen-containing heterocyclic ring and are therefore included with the alkaloids discussed in Chapter 9.) A number of aromatic amines occurring in higher plants affect the sympathetic nervous system, producing the increase in heart rate, sweating and muscular tension characteristic of adrenaline. Noradrenaline (8.60) has been isolated from bananas (*Musa*, Musaceae), spinach leaves (*Spinacea oleracea*, Chenopodiaceae), potatoes (*Solanum tuberosum*, Solanaceae) and aconite (*Aconitum napellus*, Ranunculaceae). In banana tissue, this amine is formed by hydroxylation of dopamine (3-hydroxytyramine) (8.61), a degradation product of tyrosine (*Figure 8.18*).

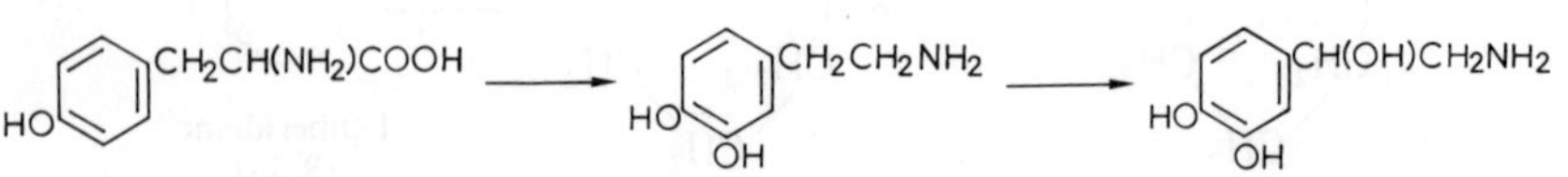

tyrosine
(8.59)

dopamine
(8.61)

noradrenaline
(8.60)

Figure 8.18 The biosynthesis of noradrenaline in banana tissue.

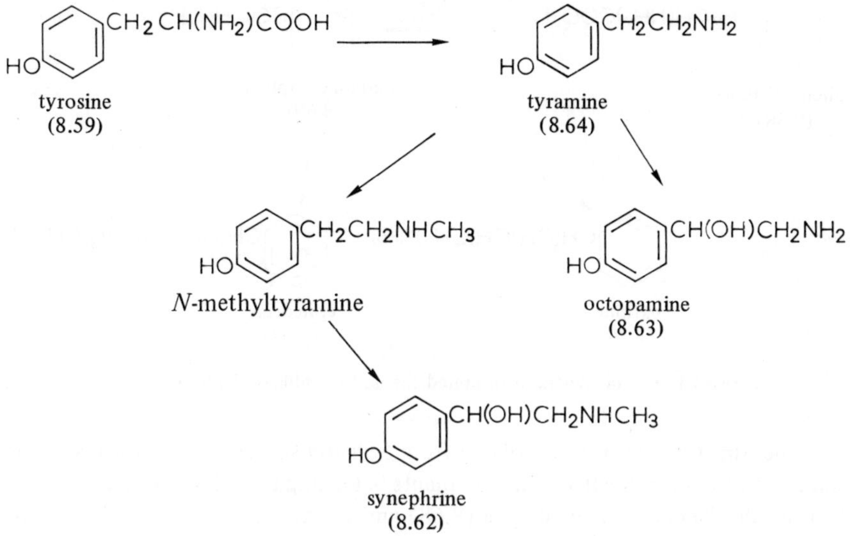

CH$_2$CH(NH$_2$)COOH

HO

tyrosine
(8.59)

CH$_2$CH$_2$NH$_2$

HO

tyramine
(8.64)

CH$_2$CH$_2$NHCH$_3$

HO

N-methyltyramine

CH(OH)CH$_2$NH$_2$

HO

octopamine
(8.63)

CH(OH)CH$_2$NHCH$_3$

HO

synephrine
(8.62)

Figure 8.19 The biosynthesis of the sympathomimetic amines.

The related sympathomimetic amines, synephrine (8.62) and octopamine (8.63), are derived from tyrosine through the intermediate formation of tyramine (8.64) (*Figure 8.19*). Synephrine occurs mainly in the Rutaceae and Amaryllidaceae, while octopamine occurs in the Rutaceae, Solanaceae, Liliaceae and Cyperaceae families.

Tyrosine is also the precursor of hordenine (8.65), the protoalkaloid character-istic of germinating barley (*Hordeum sativum*, Gramineae), and mescaline (8.66), the hallucinogen isolated from peyote or mescal buttons (*Lophophora williamsii*, Cactaceae). Dopamine (8.61) appears to be an intermediate in the biosynthesis of mescaline, with the methyl groups being donated by methionine (*Figure 8.20*).

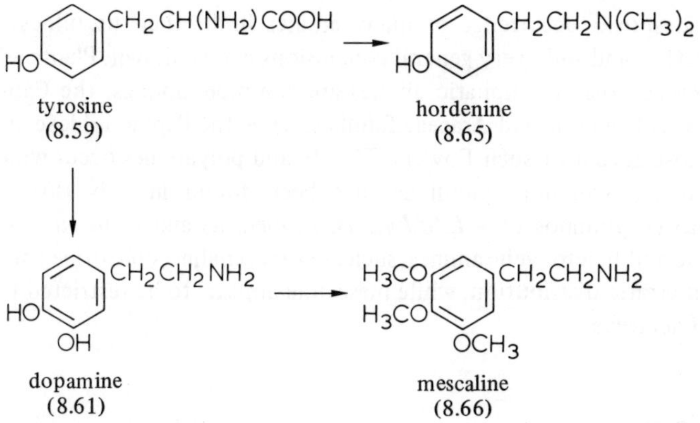

CH$_2$CH(NH$_2$)COOH

HO

tyrosine
(8.59)

CH$_2$CH$_2$N(CH$_3$)$_2$

HO

hordenine
(8.65)

CH$_2$CH$_2$NH$_2$

HO

OH

dopamine
(8.61)

H$_3$CO

H$_3$CO

CH$_2$CH$_2$NH$_2$

OCH$_3$

mescaline
(8.66)

Figure 8.20 The biosynthesis of hordenine and mescaline.

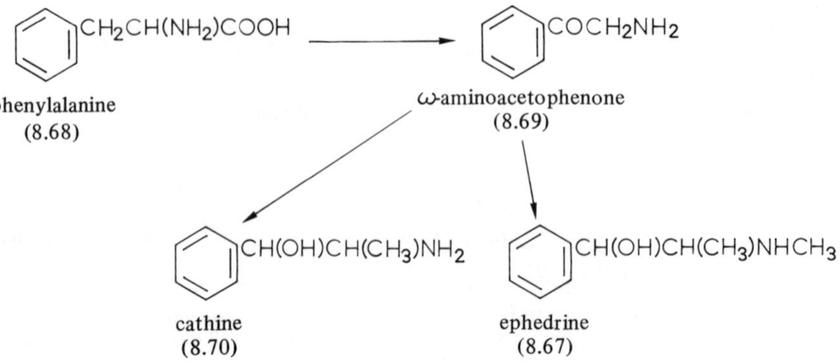

Figure 8.21 The biosynthesis of ephedrine and possible derivation of cathine.

Ephedrine (8.67) was first isolated from *Ephedra* species (Gnetaceae), and has since been found in small amounts in plants belonging to widely varying families. Isotopically labelled phenylalanine (8.68) and ω-aminoacetophenone (8.69) are converted to ephedrine in *E. distachya* (*Figure 8.21*). The C-methyl group is derived from formate and the N-methyl from methionine.

Ephedrine is used medicinally as a bronchodilator in asthma and bronchial spasm. It is also one of the pressor amines which raise the blood pressure and blood-sugar levels. Other effects of this drug are the prevention of sleep, nervousness, and dilation of the pupil of the eye.

Cathine (norpseudoephedrine) (8.70) is the active principle of the drug 'khat' or 'miraa' obtained from *Catha edulis* (Celastraceae). Cathine is a powerful exciter of the nervous system, abolishing sleep and sustaining muscular activity. *C. edulis* stems are chewed by many African tribes as a stimulant.

Cathine is derived from phenylalanine, possibly by a pathway similar to ephedrine (*Figure 8.21*).

Amines in Chemosystematics

The widespread occurrence of many amines limits their usefulness in chemosystematics, and only very general conclusions can be drawn. Plants with flowers synthesising volatile, aliphatic amines are common amongst the Caprifoliaceae, Cornaceae, Rosaceae and Araceae families, while the Papilionoideae and Labiatae are almost devoid of such flowers. The di- and polyamines occur widely but the monoamine, 3-aminopropionitrile, has been found in only three species of *Lathyrus* (Leguminosae) – *L. odoratus*, *L. hirsutus* and *L. roseus*. Some of the aromatic and heterocyclic amines, such as noradrenaline, ephedrine and histamine, have an erratic distribution, while mescaline appears to be restricted to members of the Cactaceae.

Colchicine

Because of its useful physiological properties, colchicine (8.71) is often included amongst the alkaloids. However, although biosynthesised from amino acids, this amide is not a true alkaloid, as it is neither basic nor contains a heterocyclic ring system.

Colchicine was first isolated from the autumn crocus (*Colchicum autumnale*), but has since been found in many plants belonging to the Liliaceae family. Isotopic tracer experiments have shown the tropolone ring to be derived from tyrosine (8.59) and the remainder of the molecule from phenylalanine (8.68), with acetate giving rise to the acetyl group and methionine the methyl groups. *O*-methylandrocymbine (8.72) and demecolcine (8.73) appear to be the precursors of colchicine in the autumn crocus (*Figure 8.22*).

In the last century, colchicine was much prescribed for the relief of pain due to gout, and poisoning due to overdosage was frequent: 3–5 mg of the drug is sufficient to cause death, the poison taking 3–6 h to work. Typical symptoms of colchicine poisoning are vomiting, diarrhoea, pain in the bowels, lowering of

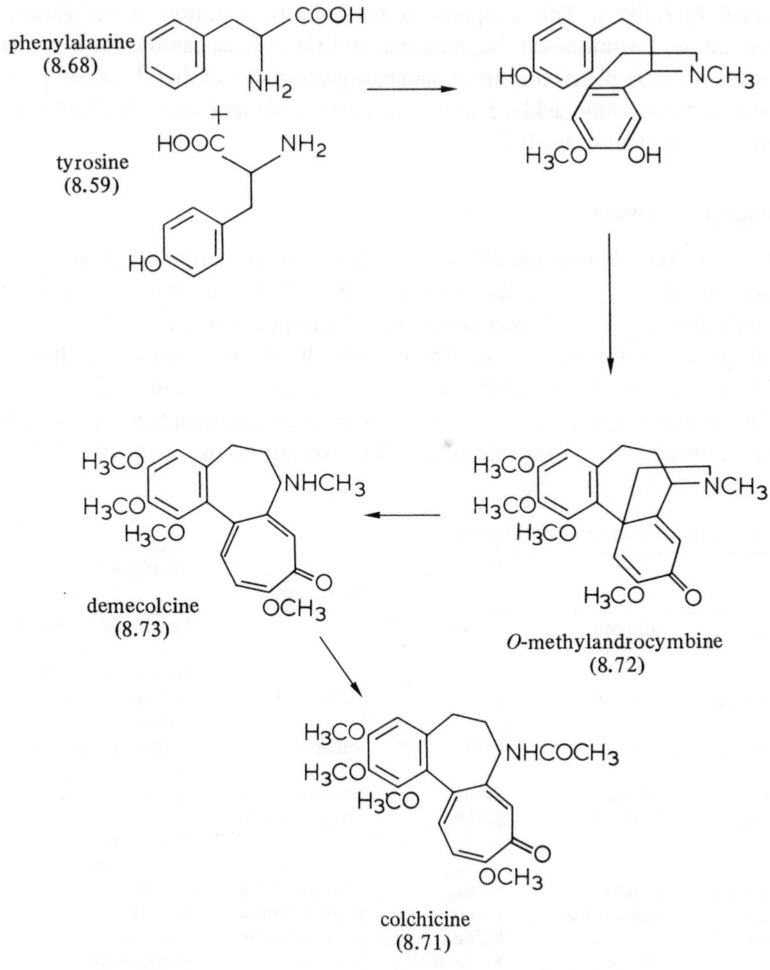

Figure 8.22 The biosynthesis of colchicine.

body temperature, circulatory collapse and death due to asphyxia. Colchicine has also caused human poisoning through its use as an abortifacient and the inadvertent inclusion of the leaves of autumn crocus in salads. In the tropics, the bulbs of *Gloriosa* species, which contain colchicine, have been used medicinally as an abortifacient or have been mistaken for edible yams. The poisoning of livestock is generally due to grazing autumn crocus or other colchicine-containing members of the Liliaceae.

Colchicine has the ability to produce polyploids by doubling the number of chromosomes, and this amide has been used experimentally in medicine, horticulture and biology to obtain mutants.

Cyanogenic glycosides and cyanolipids

Over 100 plant species have the ability to produce hydrogen cyanide by enzyme-catalysed hydrolysis. This property is particularly common in the Rosaceae, Passifloraceae, Leguminosae, Sapindaceae and Gramineae families. It is probable that most of these plants contain cyanogenic glycosides and that these compounds are derived from amino acids. The seed oils of several members of the Sapindaceae family contain cyanogenic lipids.

Cyanogenic glycosides

Work with flax (*Linum usitatissimum*, Linaceae) and sorghum (*Sorghum* sp., Gramineae), which produce linamarin (8.74a) and dhurrin (8.75a), respectively, and with cherry laurel (*Prunus laurocerasus*) and peach (*P. persica*), members of the Rosaceae, producing prunasin (8.76a), has shown that cyanogenic glycosides can be biosynthesised from amino acids by two pathways (*Figure 8.23*).

The amino acids are first converted to aldoximes, which then form cyanohydrins either via nitriles or hydroxyaldoximes. The oxidations involved are effected by

Table 8.3 Common cyanogenic glycosides.

Glycoside	Sugar	Structure	Amino acid precursor	Occurrence
linamarin	glucose	8.74a	valine	*Linum usitatissimum* *Phaseolus lunatus* *Manihot esculenta*
lotaustralin	glucose	8.74b	isoleucine	*Trifolium repens* *Lotus* sp.
acacipetalin	glucose	8.77	leucine?	*Acacia* sp. (South African)
prunasin	glucose	8.76a	phenylalanine	Rosaceae
sambunigrin	glucose	8.76b	phenylalanine	*Sambucus* sp. *Acacia* sp. (Australian)
prulaurasin	glucose	8.76c	phenylalanine	*Prunus* sp.
amygdalin	gentiobiose	8.76d	phenylalanine	Rosaceae
vicianin	vicianose	8.76e	phenylalanine	*Vicia* sp.
dhurrin	glucose	8.75a	tyrosine	*Sorghum* sp.
taxiphyllin	glucose	8.75b	tyrosine	*Taxus* sp.

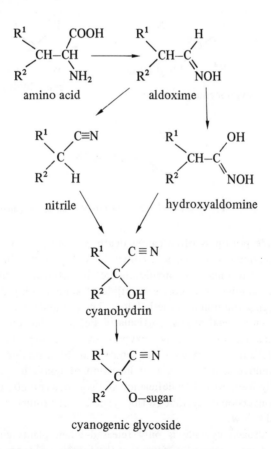

H_3C, $C \equiv N$
C
R O–glucose

HO — CH — $C \equiv N$, O–glucose

R = CH$_3$, linamarin (8.74a)
R = C$_2$H$_5$, lotaustralin (8.74b)

S-isomer = dhurrin (8.75a)
R-isomer = taxiphyllin (8.75b)

$C \equiv N$
C
O–sugar

H_3C, $C \equiv N$
C–CH
H_2C O–glucose

acacipetalin (8.77)

sugar = glucose: S-isomer = prunasin (8.76a)
                         R-isomer = sambunigrin (8.76b)
                         R, S mixture = prulaurasin (8.76c)
sugar = gentiobiose, amygdalin (8.76d)
sugar = vicianose, vicianin (8.76e)

R^1 COOH   R^1 H
CH–CH ────→ CH–C
R^2 NH$_2$ R^2 NOH
amino acid   aldoxime

R^1 C≡N
C
R^2 H
nitrile

R^1 OH
CH–C
R^2 NOH
hydroxyaldomine

R^1 C ≡ N
C
R^2 OH
cyanohydrin

R^1 C ≡ N
C
R^2 O–sugar

cyanogenic glycoside

Figure 8.23 The biosynthesis of cyanogenic glycosides.

molecular oxygen. The formation of glucosides from cyanohydrins is catalysed by β-glucosyltransferases.

The more common cyanogenic glycosides are listed in *Table 8.3*, together with their occurrence and amino acid precursors.

In the living, undamaged plant, cyanogenic glycosides are metabolised to amino acids, especially asparagine, but when the plant is damaged in any way these glycosides form free hydrogen cyanide. The first step in the degradation process (*Figure 8.24*) is the removal of the sugar and, in the case of glucose, this reaction is catalysed by β-glucosidases. The cyanohydrin which results can dissociate non-enzymatically to give hydrogen cyanide and an aldehyde or ketone, but in plants the reaction is usually catalysed by enzymes. Emulsin, an enzyme system obtained from almond kernels (*Prunus amygdalus*, Rosaceae), will catalyse both the hydrolysis of sugars and the formation of hydrogen cyanide. In amygdalin (8.76d), the gentiobiose is hydrolysed first to glucose (forming prunasin (8.76a)) and then the second glucose molecule is removed. Emulsin is specific for aromatic cyanogenic glycosides whereas linamerase separated from flax and from white clover (*Trifolium repens*, Leguminosae) will catalyse the hydrolysis of both aliphatic and aromatic glucosides but not that of diglucosides.

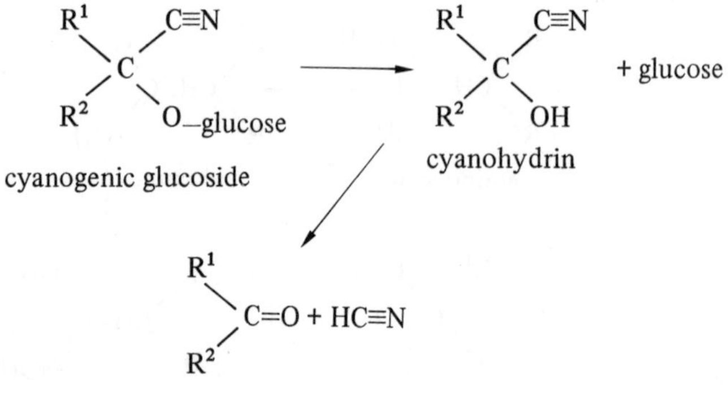

Figure 8.24 The degradation of cyanogenic glucosides.

Cyanide poisoning often causes death with no previous symptoms, although rapid breathing, vomiting and convulsions may occur. In animals, listlessness, cessation of feeding and convulsions are characteristic of this type of poisoning. Sheep are amongst the most susceptible of animals, only 2.4 mg HCN/kg body weight being the minimum lethal dose. The cyanide ion exerts its effect by complexing with metal atoms, particularly Fe, in metal-containing enzymes. This causes inactivation of many enzymes, the most important being cytochrome oxidase (the respiratory enzyme), which leads to death due to asphyxia.

In ruminants, the continuous ingestion of non-lethal quantities of cyanide eventually leads to iodine deficiency and goitrous conditions, these effects being due to conversion of cyanide to thiocyanate in the rumen. Goitrogens are further discussed below.

As hydrogen cyanide is only released when plants containing cyanogenic glycosides are damaged, it seems that the function of these compounds is to protect the plant from further damage, cyanide having an unpleasant, bitter taste. When slugs and snails were placed amongst cyanogenic and acyanogenic strains of white clover, these animals ate only the non-cyanide-producing plants.

In white clover and some *Lotus* species, the production of cyanogenic glycosides is controlled by single genes which can be bred out. Thus, acyanogenic strains have been developed for animal fodder.

Cyanolipids

The cyanolipids found in the seed oils of members of the Sapindaceae family can be divided into four types (8.78a–d). In the majority of cases, the fatty acid components are C_{20} compounds containing one unsaturated group.

Isotopic tracer experiments with *Koelreuteria paniculata* have shown the protein amino acid, leucine, to be the precursor of the cyanolipids (8.78e) and (8.78f).

$R^1 = H$, R^2 = acyl (8.78a)	$R^1 = H$, R^2 = acyl (8.78c)
R^1 = acyl-O, R^2 = acyl (8.78b)	R^1 = acyl-O, R^2 = acyl (8.78d)
	$R^1 = H$, $R^2 = CH_3(CH_2)_{18}CO$ (8.78e)
	$R^1 = CH_3(CH_2)_{18}COO$,
	$R^2 = CH_3(CH_2)_{18}CO$ (8.78f)

Cyanogenic glycosides in chemosystematics

It seems likely that all plants have the ability to synthesise cyanogenic glycosides, but that in most these compounds are metabolised and not accumulated. There appears to be some relationship between the amino acid precursors of accumulated cyanogenic glycosides and the class to which a plant belongs. Thus, the ferns synthesise glycosides from phenylalanine and the gymnosperms from tyrosine, while the angiosperms are able to synthesise all types. However, the Monocotyledones and the Polycarpicae of the order Magnoliales can only utilise tyrosine, a circumstance which lends support to the theory that the Monocotyledones evolved from the Magnoliales. Within the Dicotyledones, members of the Rosaceae are characterised by their synthesis of cyanogenic glycosides, especially amygdalin, from phenylalanine, whereas some members of the Papilionoideae subfamily derive such compounds from aliphatic valine or isoleucine. Synthesis of cyanogenic glycosides based on leucine seems to be restricted to South African *Acacia* species (Leguminosae) and members of the Sapindaceae family. The latter also synthesise the cyanolipids which are characteristic of this family.

Glucosinolates

More than 70 glucosinolates of known structure have been isolated from higher plants, the Cruciferae family being particularly rich in these compounds. Unlike

Table 8.4 Some glucosinolates found in higher plants.

Glucosinolate	R– in (8.79)	Amino acid precursor	Occurrence
glucocapparin	CH_3	alanine	Capparidaceae
sinigrin	$CH_2{=}CHCH_2$	serine	Cruciferae
gluconapin	$CH_2{=}CHCH_2CH_2$	serine	Cruciferae
glucotropaeolin	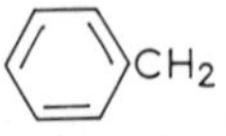 CH_2	phenylalanine	several families
sinalbin	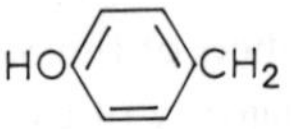 HO—CH_2	tyrosine	Cruciferae
gluconasturtiin	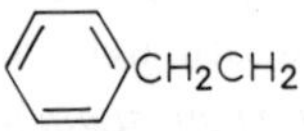 CH_2CH_2	phenylalanine	Cruciferae
glucobrassicin	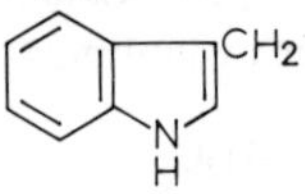 CH_2	tryptophan	*Brassica* sp.

the cyanogenic glycosides, where it is unusual for a plant to contain more than one or two such compounds, each glucosinolate-containing species synthesises several compounds. All naturally occurring glucosinolates have the general formula (8.79) and some examples are given in *Table 8.4*.

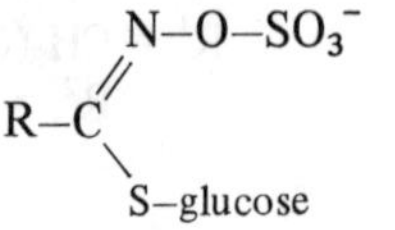

$$R-C\underset{\textstyle S-glucose}{\overset{\textstyle N-O-SO_3^-}{\Big\langle}}$$

(8.79)

It is generally accepted that the protein amino acids are the precursors of the glucosinolates and that, where necessary, these are chain-lengthened by the addition of acetate units. This process (*Figure 8.25*) is similar to that established in many plants for the biosynthesis of the higher homologues of the protein amino acids.

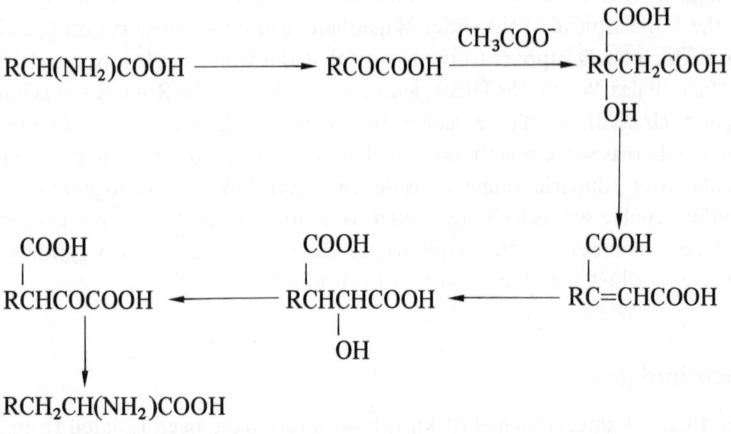

Figure 8.25 Chain lengthening of amino acids.

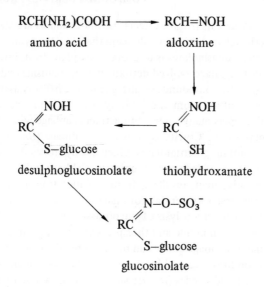

Figure 8.26 The biosynthesis of glucosinolates.

As with the cyanogenic glycosides, the glucosinolates are biosynthesised through the intermediate formation of aldoximes (*Figure 8.26*). These are converted to thiohydroxamates, the sulphur probably being derived from methionine or cysteine. A glucosyltransferase then catalyses the glucosylation of thiohydroxamates to form desulphoglucosinolates. Cell free extracts from a number of Cruciferae species contain glucosyltransferases which catalyse this reaction. *O*-sulphonation to give the glucosinolates is catalysed by sulphotransferases which utilise 3′-phosphoadenosine-5′-phosphosulphate.

Although occurring in high concentration in the intact plant, glucosinolates were not biosynthesised by tissue cultures of turnip (*Brassica napus*, Cruciferae).

Particularly high concentrations of glucosinolates are found in the flowers, seeds and roots of some plants. Although themselves odourless and tasteless, these compounds are hydrolysed to pungently odoured and tasting isothiocyanates when the plant is damaged. This hydrolysis (*Figure 8.27*) is catalysed by myrosinases (thioglucosidases). Other degradation products include thiocyanates, RSCN, nitriles, RCN, epithionitriles, $\overset{\diagdown}{\underset{S}{\diagup}}(CH_2)_nCN$ and elemental sulphur.

Isothiocyanates, which are responsible for the pungent flavours of many crucifers, are not very toxic to animals, although in high concentration they irritate, causing vomiting and diarrhoea. Mustard (*Brassica nigra*) has been known

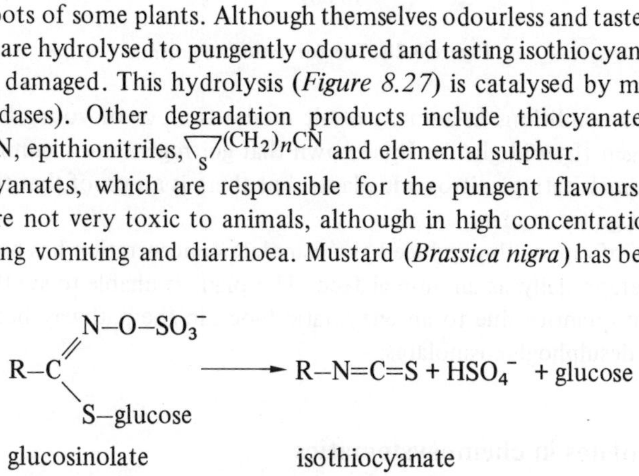

$$R-C \begin{matrix} {}^{\diagup N-O-SO_3^-} \\ {}_{\diagdown S-glucose} \end{matrix} \longrightarrow R-N=C=S + HSO_4^- + glucose$$

glucosinolate   isothiocyanate

Figure 8.27 The hydrolysis of glucosinolates by myrosinases.

to cause death through its irritant effects. Isothiocyanates are antibacterial, anti-fungal and insecticidal, and thus it would seem that, like the cyanogenic glycosides, the purpose of the glucosinolates is to protect the plant from invaders. However, some insects and fungi have evolved detoxifying mechanisms and are even stimu-lated to feed or grow by compounds resulting from the hydrolysis of glucosinolates. The mustard oil resulting from the hydrolysis of sinigrin, characteristic of the Cruciferae family, repels many insects but attracts cabbage white butterflies.

Some members of the Cruciferae, especially cabbage (*Brassica oleracea*) and rape (*B. napus*), contain glucosinolates which are hydrolysed to substances caus-ing goitrous conditions if ingested by animals. Cabbage goitre is a condition in animals and, possibly, man, resulting from the ingestion of large quantities of cabbage or kale (at least 1 kg per day for man). It is due to inorganic thiocyanate formed as a result of the hydrolysis of glucobrassicin (*Table 8.4*). The thiocyanate competes reversibly with iodide and thus prevents the accumulation of the latter in the thyroid gland. Cabbage goitre can be cured by giving iodine supplements.

The goitrous conditions caused by the seeds and roots of many *Brassica* species, especially rape seed and swedes (*B. rutabaga*), are mainly due to goitrin (8.80), a hydrolysis product (*Figure 8.28*) of progoitrin (8.81), which inhibits the for-mation of thyroxine. Other 2-thiooxazolidones have similar effects, and thyroxine inhibition cannot be cured by dosing animals with iodine.

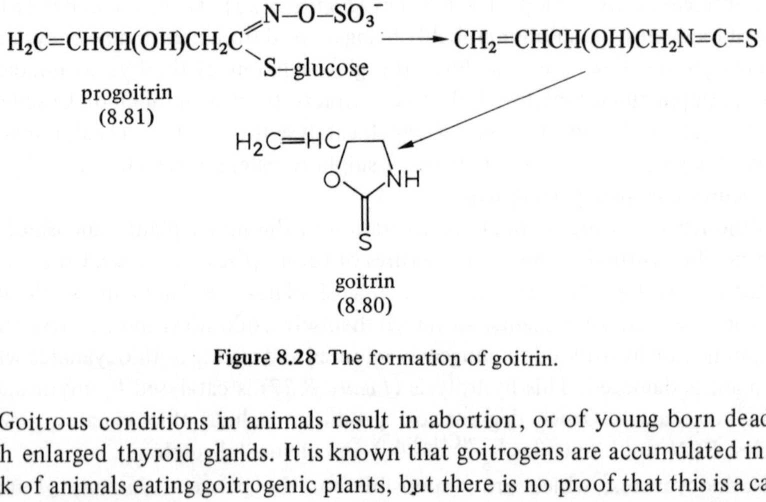

Figure 8.28 The formation of goitrin.

Goitrous conditions in animals result in abortion, or of young born dead or with enlarged thyroid glands. It is known that goitrogens are accumulated in the milk of animals eating goitrogenic plants, but there is no proof that this is a cause of endemic goitre in man.

A strain of rape with a reduced glucosinolate concentration has been bred and is used preferentially as an animal feed. The plant is unable to synthesise gluco-sinolates in quantity due to an enzymatic block in the pathway before the for-mation of desulphoglucosinolates.

Glucosinolates in chemosystematics

Glucosinolates occur abundantly in the Cruciferae and the related Capparidaceae, Resedaceae and Moringaceae families. Papaveraceae does not contain such compounds and, therefore, this family is not so closely related to Capparidaceae as was once thought on morphological grounds. The presence of glucosinolates in Tovariaceae shows that this family is more closely related to Capparidaceae than to Papaveraceae. Glucosinolates are useful taxonomic markers within the *Iberis* and *Arabis* genera (Cruciferae).

When the seeds of the cultivated mustard, *Brassica juncea*, were examined, it was found that those from the Indian subcontinent contained 3-butenylgluco-sinolate and allylglucosinolate, while those from other Asiatic countries contained only the allyl compound as the major component. *B. juncea* is thought to be a hybrid between *B. nigra*, which contains allylglucosinolate, and *B. campestris*, which contains the 3-butenyl compound. As artificially produced hybrids between these two species contain the allylglucosinolate almost exclusively, the ancestry of the Indian hybrid is in some doubt.

Glucosinolates also occur in some species belonging to the Tropaeolaceae and related Limnanthaceae, and the related Euphorbiaceae, Phytolaccaceae, Caricaceae, Salvadoraceae, Plantaginaceae and Gyrostemonaceae families. However, such compounds are not widespread within these families and, as yet, no chemosystem-atic relationships have been established. It seems possible that, like the cyanogenic glycosides, all plants have the ability to synthesise glucosinolates, but that in most they are metabolised and not accumulated.

Penicillins and cephalosporins

The penicillins and cephalosporins are important antibiotics obtained from species of fungi belonging to the *Penicillium, Cephalosporium* and *Streptomyces* genera. Both are cyclic peptides containing a β-lactam ring; the penicillins have also a five-membered sulphur-containing ring and the cephalosporins a six-membered sulphur-containing ring. It has been suggested that these cyclic peptides are sur-vivals from an early stage of evolution when amino acids condensed to polypep-tides in a specified sequence without an RNA template.

The penicillins are acyl derivatives of 6-aminopenicillanic acid (8.82a), while the cephalosporins are derivatives of δ(2-aminoadipyl)-7-aminocephalosporanic acid (deacetoxycephalosporin C) (8.83). The ring systems of both types of com-pound are biosynthesised from L-cysteine (8.10) and L-valine (8.46), these amino acids being incorporated intact (*Figure 8.29*), although L-valine changes to the D-isomer at some stage.

Although the exact mechanisms forming the β-lactam and heterocyclic rings have not yet been elucidated, the tripeptide, δ (L-2-aminoadipyl)-1-cysteinyl-D-valine, has been shown to be a precursor of penicillin N (8.82b) in a cell free ex-tract of *Cephalosporium acremonium*. This system could also synthesise the tri-peptide from 2-aminoadipic acid (8.16), cysteine and valine. 2-Aminoadipic acid is an intermediate in the biosynthesis of lysine in fungi, including *Penicillium* and *Cephalosporium* species, and thus it seems possible that penicillin N (the D-2-aminoadipyl derivative of 6-aminopenicillanic acid) or isopenicillin N (the L-2-aminoadipyl derivative) are the first cyclic peptides to be formed.

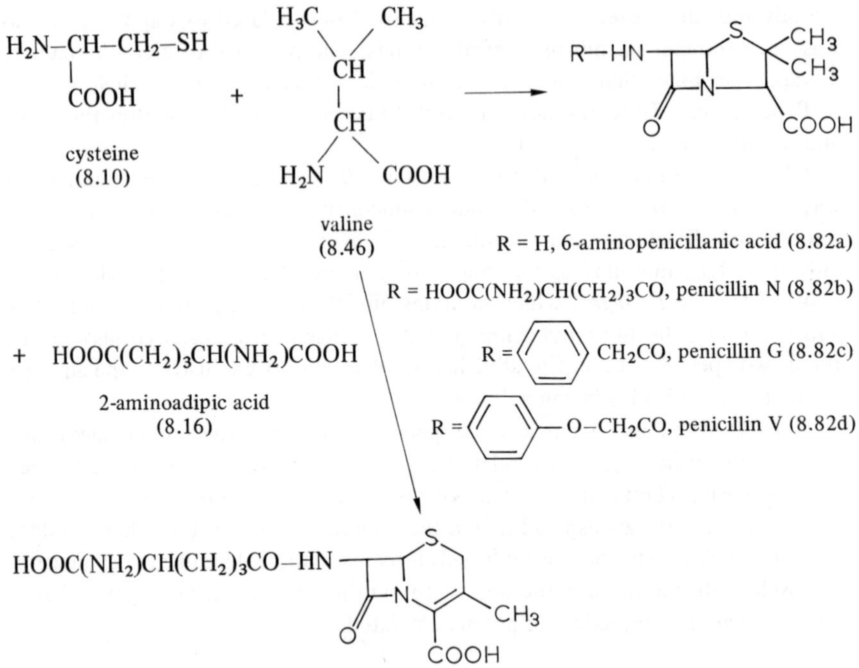

$H_2N-CH-CH_2-SH$
|
$COOH$

cysteine
(8.10)

$+$

$H_3C \quad CH_3$
\/
CH
|
CH
/\
$H_2N \quad COOH$

valine
(8.46)

$\longrightarrow$

R—HN—[β-lactam-thiazolidine ring]—CH_3, CH_3, COOH

R = H, 6-aminopenicillanic acid (8.82a)

R = $HOOC(NH_2)CH(CH_2)_3CO$, penicillin N (8.82b)

$+ \quad HOOC(CH_2)_3CH(NH_2)COOH$

2-aminoadipic acid
(8.16)

R = [phenyl]CH_2CO, penicillin G (8.82c)

R = [phenyl]$-O-CH_2CO$, penicillin V (8.82d)

$HOOC(NH_2)CH(CH_2)_3CO-HN$ —[cephalosporin ring system]— CH_3, COOH

δ(2-aminoadipyl)-7-aminocephalosporanic acid
(deacetoxycephalosporin C)
(8.83)

Figure 8.29 The biosynthesis of the penicillin and cephalosporin ring systems.

Penicillium chrysogenum contains an acyltransferase which will transfer non-polar side chains and acylate 6-aminopenicillanic acid with CoA derivatives. This *Penicillium* species also converts isopenicillin N, but not penicillin N, to benzyl-penicillin (penicillin G) (8.82c) in the presence of phenylacetyl CoA. Thus, it seems that penicillin N is the precursor of penicillins formed by *C. acremonium*, while its isomer is the precursor of those formed by *P. chrysogenum*.

Work with *C. acremonium* has shown that deacetoxycephalosporin C (8.83) is probably the first cephalosporin to be formed.

Semisynthetic penicillins can be produced by feeding *P. chrysogenum* with various substituted acetic acids. Benzylpenicillin (penicillin G) (8.82c), one of the most useful of the antibiotics, is produced naturally in small amounts, but the yield can be dramatically increased by adding phenylacetic acid to the fermen-tation medium. Penicillin G is effective against both gram positive and gram negative cocci, but, as it is hydrolysed by acid, it cannot be given by mouth. Addition of phenoxyacetic acid to the fermenting medium gives phenoxymethylpenicillin (penicillin V) (8.82d), which is more resistant to acid hydrolysis and can be given by mouth.

Synthetic penicillins with a variety of side chains are produced by acylating 6-aminopenicillanic acid. This compound is produced in quantity by a strain of

P. chrysogenum containing an amidase which catalyses the hydrolysis of the side chains of natural penicillins.

Unlike the penicillins, cephalosporins will not incorporate unnatural side chains.

The penicillins and cephalosporins are amongst the least toxic of drugs to man, but they can cause allergic reactions which range from rashes to anaphylaxis. These compounds exert their effects on bacteria by interfering with cell wall synthesis, effectively preventing viable reproduction. Mammalian cells do not have such walls.

Sulphur Compounds Derived from Cysteine

A number of sulphur compounds occurring in plants are derived from cysteine (8.10) and these are particularly plentiful in the Mimosoideae subfamily and Cruciferae and Alliaceae families. Djenkolic acid (8.84), first isolated from the djenkol bean (*Pithecolobium lobatum*), is a characteristic constituent of Mimosoideae and the *N*-acetyl derivative is useful in differentiating between the Gummiferae and other sections of the *Acacia* genus (see the section on non-protein amino acids, p. 221). Djenkolic acid is biosynthesised from cysteine, and in the plant it probably acts as a reservoir of sulphur, since an enzyme has been isolated from *A. farnesiana* which degrades djenkolic acid and a number of other cysteine

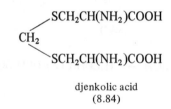

djenkolic acid
(8.84)

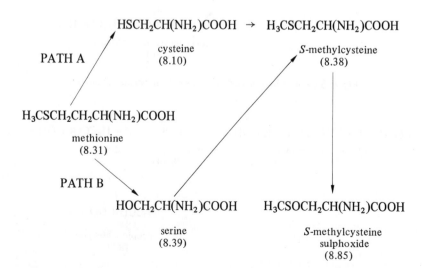

Figure 8.30 The biosynthesis of *S*-methylcysteine and *S*-methylcysteine sulphoxide.

derivatives to pyruvic acid and ammonia. Enzyme-bound dehydroalanine (8.14) is an intermediate.

Djenkolic acid is toxic to man and animals as it damages the urinary tract.

The Alliaceae contains a number of cysteine derivatives which are partly responsible for the characteristic odours and tastes of plants belonging to this family. The simplest derivative, *S*-methylcysteine (8.38), is biosynthesised in onion (*Allium cepa*) and garlic (*A. sativum*) by two pathways (*Figure 8.30*). In path A, cysteine is methylated by methionine (8.31), whereas, in path B, a methylthio radical is transferred from methionine to serine (8.39). When fed ethionine, these plants synthesised *S*-ethylcysteine, a compound which does not occur naturally. *S*-methylcysteine is oxidised to the sulphoxide (8.85) in *Allium* species and some crucifers (*Figure 8.30*). *S*-methylcysteine sulphoxide appears to function as a reservoir of sulphur which is transferred to protein amino acids as needed.

Propen-1-ylcysteine sulphoxide (8.86), a characteristic constituent of onions, is biosynthesised from cysteine and valine (8.46) (*Figure 8.31*). Valine is first converted to methacrylic acid (8.87) which adds to cysteine. Decarboxylation of the product gives propen-1-ylcysteine (8.88) which is oxidised to the sulphoxide.

CH3
CHCH(NH2)COOH
CH3
valine
(8.46)
CH3
CCOOH
CH2
methacrylic acid
(8.87)
+ HSCH2CH(NH2)COOH
cysteine
(8.10)
H3CCH=CHSCH2CH(NH2)COOH
propen-1-ylcysteine
(8.88)
H3CCH(COOH)CH2SCH2CH(NH2)COOH
H3CCH=CHSOCH2CH(NH2)COOH
propen-1-ylcysteine sulphoxide
(8.86)

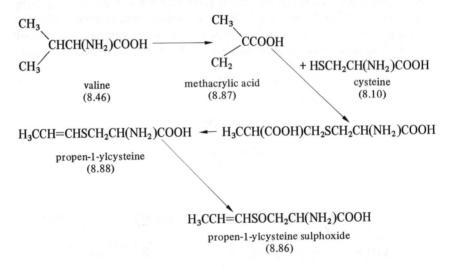

Figure 8.31 The biosynthesis of propen-1-ylcysteine sulphoxide.

H3CCH=CHSOCH2CH(NH2)COOH
propen-1-ylcysteine sulphoxide
(8.86)
H3CCH=CHSOH + H3CCOCOOH + NH3
propenylsulphenic acid
(8.89)
H3CCH2CHO + S
propionaldehyde
(8.90)

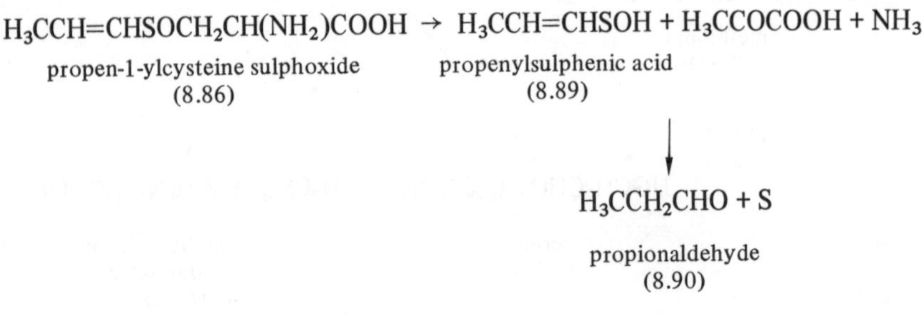

Figure 8.32 The degradation of propen-1-ylcysteine sulphoxide.

Propenylsulphenic acid (8.89), the compound responsible for the lachrymatory properties of onions, is produced by hydrolysis of propen-1-ylcysteine sulphoxide when the plants are damaged (*Figure 8.32*). Propenylsulphenic acid is rapidly decomposed to propionaldehyde (8.90).

Similar degradations occur in *Allium* species with other *S*-alkylcysteine sulphoxides, the reactions being catalysed by alliinases. In garlic, alliin (allylcysteine sulphoxide) (8.91) is hydrolysed to allicin (allylthiosulphate) (8.92) (*Figure 8.33*).

Within the *Allium* genus, each species has a characteristic pattern of sulphur-containing compounds and the patterns of related species are similar.

$$2H_2C=CHCH_2SOCH_2CH(NH_2)COOH$$

alliin
(8.91)

$$H_2C=CHCH_2SOSCH_2CH=CH_2 + 2CH_3COCOOH + 2NH_3$$

allicin
(8.92)

Figure 8.33 The degradation of alliin.

3-Indolylacetic acid

3-Indolylacetic acid (indole-3-acetic acid, IAA) (8.93) (*Figure 8.34*) is one of the auxins, which together with the gibberellins and abscisic acid (Chapter 5), cytokinins (Chapter 10) and ethylene (see below) are hormones regulating the growth and development of plants. IAA is a ubiquitous constituent of higher plants and the most important auxin. Some other, non-indolic compounds, including phenylacetic acid (8.94) biosynthesised in plants from phenylalanine (8.68), have similar properties and synthetic auxins have also been prepared.

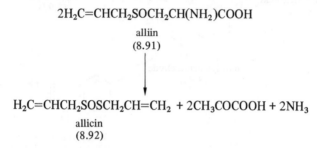

phenylacetic acid
(8.94)

In the plant, IAA conjugates with many compounds, including glucose and other sugars, and with aspartic and glutamic acids. This is probably a way of storing the hormone for future use.

IAA initiates many growth effects in plants, including geotropism and phototropism, development of the ovary, division of cells, enlargement in callus tissue, root formation and apical dominance. When fed to plants, the hormone causes growth up to a maximum, which depends on the type of tissue being fed, and thereafter inhibits further growth, probably through the formation of ethylene (see below), which is growth-inhibitory. Stem tissues tolerate the highest levels of IAA and root tissues the lowest. In the plant, the most active sites of IAA synthesis are the young, expanding leaves.

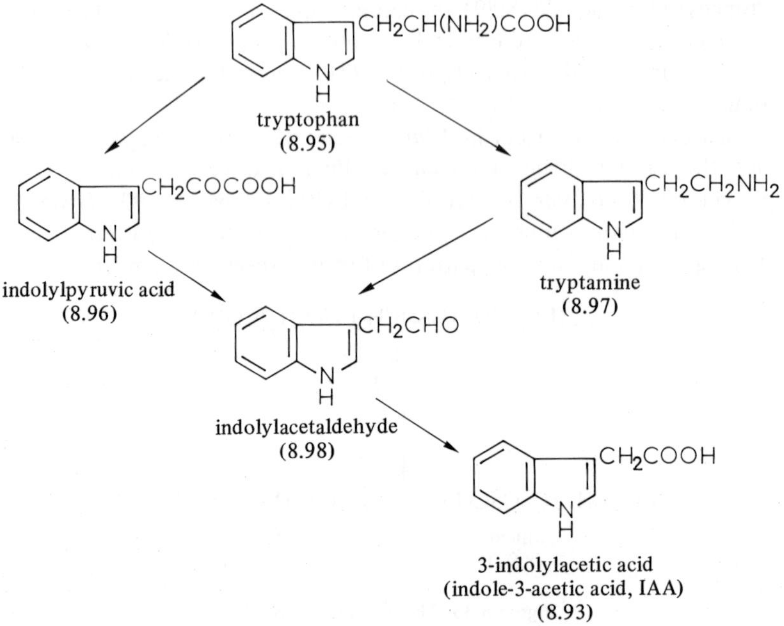

Figure 8.34 The biosynthesis of 3-indolylacetic acid.

IAA promotes growth through elongation of tissue. The principle action of this hormone is to cause a loosening of the bonds between the cellulose micro-fibrils of cell walls, which allows water to enter the vacuoles and increase the cell volume. The cell elongates longitudinally and IAA initiates the synthesis of new cell wall tissue, which is needed to surround the enlarged cell.

IAA is biosynthesised in plants from tryptophan (8.95) by two pathways (*Figure 8.34*), the indolylpyruvic acid pathway being quantitatively the more important. Experiments with tomato shoots have shown the existence of a tryptophan transaminase, which catalyses the formation of indolylpyruvic acid (8.96), and a tryptophan decarboxylase, which catalyses the formation of tryptamine (8.97). The decarboxylation of indolylpyruvic acid is catalysed by indolylpyruvate de-carboxylase, while indolylacetaldehyde dehydrogenase catalyses the oxidation of indolylacetaldehyde (8.98) to indolylacetic acid.

Members of the Cruciferae which synthesise indolylglucosinolates (see above) have a third pathway in which indolylacetonitrile (8.99) is converted to IAA.

In the intact plant and cell free systems, IAA is destroyed through oxidation to methyleneoxindole (8.100) and other products, reactions which are catalysed by indolylacetic acid oxidase. This oxidation is possibly one way of controlling growth.

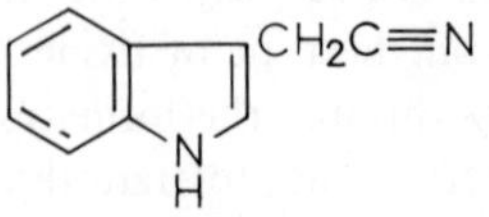

indolylacetonitrile
(8.99)

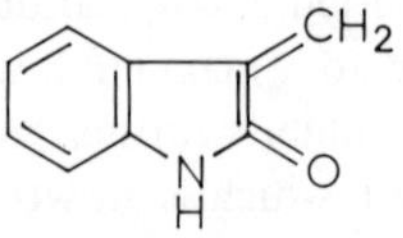

methyleneoxindole
(8.100)

Ethylene

Ethylene (8.101) is one of the most important of the compounds regulating plant growth, and it is possible that this hormone exerts an influence on all plant tissue at some stage of development. It has also been found that the production of ethylene by a plant can influence the growth of its neighbours, threshold effects in a variety of physiological responses being observed at concentrations as low as 0.01 ppm.

Ethylene production has been shown to be connected with a large number of plant growth and development effects, although the mode of action of this hormone has not been elucidated in the majority of cases. In tissue cultures, it has been observed that ethylene production reaches a maximum late in cell division and then rapidly decreases. In some instances, ethylene and the auxins (see above) have opposite effects on plant growth, ethylene inhibiting elongation of stems and roots and the expansion of new leaves. The auxins also induce the formation of ethylene, probably by initiating the synthesis of the requisite enzymes, as the induction can be blocked by inhibitors of RNA and protein synthesis.

Other growth effects of ethylene include the promotion of seed germination, flower fading, leaf senescence and fruit ripening. Its effects on flowering vary with the plant family. In most cases, ethylene inhibits flowering, but in smoke this gas is used to initiate the flowering of pineapples (*Ananas comosus*). It has a similar effect on other bromeliads. In members of the Cucurbitaceae family, ethylene can induce a change of flower sex. Carbon dioxide is a competitive inhibitor of ethylene in all its effects.

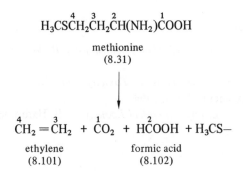

$$\overset{4}{H_3C}\overset{3}{SCH_2}\overset{2}{CH_2}\overset{}{CH(NH_2)}\overset{1}{COOH}$$

methionine
(8.31)

$$\overset{4}{CH_2}=\overset{3}{CH_2} + \overset{1}{CO_2} + \overset{2}{HCOOH} + H_3CS-$$

ethylene   formic acid
(8.101)   (8.102)

Figure 8.35 The degradation of methionine, forming ethylene.

Isotopic tracer experiments have shown methionine (8.31) to be the precursor of ethylene in plants, the reaction only taking place under aerobic conditions. As yet, the true intermediates of the *in vivo* synthesis have not been elucidated. In apple tissue (*Malus sylvestris*, Rosaceae), C-3 and C-4 of methionine are retained in ethylene (*Figure 8.35*), C-1 is converted to carbon dioxide, C-2 to formic acid (8.102) and $-SCH_3$ appears mainly as S-methylcysteine (8.38), but can also be used to methylate various compounds and the sulphur transferred to other amino acids.

Bibliography

Non-protein Amino Acids

'Non-protein Amino Acids, Cyanogenetic Glycosides and Glucosinolates', in *Biosynthesis*, Specialist Periodical Reports, vols. 2, 4 and 5, The Chemical Society, 1973, 1975 and 1977

Amino Acids, Peptides and Proteins, Specialist Periodical Reports (continuing series, vol. 1, 1968), The Chemical Society

'The Non-protein Amino Acids of Plants', in *Biosynthesis and its Control in Plants*, B. V. Milborrow (ed.), Academic Press, 1973

'Amino Acid Biosynthesis and its Control in Plants', *ibid*.

'Free Amino Acids, Bound Amino Acids, Amines and Ureides', in *Chemistry and Biochemistry of Plant Herbage*, vol. 1, G. W. Butler and R. W. Bailey (eds.), Academic Press, 1973

'Non-protein Amino Acids from Plants : Distribution, Biosynthesis and Analog Functions', in *Recent Advances in Phytochemistry*, vol. 8, V. C. Runeckles and E. E. Conn (eds.), Academic Press, 1974

'Toxic Amino Acids as Antimetabolites', L. Fowden *et al.*, in *Advances in Enzymology*, vol. 29, 1967

'Toxic Amino Acids in the Leguminosae', in *Phytochemical Ecology*, J. B. Harborne (ed.), Academic Press, 1972

'Comparative Biochemistry of Non-protein Amino Acids', in *Chemotaxonomy of the Leguminosae*, J. B. Harborne *et al.* (eds.), Academic Press, 1971

Selenoamino Acids

Organic Selenium Compounds: Their Chemistry and Biology, D. L. Klayman (ed.), Wiley, 1973

Selenium, Geobotany, Biochemistry, Toxicity and Nutrition, I. Rosenfeld and A. O. Beath, Academic Press, 1964

'Selenium Toxicity', in *Phytochemical Ecology*, J. B. Harborne (ed.), Academic Press, 1972

Amines

'The Occurrence, Metabolism and Functions of Amines in Plants', in *Biological Reviews*, vol. 46, 1971

'Free Amino Acids, Bound Amino Acids, Amines and Ureides', in *Chemistry and Biochemistry of Plant Herbage, ibid*.

Cyanogenic Glycosides

'Non-protein Amino Acids, Cyanogenetic Glycosides and Glucosinolates', in *Biosynthesis, ibid*.

'Cyanogenetic Glycosides and Glucosinolates', in *The Chemistry and Biochemistry of Plant Herbage, ibid*.

'The Biosynthesis of Cyanogenetic Glycosides and other Simple Nitrogen Compounds', in *Perspectives in Phytochemistry*, J. B. Harborne and T. Swain (eds.), Academic Press, 1969

'Naturally Occurring Cyanogenetic Glycosides', in *Progress in Phytochemistry*, vol. 4, L. Reinhold *et al.* (eds.), Pergamon Press, 1977

'Cyanogenic Glycosides and Their Function', in *Phytochemical Ecology*, J. B. Harborne (ed.), Academic Press, 1972

'Plant Toxins and Their Effects on Animals', in *Introduction to Ecological Biochemistry*, J. B. Harborne, Academic Press, 1977

'Chemistry of Geographical Races', section III, in *Chemistry in Evolution and Systematics*, T. Swain (ed.), Butterworths, 1973

Glucosinolates

'Non-protein Amino Acids, Cyanogenetic Glycosides and Glucosinolates', in *Biosynthesis, ibid.*

'Cyanogenetic Glycosides and Glucosinolates', in *Chemistry and Biochemistry of Plant Herbage, ibid.*

'The Natural Distribution of Glucosinolates: A Uniform Group of Sulphur-containing Glucosides', in *Chemistry in Botanical Classification*, G. Bendz and J. Santesson (eds.), Academic Press, 1974

'Glucosinolates in the Cruciferae', in *The Biology and Chemistry of the Cruciferae*, J. G. Vaughan *et al.* (eds.), Academic Press, 1976

'Volatile Flavour Compounds of the Cruciferae', *ibid.*

Penicillins and Cephalosporins

'Biosynthesis of Penicillins and Cephalosporins', in *Biosynthesis*, vol. 4, J. D. Bu'Lock (ed.), *ibid.*, 1976

Fungal Metabolites, W. B. Turner, Academic Press, 1971

Biogenesis of Antibiotic Substances, Z. Vareh and Z. Hostalek (eds.), Academic Press, 1965

Biosynthesis of Antibiotics (continuing series, vol. 1, 1966), Academic Press

'Chemotherapy I: Bacteria, Fungi, Viruses', in *Gaddum's Pharmacology*, 8th edn, A. S. V. Burgen and J. F. Mitchell, Oxford University Press, 1978

Sulphur Compounds

'Sulphur Compounds in Plants', in *Recent Advances in Phytochemistry*, vol. 1, T. J. Mabry *et al.* (eds.), Appleton (N.Y.), 1968

'Sulphur Compounds', in *Chemical Plant Taxonomy*, T. Swain (ed.), Academic Press, 1963

'The Distribution of Sulphur Compounds', in *Comparative Phytochemistry*, T. Swain (ed.), Academic Press, 1966

Indolylacetic Acid

'Hormones', in *Plant Biochemistry*, 3rd edn, J. Bonner and J. E. Varner (eds.), Academic Press, 1976

'The Biochemistry of the Action of Indoleacetic Acid on Plant Growth', in *Recent Advances in Phytochemistry*, vol. 7, V. C. Runeckles *et al.* (eds.), Academic Press, 1974

Ethylene

Ethylene in Plant Biology, F. B. Abeles, Academic Press, 1973
'The Biochemistry of Ethylene Biogenesis and Metabolism', in *Recent Advances in Phytochemistry*, vol. 7, *ibid*.
'Hormones', in *Plant Biochemistry, ibid*.

9 Alkaloids

Introduction

Plants synthesising alkaloids are worldwide and occur in many families, ranging from the primitive mosses to the highly complex Gramineae, although these secondary metabolites are most abundant in the Dicotyledones. Because of the proven medicinal value of a number of alkaloids and their ease of extraction, many plants have been screened for these compounds with the result that a large number of alkaloids have been identified and their structures determined.

The alkaloids are a diverse collection of compounds whose only molecular similarity is the presence of nitrogen. Those compounds occurring in plants can be divided into the true alkaloids, the protoalkaloids and the pseudoalkaloids, according to their molecular structure and biosynthetic pathways. In this chapter, our discussion is restricted to the true alkaloids, which are defined as those compounds in which nitrogen forms part of a heterocyclic ring system, and which are biosynthesised from amino acid precursors. The protoalkaloids include the physiologically active amines, and these, together with the amide, colchicine, are discussed in Chapter 8. The pseudoalkaloids contain nitrogen heterocyclic rings which are biosynthesised from precursors other than amino acids. The terpenoid and steroid pseudoalkaloids are discussed in Chapter 5, while the purine derivatives are included in Chapter 10. As the sensitivity of the methods for detecting and identifying alkaloids increases, it is becoming clear that, when the principles of chemosystematics are applied, plants should be divided into those which accumulate alkaloids and those in which they only occur in trace amounts. Thus, although many plants contain nicotine, this alkaloid is only of significance in the few plants which accumulate the compound. When structurally similar alkaloids are found in totally unrelated families, their biosynthetic pathways are usually different. Thus, it is the pathway, rather than the end-product, which is chemosystematically significant.

It is only since the advent of isotopic tracer techniques that investigations into the biosynthetic pathways of alkaloids have progressed. Although the precursors of these compounds have been identified in most cases, the detailed steps of biogenesis and the majority of the enzymes involved still await elucidation. Experiments with isotopically labelled compounds (see Chapter 1) have shown that the basic units in the biogenesis of the true alkaloids are amino acids. The non-nitrogen-containing rings or side chains are derived from terpene units and/or acetate, while methionine is responsible for the addition of methyl groups to nitrogen atoms.

The amino acids most often encountered in alkaloid biosynthesis are the aliphatic compounds, ornithine and lysine, and the aromatic nicotinic acid, phenylalanine, tyrosine and tryptophan. All these compounds originate from glucose, a product of primary metabolism. Ornithine, lysine and nicotinic acid are derived from precursors closely linked with the tricarboxylic acid cycle. Ornithine is biosynthesised from glutamic acid, while lysine and nicotinic acid are derived from aspartic acid. Tyrosine, phenylalanine and tryptophan, however, are products of the shikimic acid pathway, and their biosynthesis is described in Chapter 6.

The biosynthesis of some of the simpler alkaloids follows pathways which are analogous to those followed by some primary metabolites. For example, the pathway followed by hygrine (see below) is similar to that by which proline is derived from glutamic acid. Thus, the enzymes involved in simple alkaloid metabolism could have evolved from those of primary metabolism by simple mutation. That some, at least, of the enzymes involved in alkaloid biosynthesis are multispecific is shown by the fact that they will convert non-natural precursors to non-natural alkaloids (see, for example, the section on alkaloids biosynthesised from nicotinic acid, p. 267).

Several suggestions have been made concerning the function of alkaloids in plants, and it seems probable that these secondary metabolites are useful to the plant in several ways. As with other toxic secondary metabolites, the main function of alkaloids is probably to protect the plant against predators. Plants containing alkaloids have a selective advantage over those that contain no substances toxic to animals. In general, herbivores do not eat alkaloid-containing plants if alkaloid-free forage is available, probably because of the bitter taste of the former. It is noteworthy that the ferns, which developed several million years before the first mammals, do not contain alkaloids, whereas plants which evolved with herbivorous mammals do contain these protective compounds.

All alkaloids have some physiological action, generally on the central nervous system. This can be utilised for the benefit of man in the alkaloidal drugs. Many alkaloids, however, are too poisonous to be used medicinally and all are toxic in sufficient concentration. A few naturally occurring alkaloids cause addiction, the most potent being morphine and cocaine.

Some alkaloids have teratogenic properties, causing defects in the foetus when plants containing them are eaten by the mother. Especially potent are compounds belonging to the pyrrolizidine, quinolizidine and nicotine groups. Skeletal damage is the defect most often encountered and it has been suggested that *spina bifida* could be caused by solanine, resulting from the overeating of potatoes by the mother.

Alkaloid-containing plants probably cause more stock loss throughout the world than any other type of poisonous plant. However, the situation would be much worse if animals in general (including man) did not dislike the bitter taste of these plants. Conversely, once accustomed to the taste, stock can become so addicted to the plant that they will immediately return to it after recovering from an attack of poisoning.

The specific actions of an alkaloid often vary with the animal species. Morphine, for example, causes sleep in man and dogs, but in horses and cats it acts as a stimulant.

Even if they do not die, farm animals poisoned by alkaloid-containing plants often never completely recover and are thus an economic loss, especially as the milk and/or meat of such animals is often tainted and cannot be sold.

Plants responsible for most stock poisoning on a worldwide scale are *Senecio* and *Crotalaria* species, yew and green potatoes.

Poisoning of animals is usually accidental, except when poisoned arrow tips are used in primitive areas to kill animals for food or for their valuable skins, etc., but poisoning of man can also be deliberate, involving murder or suicide. Plants are less often involved than the pure drugs, of which nicotine, morphine, strychnine, codeine, aconitine and atropine are most often encountered by the forensic chemist. However, accidental poisoning by plants occurs distressingly often amongst children, while in adults it is most likely to occur as a result of mistaken identity, such as the use of hemlock leaves in place of parsley or monk's-hood roots instead of horseradish.

For reasons of space, we are unable to describe all the various types of alkaloid found in plants, and those discussed below have been chosen for their pharmacological, toxicological or biosynthetic importance.

Alkaloids Biosynthesised from Ornithine

Ornithine (9.1) is the precursor of the nitrogen-containing five-membered heterocyclic rings found in the pyrrolidine, tropane and pyrrolizidine alkaloids.

Pyrrolidine and Tropane Alkaloids

Tracer experiments have shown that the tropane alkaloids are biosynthesised from pyrrolidine derivatives such as hygrine (9.2). Both pyrrolidine and tropane alkaloids are abundant in the Solanaceae family and in the cocaine or coca plant (*Erythroxylum coca*, Erythroxylaceae). Although putrescine (9.3) is incorporated into the pyrrolidine and tropane alkaloids, it is not a natural precursor, as tracer experiments have shown that only the δ-nitrogen atom of ornithine is retained, and thus there can be no symmetrical intermediate. δ-*N*-methylornithine is an effective precursor for the hygrine derivative, cuscohygrine (9.4), or the tropane alkaloid, hyoscyamine (9.5), in the deadly nightshade (*Atropa belladonna*, Solanaceae), and *N*-methylputrescine was found to be a better precursor for these alkaloids than either ornithine or putrescine in both *Datura metel* and *Scopolia lurida* (Solanaceae). Thus, it seems that methylation is the first step and decarboxylation the second step in the biosynthesis of hygrine, the precursor of both the pyrrolidine and tropane alkaloids.

Tracer experiments have shown that acetate in the form of acetoacetyl CoA (9.6) is responsible for the side chain of hygrine which also forms part of the tropane ring. Thus, we can formulate a pathway for the biosynthesis of hygrine (9.2), cuscohygrine (9.4), hyoscyamine (9.5) and hyoscine (scopolamine) (9.7). As shown in *Figure 9.1*, hygrine, which is formed from ornithine, can be converted either to cuscohygrine or to the tropane alcohol, tropine (9.8), a compound which has been found in the deadly nightshade. Tropine is esterified with tropic acid (9.9) to form hyoscyamine. Tracer experiments have shown tropic acid to be derived from phenylalanine (9.10) by migration of the carboxyl group and deamination. Oxidation of hyoscyamine gives hyoscine (9.7), another tropane alkaloid found in the deadly nightshade and *Datura* species.

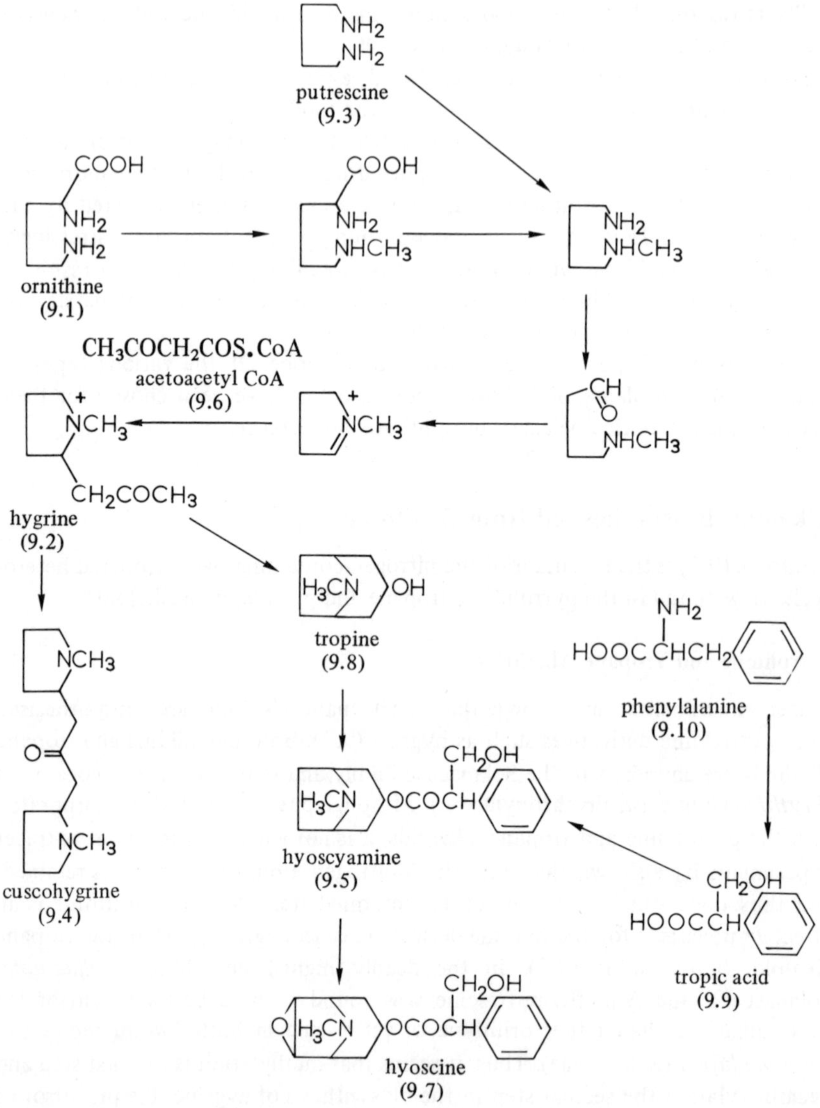

Figure 9.1 The biosynthesis of hygrine and the tropane alkaloids.

Cocaine (9.11), which occurs with hygrine in the cocaine plant, is the ester of the tropane alcohol, ecgonine (2-carboxytropine) (9.12), and benzoic acid (9.13). The biogenesis of ecgonine follows a pathway similar to that of tropine, but hygrine is not an intermediate as the carboxyl group of acetoacetate is retained and methylated, as shown in *Figure 9.2*.

The fate of pyrrolidine alkaloids in plants appears to be unknown and there has been little investigation into the metabolism of the tropane derivatives. When tritiated atropine ((±)hyoscyamine) was fed to mature *Datura innoxia* plants, it

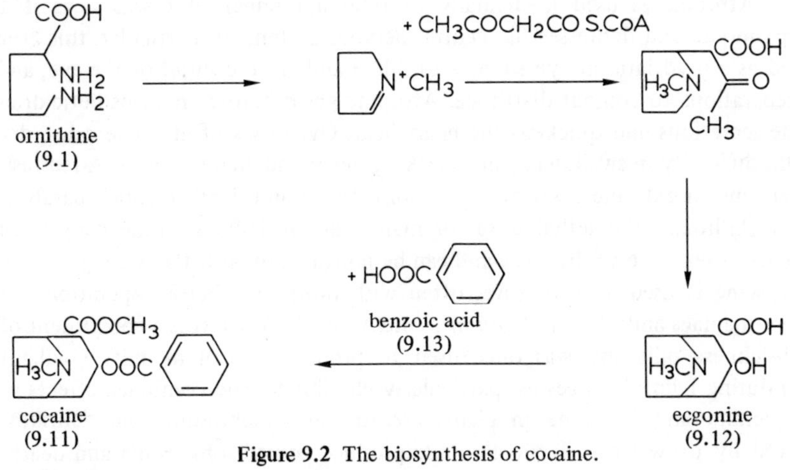

Figure 9.2 The biosynthesis of cocaine.

appeared to be metabolised to non-alkaloidal compounds, as no tritiated alkaloids were recovered. Conversely, the feeding of labelled hyoscyamine to this plant resulted in the recovery of labelled hyoscyamine, 6-hydroxyhyoscyamine (9.14) and hyoscine. 6-Hydroxyhyoscyamine is probably an intermediate in the biosynthesis of hyoscine from hyoscyamine (*Figure 9.3*).

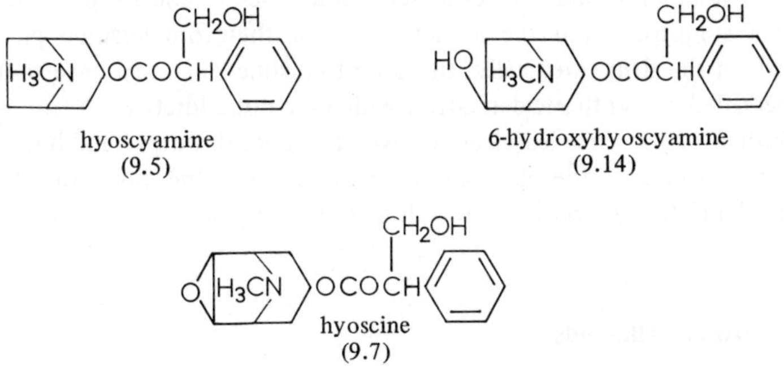

Figure 9.3 The derivation of hyoscine from hyoscyamine.

No alkaloids could be detected in leaves of the cocaine plant kept under herbarium conditions for 40 years.

Hygrine occurs fairly widely in the Angiospermae and has been detected in such unrelated families as Cruciferae, Erythroxylaceae, Solanaceae, Convolvulaceae and Orchidaceae. The simple pyrrolidine alkaloids have only a mild physiological action and are therefore unlikely to cause poisoning in animals eating plants containing these compounds. However, hygrine and its derivatives are usually intermediates in the formation of more complex alkaloids which are toxic to animals.

When isolated from the plant, (−)hyoscyamine racemises to form the drug atropine ((±)hyoscyamine) of which only (−)hyoscyamine is physiologically active. Atropine is used medicinally to treat a number of diseases, as it is antispasmodic and depresses the central nervous system. In particular, this drug is used as a mydriatic in eye surgery to dilate and fix the pupil of the eye, and in preparations to combat diarrhoea. Atropine given before an anaesthetic dries up the secretions and quickens the heart beat. Overdoses of atropine cause dry mouth, difficulty in swallowing and speaking, fever and dilated pupils. A red rash appears and in extreme cases delirium, convulsions and death through paralysis occurs. Although the lethal dose for man is about 100 mg, some people are particularly sensitive to this drug and can be poisoned by as little as 1 mg.

Hyoscine is used as a sedative, often with morphine, before operations, to quieten maniacs and to treat *delirium tremens* (alcoholism). It is a constituent of travel-sickness pills and was once used to produce loss of memory (twilight sleep) during painful processes, particularly childbirth. The combined effects of hyoscyamine and hyoscine in plants are to cause inebriation and delirium, followed by drowsiness. Toxic amounts produce convulsions, coma and death.

In animals, poisoning generally occurs through eating plants containing hyoscyamine and/or hyoscine. The susceptibility of animals varies considerably with species, pigs and cattle are easily poisoned, while rabbits can eat deadly nightshade with impunity, although subsequently their meat is poisonous to man.

Poisoning by a single dose of cocaine is rare, although it has been used as a suicidal agent. The danger of this drug lies in its addictive properties, as it probably does more harm to the brain and body than any other drug of addiction. Unlike morphine, however (see below), lack of cocaine does not cause withdrawal symptoms and the addict does not therefore become physically dependent on the drug. Medically, the use of cocaine as a local anaesthetic has been superseded by synthetic derivatives which are not addictive.

In South America, coca leaves are chewed to diminish fatigue and hunger, but although the effects on the body are less than those of the pure drug, they are still harmful and the chewer becomes addicted to the plant.

The Pyrrolizidine Alkaloids

The pyrrolizidine alkaloids are widely distributed in the plant world, but are particularly characteristic of *Senecio* species (Compositae), *Crotalaria* species (Leguminosae) and some genera belonging to the Boraginaceae family.

These alkaloids are esters formed by the combination of a necine base, which contains the pyrrolizidine ring system, with one or two necic acids. The most commonly occurring bases are retronecine and heliotridine (9.15), which differ only in the stereochemistry of the ring hydroxyl group.

Retronecine (9.15) is biosynthesised from ornithine (9.1) or putrescine (9.3) and tracer experiments have shown that the probable pathway is that shown in *Figure 9.4*.

Although the necic acids have terpene-like structures, they are not biosynthesised by the acetate–mevalonate pathway. Tracer experiments have shown that some, at least, of these acids are derived from branched-chain amino acids such as isoleucine and valine.

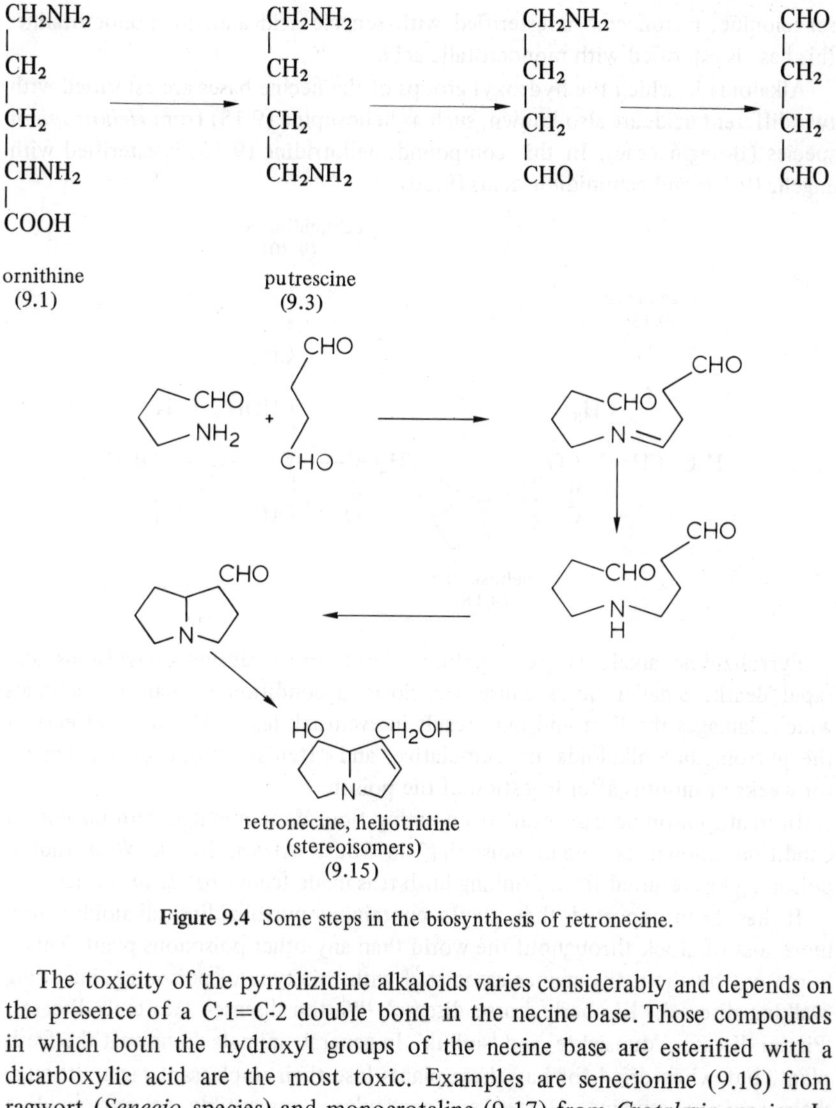

$$CH_2NH_2 - CH_2 - CH_2 - CHNH_2 - COOH$$

ornithine
(9.1)

putrescine
(9.3)

Figure 9.4 Some steps in the biosynthesis of retronecine.

retronecine, heliotridine
(stereoisomers)
(9.15)

The toxicity of the pyrrolizidine alkaloids varies considerably and depends on the presence of a C-1=C-2 double bond in the necine base. Those compounds in which both the hydroxyl groups of the necine base are esterified with a dicarboxylic acid are the most toxic. Examples are senecionine (9.16) from ragwort (*Senecio* species) and monocrotaline (9.17) from *Crotalaria* species. In

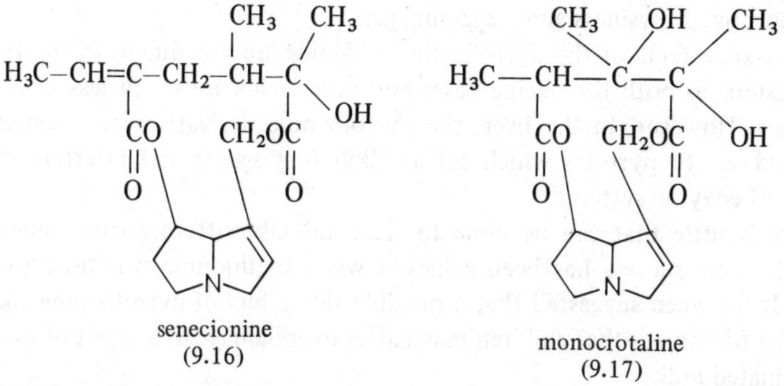

senecionine
(9.16)

monocrotaline
(9.17)

senecionine, retronecine is esterified with senecic acid and, in monocrotaline, this base is esterified with monocrotalic acid.

Alkaloids in which the hydroxyl groups of the necine bases are esterified with two different acids are also known, such as heliosupine (9.18) from *Heliotropium* species (Boraginaceae). In this compound, heliotridine (9.15) is esterified with angelic (9.19) and echimidinic acids (9.20).

heliosupine
(9.18)

Pyrrolizidine alkaloids are hepatoxic, large doses causing convulsions and rapid death. Smaller doses cause seneciosis, a condition in man and animals which damages the liver and may result in eventual death. The toxic effects of the pyrrolizidine alkaloids are cumulative, and often symptoms do not appear for weeks or months after ingestion of the poison.

In man, poisoning can result from eating flour contaminated with ragwort, a condition known as 'bread poisoning' in South Africa. In the West Indies, poisoning has resulted from drinking bush teas made from *Crotalaria* species.

It has been reported that plants containing pyrrolizidine alkaloids cause more loss of stock throughout the world than any other poisonous plant. This is borne out by the many names associated with this type of poisoning, including 'walking disease', 'Kimberly horse disease', 'Winton disease', 'Molteno disease', 'Pictou disease', 'dunsiekte' and 'suiljuk'. In general, animals do not eat the fresh plant, but when dried for hay these plants lose their unpleasant taste although their toxicity remains. Horses are particularly susceptible to pyrrolizidine poisoning, which gives rise to the typical symptoms of 'horse staggers' — nervousness, yawning, drowsiness and staggering gait.

The toxic effects of the pyrrolizidine alkaloids are not due to the hydrolysis of the esters, as both the necine bases and necic acids are much less toxic than the esters. However, in the liver, the pyrrolizidine derivatives are oxidised by liver oxidases to pyrroles which act as alkylating agents and interfere with a number of enzyme systems.

There is little that can be done to treat animals suffering from seneciosis, although some success has been achieved when methionine was used to treat horses. It has been suggested that a possible side-effect of pyrrolizidine alkaloid poisoning of cows is that children may suffer liver damage as a result of drinking contaminated milk.

Although toxic to animals, plants containing pyrrolizidine alkaloids attract certain butterflies (*Danais* spp.), which are able to convert the alkaloids to an insect pheromone which promotes mating.

Alkaloids Biosynthesised from Lysine

Lysine (9.21) is the precursor of the nitrogen-containing, six-membered heterocyclic ring systems found in a number of alkaloids, including the piperidine derivatives and the lupin alkaloids.

The Piperidine Alkaloids

The piperidine alkaloids occur in a number of unrelated families including Moraceae, Chenopodiaceae, Piperaceae, Crassulaceae, Caricaceae, Punicaceae, Solanaceae and Lobeliaceae.

It has been shown that the nitrogen atom of the heterocyclic ring of piperidine alkaloids, such as pelletierine (9.22) and anaferine (9.23), is derived solely from the ϵ-amino group of lysine. In *Sedum sarmentosum* (Crassulaceae), ϵ-(*N*-methyl)lysine is not an intermediate even for *N*-methyl derivatives, and therefore a pathway analogous to the biosynthesis of hygrine (*Figure 9.1*) is

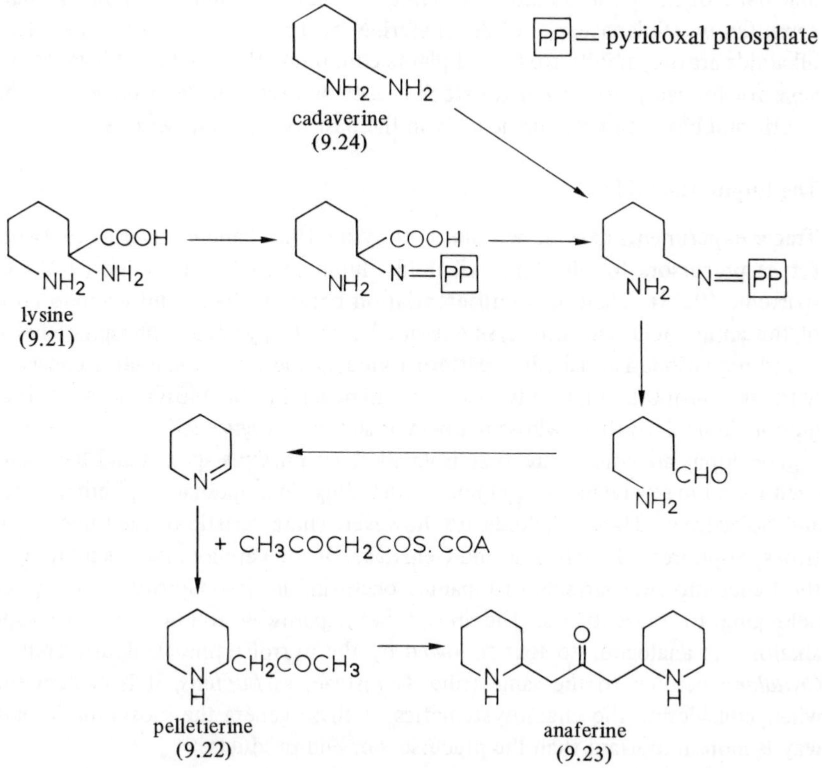

Figure 9.5 The biosynthesis of pelletierine and anaferine.

ruled out. Furthermore, cadaverine (9.24) can act as a precursor for pelletierine alkaloids present in *S. sarmentosum,* and this compound is both present in the plant and produced from lysine when the latter is fed to this *Sedum* species. To reconcile these facts, it has been suggested that lysine or cadaverine are first bound to pyridoxal phosphate (PP) through the α-amino group and that the bound compound then follows the pathway shown in *Figure 9.5.*

Tracer experiments have shown that the side chain of pelletierine is derived from acetate, probably in the form of acetoacetyl CoA, and that pelletierine is the precursor of anaferine (*Figure 9.5*).

Although coniine (9.25), the alkaloid responsible for the toxicity of hemlock (*Conium maculatum,* Umbelliferae), is a piperidine derivative, it is biosynthesised from acetate and is therefore a pseudoalkaloid.

$$\text{N} \quad CH_2\ CH_2\ CH_3$$

coniine
(9.25)

Pelletierine has been isolated from *Sedum* species (Crassulaceae), the roots and bark of the pomegranate tree (*Punica granatum,* Punicaceae) and *Withania somniferum* (Solanaceae), while anaferine occurs in *W. somniferum.* These alkaloids are only mildly toxic, and plants containing them have not been responsible for human poisoning or for stock loss on any scale. Pelletierine is an anthelmintic and has been used medicinally in the past to expel tapeworms.

The Lupin Alkaloids

Tracer experiments have shown that both lysine (9.21) and cadaverine (9.24) can act as precursors for the lupin alkaloids, lupinine (9.26), lupanine (9.27) and sparteine (9.28). There is no differentiation between the α- and ϵ-amino group of the amino acid, and thus lysine is not bound to pyridoxyl phosphate before decarboxylation. The labelling pattern found in the lupin alkaloids is consistent with the proposal that only lysine is involved in the biosynthesis of these quinolizidine derivatives whose pathway is shown in *Figure 9.6.*

The lupin alkaloids were first isolated from *Lupinus* species and have since been found in several unrelated families including Chenopodiaceae, Berberidaceae and Solanaceae. These alkaloids are, however, characteristic of the three related tribes, Sophoreae, Podalyrieae and Genisteae, of the Papilionoideae subfamily of the Leguminosae, sparteine or lupanine occurring in the majority of the genera belonging to these tribes. The biosynthetic pathway followed by the lupin alkaloids is analogous to that followed by the pyrrolizidine alkaloids. Thus, as *Crotalaria* belongs to the same tribe, Genisteae, as *Lupinus,* it is evident that when considering the chemosystematics of these genera the biosynthetic pathway is more important than the precursors or end-products.

The lupin alkaloids are toxic, causing nausea, vomiting, convulsions and death due to respiratory failure. Poisoning of man can occur when the seeds are used

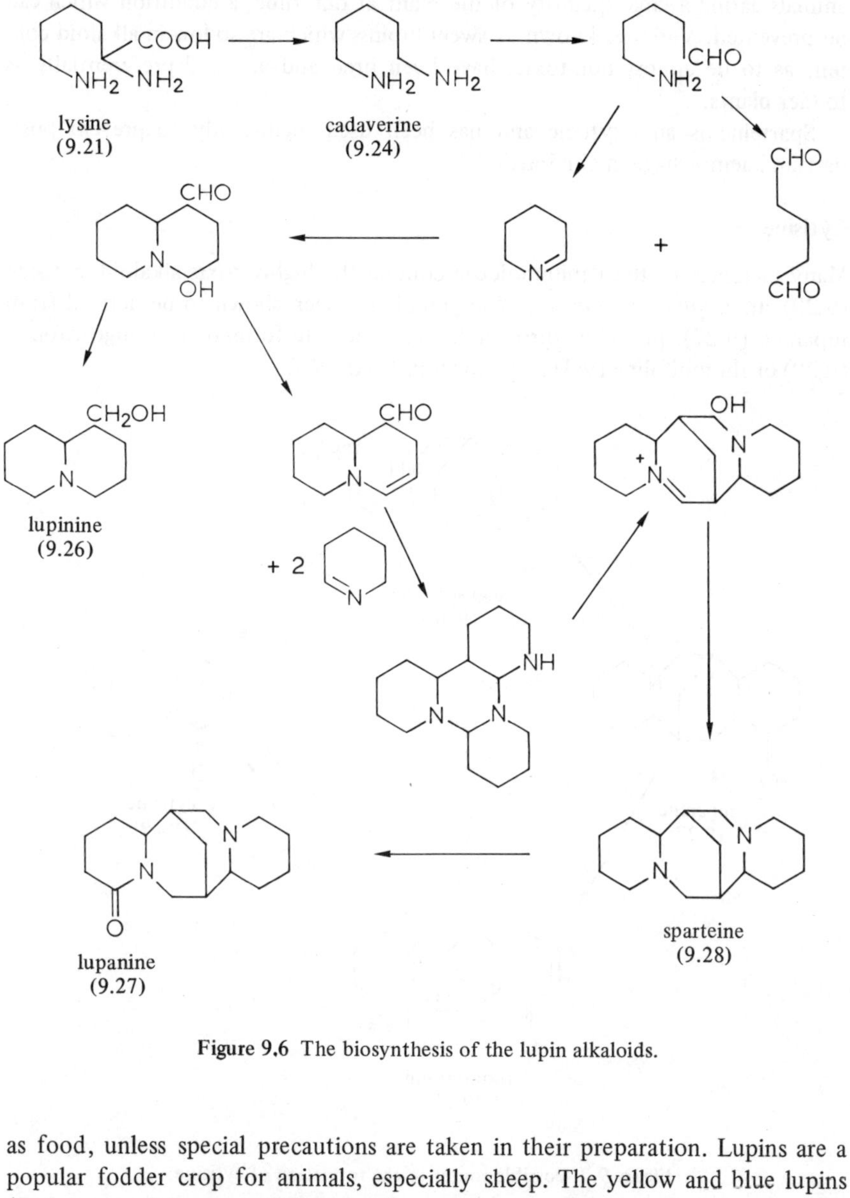

Figure 9.6 The biosynthesis of the lupin alkaloids.

as food, unless special precautions are taken in their preparation. Lupins are a popular fodder crop for animals, especially sheep. The yellow and blue lupins (*L. luteus* and *L. angustifolius*) are the species generally used for this purpose in Europe. Poisoning of animals, known as lupinosis, can occur from a fungus which infests the plants and whose effects are cumulative. Lupinosis causes extensive liver damage and has resulted in the loss of many sheep in Europe and Australia.

Poisoning of animals due to the alkaloid content of the plants can also occur, particularly from *L. sericeus*, *L. leucophyllus*, *L. perennis*, *L. argenteus* and *L. caudatus*. The plants are most dangerous when in seed, and drying for hay has no effect on toxicity. However, such poisoning is not cumulative and is due to animals eating a large quantity of the plant at one time, a condition which can be prevented. Varieties known as sweet lupins, which are so low in alkaloid content as to be almost non-toxic, have been bred and are used preferentially as fodder plants.

Sparteine is an oxytocic and has been used medicinally to prevent post-partum haemorrhage in childbirth.

Cytisine

Many members of the Papilionoideae contain the highly toxic alkaloid, cytisine (9.29). In *Lupinus luteus,* this compound has been shown to be derived from lupanine (9.27), probably through the intermediate formation of angustifoline (9.30) or rhombifoline (9.31), as shown in *Figure 9.7.*

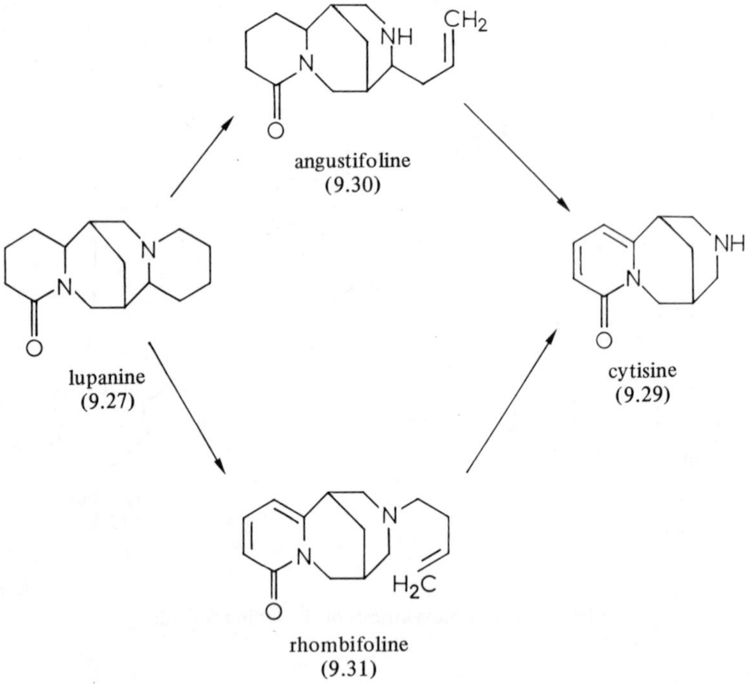

angustifoline
(9.30)

lupanine
(9.27)

cytisine
(9.29)

rhombifoline
(9.31)

Figure 9.7 Possible routes to the formation of cytisine.

Cytisine has a physiological action similar to that of nicotine (see below). Poisoning in Europe is generally from laburnum trees (*Laburnum anagyroides,* Leguminosae), while, in the tropics and America, *Sophora* species (Leguminosae) are often the cause. The alkaloid occurs in all parts of laburnum, but is most concentrated in the seeds, which can be mistaken for peas by children. Cytisine does not always affect cattle, but this compound can be excreted in milk and so poison children.

Alkaloids Biosynthesised from Nicotinic Acid

The Tobacco Alkaloids

The tobacco alkaloids, nicotine (9.32), anabasine (9.33) and anatabine (9.34), are biosynthesised from nicotinic acid (9.35) (*Figure 9.8*), anatabine being derived from two identical nicotinic acid metabolites (see Chapter 1). It is interesting that the biosynthesis of the pyrrolidine ring of nicotine involves free putrescine (9.3), unlike hygrine (9.2) or the tropane alkaloids, whereas the genesis of the piperidine ring in anabasine does not involve free cadaverine (9.24), but follows a pathway similar to that of pelletierine (9.22).

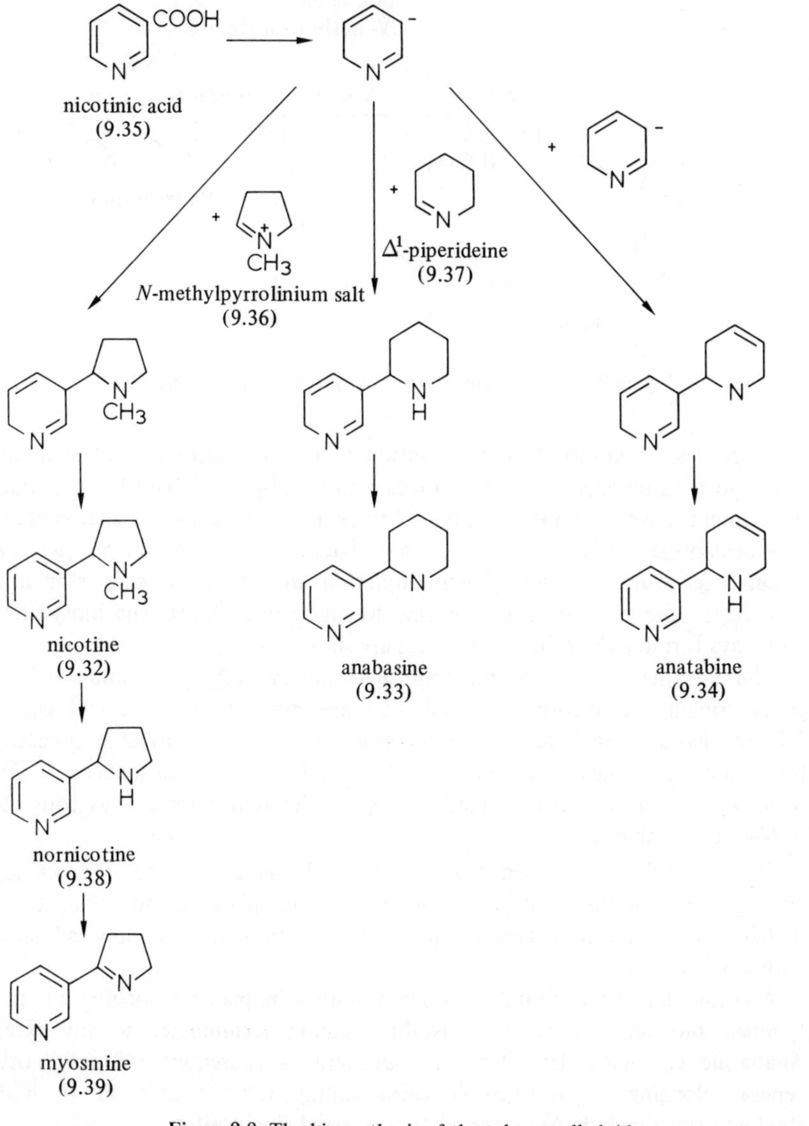

Figure 9.8 The biosynthesis of the tobacco alkaloids.

Much effort has been expended in determining the sequence of reactions and the enzymes involved in the formation of the pyrrolidine ring of nicotine, which originates from ornithine (9.1), through the formation of an *N*-methylpyrrolinium salt (9.36), as shown in *Figure 9.9*.

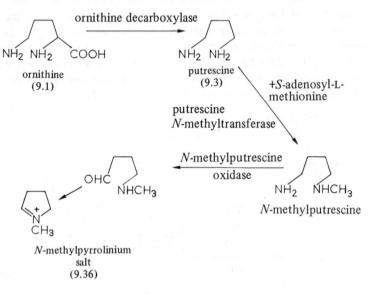

Figure 9.9 The biosynthesis of the pyrrolidine ring of nicotine.

Much less is known of the derivation of the piperidine ring of anabasine, although the immediate precursor appears to be Δ^1-piperideine (9.37), as tracer experiments have shown this compound to be a good precursor of anabasine.

Nornicotine (9.38) is derived in tobacco plants (*Nicotiana tabacum,* Solanaceae) from nicotine by demethylation and this compound is an intermediate in the conversion of nicotine to myosmine (9.39). The biosynthetic pathways forming the tobacco alkaloids are shown in *Figure 9.8*.

The enzymes converting nicotinic acid and *N*-methylpyrrolinium salts or Δ^1-piperideine to nicotine or anabasine are multispecific and will convert 5-fluoronicotinic acid to 5-fluoronicotine or various methylated *N*-methylpyrrolinium chlorides to the corresponding methylnicotines in tobacco. In both *N. tabacum* and *N. glauca,* *N*-methyl-Δ^1-piperideinium chloride was converted to *N*-methylanabasine.

It has already been shown that nicotine can be metabolised to other alkaloids (*Figure 9.8*), but further degradation can also take place, as the administration of labelled nicotine to *N. tabacum* resulted in the formation of many radioactive amino acids.

Nicotine has been found in trace amounts in plants belonging to many families, but only in *Nicotiana* is this alkaloid accumulated to any extent. Anabasine was first isolated from *Anabasis* and was subsequently found in other genera belonging to the Chenopodiaceae family. Other families in which this alkaloid occurs include Araliaceae, Lauraceae and Compositae.

Pure nicotine produces instantaneous death when ingested in only small quantities, the fatal dose for man being 40–60 mg. Symptoms produced by plant poisoning are irregular, weak heart action, vomiting and unconsciousness, followed by death. The alkaloid can also be absorbed through the skin and there have been cases of nicotine poisoning caused by prolonged contact of tobacco leaves with damp skin. Children playing with old tobacco pipes have also suffered nicotine poisoning.

Due to its popularity as a horticultural and agricultural insecticide, nicotine was once one of the most common causes of accidental poisoning by alkaloids. It has also been used to commit suicide and as a homicidal agent. The use of nicotine-containing medicaments on animals has caused many deaths. However, this alkaloid is being replaced by safer pesticides, and its use discouraged.

Animals do not usually have access to tobacco fields, but there have been instances of poisoning, especially of pigs. Rabbits, however, can eat 500 g of fresh tobacco leaves per week with no ill effects.

Anabasine has similar physiological effects to nicotine.

Alkaloids Biosynthesised from Tyrosine

Tyrosine (9.40) is the precursor of many alkaloids of which the benzylisoquinoline derivatives are pharmacologically the most important.

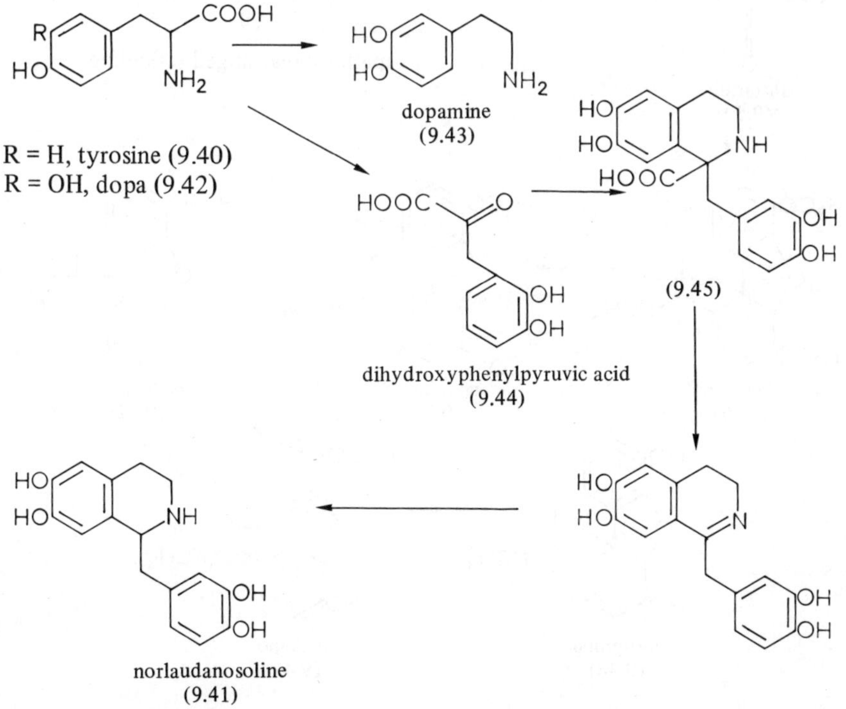

Figure 9.10 The biosynthesis of norlaudanosoline.

ALKALOIDS

The Benzylisoquinoline Alkaloids

Numerous benzylisoquinoline alkaloids occur in the related orders, Magnoliales, Ranunculales, Aristoliales and Papaverales. Tracer work has shown that norlaudanosoline (9.41) is the precursor of these alkaloids and that this compound is biosynthesised entirely from tyrosine (9.40). The amino acid is first hydroxylated to dopa (2,3-dihydroxyphenylalanine) (9.42), which is then

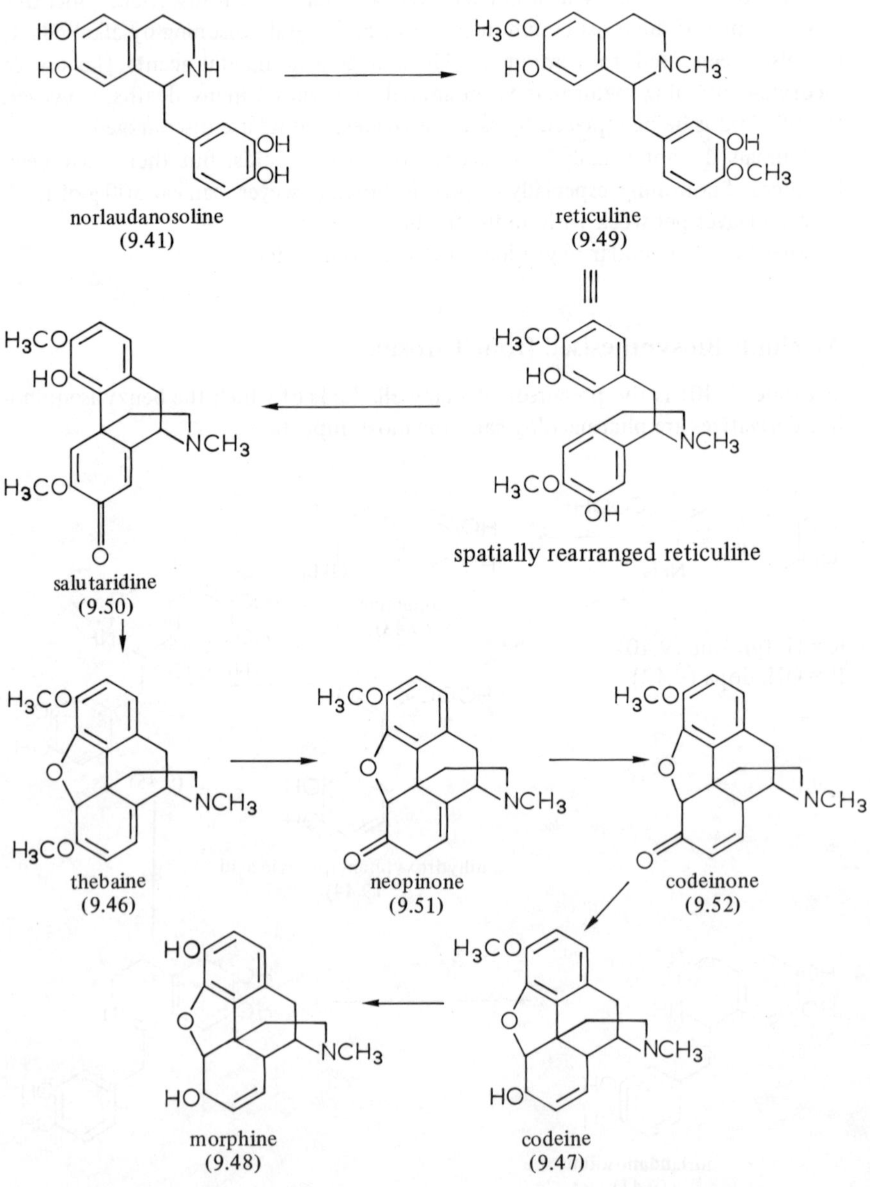

Figure 9.11 The biosynthesis of morphine, codeine and thebaine.

converted to both dopamine (9.43) and dihydroxyphenylpyruvic acid (9.44), compounds which link together to form the carboxylic acid (9.45). Norlaudanosoline is derived by decarboxylation of (9.45) as shown in *Figure 9.10*.

The opium alkaloids, thebaine (9.46), codeine (9.47) and morphine (9.48), are biosynthesised from norlaudanosoline through the intermediate formation of reticuline (9.49) and salutaridine (9.50). Thebaine is the precursor of codeine through conversion to neopinone (9.51) and codeinone (9.52), while morphine is derived from codeine. The biosynthesis of the opium alkaloids is summarised in *Figure 9.11*.

Papaverine (9.53), another opium alkaloid, is also derived from norlaudanosoline, but not through the formation of reticuline, as tracer work has shown this compound to be a poor precursor of the alkaloid. Presumably demethylation of reticuline does not take place easily as norreticuline (9.54), derived from norlaudanosoline, is a good precursor of papaverine. Tracer work has shown that the order in which the hydroxyl groups are methylated can be varied, but all the hydroxyl groups of norlaudanosoline are methylated before aromatisation of the heterocyclic ring, tetrahydropapaverine (9.55) being the immediate precursor of papaverine, as shown in *Figure 9.12*.

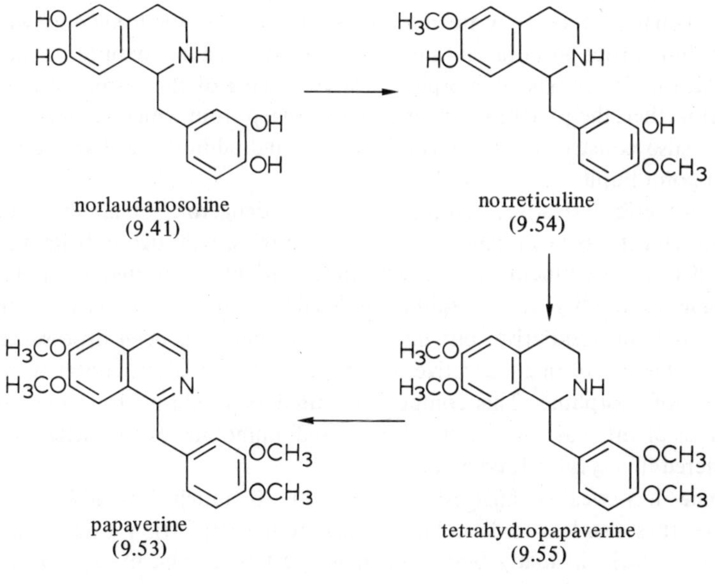

norlaudanosoline
(9.41)

norreticuline
(9.54)

papaverine
(9.53)

tetrahydropapaverine
(9.55)

Figure 9.12 Some steps in the biosynthesis of papaverine.

Although morphine accumulates in the latex of the opium poppy (*Papaver somniferum*, Papaveraceae), particularly in the unripe capsule, its concentration has been found to vary considerably and often depends on the time of day, being greatest in the early morning. Tracer experiments have shown that this alkaloid is metabolised to non-alkaloidal compounds and that demethylation to normorphine is not a reversible reaction. It was also found that isolated latex from the capsule would convert dopa to morphine but that latex from the stem would not. The opium poppy can convert various synthetic codeine derivatives into morphine derivatives which do not occur naturally in the plant.

Morphine has only been isolated from the closely related *P. somniferum* and *P. setigerum,* while codeine also occurs in another related species, *P. paeoniflorum.* Thebaine and papaverine, however, occur much more widely in the *Papaver* genus, indicating that their biosynthetic pathways evolved before that by which thebaine is converted to morphine.

Morphine, codeine and papaverine are important as drugs, while thebaine, although of no pharmacological use itself, is easily converted to codeine, and so helps to fill the demand for this drug by the pharmaceutical industry.

In man, morphine is both narcotic and analgesic. It is used to relieve prolonged pain, especially that of terminal patients. It has been discovered that the brain contains discrete opiate receptors and that morphine imitates the action of painkillers, known as enkephalins, which are naturally produced by the brain. This drug is also used to treat many diseases and is included in prescriptions for unproductive coughs and diarrhoea.

Morphine affects the central nervous system, causing reduced powers of concentration and lessening feelings of fear, anxiety and hunger, thus leading to a state of contentment. Addicts of this drug become both tolerant and physically dependent on its effects. They are able to take many times the toxic dose without ill effect, but abrupt withdrawal causes extremely unpleasant symptoms and may lead to death. Overdoses of morphine cause slowing of the respiration until it ceases altogether, the patient falling into a deep sleep from which he never wakes. Morphine also constricts the pupils of the eyes and addicts are characterised by their pin-point pupils.

The synthetic derivative, heroin, is the most dangerous of all the drugs of addiction, and it has been banned in many countries, although in Britain its use is allowed in the treatment of terminal cancer patients. Repeated attempts have been made to modify the morphine molecule so that it retains its analgesic powers but is not addictive, but unfortunately there has been little success in this field. One interesting derivative, etorphine, has an analgesic potency 10 000 times that of morphine. This compound is used as a sedative for large animals such as elephants, as they can be rendered immobile by as little as 2 mg administered in a syringe fired as a dart.

Codeine has weaker analgesic powers but is much less addictive than morphine. It is also less toxic and does not slow the respiration. Codeine is combined with aspirin in many brands of analgesic tablets, and huge quantities of the drug are required for this purpose every year. Codeine is also included in some cough mixtures as it supresses the urge to cough.

Papaverine is a muscle relaxant used to treat spasms and asthma. It is a coronary vasodilator and is prescribed in some forms of heart disease.

Although benzylisoquinoline alkaloids occur in all *Papaver* species, animals generally do not have access to sufficient of the plants to cause poisoning. Symptoms which have been observed in poppy poisoning are gastroenteritis and nervous excitement. Morphine has been used to dope horses, as it acts as a stimulant in these animals.

Alkaloids Biosynthesised from Tyrosine and Phenylalanine

The Phenanthridine Alkaloids

Tyrosine (9.40) and phenylalanine (9.56) are the precursors of the phenanthridine alkaloids which are characteristic of the Amaryllidaceae family. Tracer experiments have shown that norbelladine (9.57) is the precursor of these alkaloids and that tyrosine is incorporated into norbelladine as tyramine (9.58), while phenylalanine is first deaminated and then converted to protocatechualdehyde (9.59), as shown in *Figure 9.13.*

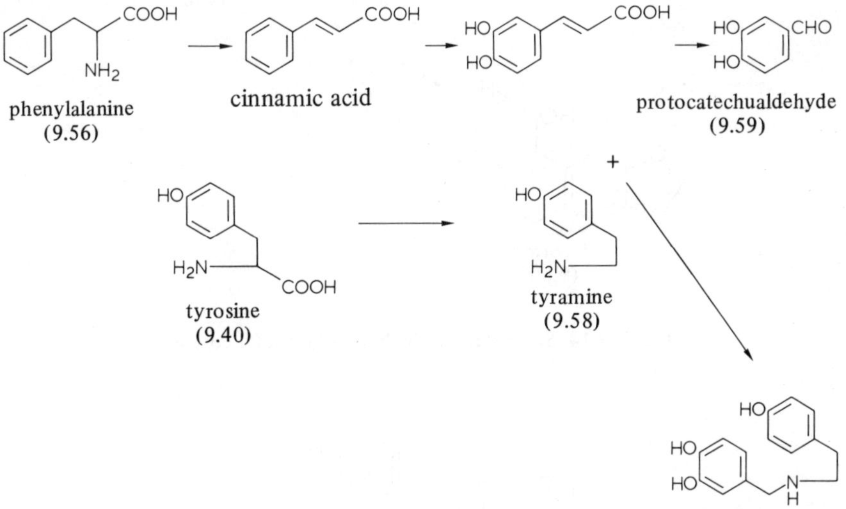

phenylalanine (9.56); cinnamic acid; protocatechualdehyde (9.59); tyrosine (9.40); tyramine (9.58); norbelladine (9.57)

Figure 9.13 The biosynthesis of norbelladine.

Lycorine (9.60), the most common of the Amaryllidaceae alkaloids, occurs in at least 24 genera. It is derived from norbelladine through methylation to form *O*-methylnorbelladine (9.61) and ring closure to give norpluviine (9.62). Hydroxylation and oxidation of norpluviine then gives lycorine, as shown in *Figure 9.14*. The methylenedioxy group occurs in many of the Amaryllidaceae alkaloids and is formed by direct ring closure between the methoxyl and adjacent hydroxyl groups.

Several alkaloids can be synthesised from *O*-methylnorbelladine by twisting the tyrosine-derived aromatic ring as shown in *Figure 9.15*. The formation of galanthamine (9.63), haemanthamine (9.64) and pretazettine (9.65) illustrates the diversity of compounds which can be derived from a norbelladine precursor.

Although alkaloids occur in the leaves, stems and flowers of some members of the Amaryllidaceae family, it is generally the bulbs which are the most toxic part of the plant. Human poisoning has occurred through the bulbs of daffodils (*Narcissus* species) being mistaken for onions, while pigs and cattle have been

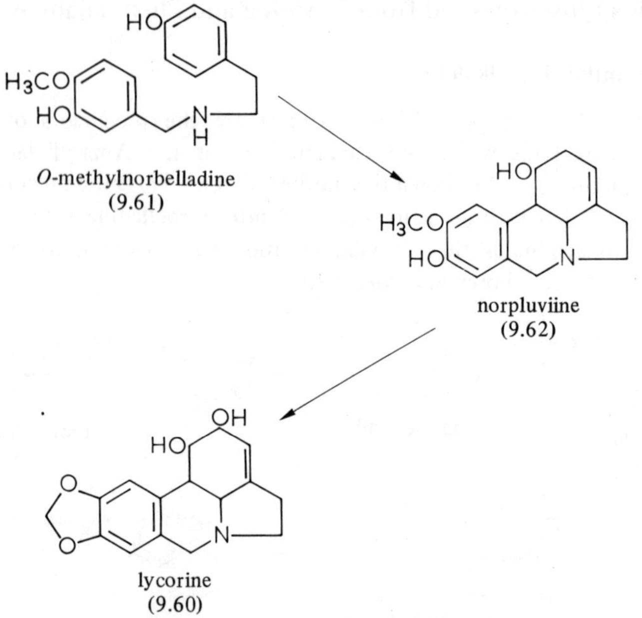

O-methylnorbelladine
(9.61)

norpluviine
(9.62)

lycorine
(9.60)

Figure 9.14 Some steps in the biosynthesis of lycorine.

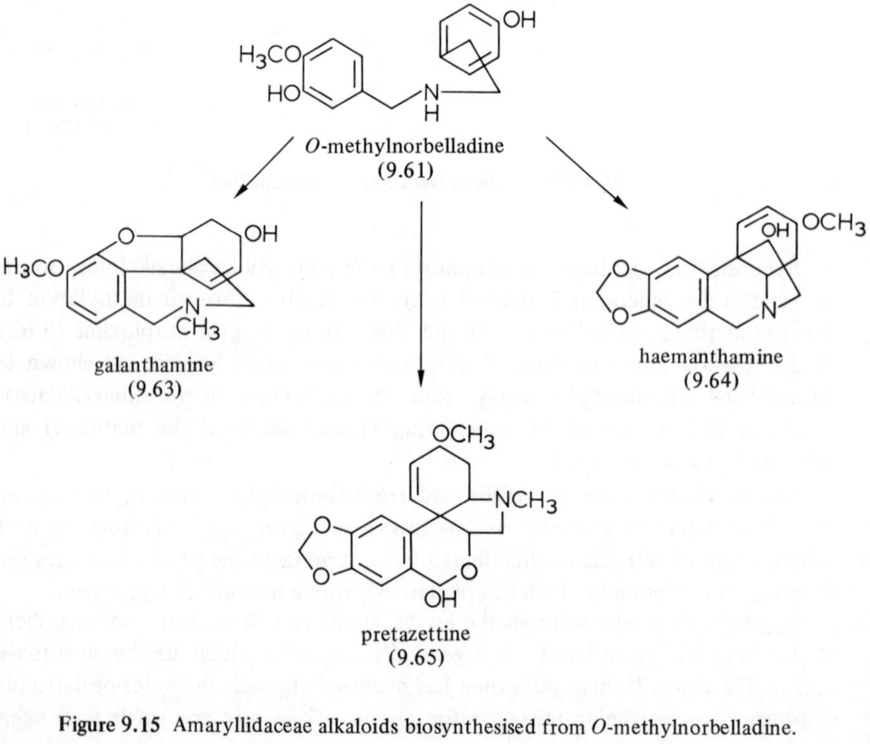

O-methylnorbelladine
(9.61)

galanthamine
(9.63)

haemanthamine
(9.64)

pretazettine
(9.65)

Figure 9.15 Amaryllidaceae alkaloids biosynthesised from O-methylnorbelladine.

poisoned when fed the bulbs of daffodils and snowdrops (*Galanthus* species). Some species of *Amaryllis, Crinum, Haemanthus* and *Nerine* have caused the loss of sheep and goats in South Africa, while the highly toxic *Buphane disticha* has been used to prepare arrow poisons in Africa.

Lycorine causes vomiting, diarrhoea, general collapse and death due to paralysis of the central nervous system.

Alkaloids Biosynthesised from Tryptophan

Tryptophan (9.66) is the precursor of a large number of indole alkaloids, which range from simple compounds such as serotonin (9.67) to the highly complex dimers, such as vincaleucoblastine (9.88).

The Simple Indole Alkaloids

The simple indole alkaloids occur in a large number of plants belonging to widely varying families. The majority have such a low toxicity that they can be considered non-poisonous, especially in the concentrations occurring in plants. Many of the simple indole alkaloids are tryptamine (9.68) derivatives, and one of the simplest, serotonin (9.67) (enteramine, thrombocytin, 5-hydroxy-tryptamine), is thought to act with histamine in producing the intense pain and itching caused by the stinging hairs of several plants, including stinging nettles (*Urtica* species, Urticaceae). The biosynthesis of serotonin is shown in *Figure 9.16*.

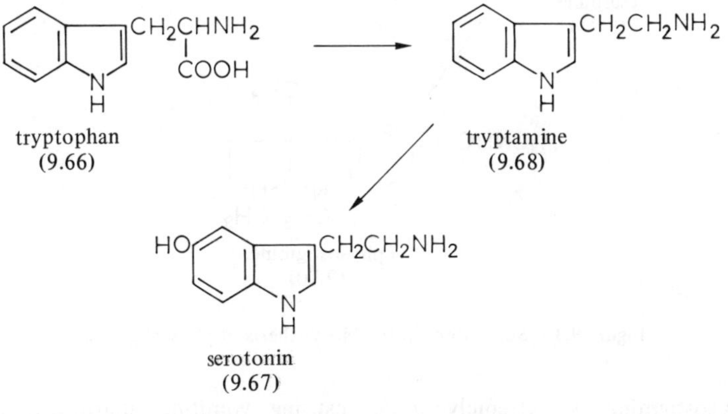

tryptophan
(9.66)

tryptamine
(9.68)

serotonin
(9.67)

Figure 9.16 The biosynthesis of serotonin.

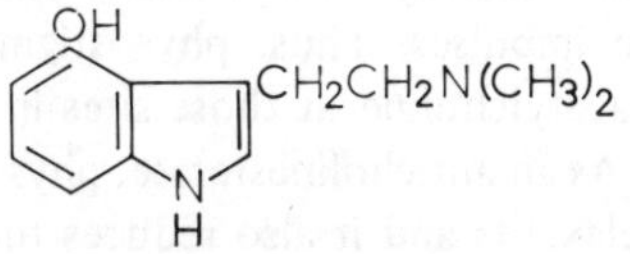

psilocin
(9.69)

Serotonin is produced naturally in the body and it is one of the compounds in the brain which influences behaviour. It has been suggested that lack of serotonin causes depression. Other tryptamine derivatives are hallucinogens, including psilocin (9.69), found in some mushrooms, especially the South American *Psilocybe* (Agaricaceae).

PHYSOSTIGMINE

Physostigmine (9.70) is the main alkaloid of the Calabar bean (*Physostigma venenosum, Leguminosae*). Biosynthesis is from tryptophan (9.66) with eseroline (9.71) the immediate precursor of physostigmine, as shown in *Figure 9.17*. Physostigmine is also present in some other *Physostigma* species and in *Hippomane* (Euphorbiaceae).

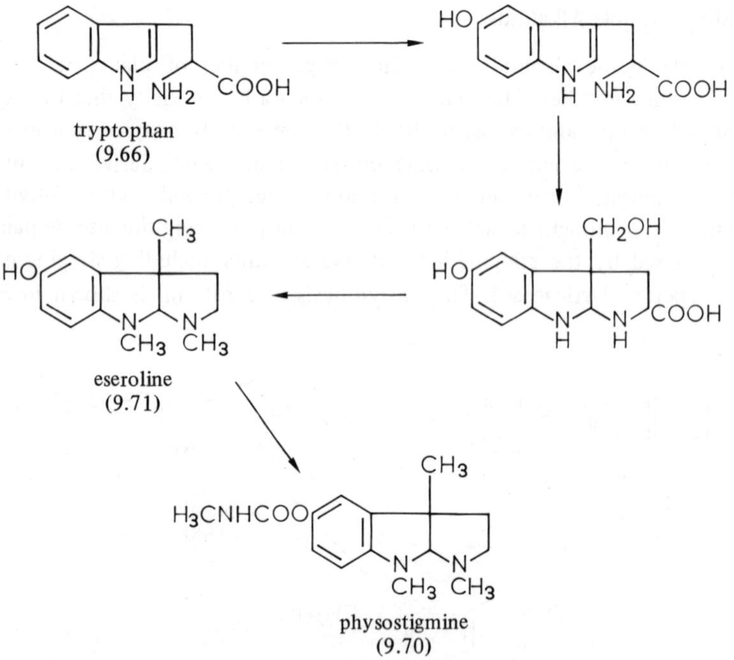

Figure 9.17 Some steps in the biosynthesis of physostigmine.

Physostigmine is extremely toxic, causing vomiting, diarrhoea, spasms, circulatory collapse and paralysis of the lower limbs. Death is due to respiratory failure or paralysis of the heart. This alkaloid is an inhibitor of acetylcholinesterase, an enzyme which destroys acetylcholine once this compound has been used to transmit nerve impulses. Thus, physostigmine causes an abnormally high concentration of acetylcholine at those sites in the nervous system where it is normally released. As an anticholinesterase, physostigmine is used to reverse the effects of muscle relaxants and it also reduces the blood pressure and slows the pulse. Physostigmine contracts the pupil of the eye, and dilute solutions are used to treat glaucoma.

The seeds of the Calabar bean constitute one of the most notorious ordeal poisons of West Africa. Accidental poisoning of man has taken place when overdoses of the drug have been taken either as orthodox or as native medicines. There seems to be no record of animal poisoning by this alkaloid.

THE ERGOT ALKALOIDS

The ergot alkaloids or ergolines occur in *Claviceps* species, fungi which infest cereals and grasses, especially rye (*Secale cereale*, Gramineae). Tracer experiments have shown that chanoclavine I (9.72) is the precursor of the more complex alkaloids and that this compound is biosynthesised from tryptophan (9.66) and mevalonic acid (9.73), through the intermediate formation of dimethylallyltryptophan (9.74), as shown in *Figure 9.18*. Before reaction, mevalonic acid is converted to its pyrophosphate, and an enzyme has been isolated which catalyses the formation of dimethylallyltryptophan.

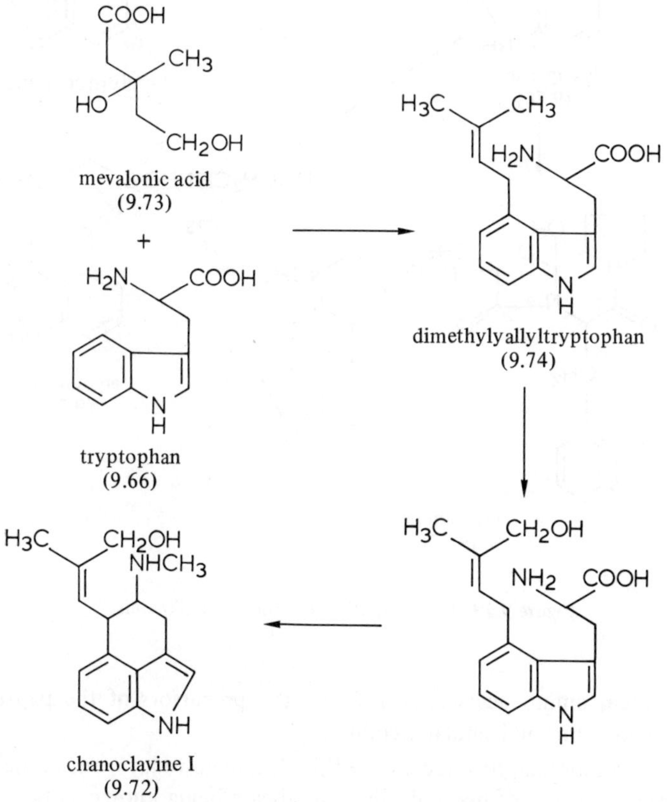

Figure 9.18 The biosynthesis of chanoclavine I.

The more complex alkaloids can be divided into two types: the clavines, such as agroclavine (9.75), and the lysergic acid (9.76) derivatives, such as ergotamine (9.77) and ergometrine (9.78). Agroclavine (9.75), biosynthesised from chanoclavine I (9.72), is the precursor of the lysergic acid derivatives (*Figure 9.19*).

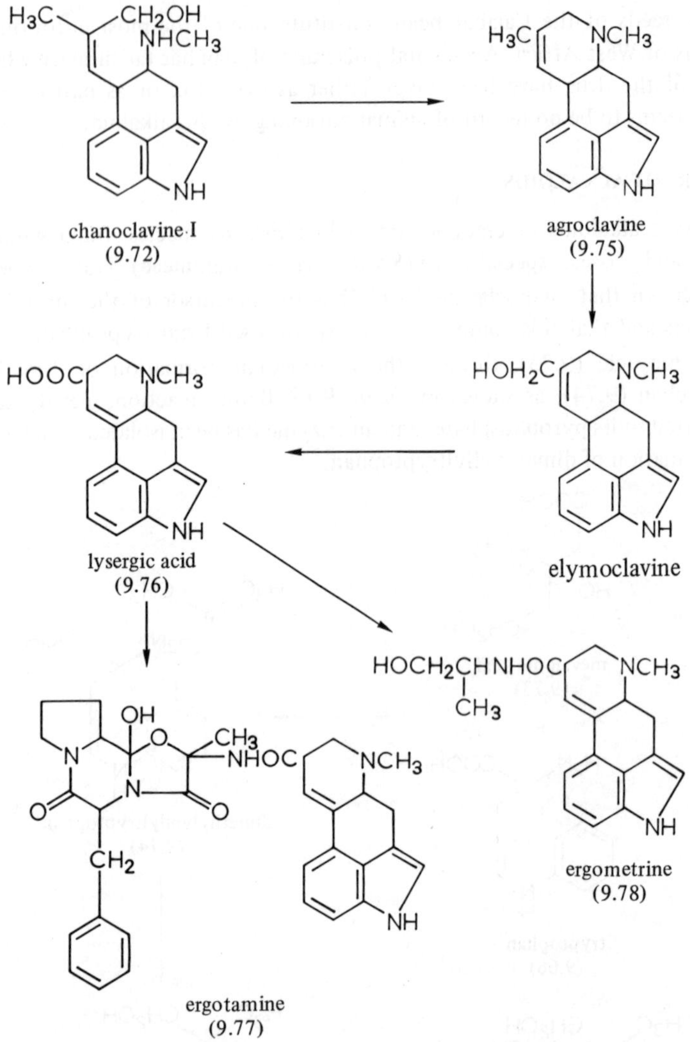

Figure 9.19 The biosynthesis of the ergot alkaloids.

The protein amino acids appear to be the precursors of the peptide side chains of ergotamine and similar ergolines.

Ergot poisoning due to infected rye flour is rare nowadays, but in the Middle Ages it was the cause of many deaths, the disease being known as St Antony's Fire. Poisoning can be of two types: the rarer form affects the brain, causing mental disturbance and convulsions, while the more common form causes severe constriction of the blood vessels, resulting in gangrene. Today, ergot poisoning is most likely to occur in women who take ergot preparations to produce abortion.

Poisoning of animals eating infected grasses is much more common, especially amongst cattle, and the animals exhibit both types of symptoms.

Ergometrine stimulates smooth muscle, particularly of the uterus. It is widely used in obstetrics to expel the placenta and blood clots and to reduce the risk of haemorrhage. Ergotamine is used to treat migraine, which is due to dilation of the blood vessels in the head. The alkaloid acts by constricting these blood vessels but it also causes unpleasant side-effects such as nausea, vomiting, diarrhoea, dizziness and confusion, and can thus only be used in small concentrations. Ergotamine increases the severity of non-migraine headaches which are due to constriction of the blood vessels.

The Complex Indole Alkaloids

Over 600 bases classified as complex indole alkaloids have been isolated from plants belonging to the Apocynaceae, Loganiaceae and Rubiaceae families. Where tracer experiments have been carried out, they all show the indole nucleus to be derived from tryptophan (9.66), as with the simple indole alkaloids. The remainder of the complex indole molecule is derived from mevalonic acid (9.73) and falls into one of three structural types denoted *Corynanthe, Aspidosperma* or *Iboga* (*Figure 9.20*).

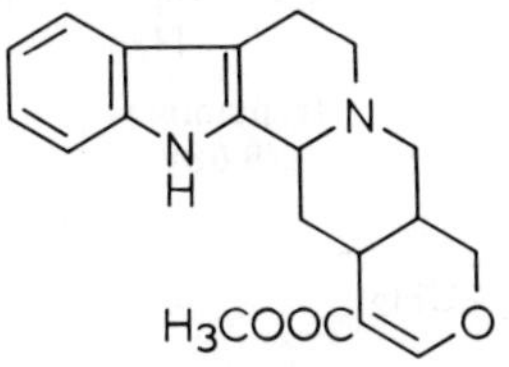

Corynanthe type

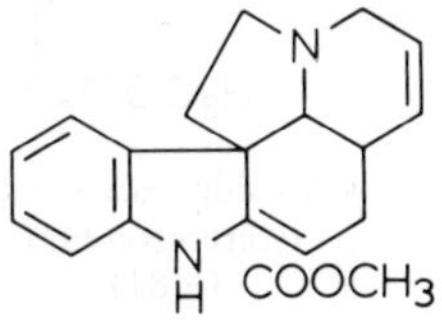

Aspidosperma type

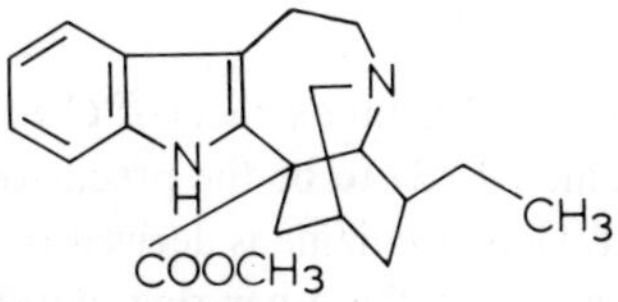

Iboga type

Figure 9.20 The complex indole alkaloids.

Isotopic tracer experiments with the Madagascar periwinkle (*Catharanthus roseus* syn. *Vinca rosea,* Apocynaceae), a plant which contains all three types of alkaloid, have shown that loganin (9.79) is a precursor of the complex indole alkaloids. Loganin, which occurs in several plants containing these alkaloids, is derived from geraniol and is therefore a product of the acetate–mevalonate pathway, and its biosynthesis is described in Chapter 5.

Before reacting with tryptamine (9.68), loganin is converted to secologanin (9.80), a compound found in minute quantities in *C. roseus*. The first products of the reaction of secologanin with tryptamine (formed by decarboxylation of tryptophan) are the stereoisomers, vincoside (9.81) and isovincoside, the C-5 epimer (*Figure 9.21*).

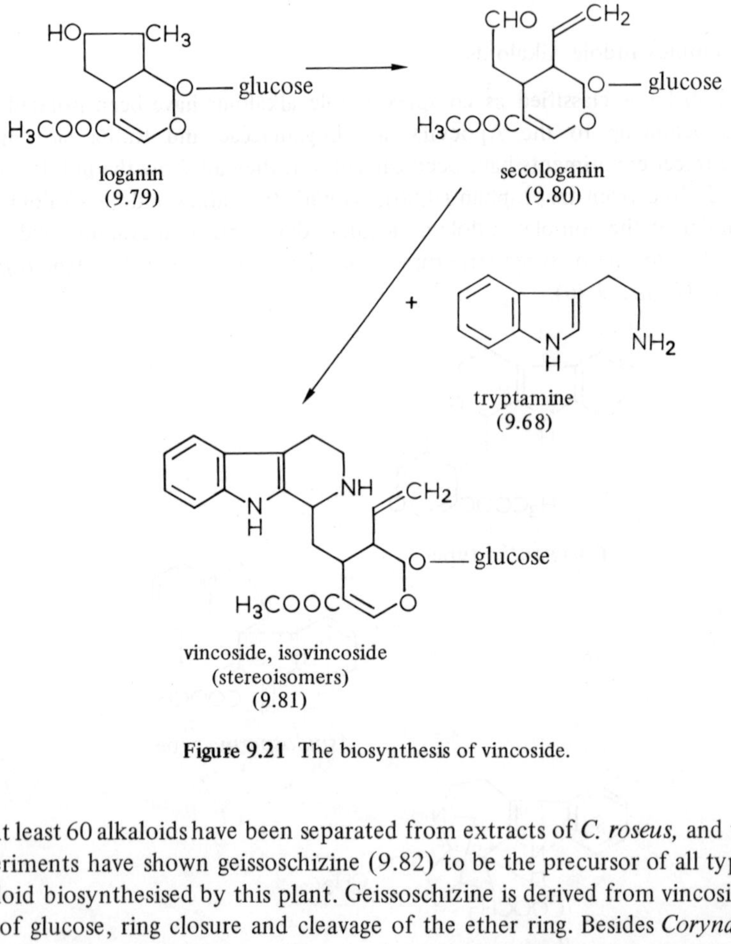

Figure 9.21 The biosynthesis of vincoside.

At least 60 alkaloids have been separated from extracts of *C. roseus,* and tracer experiments have shown geissoschizine (9.82) to be the precursor of all types of alkaloid biosynthesised by this plant. Geissoschizine is derived from vincoside by loss of glucose, ring closure and cleavage of the ether ring. Besides *Corynanthe*-type, *Aspidosperma*-type and *Iboga*-type alkaloids, *C. roseus* also synthesises alkaloids of the strychnine type and *bis*-indole dimers. *Figure 9.22* shows some steps in the biogenesis from geissoschizine (9.82) of ajmalicine (9.83) (*Corynanthe* type), vindoline (9.84) (*Aspidosperma* type), catharanthine (9.85) (*Iboga* type) and akuammicine (9.86) (strychnine type) which appear to be end-products, while tabersonine (9.87) is actively metabolised. The *bis*-indole dimers, such as vincaleucoblastine (9.88) and leurocristine (9.89), are formed by the linkage of an *Iboga*-type alkaloid to an *Aspidosperma* type, vincaleucoblastine being evolved from catharanthine and vindoline.

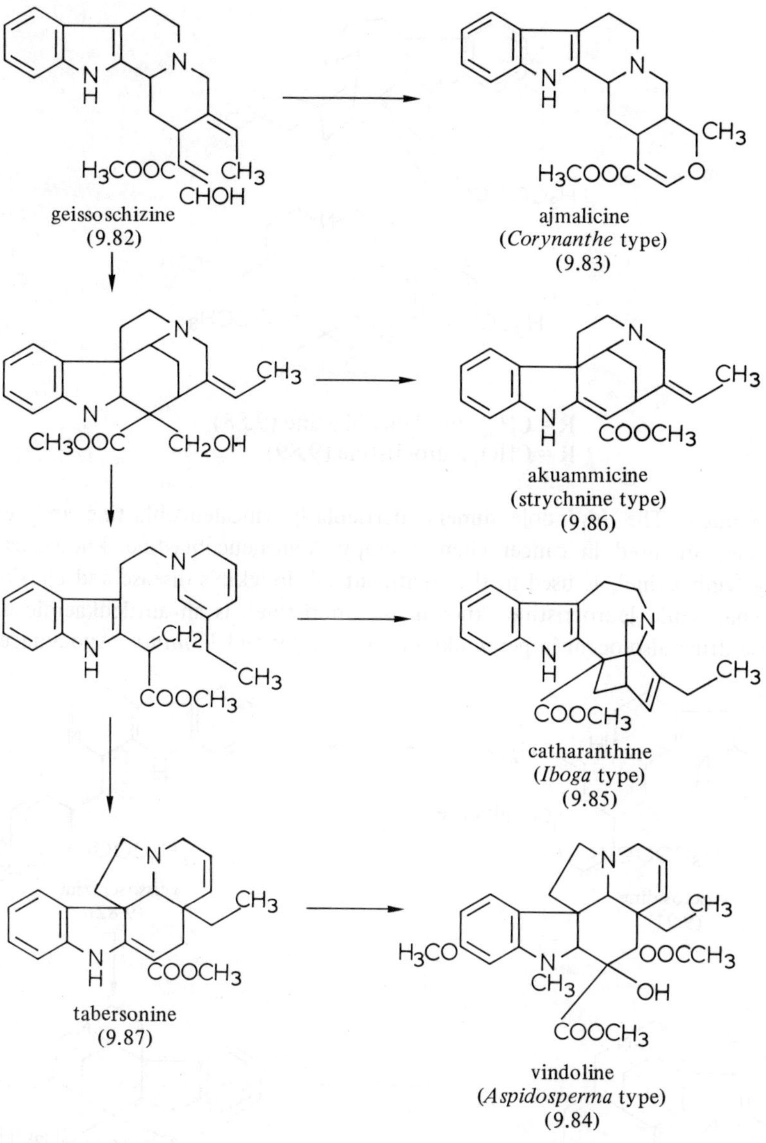

Figure 9.22 The biosynthesis of some alkaloids in *Catharanthus roseus*.

A few of the enzymes involved in the preliminary reactions by which the complex indole alkaloids in *C. roseus* are biosynthesised have been isolated, including a tryptophan decarboxylase which catalyses the formation of tryptamine.

The alkaloids in *C. roseus* have only a low toxicity and extracts have for long been used in native medicines. These extracts are hypotensive and have a limited antibiotic activity. Claims that they are also hypoglycaemic appear to be

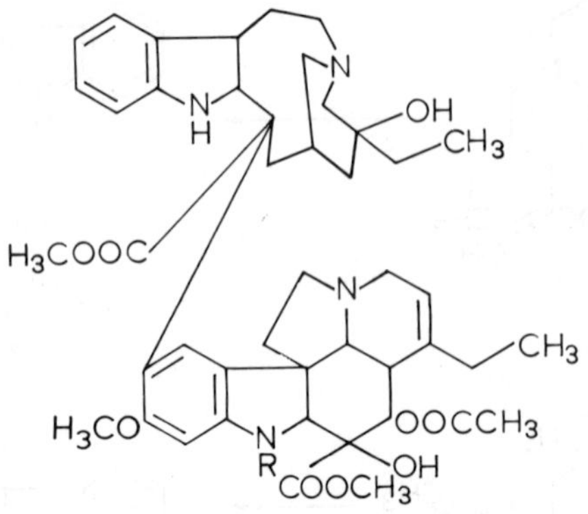

R = CH₃, vincaleucoblastine (9.88)
R = CHO, leurocristine (9.89)

unfounded. The *bis*-indole dimers, particularly vincaleucoblastine and leuro-cristine, are used in cancer chemotherapy. Vincaleucoblastine, known as the drug 'vinblastine', is used in the treatment of Hodgkin's disease and choriocar-cinoma, while leurocristine, known as 'vincristine', is an antileukaemic drug. These drugs also occur in periwinkles (*Vinca major* and *V. minor*, Apocynaceae).

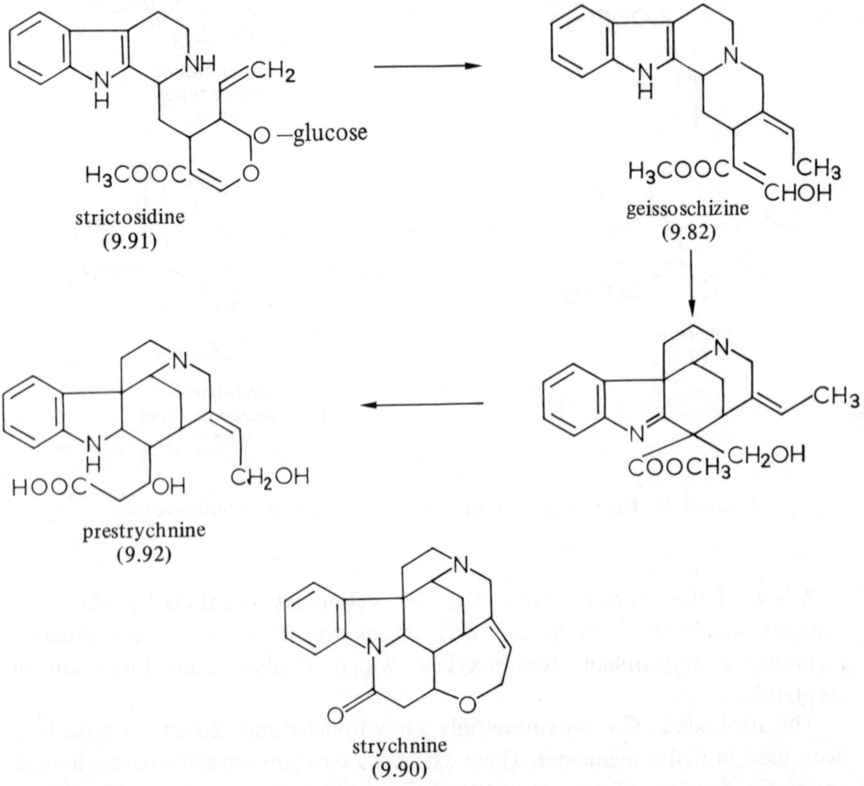

strictosidine
(9.91)

geissoschizine
(9.82)

prestrychnine
(9.92)

strychnine
(9.90)

Figure 9.23 Some steps in the biosynthesis of strychnine.

STRYCHNINE

Strychnine (9.90), found in some *Strychnos* species (Loganiaceae), is biosynthesised from tryptophan (9.66) and loganin (9.79) through the formation of
the unstable strictosidine (9.91). The latter compound has recently been shown
to be identical with isovincoside (9.81). Tracer experiments have shown geissoschizine (9.82) to be readily converted to strychnine, while prestrychnine (9.92)
appears to be the immediate precursor of strychnine. It is possible that the
formation of this alkaloid follows a pathway similar to that by which strychnine-
type compounds are formed in *C. roseus* (*Figure 9.22*). A possible biosynthetic
pathway for strychnine is given in *Figure 9.23*.

Strychnine is a highly toxic alkaloid, 30–60 mg being sufficient to kill a man.
It is a stimulant, producing mental and muscular activity and overdoses cause
characteristic violent convulsions, five or six of which are sufficient to cause
death, which is due to respiratory failure. However, strychnine is one of the few
poisons whose effects can be counteracted after absorption. It is important to
keep the patient absolutely still and quiet and one of the barbiturate drugs is
usually given as an antidote. Strychnine is used medicinally to combat morphine
stupour and overdoses of depressant drugs, and sometimes in the treatment of
alcoholism (*delirium tremens*).

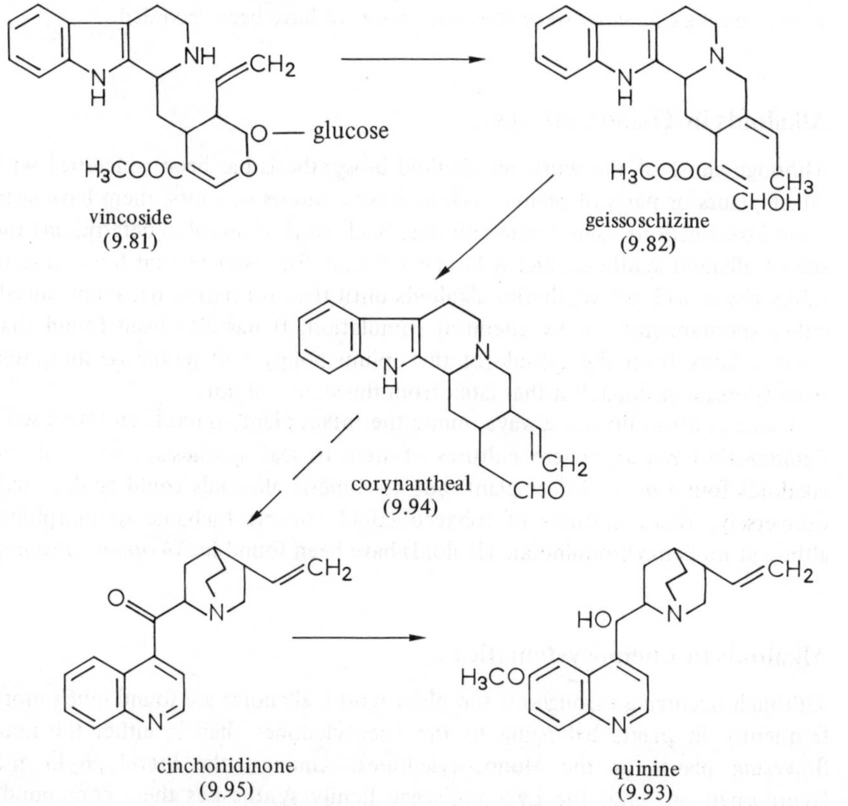

vincoside
(9.81)

geissoschizine
(9.82)

corynantheal
(9.94)

cinchonidinone
(9.95)

quinine
(9.93)

Figure 9.24 Some steps in the biosynthesis of quinine.

QUININE

Quinine (9.93) occurs in some *Cinchona* species (Rubiaceae) and, although this alkaloid is a quinoline derivative, it is derived from tryptophan (9.66) and loganin (9.79) and is thus biosynthetically related to the complex indole alkaloids. Tracer experiments have shown both vincoside (9.81) and the *Corynanthe*-type alkaloid, corynantheal (9.94), to be early intermediates in the biosynthesis of quinine, while cinchonidinone (9.95) was also converted to quinine and has been found in *Cinchona ledgeriana* by dilution analysis. The steps by which *Corynanthe*-type indole alkaloids are converted to quinoline derivatives have not yet been elucidated. Thus, only a partial biosynthetic pathway can be written for quinine (*Figure 9.24*).

Quinine is one of the least toxic of all alkaloid drugs, but it can cause deafness and blindness, which may be permanent, and also produce allergies. The main use of this drug is as an antimalarial, particularly in resistant cases which do not respond to treatment with synthetic drugs. Non-fatal cases of quinine poisoning have very occasionally resulted from its use in malaria. This drug is also an oxytocic, stimulating the uterus, and fatal poisoning has occurred when quinine has been used as an abortifacient. Quinine has been used in doping animals, as it acts as a stimulant in the horse and a sedative in the greyhound. Poisoning of animals eating *Cinchona* trees does not appear to have been recorded.

Alkaloids in Tissue Cultures

Although most of the work on alkaloid biosynthesis has been concerned with intact plants or parts of plants, such as leaves, shoots or roots, there have been some investigations using tissue cultures. Such work is useful in determining the site of alkaloid synthesis, and it has been found, for instance, that tobacco stem callus tissue will not synthesise alkaloids until root formation has commenced, either spontaneously or by chemical stimulation. It has also been found that isolated latex from the capsule of the opium poppy will synthesise morphine from tyrosine or dopa, but that latex from the stem will not.

Tissue cultures do not always mimic the intact plant, as has been found with *Catharanthus roseus,* where cultures of stem or leaf synthesised some of the alkaloids found in the intact plant, but no dimeric alkaloids could be detected. Conversely, tissue cultures of tobacco could convert thebaine to morphine, although no benzylisoquinoline alkaloids have been found in *Nicotiana tabacum.*

Alkaloids in Chemosystematics

Although occurring throughout the plant world, alkaloids are found much more frequently in plants belonging to the Dicotyledones than in either the non-flowering plants or the Monocotyledones. Amongst the Pteridophyta and Gymnospermae, only the Lycopodiaceae family synthesises these compounds to any extent. The *Lycopodium* alkaloids are quinolizidine derivatives and have the ring structure shown by lycopodine (9.96).

lycopodine
(9.96)

Alkaloids are rare in dicotyledon orders classified before the Centrospermae, thus linking these simpler flowering plants with the Gymnospermae. Amongst the remaining orders, the occurrence of alkaloids is very uneven, and an order rich in such compounds may be both preceded and followed by an order in which alkaloids do not appear to be synthesised at all. On a family level, such erratic distribution is even more apparent. Thus, alkaloids by themselves are useless in determining phylogenetic relationships between orders or between families within an order, except in a few special cases, such as that described below for the Centrospermae.

Those orders rich in alkaloids are the Centrospermae, Magnoliales, Ranunculales and Gentianales in the Dicotyledones, and the Liliales and Orchidales in the Monocotyledones. Of the individual families, probably the Papaveraceae and the totally unrelated Apocynaceae will prove to contain the greatest number of these secondary metabolites. Every species belonging to the Papaveraceae so far examined contains alkaloids, while the Apocynaceae exhibits a great diversity of complex indole alkaloids. Other families in which alkaloids frequently occur include the Ranunculaceae, Leguminosae, Rutaceae, Loganiaceae, Rubiaceae, Solanaceae, Compositae, Amaryllidaceae, Liliaceae and Orchidaceae. The Amaryllidaceae alkaloids are specific to that family, which contains no other type.

When applying chemosystematics to the classification of plants, the biosynthetic pathway is more important than the end-product. Thus, quinine, a quinoline derivative found in the Rubiaceae, is biosynthesised from tryptophan by a pathway similar to that forming the complex indole alkaloids characteristic of the family. Quinoline derivatives also occur in the Rutaceae, but here they are biosynthesised from anthranilic acid and by a pathway which is specific to this family.

The application of alkaloid biosynthesis to classification is also complicated by convergence and divergence. The synthesis of tropane alkaloids from hygrine is an example of convergence. This takes place in such unrelated families as Convolvulaceae (*Convolvulus* species), Cruciferae (*Cochlearia arctica*), Erythroxylaceae (*Erythroxylum coca*), Euphorbiaceae (*Phyllanthus discoideus*) and Solanaceae (several genera). However, the recent removal of the tropane alkaloid-containing *Anthocercis* from Solanaceae, a family rich in such compounds, to the non-tropane alkaloid-containing Scrophulariaceae, would not seem to be justified on chemosystematic grounds. Within the Papilionoideae are many examples of the divergence which can be shown by related plants, as members of this subfamily synthesise a wide range of secondary metabolites.

Even the Papilionoideae alkaloids show little relationship either in molecular structure or in their biosynthetic pathways. Thus, the spiroamine alkaloids found in *Erythrina* species have no structural relationship to the pyrrolizidine alkaloids characteristic of *Crotalaria* species and are biosynthesised from tyrosine, while the pyrrolizidine alkaloids originate from ornithine. (The biosynthesis of pyrrolizidine alkaloids in both *Crotalaria* and the totally unrelated *Senecio* (Compositae) is another example of convergence.) However, the quinolizidine alkaloids characteristic of the Sophoreae, Podalyrieae and Genisteae tribes are biosynthesised by a pathway similar to the pyrrolizidine alkaloids in the related *Crotalaria*.

The similarity of the quinolizidine alkaloids in the Papilionoideae and Ranunculaceae indicates a common ancestor, a suggestion strengthened by the similarity of the amino acid sequences of cytochrome c in *Phaseolus aureus* (Papilionoideae) and *Nigella damascena* (Ranunculaceae).

There has been much controversy amongst botanists over the families which should be included in the order Centrospermae. Although not all systematists agree, it has been suggested that, of the original 12 families considered on morphological grounds to belong to this order, only the 10 containing betalains are phyletically related. The betalains are red or yellow alkaloids which take the place of the more usual anthocyanins. An example is the glycoside, betanin (9.97), which gives the red colour to beetroot *(Beta vulgaris* forma *rubra,* Chenopodiaceae).

betanin
(9.97)

The two families, Molluginaceae and Caryophyllaceae, which are usually included in the Centrospermae, synthesise only anthocyanins and not betalains. Thus, it seems reasonable to exclude these families, especially as the morphological evidence for their inclusion is slight. The specificity of the betalains to the Centrospermae, and the total lack of anthocyanins in these families, indicates an early evolutionary divergence of the order from the remaining angiosperms.

There are many instances in which the species within a genus can be arranged according to their alkaloid content, and it is at this level that alkaloid biosynthesis is most useful to systematics. For instance, in the genus *Argemone,* which is difficult to classify on purely morphological grounds, the species can be arranged in three groups according to their alkaloid content. Group 1 contains alkaloids of the protopine and berberine types but not of the pavine type; Group 2 contains all three alkaloid types; while Group 3 contains only the pavine type.

Although specific alkaloids can be said to be characteristic of an order, family or genus, it does not follow that the same alkaloids are not also synthesised in plants belonging to totally different orders, families or genera. However, the frequency of occurrence will be much higher in the former. Some characteristic alkaloids are the betalains of the Centrospermae, the protoberberines of the Ranunculales, lycorine of the Amaryllidaceae family, the spiroamine alkaloids of *Erythrina* and the α,α'-substituted piperidines characteristic of *Lobelia* species (Lobeliaceae). Such characterisations must be applied with caution, however, as alkaloid types previously thought to be unique to a particular family, genus or species have now been found in totally unrelated plants. Such a trend is sure to continue, and even morphine, the once classical example of an alkaloid unique to a single species, has been found in a second *Papaver* species (*P. setigerum*), albeit one that is closely related to *P. somniferum*.

Bibliography

General

The Alkaloids, R. H. F. Manske (ed.) (continuing series, vol. I, 1960; cumulative index, vol. XI, 1968), Academic Press. (The reader should note that some earlier volumes contain structures and biosynthetic pathways which have subsequently been found to be incorrect)
The Alkaloids, Specialist Periodical Reports (continuing series, vol. 1, 1971), The Chemical Society
Alkaloid Biology and Metabolism in Plants, G. R. Waller and E. K. Nowacki, Plenum, 1978

Biosynthesis

'Biosynthesis of Alkaloids', in *Biosynthesis,* Specialist Periodical Reports (continuing series, vol. 1, 1972), The Chemical Society
'Alkaloid Biogenesis', in *Biogenesis of Natural Compounds,* 2nd edn, P. Bernfeld (ed.), Pergamon Press, 1967
'Alkaloid Biosynthesis', in *Organic Chemistry of Secondary Plant Metabolism,* T. A. Geissman and D. H. G. Crout, Freeman, Cooper and Co., 1969
'Alkaloids from Aromatic Amino Acids', *ibid.*
'Oxidative Phenol Coupling in Alkaloid Biosynthesis', *ibid.*
'Alkaloids of Mixed Amino Acid Mevalonate Origin', *ibid.*
'The Secondary Metabolism of Amino Acids', in *Secondary Metabolism,* J. Mann, Oxford University Press, 1978
'Regulatory Control in Alkaloid Biosynthesis', in *Recent Advances in Phytochemistry,* vol. 8, V. C. Runeckles and E. E. Conn (eds.), Academic Press, 1974

'Alkaloids', in *Chemistry and Biochemistry of Plant Herbage,* vol. 1, G. W. Butler
and R. W. Bailey (eds.), Academic Press, 1973
The Alkaloids: The Fundamental Chemistry – A Biogenetic Approach, Studies
in Organic Chemistry, vol. 7, D. R. Dalton, Dekker, 1979

Toxicity

'Toxicity and Metabolism of *Senecio* Alkaloids', in *Phytochemical Ecology,*
J. B. Harborne (ed.), Academic Press, 1972
'Alkaloids', in *Chemistry and Biochemistry of Plant Herbage, ibid.*
'Plant Toxins and Their Effects on Animals', in *Introduction to Ecological
Biochemistry,* J. B. Harborne, Academic Press, 1977

Uses of Alkaloids as Drugs

'Drugs', in *Plant Products of Tropical Africa,* M. L. Vickery and B. Vickery, The
Macmillan Press, 1979
Gaddum's Pharmacology, 8th edn, A. S. V. Burgen and J. F. Mitchell, Oxford
University Press, 1978. (Alkaloids are described in several chapters)
The Isoquinoline Alkaloids, Chemistry and Pharmacology, M. Shamma,
Academic Press, 1972

Chemotaxonomy

'Plant Systematics and Alkaloids', in *The Alkaloids,* vol. XVI, *ibid.,* 1977
'The Taxonomic Significance of Alkaloids', in *Chemical Plant Taxonomy,*
T. Swain (ed.), Academic Press, 1963
'The Distribution of Alkaloids in the Rutaceae', *ibid.*
'Alkaloid Chemistry and the Systematics of *Papaver* and *Argemone',* in *Recent
Advances in Phytochemistry,* vol. 1, T. J. Mabry *et al.* (eds.), Appleton (N.Y.),
1968
'The Quinolizidine Alkaloids', in *Chemistry in Botanical Classification,* G. Bendz
and J. Santesson (eds.), Academic Press, 1974
'Chemistry of Geographical Races', section III, in *Chemistry in Evolution and
Systematics,* T. Swain (ed.), Butterworths, 1973
'Senecioneae – Chemical Review', in *The Biology and Chemistry of the
Compositae,* vol. 2, V. H. Heywood *et al.* (eds.), Academic Press, 1977
'Cynareae – Chemical Review', *ibid.*
'Alkaloids in the Leguminosae', in *Chemotaxonomy of the Leguminosae,*
J. B. Harborne *et al.* (eds.), Academic Press, 1971
'Comparative Phytochemistry of the Alkaloids', in *Comparative Phytochemistry,*
T. Swain (ed.), Academic Press, 1966
The Catharanthus Alkaloids, W. I. Taylor and N. R. Farnsworth (eds.), Dekker,
1975

10 Porphyrins, Purines and Pyrimidines

Introduction

The porphyrins, purines and pyrimidines are compounds with nitrogen-containing heterocyclic ring systems essential to the primary biochemical processes which maintain life. Without the porphyrin derivative, chlorophyll, there would be no *de novo* synthesis of sugars and therefore no energy to carry out the biochemical reactions of higher plants and animals. The nucleic acids, derivatives of purines and pyrimidines, are even more important to life, as without these compounds there would be no replication of protein molecules and no transmittance of heredity. Porphyrins, purine and pyrimidine derivatives are also involved in the prosthetic groups of many enzymes. A prosthetic group is the non-protein component of an enzyme which is firmly bound to the protein or apoenzyme, the complete complex forming a haloenzyme. More loosely bound non-protein components are known as coenzymes. A purine derivative is involved in the important complex, coenzyme A.

Porphyrins

The porphyrin ring system of four pyrrole units (*Figure 10.1*) is widely distributed in nature and is usually found bound to metal atoms. Non-metal-containing porphyrins are either precursors or degradation products of the metal-containing compounds.

(*a*) Porphyrin ring system   (*b*) Corrin ring system

Figure 10.1 The porphyrin and corrin ring systems.

The most common metal bound to a porphyrin ring system is iron, found as Fe(II) in haeme, haemoglobin and oxyhaemoglobin, and Fe(III) in haemin, peroxidase and catalase. The cytochromes contain iron with variable valency, depending on whether they are in the oxidised or reduced form. In haemoglobin, oxyhaemoglobin, peroxidase, catalase and the cytochromes, the iron atom is also coordinated to protein. Other metals can occur bound to porphyrins, including magnesium in the chlorophylls *a* and *b*.

The corrin ring system found in the cobalamins differs from the porphyrins in having lost a methylene group (*Figure 10.1*). The corrins are bound to cobalt in the cobalamins, which include vitamin B_{12}.

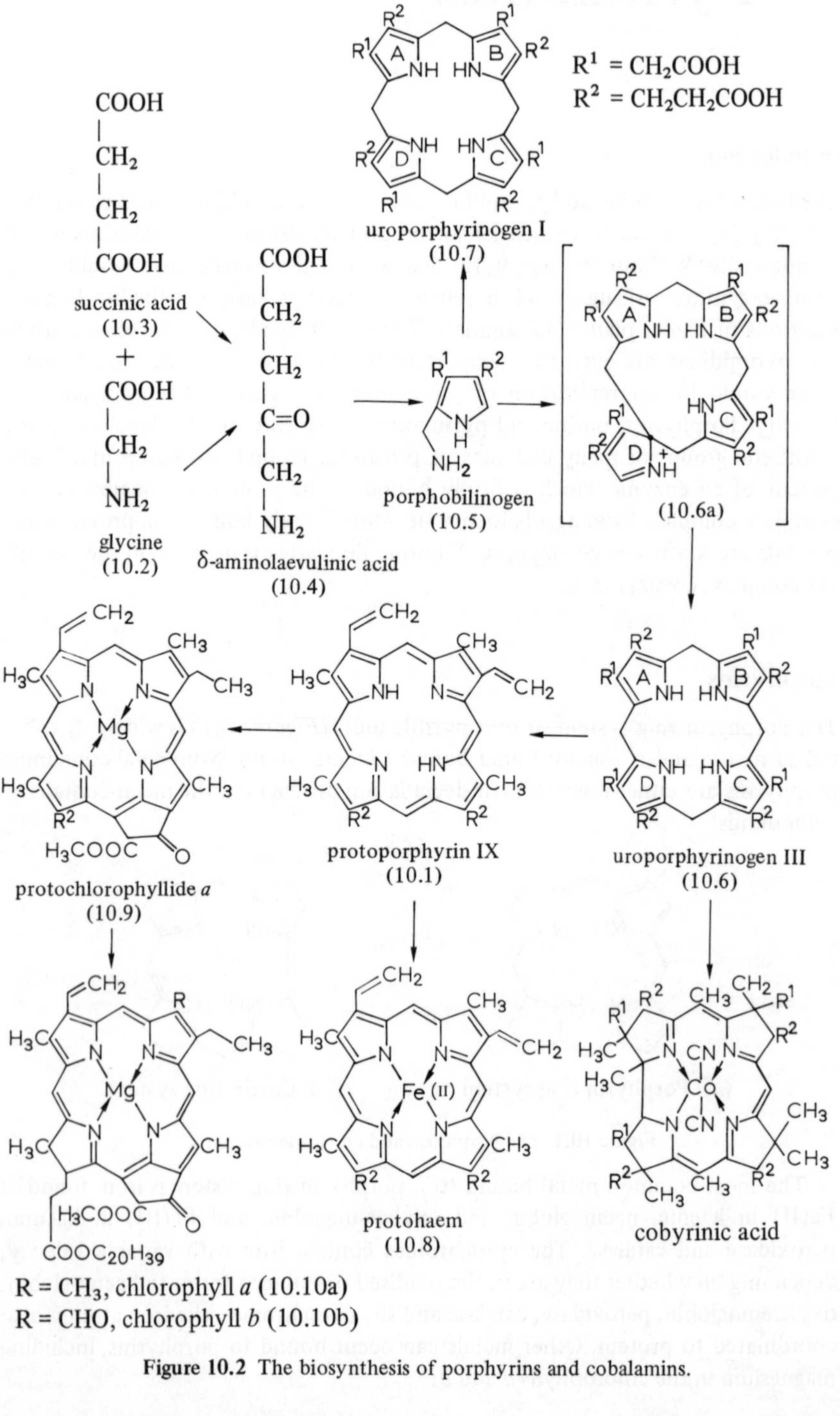

$R^1 = CH_2COOH$
$R^2 = CH_2CH_2COOH$

uroporphyrinogen I
(10.7)

succinic acid
(10.3)

glycine
(10.2)

δ-aminolaevulinic acid
(10.4)

porphobilinogen
(10.5)

(10.6a)

protochlorophyllide *a*
(10.9)

protoporphyrin IX
(10.1)

uroporphyrinogen III
(10.6)

$R = CH_3$, chlorophyll *a* (10.10a)
$R = CHO$, chlorophyll *b* (10.10b)

protohaem
(10.8)

cobyrinic acid

Figure 10.2 The biosynthesis of porphyrins and cobalamins.

Protoporphyrin IX (10.1) is the key intermediate in the biosynthesis of many porphyrins. It originates from the protein amino acid, glycine (10.2), which condenses with succinic acid (10.3), in the form of succinyl CoA, to form δ-aminolaevulinic acid (10.4) (*Figure 10.2*). Condensation of two molecules of this amino acid gives the pyrrole derivative, porphobilinogen (10.5), four molecules of which link up one at a time with the elimination of ammonia to give uroporphyrinogen III (10.6). In uroporphyrinogen III, one of the pyrrole rings (ring D, *Figure 10.2*) has rearranged intramolecularly. It has been found in microorganisms that two enzymes are concerned in the head-to-tail linking of the four pyrrole units. If uroporphobilinogen I synthase is used alone, uroporphyrinogen I (10.7) is produced, in which all four pyrrole rings retain their original configuration. If this synthase is combined with uroporphyrinogen III cosynthase, uroporphyrinogen III results. However, neither cosynthase alone nor in combination with the synthase can convert the I isomer to the III. Thus, the rearrangement of ring D takes place at some stage prior to ring closure. Recent work has shown that this rearrangement takes place after the four pyrrole units have linked up but before ring closure. A suggested intermediate (10.6a) is shown in *Figure 10.2*.

The pathway to the cobalamins (*Figure 10.2*) is thought to branch at the uroporphyrinogen III stage — one of the methylene groups being removed to give the corrin ring system.

Conversion of uroporphyrinogen III to protoporphyrin IX takes place by stepwise decarboxylation and dehydrogenation. The various iron complexes are derived from protohaem (10.8), which is formed from protoporphyrin IX and iron, a reaction catalysed by ferrochelatase. Protohaem is both the prosthetic group of haemoglobin, oxyhaemoglobin, peroxidase, catalase and cytochrome b, and the precursor of cytochromes a and c.

Modification of the substituents on rings C and D of protoporphyrin IX leads to protochlorophyllide *a* (10.9) in plants. The formation of protochlorophyllide *a* is the final reaction which can take place in the dark. Under the influence of light, the proto compound is hydrogenated in ring D to form chlorophyllide *a* and esterified to give chlorophyll *a* (10.10a). It is not yet known if esterification with phytol is direct or if chlorophyllide *a* is esterified with geranylgeraniol, which is then reduced to phytol. (The biosynthesis of geranylgeraniol is described in Chapter 5.) Chlorophyll *b* (10.10b) is probably synthesised from chlorophyll *a* by oxidation of the methyl group to an aldehyde group.

Chlorophylls *a* and *b*, found in the chloroplast lamellae, are the only green pigments of plants. Their light-absorbing properties are essential to the photosynthetic processes for the production of sugars from carbon dioxide and water. Chlorophyll synthesis is dependent on adequate levels of nitrogen, potassium, magnesium, iron and manganese within the plant, and leaves suffering a deficiency of these elements show symptoms of chlorosis. Conversely, excess concentrations of cobalt, copper, manganese, nickel and zinc also produce chlorosis, probably through interference with iron-requiring enzymes. During senescence and storage, chlorophyll is rapidly destroyed by oxidative degradation. The synthesis of chlorophyll is controlled by genes, and mutants have been bred which do not contain this pigment.

Haemoglobin and oxyhaemoglobin are important to the respiratory processes of animals, but they have not been found as true plant constituents, although haemoglobin occurs in the root nodules of some legumes. Peroxidase and catalase, however, occur throughout the plant kingdom. Peroxidase catalyses reactions involving oxidation by hydrogen peroxide, while catalase is responsible for the destruction of excess hydrogen peroxide, which is converted to water and oxygen. Horseradish (*Armoracia lapathifolia*, Cruciferae) is particularly rich in peroxidase.

The cytochromes are pigments which act as catalysts in electron transport reactions. They are divided into three groups — a, b and c — depending on their porphyrin substituents. In cytochrome c, the iron atom is coordinated to two protein amino acid residues. As all the coordination sites are occupied, cytochrome c cannot be poisoned by inhibitors such as cyanate or carbon monoxide.

Vitamin B_{12} (10.11) occurs in all plants but in concentrations much less than are found in raw animal liver. Thus, people living on a vegetarian diet can suffer from pernicious anaemia. In animals and microorganisms, vitamin B_{12} is concerned in the transfer of methyl groups, the methyl group to be donated being attached to the cobalt atom ($R = CH_3$ in (10.11)). This vitamin also takes part in the synthesis of new methyl groups. Coenzymes containing vitamin B_{12} have been isolated from plants, indicating that similar processes take place in these organisms.

$R^1 = CH_2CONH_2$
$R^2 = CH_2CH_2CONH_2$

vitamin B_{12}
(10.11)

The ubiquitous occurrence of porphyrin derivatives makes them of little value in chemosystematics.

Purine and Pyrimidine Derivatives

Purine and pyrimidine derivatives are essential constituents of all living organisms. An enormous number of biochemical reactions depends on these derivatives, including the transfer of energy, electrons and genetic information. The purines most commonly encountered are the bases, adenine (10.12), usually abbreviated to A, and guanine (G) (10.13), while uracil (U) (10.14), cytosine (C) (10.15) and thymine (T) (10.16) are the most common pyrimidine bases.

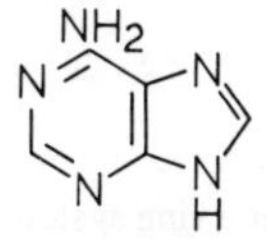

adenine
(10.12)

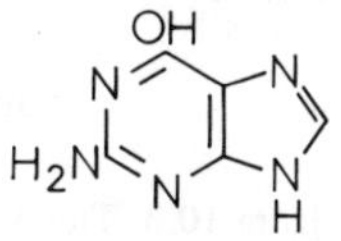

guanine
(10.13)

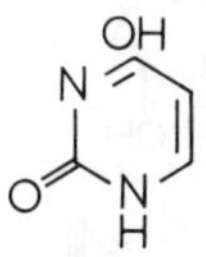

uracil
(10.14)

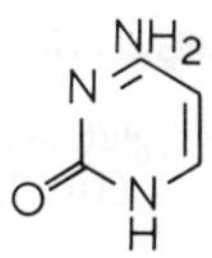

cytosine
(10.15)

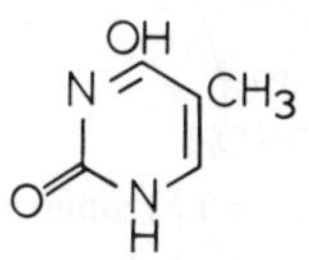

thymine
(10.16)

When purine or pyrimidine bases are in combination with the sugars, ribose or deoxyribose, the compounds are known as nucleosides. The addition of phosphate (abbreviated to P) to a nucleoside gives a nucleotide. Monophosphates (MP), diphosphates (DP) and triphosphates (TP) of nucleosides occur in all living organisms. Some common ribose nucleosides and nucleotides and their bases are given in *Table 10.1*.

Table 10.1 Some commonly occurring ribonucleosides and nucleotides.

Nucleoside	Nucleotides	Base
adenosine	AMP (adenylic acid), ADP, ATP	adenine (10.12)
guanosine	GMP (guanylic acid), GDP, GTP	guanine (10.13)
inosine	IMP (inosinic acid) (10.21)	hypoxanthine
uridine	UMP (uridylic acid), UDP, UTP	uracil (10.14)
cytidine	CMP (cytidylic acid),CDP, CTP	cytosine (10.15)
thymidine	TMP (thymidylic acid),TDP, TTP	thymine (10.16)

The Biosynthesis of Purines and Pyrimidines

Little is known of the biosynthetic pathways of either the purines or the pyrimidines in higher plants. Detailed information concerning intermediates and the enzymes involved has been obtained for avian liver and some microorganisms (see, for example, *Metabolic Pathways* listed in the bibliography), and there are

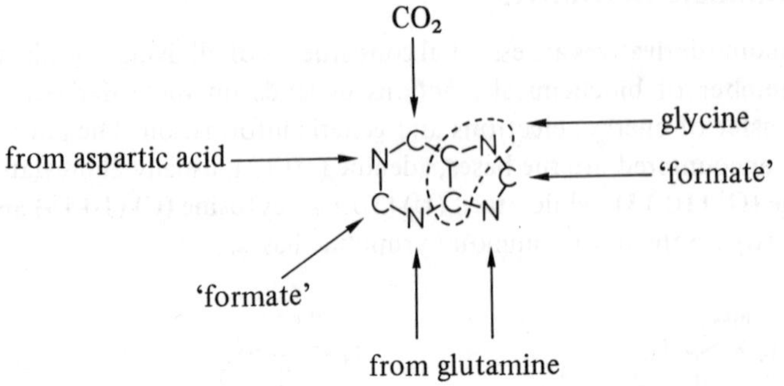

Figure 10.3 The derivation of the purine ring system.

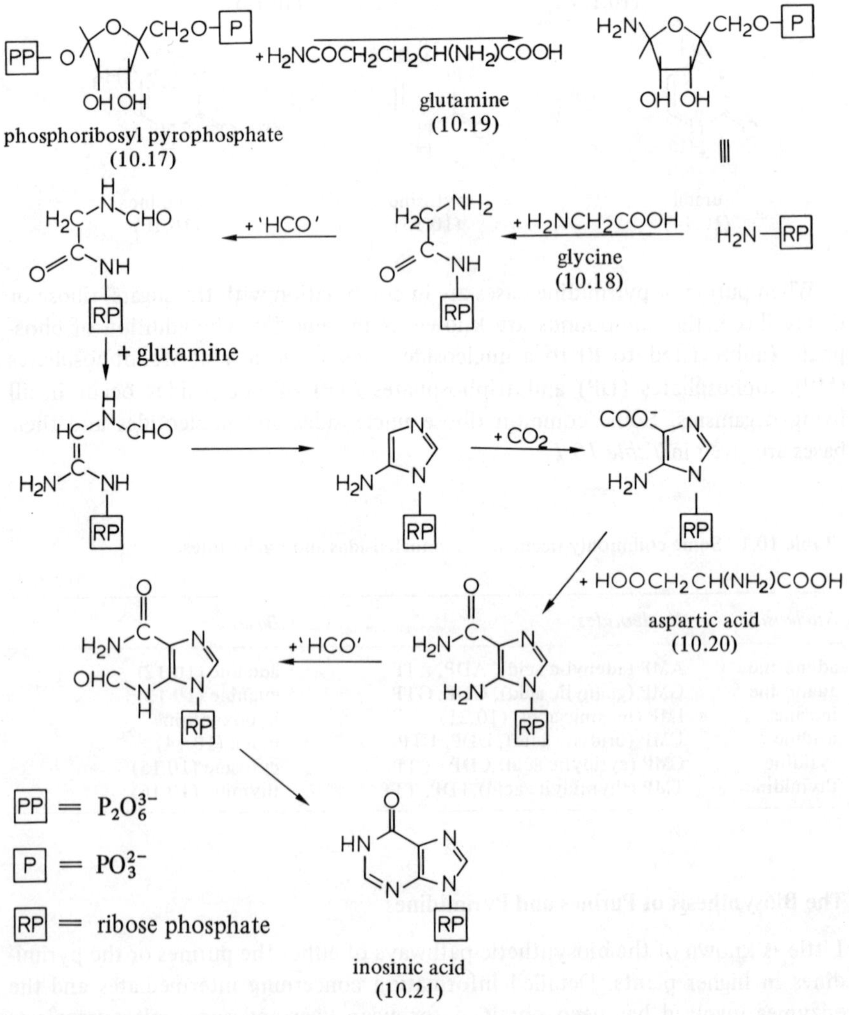

Figure 10.4 The biosynthesis of the purine ring system.

indications that similar pathways are followed in plants. Only an outline will be given here, as experience has shown that extreme caution should be used when animal biosynthetic pathways are extrapolated to higher plants.

In avian liver and yeast, it has been shown that the purine ring system is built onto the sugar, phosphoribosyl pyrophosphate (PRPP) (10.17) (*Figure 10.4*) and that this ring system is derived from glycine (10.18), glutamine (10.19), aspartic acid (10.20), carbon dioxide (probably donated by carboxylated biotin), and formate (donated by N^5, N^{10}-methylenyltetrahydrofolate) as shown in *Figure 10.3*. The incorporation of isotopically labelled glycine into the purine derivatives of some plants has been reported.

The first step in the biosynthetic pathway (*Figure 10.4*) is the addition of an amino group, donated by glutamine, to PRPP, forming an amino sugar. Glycine is added and then a formyl group. The transfer of a second amino group from

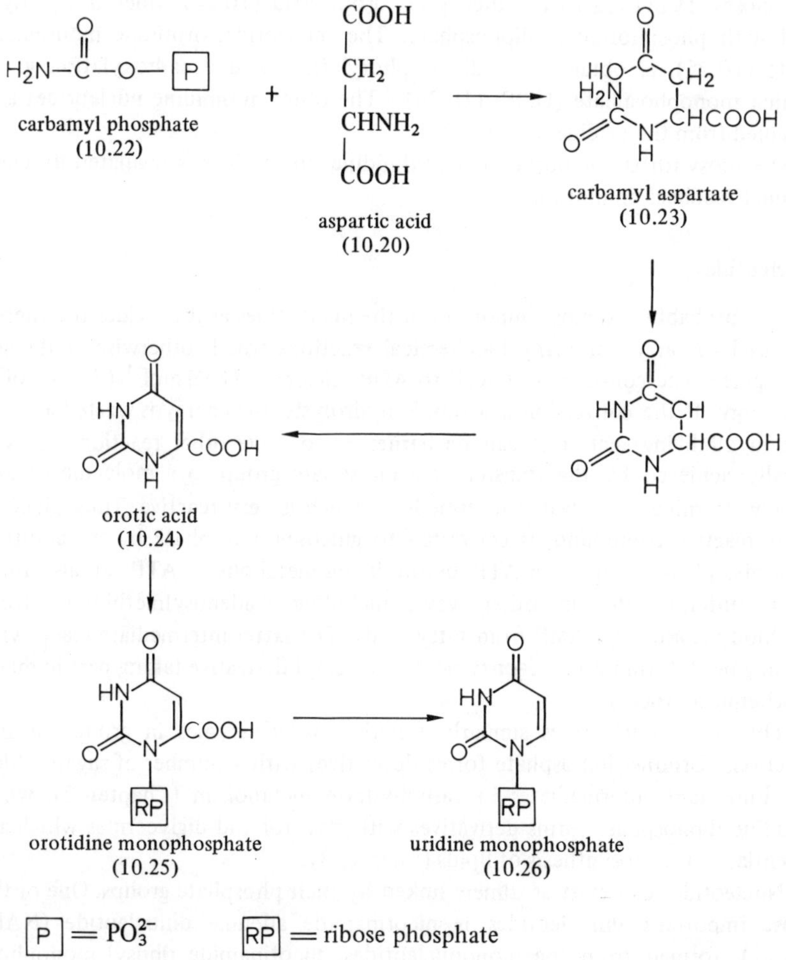

Figure 10.5 The biosynthesis of the pyrimidine ring system.

glutamine followed by ring closure gives an imidazole derivative. Carboxylation by carbon dioxide then takes place, and the carboxylic acid is converted to an amide by donation of an amino group from aspartic acid. The final carbon atom is added as a formyl group and ring closure completes the purine ring system, giving the nucleotide, inosinic acid (10.21).

Inosinic acid is the key intermediate from which the other purine nucleotides are derived. The monophosphates are converted to the di- and triphosphates by addition of phosphate, reactions which are catalysed by kinases.

Free purine bases, which are degradation products of nucleic acids (see below), can be reconverted to nucleosides or nucleotides by reaction with ribose phosphate and phosphoribosyl pyrophosphate, respectively.

The pyrimidine bases are synthesised in animal liver and microorganisms from aspartic acid (10.20) and carbamyl phosphate (10.22) (*Figure 10.5*). Addition of a carbamyl residue to aspartic acid gives carbamyl aspartate (10.23), which ring closes. Dehydrogenation then gives orotic acid (10.24) which is glycosylated with phosphoribosyl diphosphate. The nucleotide, orotidine monophosphate (10.25), is obtained by dephosphorylation, and decarboxylation gives uridine monophosphate (UMP) (10.26). The other pyrimidine nucleotides are obtained from the uridine derivatives.

The biosynthesis of purine and pyrimidine nucleotides is regulated by end-product feedback inhibition.

Nucleotides

ATP is probably the most important of the nucleotides as it provides the energy required to carry out many biochemical reactions which otherwise could not take place. The conversion of ATP to ADP releases $7.9 \, kcal \, mol^{-1}$ ($33 \, kJ \, mol^{-1}$) of energy. If the conversion is a simple hydrolysis, the energy is wasted as heat, but in all living matter it can be harnessed to drive other reactions. This is usually achieved by the transfer of a phosphate group to a molecule of low energy, forming an activated intermediate which is very reactive. Thus, glucose, an unreactive compound, is converted to glucose-6-phosphate by the addition of a phosphate group from ATP, before being metabolised. ATP can also form active intermediates in other ways, including *S*-adenosylmethionine from methionine, and acyl AMP from fatty acids. The latter intermediate reacts with coenzyme A forming acyl coenzyme A, the acetyl derivative taking part in many biochemical processes.

The other nucleotides similarly function as coenzymes in group transfer reactions. Uridine diphosphate forms derivatives with a number of sugars which are important intermediates in carbohydrate metabolism (Chapter 2), while cytidine diphosphate forms derivatives with glycerol and diglycerides which are essential to the biosynthesis of lipids (Chapter 3).

Nucleotides can exist as dimers linked by their phosphate groups. One of the most important dinucleotides is nicotinamide adenine dinucleotide (NAD) (10.27), formed from the mononucleotides, nicotinamide ribosyl monophosphate and adenosine monophosphate. NAD is important to many oxidation reactions and can exist in the oxidised NAD^+ form, when the nicotinamide

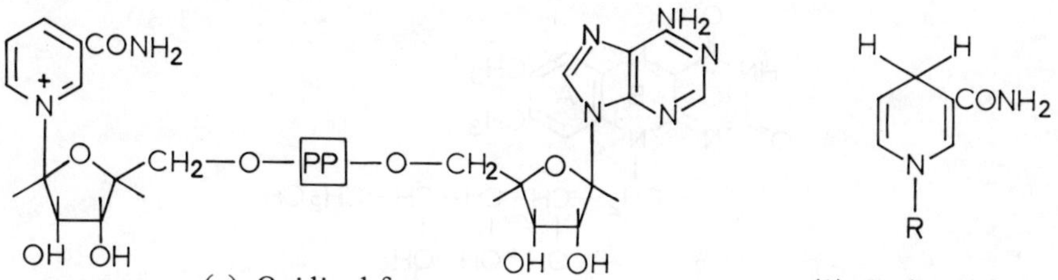

(a) Oxidised form   (b) Reduced form

$\boxed{PP} = H_2P_2O_5$ nicotinamide adenine dinucleotide
(10.27)

Figure 10.6 The oxidised and reduced forms of nicotinamide adenine dinucleotide.

nitrogen is tetravalent, or the reduced NADH form, when this nitrogen is trivalent and a hydrogen atom has added to the heterocyclic ring (*Figure 10.6*).

Nicotinamide (10.28) is one of the B vitamins. In animals, this compound is a degradation product of tryptophan, but in plants it is synthesised from aspartic acid (10.20) and glyceraldehyde phosphate (10.29) through the intermediate formation of quinolinic acid (10.30) (*Figure 10.7*). Glycosylation with phosphoribosyl pyrophosphate gives the nucleotide.

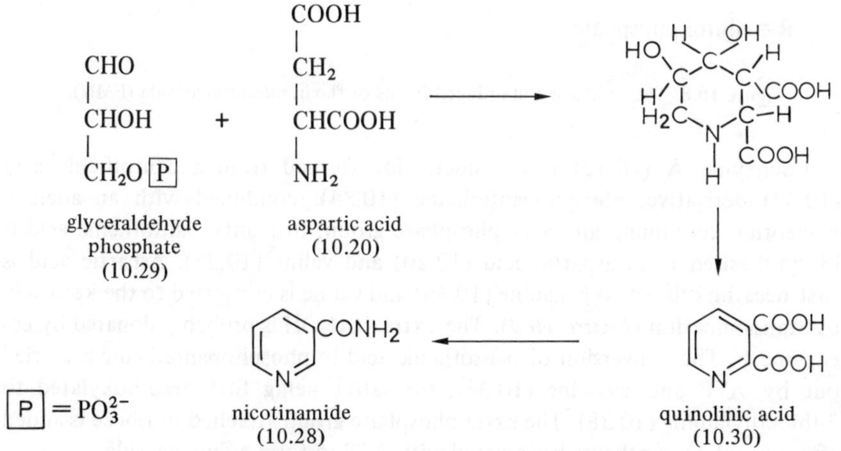

Figure 10.7 The biosynthesis of nicotinamide.

Addition of phosphate to NAD gives NADP (nicotinamide adenine dinucleotide phosphate) which can exist in similar oxidised (NADP$^+$) or reduced (NADPH) forms.

The coenzyme, flavin mononucleotide (FMN), which oxidises NAD and NADP, contains riboflavin (10.31), another of the B vitamins. The detailed biosynthesis of riboflavin in plants is not known, but the pyrimidine ring (ring C) can be derived from adenine (10.12) or guanine (10.13), while ring A can be derived from acetate and ring B from glycine.

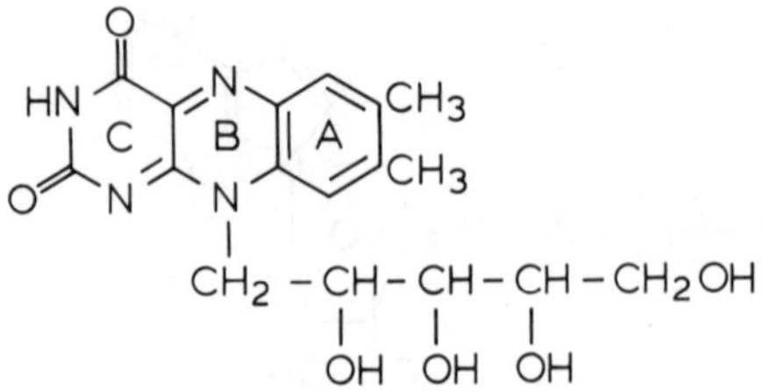

riboflavin
(10.31)

Riboflavin also combines with AMP to give the dinucleoside, flavin adenine dinucleotide (FAD). This coenzyme takes part in many oxidation reactions, where it acts as a carrier of electrons (*Figure 10.8*).

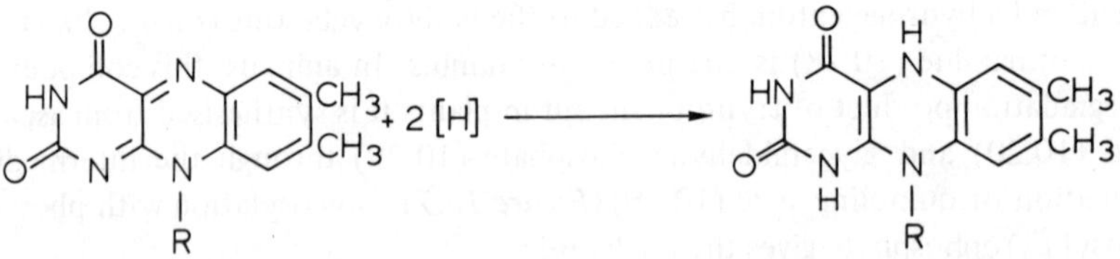

(*a*) Oxidised form   (*b*) Reduced form

R = ribitol phosphate

Figure 10.8 The oxidised and reduced forms of flavin mononucleotide (FMN).

Coenzyme A (10.32) is a dinucleotide formed from a pantothenic acid (10.33) derivative, phosphopantetheine (10.34), combined with an adenine nucleotide containing an extra phosphate group. In plants, pantothenic acid is biosynthesised from aspartic acid (10.20) and valine (10.35). Aspartic acid is first decarboxylated to β-alanine (10.36) and valine is converted to the keto acid by transamination (*Figure 10.9*). The extra –CH$_2$OH is probably donated by co-enzyme F. The conversion of pantothenic acid to phosphopantetheine is carried out by ATP and cysteine (10.37), the latter being first decarboxylated to 2-thioethylamine (10.38). The extra phosphate group attached to ribose is added after phosphopantetheine has reacted with ATP to form a dinucleoside.

Coenzyme A is essential to the transfer of acyl groups in biological systems. It is responsible for the biosynthesis and metabolism of fatty acids (Chapter 3) and 'feeds' the tricarboxylic acid cycle with acetate derived from the primary process of glycolysis. It seems that coenzyme A is always involved in the building of molecules from acetate units, examples of which occur throughout this book.

We have only considered the major nucleotides, but many derivatives of the bases, adenine, guanine, uracil and cytosine, occur in plants, including methylated derivatives which are further discussed below.

The deoxyribonucleotides occurring in deoxyribonucleic acid (DNA) are obtained from the ribonucleotides discussed above by reduction of the sugar

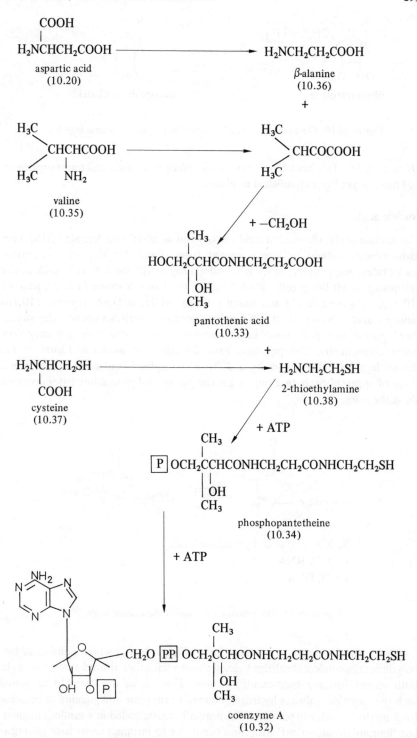

Figure 10.9 The biosynthesis of coenzyme A.

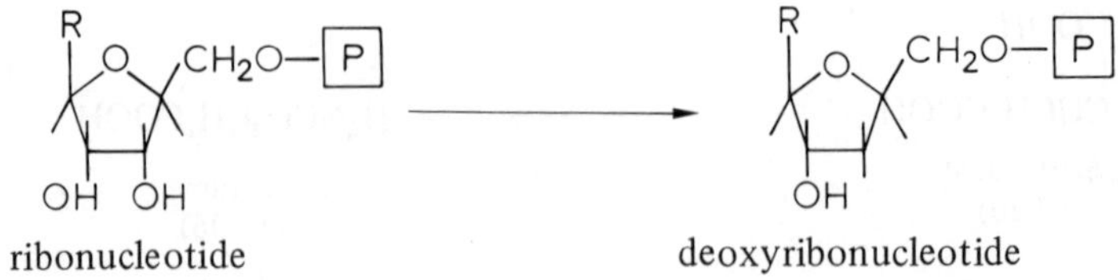

ribonucleotide   deoxyribonucleotide

Figure 10.10 The conversion of ribonucleotides to deoxyribonucleotides.

(*Figure 10.10*). This has been shown to take place in animals and microorganisms, but has not yet been established in plants.

Nucleic acids

The nucleic acids, ribonucleic acid (RNA) and deoxyribonucleic acid (DNA) are high-molecular-weight polymers formed from ribonucleotides and deoxyribo-nucleotides, respectively. DNA is found mainly in the nucleus and RNA in the cytoplasm of all living cells. RNA contains the bases, adenine (10.12), guanine (10.13), cytosine (10.15) and uracil (10.14), while, in DNA, thymine (10.16) replaces uracil. Some plant DNAs also contain 5-methylcytosine. The substi-tuted purine and pyrimidine bases sometimes occur but they are only very minor constituents. The polymers have the structure shown in *Figure 10.11*, the nucleotides being joined by a 3′ to 5′ phosphate bridge. This results in a chain of sugar–phosphate groups with the purine and pyrimidine bases arranged along the sides.

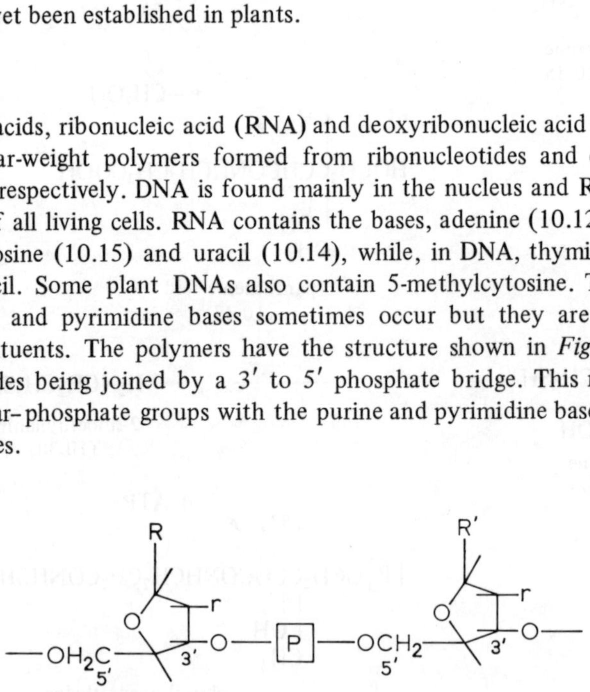

R, R′ = purine or pyrimidine bases
r = OH, RNA
r = H, DNA

Figure 10.11 The primary structure of the nucleic acids.

DNA normally exists as a double helix (*Figure 10.12*) which consists of two polydeoxyribonucleotide chains twisted about each other around a common axis, both chains forming right-handed helices. This allows the bases to be paired such that adenine is always hydrogen-bonded to thymine and guanine to cytosine (or 5-methylcytosine in plants). RNA normally exists coiled in a random manner, but 'hairpin' or 'cloverleaf' structures occur due to intramolecular base pairing as in DNA (*Figure 10.12*).

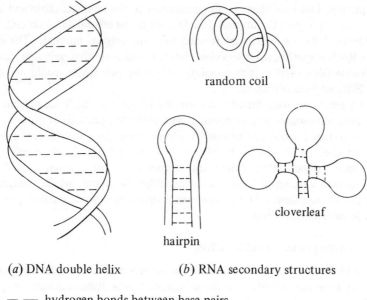

random coil

cloverleaf

hairpin

(*a*) DNA double helix   (*b*) RNA secondary structures

— — hydrogen bonds between base pairs

Figure 10.12 The secondary structures of RNA and DNA.

The nucleic acids are essential to life as they are the key to the reproduction of proteins. DNA is the centre of genetic information, while RNA can exist in three forms – ribosomal or rRNa, transfer or tRNA and messenger or mRNA. The parts played by DNA and the RNAs in the biosynthesis of proteins will not be described here, as they can be found in any good textbook of biochemistry (see bibliography for examples).

Little work on the biosynthesis of RNA or DNA in plants has been carried out, but an enzyme similar to the DNA polymerases of microorganisms has been isolated from maize (*Zea mays*, Gramineae). This enzyme requires the deoxyribose triphosphates of adenine (dATP), guanine (dGTP), cytosine (dCTP) and thymine (dTTP) as substrates, and a small amount of DNA to act as primer is essential. Many studies carried out on chloroplasts show that these cells are able to replicate DNA.

DNA partakes both in its own synthesis and acts as a template for the production of RNA, catalysed by RNA polymerase, an enzyme which has been isolated from higher plants. It is present in dry seeds and one of the earliest reactions of germination is the synthesis of RNA. The sequence of bases in some 60 tRNAs of microorganisms including yeast have been determined. These tRNAs all exist in the cloverleaf form shown in *Figure 10.12*.

The mechanisms involved in the replication of DNA and RNA lie outside the scope of this book (for details, see *The Biochemistry of the Nucleic Acids* listed in the bibliography), but the process can be explained simply as follows. When a cell divides, the DNA double helix unwinds so that each daughter cell contains one polynucleotide chain. A new chain is then formed in each daughter cell by base pairing. This new chain is complementary to the template chain and identical to the chain from the original double helix in the second daughter cell. Thus, two identical double helices are formed from the original template. The way in which RNA is synthesised is less clear, but it would seem that only one strand of the double DNA helix acts as template, otherwise two RNAs would be formed with different base sequences.

DNA combines with protein to form the ribosomes which are present in all living cells. Ribosomes play an essential part in the biosynthesis of proteins when they are attached to a strand of mRNA, forming complexes known as polysomes. DNA attached to protein of the histone type is present in the chromatin of cell nuclei and is responsible for the transmittance of heredity. When a cell divides, the dispersed chromatin forms first threadlike chromatid, which contracts to form the chromosomes. The chromosomes divide into equal portions, one half going to each daughter cell.

THE CATABOLISM OF NUCLEIC ACIDS

DNA and RNA can be degraded to nucleotides and nucleosides by enzymes which catalyse the hydrolysis of the phosphate bonds. Ribonucleases are specific to RNA and deoxyribonucleases to DNA, while multispecific nucleases catalyse the hydrolysis of both types of nucleic acid. Nucleotides are catalytically dephosphorylated to nucleosides by pyrophosphatases and phosphatases and the free bases are obtained by hydrolysis of the sugars, a reaction catalysed by multispecific nucleoside hydrolases. Free hypoxanthine and guanine, but not adenine, are also obtained by reaction of the nucleosides with phosphate giving the base and a sugar phosphate. These reactions are catalysed by nucleoside phosphorylases.

All these degradations are reversible and the free bases can be reassembled into RNA and DNA, processes known as 'salvage pathways'. However, the salvage pathways account for only a minor portion of total nucleic acids, the *de novo* synthesis of purines from phosphoribosyl pyrophosphate and pyrimidines from orotic acid being much more important.

In general, DNA is conserved and there is little breakdown of this nucleic acid. Some types of RNA, however, have a rapid turnover.

About 20 different nucleosides with antibiotic properties have been isolated from fungi, especially *Streptomyces* species. These are mostly purine derivatives, but a few pyrimidine derivatives have also been obtained. The purpose of these

crotonoside
(10.39)

compounds appears to be to protect the fungus against bacterial attack. The nucleoside, crotonoside (10.39), is characteristic of *Croton tiglium* (Euphorbiaceae).

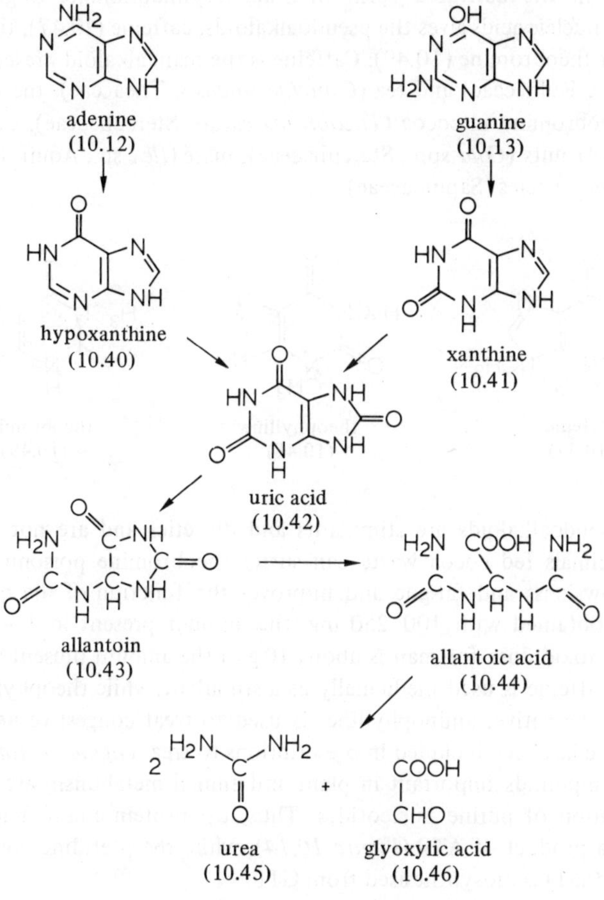

Figure 10.13 The degradation of purines.

Adenine (10.12) and guanine(10.13), in the form of the free base, nucleoside or nucleotide, can be deaminated to give hypoxanthine (10.40) and xanthine (10.41), respectively (*Figure 10.13*). Oxidation, catalysed by xanthine oxidase, gives uric acid (10.42), a compound detected in many plants. Further oxidation leads to allantoin (10.43), the reaction being catalysed by urate oxidase (uricase), an enzyme present in many seeds. The concentration of allantoin often increases on germination and decreases as the plant matures. Some plants, especially comfrey (*Symphytum officinale*) and other members of the Boraginaceae family, accumulate allantoin, while in other plants it is further metabolised to allantoic acid (10.44) and thence to urea (10.45) and glyoxylic acid (10.46). Tracer experiments have shown the conversion of adenine to allantoin, allantoic acid and urea in *Acer* sp. (Aceraceae). The enzyme, allantoinase, catalysing the hydrolysis of allantoin to allantoic acid, has been isolated from many plants.

Methylation of purine derivatives in plants takes place at the nucleic acid stage and is catalysed by methyltransferases. The methyl groups are donated by methionine in the activated form of *S*-adenosylmethionine. Degradation of methylated nucleic acids gives the pseudoalkaloids, caffeine (10.47), theophylline (10.48) and theobromine (10.49). Caffeine is the main alkaloid present in coffee (*Coffea* spp., Rubiaceae) and tea (*Camellia sinensis,* Theaceae), theophylline in tea and theobromine in cocoa (*Theobroma cacao*, Sterculiaceae). Caffeine also occurs in kola nuts (*Cola* spp., Sterculiaceae), maté (*Ilex* sp., Aquifoliaceae) and some *Paullinia* species (Sapindaceae).

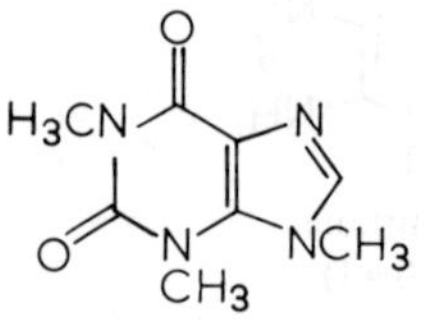

caffeine
(10.47)

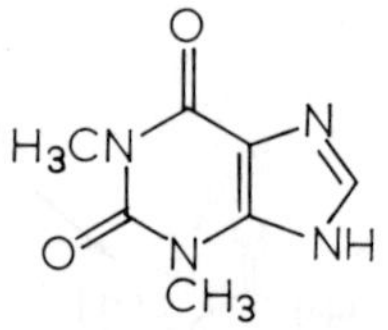

theophylline
(10.48)

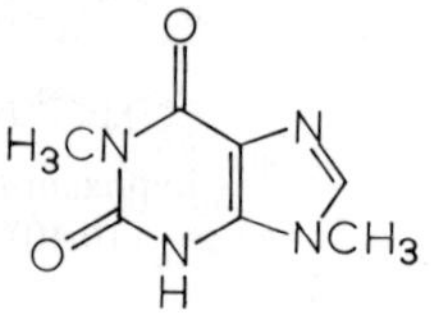

theobromine
(10.49)

These pseudoalkaloids are stimulants and diuretics and are not very toxic, although animals fed cocoa waste can suffer theobromine poisoning. Caffeine reduces drowsiness and fatigue and improves the function of the brain. These effects are obtained with 100–250 mg, the amount present in 1 – 2 cups of coffee. The toxic dose for man is about 10 g or the amount present in 100 cups of coffee. Caffeine is used medicinally as a stimulant, while theophylline, in the form of its derivative, aminophylline, is used to treat congestive heart failure. Theobromine has been included in prescriptions to treat *angina pectoris*.

Some compounds important in plant and animal metabolism are formed by the degradation of purine nucleotides. Thus, the protein amino acid, histidine (10.50), is a product of ATP (*Figure 10.14*), while the pteridine component of folic acid (10.51) is biosynthesised from GTP.

A pathway from GTP (10.52) to the pteridine derivative (10.53) has been suggested for animals, the remainder of the folic acid molecule being derived

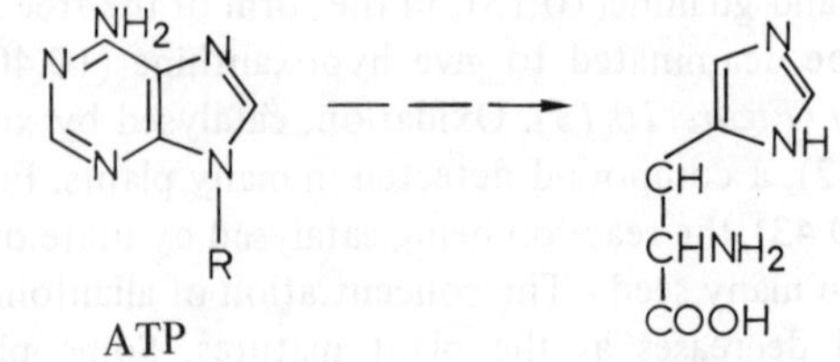

ATP

histidine
(10.50)

Figure 10.14 The biosynthesis of histidine.

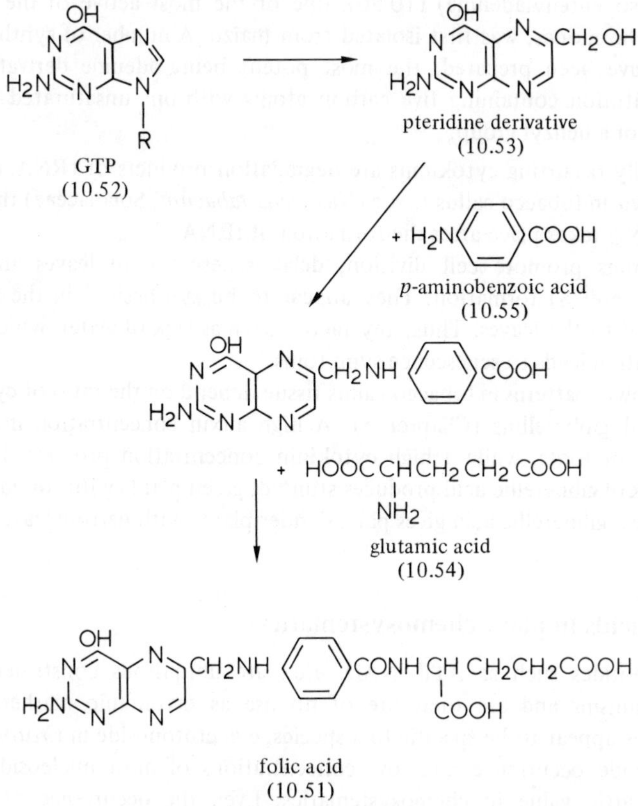

Figure 10.15 The probable biosynthesis of folic acid in plants.

in both plants and animals from glutamic acid (10.54) and *p*-aminobenzoic acid (10.55), a product of the shikimic acid pathway (Chapter 6) (*Figure 10.15*).

Folic acid (sometimes known as citrovorum factor or vitamin M) is now usually included in the vitamin B complex. It is one of the factors involved in the transfer of methyl groups (other methylating agents being methionine, choline, betaine, serine and vitamin B_{12}). When folic acid is acting as a methylating agent, the methyl group is attached to N^5. Both folic acid and vitamin B_{12} are involved in the synthesis of new methyl groups.

The cytokinins, growth hormones found in plants, are purine bases, the most potent being adenine derivatives substituted on the amino group. Zeatin

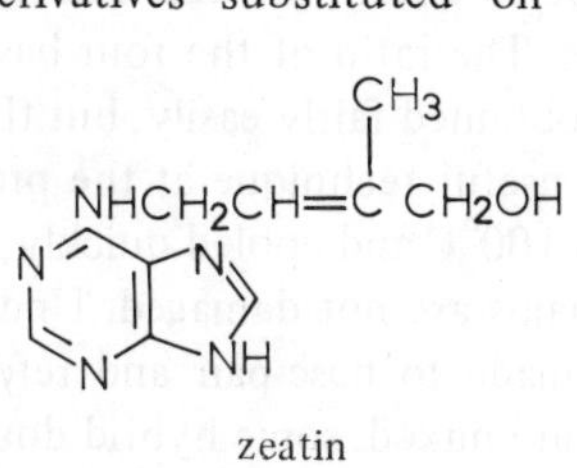

zeatin
(10.56)

(hydroxyisopentenyladenine) (10.56), one of the most active of the naturally occurring hormones, was first isolated from maize. A number of synthetic compounds have been prepared, the most potent being adenine derivatives with N-6 substitution containing five carbon atoms with one unsaturated group (as in zeatin) or a benzyl group.

Naturally occurring cytokinins are degradation products of tRNA, and it has been shown in tobacco callus tissue (*Nicotiana tabacum*, Solanaceae) that substitution at N-6 takes place after the formation of tRNA.

Cytokinins promote cell division, delay senescence of leaves and induce auxin (Chapter 8) formation. They appear to be synthesised in the roots and transported to the leaves. Thus, any factor, such as lack of water, which hinders transportation leads to senescence symptoms.

The growth patterns in tobacco callus tissue depend on the ratio of cytokinins, auxins and gibberellins (Chapter 5). A high auxin concentration induces the formation of roots, while a high cytokinin concentration produces leaves and buds. Lack of gibberellic acid produces stunted, green plants with rounded leaves, while excess gibberellic acid gives pale, slender plants with narrow leaves.

Nucleic acids in plant chemosystematics

The nucleotides such as ATP, NAD, etc., are ubiquitous constituents of all living organisms and are therefore of no use as taxonomic markers. A few nucleosides appear to be specific to a species, e.g. crotonoside in *Croton tiglium*, but the wide occurrence and low concentrations of most nucleosides makes them of little value in chemosystematics. Even the occurrence of the free xanthine bases has little systematic relevance — caffeine having an erratic distribution in higher plants.

Theoretically, the study of plant nucleic acids, especially DNA, should be capable of solving many phylogenetic problems, as not only does each organism have DNA with a unique base sequence but the theory of evolution is based on the premise that related organisms should show similarities in their DNAs which are not shown by unrelated species. Thus, if the base sequences of all plant DNAs were known, it should be theoretically possible to trace the ancestry of every species. The determination of plant evolution is hampered at present by lack of fossil evidence.

Unfortunately, although it is not difficult to isolate DNA from plants, determining the base sequences by present methods is such a long and laborious process that it is impossible to obtain information from enough species to draw any systematic conclusions. The ratio of the four bases, adenine, guanine, cytosine and thymine, can be obtained fairly easily, but these are of little taxonomic value. Probably the most useful technique at the present time is DNA hybridisation. If DNA is heated to $100°C$ and cooled quickly, the double helices unwind but the polynucleotide chains are not damaged. Under suitable conditions, the separated chains can be made to base-pair and reform double helices. If the DNAs of two plant species are mixed, some hybrid double helices will be formed, the number depending on the closeness of the two DNA base sequences.

Unfortunately, experimental techniques have not yet been perfected and variable results, depending on experimental conditions, are obtained. However, enough useful results have been obtained to show the potential value of this method. A comparison of the DNAs of the Gramineae members, wheat (*Aestivum sativum*), barley (*Hordeum vulgare*), oats (*Avena* spp.) and rye (*Secale cereale*), has shown that wheat is closely related to rye and both are related to barley but not to oats. This agrees with the classification on morphological grounds of wheat, rye and barley in the tribe Triticeae and oats in Aveneae. The close resemblance of wheat and rye DNA is also shown in the ability of these cereals to form the intergeneric hybrids known as *Triticales*.

Evidence supporting morphological classification was also obtained from the *Cucurbita* genus (Cucurbitaceae), where DNA hybridisation percentages showed that *C. lundelliana, C. pepo* and *C. maxima* have very similar DNAs, lending support to the suggestion that *C. pepo* and *C. maxima* were originally derived from *C. lundelliana*. As would be expected, the xerophytic *C. palmata* was only distantly related to these three species.

Bibliography

Porphyrins

'Tetrapyrroles and Their Biosynthesis', D. G. Buckley in *Annual Reports,* vol. 74, section B, The Chemical Society, 1977
'Chlorophyll', in *Chemistry and Biochemistry of Plant Herbage,* G. W. Butler and R. W. Bailey (eds.), Academic Press, 1973
'Chlorophyll Biosynthesis and its Control', in *Progress in Phytochemistry*, vol. 5, L. Reinhold *et al.* (eds.), Pergamon Press, 1978
'Chlorophyll Biosynthesis', in *Chemistry and Biochemistry of Plant Pigments*, vol. 1, T. W. Goodwin (ed.), Academic Press, 1976

Purines and Pyrimidines

'Purines, Pyrimidines and Their Derivatives', in *Phytochemistry*, vol. 2, L. P. Miller (ed.), Van Nostrand Reinhold Co., 1973
The Alkaloids, R. H. F. Manske (ed.) (continuing series, vol. 1, 1960), Academic Press
'Biosynthesis of Alkaloids', in *Biosynthesis*, Specialist Periodical Reports (continuing series, vol. 1, 1972), The Chemical Society

Cytokinins

'Chemistry of the Cytokinins', in *Recent Advances in Phytochemistry*, vol. 7, V. C. Runeckles *et al.* (eds.), Academic Press, 1974
'Plant Growth Substances', in *Plant Biochemistry*, Biochemistry Series One, vol. 11, D. H. Northcote (ed.), Butterworths, 1974
'Hormones', in *Plant Biochemistry,* 3rd edn, J. Bonner and J. E. Varner (eds.), Academic Press, 1976

Nucleic Acids

The Biochemistry of the Nucleic Acids, 8th edn, J. N. Davidson, Chapman and Hall, 1976

'Carbohydrates and Nucleotides', in *Cellular Biochemistry and Physiology,* N. A. Edwards and K. A. Hassall, McGraw-Hill, 1971

'Polynucleotides and Protein Synthesis', *ibid.*

'Proteins and Nucleic Acids', in *Chemistry and Biochemistry of Plant Herbage, ibid.*

Metabolic Pathways, 3rd edn, vol. IV, D. M. Greenberg (ed.), Academic Press, 1970

'Protein Biosynthesis', in *Plant Biochemistry, ibid.*

'Nucleic Acid Metabolism', in *Plant Biochemistry, ibid.*

'The Regulation of Ribosomal RNA Synthesis', in *Biosynthesis and its Control in Plants,* B. V. Milborrow (ed.), Academic Press, 1973

Chemosystematics

'Nucleotide Variation in Plant Taxonomy', in *The Chemotaxonomy of Plants,* P. M. Smith, Edward Arnold, 1976

'Is the Order Centrospermae Monophyletic?', in *Chemistry in Botanical Classification,* G. Bendz and J. Santesson (eds.), Academic Press, 1974

'Biochemical Systematics', in *Plant Taxonomy,* 2nd edn, V. H. Heywood, Edward Arnold, 1976

'Heliantheae – Chemical Review', in *The Biology and Chemistry of the Compositae,* vol. 2, V. H. Heywood *et al.* (eds.), Academic Press, 1977

General Index

Abortifacients, 238, 244, 278, 284
Abscission, 50, 124
Acetate–malonate pathway, 1, 2, 56–85, 88, 91, 97, 98, 110, 157, 183, 206, 208, 212, 216, 217
Acetate–mevalonate pathway, 1, 2, 112–54, 174, 183, 280
Acetyl CoA
  acetyltransferase, 113
  acyltransferase, 63
  carboxylase, 58
  synthetase, 57
Acetylcholinesterase, 276
Acetyltransferase, 59
ACP, 59
  acetyltransferase, 60
  acyltransferases, 58
  malonyltransferase, 59
Acyl carrier protein, see ACP
Acyl CoA
  dehydrogenase, 63
  fatty alcohol acyltransferase, 76
  synthetase, 63
Acyltransferase, 246
Advanced characters, 152
Akee-akee, 227
Alkali disease, 229
Allantoinase, 304
Allelopaths, 149, 179, 212
Allergens, 234, 247, 275, 284
Allinases, 249
Aloes, 8, 100
Amidase, 247
Amine oxidases, 233
Aminoacyl tRNA synthetases, 223, 229
Aminotransferases, 220
Amoebicides, 8, 141
α-Amylase, 42, 43, 124, 127
β-Amylase, 43
Anaesthetics, 8, 260
Analgesics, 8, 167, 272
Anaphylaxis, 234, 247
Angina pectoris, 304
Anthelmintics, 96, 119, 122, 208, 264
Anthranilate
  phosphoribosyltransferase, 160
  synthase, 160
Antibiotics, 5, 29, 81, 84, 93, 96, 104, 107, 108, 109, 130, 131, 140, 177, 203, 214, 244, 245–7, 281, 303
Anticholinesterases, 276
Anticoagulants, 173
Antihaemorrhagic agents, 215, 266, 279
Anti-inflammatory agents, 131
Antileukaemic drugs, 282

Antilipemic agents, 131
Antimalarials, 284
Antimetabolites, 73, 221–3, 226
Antimitotics, 150, 282
Antioxidants, 171, 212
Antipyretics, 167
Antispasmodics, 8, 260
Antistiffness factor, 151
Antitumour agents, 104, 108, 109, 122, 131, 150, 171, 214, 282
Antiviral agents, 108, 179, 203
Apoenzymes, 289
Arginase, 226
Arginine decarboxylase, 233
Arrow poisons, 6–7, 9, 143, 257, 275
ATPase, 144

Bactericides, 143, 150, 303
Bakanae disease, 127
Bees, 69, 77, 196
Bioflavonoids, 203
Birds, 5, 151, 196, 203
Blind staggers, 229
Bloat, 131
Bombax moi, 151
Bread poisoning, 262
Bronchodilators, 8, 236
Buffa, 141
Butterflies, 6, 151, 244, 263

Cabbage goitre, 244
Calvin cycle, 20–1, 23
Capillary resistance, 203–4
Carbamylputrescine amidohydrolase, 233
Carcinogens, 93, 103
Carminatives, 119
Cascara sagrada, 100
Castor oil, 8, 70, 78
Catabolism, 57, 62–7, 76, 101, 132–3, 177–8, 202–3, 267, 268, 302–6
Catalase, 291, 292
Cathartics, 8, 100
Cell suspension cultures, see Tissue cultures
Cell walls, 28, 46, 50, 51, 52, 168, 250
  primary, 46, 50
  secondary, 46, 50, 168
Cellobiose phosphorylase, 47
Chalcone–flavanone isomerase, 187
Chemosystematics, 11–12, 53–4, 79–81, 84–5, 109–10, 151–4, 180–1, 204–5, 207, 216–7, 230–1, 236, 241, 244–5, 284–7, 306–7
Chlorophyllase, 125
Chlorosis, 291
Cholera, 107

Choriocarcinoma, 282
Chorismate
  mutase, 160
  synthase, 159
Chromosomes, 302
Chrysolina brunsvicensis, 101
Cinnamyl alcohol oxidoreductase, 169
Citrovorum factor, 305
CMR spectroscopy, 13, 15, 17
Coenzymes, 289, 297–8
Condensing enzyme, 59
Coniferyl alcohol dehydrogenase, 169
Convergence, 12, 285
Copigmentation, 197
Cotton, 47
Curare, 8
β-Cyanoalanine
  hydrolase, 224
  synthase, 224
γ-Cystathionase, 224
Cystathioninuria, 224
Cytochrome oxidase, 240
Cytotoxins, 131, 150, 171

DAHP synthase, *see* Phospho-2-oxo-3-de-
  oxyheptonate aldolase
Danais, 263
Dark reactions, 20
Decarboxylase, 75
Degree of incorporation, 14
Dehydrogenases, 169
Dehydroquinate dehydratase/shikimate de-
  hydrogenase, 159
Dehydroquinate synthase, 157
Delerium tremens, 260, 283
Deoxyribonucleases, 302
Dermatitis, 107, 131, 208
Derris, 201
Desaturases, 68
Diabetes, 39
Digitalis, 143
Dilution value, 14
Dimethylallyl transferase, 117
Diuretics, 304
Divergence, 285
DNA polymerases, 301
Dormancy, 124, 127, 179, 203
Drosophila pachea, 151
Drugs, 6–8, 94, 138, 140, 144, 145, 208,
  236, 237, 255, 256, 257, 259–60, 272,
  284
  addiction, 256, 260, 272
Dunsiekte, 262
Dyes, 10, 214
Dysentery, 141

Electron transport reactions, 210, 211, 215,
  298
Elongase, 62
Emetics, 8
Emulsin, 40, 240
End-product feedback inhibition, 160, 296

Enolpyruvylshikimate phosphate synthase,
  159
Enoyl ACP reductase, 59
Enoyl CoA hydratase, 63
Epimerase, 52, 63
Esterases, 51
Ethanolamine phosphotransferase, 75
Evolution, 11, 53, 151–2, 180, 204, 306
Expectorants, 119, 131

Fatty acid
  elongation systems, 60–1
  synthesis control, 58
  synthetase complex, 59, 62
  transferases, 73
Fatty alcohol acyltransferase, 76
Ferrochelatase, 291
Fish poisons, 8–9, 131, 174, 202, 204
Flavanone
  oxidase, 187
  synthase, 183
Flavourings, 9–11, 119
Flies, 77
Fungicides, 5, 78, 94, 108, 109, 131, 140,
  149, 150, 179, 203, 244

Galactokinase, 40
Galactolipase, 76
Galactosidase, 40
Geotropism, 249
Geraniol hydratase, 119
Geranylgeranyl pyrophosphate synthase, 124
Githagism, 132
Glaucoma, 8, 276
Glucan branching enzyme, 42
Glucan synthase, 42
Glucosidases, 37, 51, 172, 240
Glucosyltransferases, 240, 243
Glutamate
  dehydrogenase, 220
  synthase, 220
Glutamine synthase, 220
Glycolysis, 57, 157, 221
Glycosidases, 51
Glycosuria, 39
Glyoxalate pathway, 64, 65, 77
Glyoxysomes, 64
Goitrogens, 240, 244
Gout, 8, 237
Growth
  excessive, 127
  inhibition, 122, 124, 150, 179, 206, 249,
  251
  regulators, 179, 203, 232–3, 249–50,
  251, 305–6
  stimulation, 126, 127, 150, 232, 249,
  251, 306
Gum
  arabic, 45
  cholla, 46
  ghatti, 45
  Indian, 45

Gum (contd)
  karaya, 46
  khaya, 46
  mesquite, 46
  tragacanth, 46
Gums, 9, 44, 45–6, 51, 52, 54
  seaweed, 51–2

Haemolytic agents, 131
Hallucinogens, 235, 276
Haloenzymes, 289
Hardwoods, 168, 180
Heart disease, 8, 78, 144, 304
Henna, 214
Hepatoxins, 262
Heredity, 302
Hess's cinnamic acid starter hypothesis, 184
Hexose monophosphate shunt, 20, 157
HMG CoA reductase, 113
HMG CoA synthase, 113
Hodgkin's disease, 8, 282
Hormones, 4, 6, 43, 69, 77, 122, 126, 138,
    144, 145, 149, 150, 151, 179, 203,
    220, 249, 251, 305–6
Horse staggers, 262
Hydrolases, 51
Hydroperoxyisomerases, 66
Hydroxyacyl ACP hydratase, 59
Hydroxyacyl CoA dehydrogenase, 63
Hydroxybenzoate polyprenyltransferase, 210
Hypoglycaemia, 227, 281
Hypotensive agents, 140, 141, 281

Indolylacetaldehyde dehydrogenase, 250
Indolylacetic acid oxidase, 250
Indolylglycerol phosphate synthase, 160
Indolylpyruvate decarboxylase, 250
Indolylpyruvic acid pathway, 250
Insect
  attractants, 5, 6, 149, 151, 244, 263
  juvenile hormones, 6, 149, 203, 231
  moulting hormones, 6, 137, 151
  pheromones, 263
  repellents, 5, 6, 149, 151, 244
  vitamins, 6, 150
Insecticides, 5, 9, 96, 133, 150, 174, 201–2,
    203, 244
Invertase, 40
Iodine deficiency, 240, 244
Iridomyrex, 119
Irritants, 38, 107, 244
Isomerases, 63, 66
Isopentenyl pyrophosphate isomerase, 115
Isoprene rule, 112, 129
Isotopes, 13–17
Isotopic tracer analysis, 13–17, 30, 38, 40,
    103, 122, 129, 162–3, 183, 189, 192,
    208, 236, 237, 241, 251, 255, 257,
    258, 260, 264, 268, 270, 271, 273,
    277, 280, 283, 284, 295, 303
Isozymes, 162

Juvenile pigmentation, 202

Kaurene synthetase, 127
Ketoacyl CoA thiolase, 63
Khat, 236
Kimberly horse disease, 262
Kinases, 296
Kwashiorkor, 176

Lathyrism, 225, 232
Laxatives, 100, 131
Lepidoptera, *see* Butterflies and Moths
Leprosy, 79, 108, 131
Leukaemia, 8, 282
Light reactions, 20
Lignification, 168, 169
Limit dextrinase, 43
Limit dextrins, 42
Linamerase, 240
Lipases, 76
Lipoxygenase, 67
  catalysed oxidation, 67
Liquorice, 131
Lupinosis, 265
Lyases, 51

Major fatty acids, 68, 80
Malonyl transferase, 59
Margarine, 78
Medicines, *see* Drugs
Membranes, 77, 137, 151
Methylputrescine oxidase, 268
Methylsalicylic acid synthase, 91
Methyltransferase, 120, 164, 169, 304
Mevalonic acid kinase, 114
Milk sickness, 5
Minor fatty acids, 68, 80
Miraa, 236
Molteno disease, 262
Monoterpenoid reductase, 118
Moths, 151, 196
Multiple labelling, 15–16
Muscle relaxants, 8
Mydriatics, 8, 260
Myo-inositol-1-phosphate synthase, 25
Myrosinase, 243

Narcotics, 8, 272
Neurolathyrism, 225–6
Night blindness, 148
NIH shift, 162–3
NMR spectroscopy, 17
Nucleases, 302
Nucleoside
  hydrolases, 302
  phosphorylases, 302

Oestrogens, 198–9, 203
Oil bodies, 77
Oils
  essential, 10–11, 119
  eucalyptus, 119

Oils (contd)
  lemon grass, 119
  neroli, 119
  peppermint, 119
  rose, 17, 119
Oleosomes, 77
Oligo-1,6-glucosidase, 43
OMT, 169
Opiate receptors, 272
Opium, 271–2
Ordeal poisons, 277
Ornithine decarboxylase, 268
Osteolathyrism, 232
Oxidases, 262, 268
Oxidative pentose phosphate cycle, 20, 157
Oxoacyl ACP reductase, 59
Oxoacyl ACP synthase, 59
Oxygenases, 26, 148, 162, 178
Oxytocics, 266, 279, 284

PAL, *see* Phenylalanine ammonia lyase
Pentose phosphate pathway, *see* Oxidative pentose phosphate cycle
Perfumes, 9, 119
Pernicious anaemia, 292
Peroxidases, 170, 291, 292
Phenolases, 162, 171, 194
Phenylalanine
  ammonia lyase, 162
  aminotransferase, 161
Phosphatases, 302
Phosphoinositol kinase, 27
Phospholipase, 76
Phosphomevalonate
  decarboxylase, 115
  kinase, 114
Phospho-2-oxo-3-deoxyheptonate aldolase, 157
Phosphoribosylanthranilate isomerase, 160
Phosphorylase, 43
Photodynamic activity, 174–5
Photosensitisation, 101
Photosynthesis, 20–3, 27, 39, 40, 151, 291
Photosynthetic electron transport system, 77
Phototropism, 249
Phytases, 27
Phytoalexins, 5, 149, 179, 201, 203
Phytoecdysones, 137
Pictou disease, 262
Pollen, 69, 77, 151
Polyprenyl pyrophosphate synthases, 210
Polysaccharide synthases, 48
Polysomes, 302
Porphobilinogen I synthase, 291
Prenyl transferase, 122
Prephenate
  dehydratase, 161
  dehydrogenase, 161
Primary metabolism, 1–3, 20–3, 57–9, 112–5, 157–61, 289

Primitive characters, 53, 109, 113, 151–2, 161, 180, 204, 205, 217, 245
Prolyl tRNA synthetase, 223, 228
Prosthetic groups, 289, 291
Protease, 124, 127
Purgatives, 70, 79, 100
Putrescine *N*-methyltransferase, 268
Pyrophosphatases, 302
Pyruvate–malate pathway, 20

Q-enzyme, 42
Queen substance, 69
Quinate transferase, 165

R-enzyme, 43
Randomisation of labelling, 13
Rat poisons, 8, 77
Reductase, 91
Reductive pentose phosphate cycle, 20–1
Respiratory enzyme, 240
Ribonucleases, 302
RNA polymerase, 301
Rubber, 148–9

St Antony's fire, 5, 278
Salvage pathways, 302
Saponification, 79
Scurvy, 31
Sedatives, 8, 212, 260, 284
Seed oils, 70, 71, 77, 78, 80, 107, 238, 241
  economic importance, 78–9
Seneciosis, 262
Senna, 8, 100
  Alexandrian, 100
  Indian, 100
  Italian, 100
Sequential analysis, 13
Shikimate
  dehydrogenase, 157
  kinase, 159
Shikimic acid pathway, 1, 23, 98, 110, 157–81, 183, 206, 212, 217, 256, 305
  catabolism of metabolites, 177–8
Shikimic acid – *o*-succinoylbenzoic acid pathway, 212, 216
Shorthand notation for
  fatty acids, 59
  sugars, 23
Silkworms, 151
Slack and Hatch pathway, 20
Soap, 78, 79
Softwoods, 168, 169, 180
Spherosomes, 77
Spices, 10 – 11
Spina bifida, 256
Squalene
  epoxidase, 127
  synthase, 127
Squill, 8, 144
Starch synthase, 42
Stearyl ACP desaturase, 60

Stimulants, 8, 236, 256, 272, 283, 284, 304
Sucrose
  phosphatase, 40
  phosphate synthase, 40
  synthase, 40
Suiljuk, 262
Sulpholipase, 76
Sulphotransferases, 243
Sympathomimetic amines, 234–6

Taxonomy, 11–12
  see also Chemosystematics
Teratogens, 256
Thioglucosidases, 243
Tissue cultures, 12–13, 25, 46, 118, 136, 145–6, 160, 164, 166, 169, 171, 202, 203, 220, 243, 251, 284, 306
Toad poisons, 141–2
Toxins, 5, 9, 33, 38, 77, 84, 103, 107, 109, 130–1, 132, 133, 137, 140, 141, 143, 149, 150, 173, 179, 221, 222, 224, 225, 226, 229, 234, 237–8, 240, 248, 256, 259, 260, 261–2, 263–5, 266, 269, 272, 273, 276–7, 278, 283, 304
Tracer analysis, *see* Isotopic tracer analysis
Tranquillisers, 8
Transaminases, 220, 250
Transamination, 138, 140, 161, 220, 226, 232, 250
Transeliminases, 51

Transesterification, 166
Transferases, 29
Transglucosylation, 42, 172
Transglycosylation, 37
Trial by ordeal, 9
Tricarboxylic acid cycle, 57, 64, 77, 221, 256, 298
Tryptophan
  aminotransferase, 250
  decarboxylase, 250, 281
  synthase, 160
Tuberculosis, 30, 108
Turkey X disease, 103
Turmeric, 11
Tyrosine aminotransferase, 161

UDP-glucuronate cyclase, 27
Urate oxidase, 303
Uricase, 303
Uroporphobilinogen I synthase, 291
Uroporphyrinogen III cosynthase, 291

Vegetable oils, 10, 13, 68
Venereal diseases, 107
Vesicants, 107, 110
Viruses, 179, 203

Walking disease, 262
Winton disease, 262

Xanthine oxidase, 303

Chemical Index

Page numbers in italics indicate chemical formulae

Abietic acid, *125*
Abscisic acid, *4*, 122, *123*, 124, 127, 149, 249
Acacipetalin, 238, *239*
Acetate, 57, 59, 89, 95, 97, 98, 103, 143, 167, 179, 183, 207, 227, 237, 242, 255, 257, 264, 297, 298
Acetic acid, 14, 71
Acetoacetate, 258
Acetoacetyl ACP, *58*, 59
Acetoacetyl CoA, 113, *114*, 257, *258*, *259*, *264*
Acetyl ACP, *58*, 59, 62
Acetyl CoA, 24, 56–7, *58*, 59, 62, 63, 64, 77, 88, 113, *114*, 296
Acetylcholine, 276
Acetyldjenkolic acid, 231, 247
Acetylenic acids, 68, 70
Acetylsalicylic acid, 167
Acetylserine, 221, *222*, *223*, *224*, 225, *229*
Aconitine, *150*, 153, 257
Acyl AMP, 296
Acyl CoA, 296
Adenine, *293*, 297, 298, 300, 301, 302, 303, 305, 306
Adenosine, 293
  monophosphate, 296
Adenosylmethionine, 29, 51, 75, 120, 134, 268, 296, 304
Adenylic acid, *see* AMP
ADP, 293, 296
  glucose, 40, 42
Adrenaline, 234
Aesculetin, *172*
Aesculin, *172*
Aflatoxin B$_1$, *102*, 103
Aflatoxins, 102, 103
Agar, 52
Agaran, 52
Agaropectin, 52
Agarose, 52
Agathic acid, *125*
Agmatine, *233*
Agroclavine, 277, *278*
Ajmalicine, 280, *281*
Akuammicine, 280, *281*
Alanine, *222*, 242, 298, *299*
Alcohols, 76, 82
Aldoximes, 238, *239*, *243*
Aleprolic acid, 71, *72*
Alginates, 52

Alginic acid, 52
Alizarin, 9, *10*, *98*, *216*
Alkaloids, 5, 6, 8, 9, 12, 14, 57, 120, 138–40, 150, 220, 231, 234, 237, 255–87
  benzylisoquinoline, 269, 270–2, 285
  chemosystematics, 284–7
  diterpenoid, 150
  ergot, 277–9
  function, 256
  furanoquinoline, 176
  indole, 275–83
  lupin, 263, 264–6
  lycopodium, 284–5
  nicotine, 256, 267–9
  phenanthridine, 273–5
  physiological action, 256
  piperidine, 263–4, 287
  pyrrolidine, 257–60
  pyrrolizidine, 5, 12, 220, 256, 257, 260–3, 286
  quinoline, 284
  quinolizidine, 256, 264–6, 284, 286
  spiroamine, 286, 287
  steroidal, 138–40
  tobacco, 267–9
  tropane, 257–60, 267, 285
Alkannin, 213, *214*, 217
Alkylcysteine sulphoxides, 249
Allantoic acid, *303*, 304
Allantoin, *303*, 304
Allicin, *249*
Alliin, *249*
Allylcysteine sulphoxide, *see* Alliin
Allylglucosinolate, 245
Allylthiosulphate, *see* Allicin
Aloe-emodin, 99, 100
Aluminium in flower pigmentation, 197
Ambrosanolides, *154*
Amines, 220, 231–6, 255
  chemosystematics, 236
Amino acids, 24, 57, 116, 162, 220, 221–31, 238, 240, 242, 243, 245, 248, 255, 268, 278
  chain lengthening, 242
  non-protein, 221–31
Amino sugars, 29, 30
Aminoacetophenone, *236*
Aminoadipic acid, *224*, 225, 245
δ-(2-Aminoadipyl)-7-aminocephalosporanic acid, 245, *246*

δ-(2-Aminoadipyl)-1-cysteinylvaline, 245
3-Aminobenzoic acid, *108, 177*
p-Aminobenzoic acid, *158,* 177, *305*
Aminodihydrodiosgenin, 138, *139*
Aminoguanidoxybutyric acid, *see* Canavanine
Aminolaevulinic acid, *290,* 291
Amino-oxobutyric acid, *228*
Aminopenicillanic acid, 245, *246*
Aminophylline, 304
Aminopropionitrile, *232, 236*
  -glutamyl derivative, 232
Amorphogenin, *201*
AMP, 293, 298
Amygdalin, 35, 53, 238, *239,* 240, 241
Amylopectin, *41,* 42, 43
Amylose, *41,* 42, 43
α-Amyrin, *130,* 131
β-Amyrin, *129, 130*
Anabasine, *15, 267,* 268, 269
Anacardic acid, 105, *106,* 107
Anaferine, *263,* 264
Anatabine, 14, *15, 267*
Androgens, 144
Anemonin, 38, *39*
Anethole, *164*
Angelic acid, *71, 262*
Angelicin, *174*
Angustifoline, *266*
Anhydrogalactose, 52
Ansamycins, 108, 109
Anthocyanidins, 192, *195*
Anthocyanins, 35, 184, 194–7, 203, 204, 205, 286
  pigmentation, 196–7
Anthranilic acid, *158,* 160, 285
Anthranols, 100
Anthraquinones, 8, 98–101, 109, 110, 116, 215–16, 217
  glycosides, 100
Anthrones, 99, 100, 101
Apigenin, 187, *188,* 189
Apiin, 27
Apiose, 26, *27,* 188
Arabans, 50
Arabinogalactans, 48, 49, 50, 54
Arabinose, *22,* 23, 33, 35, 37, 45, 46, 48, 49, 50, 188
Arachidic acid, *68*
Arachidonic acid, *68*
Arbutin, 180, *208,* 215
Arginine, 29, *222, 226,* 232, *233*
Artemisia ketone, *122*
Artocarpesin, *189*
Artocarpin, *189*
Arylchromenes, *see* Neoflavonoids
Arylcoumarins, *see* Neoflavonoids
Ascaridole, *117,* 119
Ascorbic acid, 6, 7, 20, 25, 30–2, *30*
Asiaticoside, *131*
Asparagine, *222,* 224, 231, 232, 240
Aspartic acid, *222, 224,* 225, 249, 256, *294, 295,* 296, 297, *298*

Aspidinol, *95*
Aspidosperma-type alkaloids, *279*
Aspirin, 167, 272
Astragalin, 191, *192*
ATP, 293, 296, 298, 304, 306
Atropine, 257, 258, 260
Aucubin, *153*
Aureomycin, *see* Chlorotetracycline
Aureusidin, 186, *187,* 197
Aureusin, 186, *187*
Aurones, 184, 186–7, 196, 197, 204, 205
Auxins, *4,* 28, 249, 251, 306
Averufin, *102,* 103
Azetidine-2-carboxylic acid, *1, 223,* 227–8, *228,* 230

Balata, 9, 148, 149
Barbaloin, *36, 100*
Barbiturates, 283
Behenic acid, *68*
Benzoic acid, 166, *167,* 207, *259*
  derivatives, 166–8, 202, 207
Benzoquinones, 97, *208,* 215, 217
Benzotropolones, 194
Benzylpenicillin, *see* Penicillin G
Berberine, 286
Bergapten, *116*
Betacyanins, 204
Betaine, 305
Betalains, 286, 287
Betanin, *286*
Betaxanthins, 204
Biotin, 295
Bixin, 9, *10*
Bolekic acid, 70
Boviquinones, 208, *209*
Bufadienolides, 134, 141, *142,* 144
Butein, *185*
Butenylglucosinolate, 245
Butyrate, 59, 94
Butyric acid, 62
Butyryl ACP, *58, 59*
Butyrylfilicinic acid, 94, *95*

Cadaverine, *233, 263,* 264, *265,* 267
Caffeic acid, *161,* 162, 164, *165,* 169, 172, 180, 195
Caffeine, 8, *304,* 306
Caffeyl CoA, *165*
Caffeylquinic acid, *see* Chlorogenic acid
Camphor, *117,* 149
α-Camphorene, 124, *125*
Canaline, *226*
Canavanine, *226,* 229, 230, 231
Capric acid, *68*
Caproic acid, *68*
Caprylic acid, *68*
Carbamyl
  aspartate, *295,* 296
  phosphate, *226, 295,* 296
Carbamylputrescine, *233*
Carbohydrates, 6, 20–54

Carboxymethylcellulose, 48
Carboxyphenylalanine, 231
Carboxyphenylaminodeoxyribulose-5-phosphate, *158*, 160
Carboxytropine, *see* Ecgonine
Cardenolides, 134, 141, *142*, 144, 146
Cardiolipin, 74, 75
Cardol, 105, *106*, 107
Carlina oxide, *81*
Carnitine, 227
α-Carotene, *147*
β-Carotene, 6, *7*, 78, 119, *147, 148*
Carotenes, *146, 147*
Carotenoids, 2, 5, 113, 123, 146–8, 151, 152, 204, 205
Carrageenan, 52
Carthamin, 9, *10*
Caryatin, 205
Catalase, 289, 291, 292
Catalpic acid, *69, 70*
Catechin, 192
Catechins, 192, 204
Catechol, *178*
Catharanthine, 280, *281*
Cathine, *236*
CDP-choline, 75
CDP-diglyceride, 75
CDP-ethanolamine, 75
CDP-glucose, 40
Cellobiose, 47, 48
Cellulose, 9, 20, 46, 47–8, *47*, 168, 250
  acetate, 48
  acetate butyrate, 48
  amorphous, 47
  crystalline, 47
  trinitrate, 48
Cembrene, 124, *125*
Cephalosporins, 245–7
Cerebrosides, *76*
Cernuoside, 186, *187*
Chalcones, 184–6, *187*, 189, 190, 196, 197, 202, 203, 205
Chanoclavine I, *277, 278*
Chaulmoogric acid, 79
Chicle, 148, 149
Chimaphilin, 213, *214*
Chinese gallotannin, *175*
Chitin, *48*
Chlorflavonin, 191, *192*, 204
Chlorogenic acid, *165*, 166
Chlorophyll, 124, 125, 220, 289
  a, *20*
Chlorophyllide
  a, 124, *290*, 291
  b, 290, *291*
Chloroplumbagin, *98*, 109
Chlorotetracycline, *106*, 107, 109
Cholesterol, 134, *136*, 137, 138, *140*, 145
Choline, 305
Chorismic acid, *3*, *158*, 159, 160, 177
Chromatid, 302
Chromatin, 302

Chrysophanol, *99*, 100
Cichoriin, *172*
Cicutol, *84*
Cicutoxin, *84*
Cinchonidinone, 283, *284*
Cineol, *117*, 119, 149
Cinnamaldehyde, 10, *11*, 169
Cinnamic acid, 105, *161, 162, 163*, 164, *165, 166*, 169, *171, 176*, 183, *207, 273*
  derivatives, 37, 161–6, 168, 176, 202
  hydroxylation, 162–3
Cinnamyl alcohol, 168, 169
Cinnamylquinic acid, *165*
Cis-aconitate, 57, 65
Cis-9-ethylenic acids, 69–70, 80
Cis-12-ethylenic acids, 69–70
Cis-trans-farnesyl pyrophosphate, *122*
Cis-vaccenic acid, 79
Citral, *117*, 118, 119
Citrate, 57, 65
Citronellol, *118*
Citrulline, 221, *222*, 226, *233*
Clavines, 277, 278
CMP, 293
Cobalamins, 290, 291
Cobalt, 290, 292
Cobyrinic acid, *290*, 291
Cocaine, 8, 256, 258, *259*, 260
Codeine, 8, 257, *270*, 271, 272
Codeinone, *270*, 271
Coenzyme A, 58, 166, 289, 296, 298, *299*
  esters, 60, 88, 169, 246
Coenzyme F, *177*, 298
Coenzyme Q, *see* Ubiquinones
Colchicine, 8, 236–8, *237*, 255
Collagen, 228, 232
Columbianetin, *174*
Complex indole alkaloids, 120, 279–84, 285
Conessine, 140, *141*
Coniferin, 15, *16*, *164*
Coniferyl alcohol, 168, *169*
Coniine, *264*
Conjugated ethylenic acids, 68, 69
Convallatoxin, 143
Copalyl pyrophosphate, *126*, 127
Coreopsin, *185*
Corilagen, *176*
Corrins, 289
Corticosteroids, 145
Corynanthe-type alkaloids, *279*
Corynantheal, *283, 284*
Coumaranones, *186*
o-Coumaric acid, *163, 167, 173*
  glucoside, 172, *173*
p-Coumaric acid, *105, 161, 162, 163*, 164, *165, 166, 167, 169, 171, 172*, 187, 195
Coumarin, *171*, 172, *173*, 179, 180
Coumarins, 37, 171–5, 180, 181
Coumaryl
  alcohol, 168, *169*, 180
  CoA, *165*
Coumestans, 199, *200*, 203

Coumestrol, 199, *200*
Crepenynic acid, *70,* 81, 82, 85
Cresol, 91, *92*
Crocetin, 9, *10*
Crotonoside, *302,* 303, 306
Crotonyl ACP, *58, 59*
CTP, 75, 293
Cucurbitacins, *132,* 133, 153
Curcumin, 9, *10, 105*
Cuscohygrine, 257, *258*
Cutin, 76
Cyanidin, 194, *195,* 196
Cyanin, 197
Cyanoalanine, 223, *224,* 231, 232
  glutamyl derivative, 224
Cyanogenic glycosides, 5, 33, 38, 53, 220,
    238–41, 243
  chemosystematics, 241
  function, 240
Cyanogenic lipids, 238, 241
Cyanohydrins, 238, *239,* 240
Cyclitols, 53–4
Cycloartenol, *133, 134,* 135, *136,* 141
Cyclobuxamine, *141*
Cycloeucalenol, *135*
Cyclopentenes, 71, 80
Cyclopropenes, 80
Cymarose, *33, 34*
Cyperaquinone, 208, *209,* 217
Cystathionine, *230*
Cysteine, *222, 223, 224, 230,* 243, 245,
    246, 247, 248, 298
  sulphur compounds derived from, 247–9
Cystine, *222*
Cytidine, 293
  diphosphate, 296
Cytidylic acid, *see* CMP
Cytisine, *266*
Cytochrome
  a, 291, 292
  b, 211, 212, 291, 292
  c, 286, 291, 292
Cytochromes, 289, 292
Cytokinins, 4, 249, 305–6
Cytosine, *293,* 298, 300, 301, 306

DAHP, *see* Deoxyarabinoheptulosonic acid-
    7-phosphate
Daidzein, *198,* 199, *200*
  C-glucoside, *198*
Dalbergione, 208, *209*
Dammarene, 129, 130
Dammarenediol, *129*
Daphnetin, 172, 181
  7-glucoside, *37*
  8-glucoside, *37, 172*
Daphnin, *37, 172*
dATP, 301
Deacetoxycephalosporin C, 245, *246*
dCTP, 470
Dehydroalanine, *224,* 225, 248
Dehydroascorbic acid, 31, *32*

Dehydrodianthrone derivatives, *101*
Dehydrogriseofulvin, *93,* 94
Dehydromatricaria ester, *81*
Dehydropterocarpans, *200*
Dehydroquinic acid, 157, *158, 159*
Dehydroshikimic acid, 157, *158,* 159, 168,
    176, 180
Dehydrosphingosine, 76
Delphinidin, *195,* 196, 197
Demecolcine, *237*
Demethylsuberosin, *116, 174*
Dendrobine, 149, *150*
Deoxy-sugars, 33, 34, 53
Deoxyanthocyanins, 204
Deoxyarabinoheptulosonic acid-7-phosphate,
    157, *158, 159*
Deoxyarabinohexose, *33*
Deoxyglucose, *33*
Deoxyloganic acid, 120, *121*
Deoxyloganin, 120, *121*
Deoxymethylpentoses, 33, 53
Deoxyribonucleic acids, 23, 298, *300*
Deoxyribonucleotides, 298, *300*
Deoxyribose, 36, *37,* 293
  triphosphates, 301
Depsides, 96, 97, 109, 175
Depsidones, 97, 109
Desmosine, 232
Desulphoglucosinolates, *243,* 244
Dextrins, 42
dGTP, 470
Dhurrin, 238, *239*
Diacylglycerol-3-phosphate, *see* Phosphatidic
    acid
Diaminobutyric acid, *228,* 231
Diaminopropionic acid, *225*
Dianthrones, 110
  glucosides, 100
Dicoumarol, *173*
Dictamnine, *116*
Dideoxy-sugar methylethers, 33
Dideoxyhexosone, *34, 35*
Diethylstilboestrol, *199*
Diffractic acid, *97*
Digalactosyldiglyceride, *72,* 73, 74, 76, 77
Digicitrin, 190, *191*
Diginose, *34*
Digipurpurogenin, *140,* 142
Digitogenin, 137, *138*
Digitoxigenin, *142,* 143
Digitoxin, 8, 143, *145,* 146
Digitoxose, 143, *144*
Digoxin, 144, *145,* 146
Dihydrochalcones, 39, 184, 186
Dihydrodaidzein, *198, 200*
Dihydroflavonols, 184, 190, 192, 196, 205
Dihydrogeodin, 101, *102,* 103
Dihydrokaempferol, 192, *193, 195,* 196
Dihydroquercetin, *195,* 196
Dihydroxyacetone phosphate, *21*
Dihydroxybenzaldehyde, 179
2,3-Dihydroxybenzoic acid, 180

3,5-Dihydroxybenzoic acid, 180
Dihydroxycinnamic acid diglucoside, *38*
Dihydroxyflavanones, 202
Dihydroxypalmitic acid, 70
Dihydroxyphenylacetic acid, *178*
Dihydroxyphenylalanine, *see* Dopa
Dihydroxyphenylpyruvic acid, *269, 271*
Dihydroxypyridine, *224,* 225
Dihydroxystearic acid, 70
5,6-Dimethoxy-3-methyl-2-polyprenyl-1,4-
    benzoquinone, *see* Ubiquinones
Dimethylallol, 118
Dimethylallyl pyrophosphate, *104,* 113,
    115, 116, 117, 118, *174, 189, 201,*
    213, *214, 216*
Dimethylallyltryptophan, *277*
Dimethylbenzoquinones, *see* Plastoquinones
Dimethyldiselenide, *230*
Dimethylselenide, *230*
α-Dimorphecolic acid, *69,* 70
β-Dimorphecolic acid, *69,* 70
Dinucleotides, 296
Diosgenin, *12,* 13, 137, *138, 139,* 145
Dioxogulonic acid, 31, *32*
Diphosphatidylglycerol, *see* Cardiolipin
Diphosphoglycerate, *21*
Diploicin, *97*
Disaccharides, 34–5, 39
Diterpenoids, 2, 113, 124–7, 150, 153
Djenkolic acid, *247,* 248
DNA, 298, 300, 302, 306, 307
    double helix, 300, *301,* 307
    hybridisation, 307, 308
    replication, 301, 302
    synthesis, 124, 301
Dodecendioic acid, *see* Traumatic acid
Dopa, *269,* 270, 272, 284
Dopamine, *234, 235, 269,* 271
dTTP, 301

Ecgonine, 258, *259*
Echimidinic acid, *262*
Echinatin, *185*
Elastin, 232
Eleostearic acid, *69,* 78
Ellagic acid, 175, *176,* 180
    derivatives, 176
Ellagitannins, *176*
Elymoclavine, *278*
Embelin, 208, *209,* 217
Emetine, 8
Emodin, 8, 98, *99,* 100, 103, 109, *216,* 217
Emodinanthrone, *101*
Emodin-2-carboxylic acid, *see* Endocrocin
Endocrocin, 99
Enolpyruvate, 159
Enolpyruvylshikimic acid-5-phosphate, *158,*
    159
Enteramine, *see* Serotonin
Ephedrine, 8, *236*
Epi-catechin, 194
    3-gallate, 194

Epi-catechins, 192
Epi-gallocatechin, *194*
    3-gallate, *194*
Epi-thionitriles, 243
Epoxyfatty acids, 71, 80
Epoxyoleic acid, *see* Vernolic acid
Equol, *199*
Eremophilanolides, 153, *154*
Ergocalciferol, *133*
Ergolines, *see* Alkaloids, ergot
Ergometrine, 277, *278,* 279
Ergosterol, 6, *7, 133*
Ergotamine, 277, *278,* 279
Eriodictyol, *195,* 196
Erucic acid, *68,* 78
Erythromycin A, 107, 108
Erythromycins, 107, 109
Erythronolide A, *107*
Erythrose, 20
    4-phosphate, *21,* 23, 24, 157, *158*
Eseroline, *276*
Ethanol, 43
Ethanolamines, 74, 75
Ethionine, 248
Ethylcysteine, 248
Ethylene, *4, 67,* 77, 220, 249, *251*
Ethylenelophenol, *135*
Etorphine, 272
Eudesmol, *126*
Eugenol, 10, *11,* 15, *16, 164*
Eugenone, 95
Euphol, *130*
Exocarpic acid, 70

FAD, *see* Flavin adenine dinucleotide
Falcarinone, 84, *85*
Farnesol, *112*
Farnesyl pyrophosphate, 2, *3,* 113, 114,
    122, 123, 124, 127, *128,* 149
Fatty acids, 56, 58–73, 88, 296, 298
    oxidation, 62–7, 79, 82, 227
    synthesis control, 58
Ferulic acid, *105, 161,* 162, 164, *169,*
    171, 179, 180, 183, 195
Flavandiols, 184, 192, *193,* 203
Flavanols, 185, 192, 193
    gallate esters, 193
Flavanones, 184, 187, 197, 202, 203
    glycosides, 187
Flavin adenine dinucleotide, 298
Flavin mononucleotide, 211, 297, *298*
Flavolans, *see* Proanthocyanidins
Flavones, 184, 187, *188,* 189, 196
    C-glycosides, 36, 189, 197, 204
Flavonoids, *5,* 35, 78, 105, 152, 162, 164,
    179, 183–205, 217
    catabolism, 202–3
    chemosystematics, 204–5
    function, 203–4
    glycosides, 53
    physiological effects, 203–4

Flavonols, 35, 188, 190-2, *193,* 196, 197, 202, 204, 205
Flavylium ion, 194, *195*
Fluoroacetic acid, 73, 80
Fluoroacetyl CoA, 73
Fluorofatty acids, 77, 80
Fluoronicotine, 268
Fluoronicotinic acid, 268
Fluorooleic acid, 73
Fluoropalmitic acid, 73
FMN, *see* Flavin mononucleotide
FMNH$_2$ *see* Flavin mononucleotide
Folic acid, 177, 304, *305*
Formate, 236, 294, 295
Formic acid, *251*
Formononetin, *199, 201*
Fragilin, 215, *216*
Fructans, 43–4, 54
Fructofuranose, *22, 23*
Fructosans, 39
Fructose, 3, *4,* 20, *22,* 23, 33, 37, 39, 40, 41, 43, 53
  diphosphate, 21
  phosphate, 20, *21,* 23, 24, 25, 39, 40
Fucose, *34,* 35, 50
  methyl, 50
Fumarate, 57
Furan ring, 116
Furanocoumarins, 103, 116, 173–5
Furanoquinoline alkaloids, 116
Furanose ring, *22,* 23, 49
Fusidic acid, *130*

Galactans, 50
Galactinol, *28,* 29, 40
Galactoarabans, 50
Galactoglucomannans, 49
Galactomannans, 44
Galactonolactone, 30, *31*
Galactopyranosyl-*myo*-inositol, *28, 29*
Galactose, *28,* 29, 30, *31,* 33, 35, 37, 40, 44, 45, 46, 48, 49, 50, 52, 76, 188
Galactosylglucopyranose, *46*
Galacturonic acid, *35,* 48, 50
Galangin, *190,* 191
Galanthamine, 273, *274*
Gallic acid, *158, 168,* 175, 176, 179, 180
Gallotannin, *175,* 180
Gardenin
  A, *188*
  C, *188*
  E, *188*
GDP, 293
  galactose, 25
  glucose, 25, 40
  guluronic acid, 52
  mannose, 25
  mannuronic acid, 52
Geissoschizine, 280, *281, 282, 283*
Gentamicin, 30
Gentianine, 120, *121*
Gentianose, 40, 53

Gentiobiose, *34,* 35, 37, 41, 53, 238, 239, 240
Gentiobiosides, 37
Gentiopicroside, 120, *121*
Gentioside, *121*
Gentisein, *206,* 207
Gentisic acid, 97, *98*
Gentisin, *206*
Gentisyl alcohol, 91, *92*
Gentisylaldehyde, 91, *92,* 97
Geodin, 101, *102*
Geraniol, *112, 117, 118,* 119, *121,* 279
Geranyl pyrophosphate, 2, *3,* 113, 114, *117, 118, 122,* 149, 213, *214,* 217
Geranylgeraniol, *112, 125,* 291
Geranylgeranyl pyrophosphate, 2, *3,* 113, 114, *124,* 127, *146*
Germacranolides, 153, *154*
Gibbane ring, *126*
Gibberellic acid, *4,* 43, *125,* 126, 127
Gibberellin
  A$_{12}$, *126,* 127
    aldehyde, *126,* 127
  A$_3$, *see* Gibberellic acid
  A$_{14}$, *126,* 127
Gibberellins, 4, *124,* 126, 127, 150, 249, 306
Gitogenin, 137, *138*
Glucan, 42, 47
Glucobrassicin, *242,* 244
Glucocapparin, *242*
Glucoferulic acid, 15, *16,* 164
Glucomannans, 44, 49
Gluconapin, *242*
Gluconasturtiin, *242*
Glucopyranose, *22,* 23
Glucopyranosylfructofuranoside, *22,* 23
Glucose, 3, *4,* 20, *22,* 23, *30,* 31, 32, 33, 35, 36, 37, 39, 40, 41, 42, 43, 47, 48, 49, 50, 53, 57, 64, 65, 76, 172, 188, 195, 238, 239, 240, 249, 255, 296
  1-phosphate, 25, 43
  6-phosphate, 23, 24, 25, *26,* 56, 157, 296
Glucosides, 37, 119, 145, 164, 167, 171, 179, 180, 185, 186, 187, 198, 208, 212, 240
Glucosinolates, 6, 33, 38, 220, 241–5, *242, 243*
  chemosystematics, 244–5
Glucotropaeolin, *242*
Glucuronic acid, *26, 27, 30,* 35, 45, 46, 48, 49, 52, 179, 188
Glutamate, 212, *220, 221*
Glutamic acid, *177,* 222, 249, 256, *305*
Glutamine, 160, 177, 220, *221, 222, 294,* 295
Glutaric acid, 24
Glutaryl phosphate, 220
Glyceraldehyde, 20
  phosphate, *21,* 23, *27, 160, 297*
Glycerate, 20

Glycerides, 73, 75, 76, 77, 296
Glycerol, 73, 296
  phosphate, 73, *74, 75*
Glycine, *222, 290,* 291, *294,* 295, 297
Glycogen, 227
Glycosides, 37–9, 53, 130–2, 137–8, 143,
    166, 187, 188, 191, 193, 194, 195,
    198, 202, 203, 204, 215
  C- , 33, 36
  N- , 33, 36–7
  O- , 33–6
  S- , 33
Glycyrrhizin, *131*
Glyoxalate, 65
Glyoxylic acid, *303*
GMP, 293
Goitrin, *244*
Gossypetin, *191,* 204, 205
Griseofulvin, *16,* 17, *93,* 94, 109
Griseophenone
  A, *93,* 94
  B, *93,* 94
  C, *93,* 94
GTP, 293, 304, *305*
Guaianolides, 153, *154*
Guanine, *293,* 297, 298, 300, 301, 302,
    303, 306
  diphosphate, 25
Guanosine, 293
Guanylic acid, *see* GMP
Gulonic acid, *30*
Guluronic acid, *52*
Gutta percha, 9, 148, 149, 151
Gypsoside A, *35,* 36, 131

Haem, 289
Haemanthamine, 273, *274*
Haemin, 289
Haemoglobin, 289, 291, 292
Hamamelose, 27
Hecogenin, 137, *138*
Hederacosides, 131
Hederagenin, 131, *132*
Helicin, *167*
Heliosupine, *262*
Heliotridine, 260, *261,* 262
Hellebrigenin, *144*
Hemicelluloses, 46, 48–9, 168, 170
Hemiterpenoids, 113, 115, 116
Herbacetin, 204
Heroin, 272
Hesperidin, 204
Hexadecenoic acid, 68, 73
  phosphatidyl ester, 68, 73
Hexahydroxydiphenic acid, *176*
Hexosans, 49
Hispidol, *186*
Histamine, *234,* 236, 275
Histidine, *222, 234, 304*
HMG CoA, *see* Hydroxymethylglutaryl CoA
Holaphyllamine, 140, *141*
Holarrhimine, 140, *141*

Homoarbutin, *213*
Homoarginine, 231
Homogentisic acid, *211,* 213, *214*
Homoserine, *226, 228*
Hordenine, *235*
Humulene, *123*
Humulone, 95, *96*
Hydnocarpic acid, 71, *72,* 79, 80
Hydrangenol, *207*
Hydrogen cyanide, 38, 238, *240*
  poisoning, 240
Hydrogen peroxide, 170, 292
Hydroperoxides, 67, 70, 71, 76
Hydroperoxy acids, 65, 66
Hydroquinone, 180, 212
Hydroxy acids, 62
Hydroxyaldoximes, *239*
Hydroxybenzoic acid, *163, 166, 167,* 180,
    208, *210,* 213, *214*
Hydroxybenzoic acids, 163, 166, 179, 202,
    217
Hydroxybenzyl alcohol, 91, *92*
Hydroxybutyryl ACP, *58, 59*
Hydroxychalcones, 184, *185,* 190
Hydroxycholesterol, 137, *138*
Hydroxycinnamic acid, 37, *38*
Hydroxyfatty acids, 65, 67, 70, 76
Hydroxyferulic acid, *161, 162,* 164, 169
Hydroxyflavonols, 191
Hydroxygeraniol, 119, *121*
Hydroxyhyoscyamine, *259*
Hydroxyisopentenyladenine, *see* Zeatin
Hydroxykaempferol, *see* Herbacetin
Hydroxykaurenoic acid, *126,* 127
Hydroxymatairesinol, 179
Hydroxymethoxybenzoic acids, 180
Hydroxymethylglutaryl CoA, 113, *114*
Hydroxynaphthoquinones, *see* Juglone and
    Lawsone
Hydroxynerol, 120, *121*
2-Hydroxyphenylacetic acid, 178, 180
4-Hydroxyphenylacetic acid, 180
Hydroxyphenylpyruvic acid, *158,* 161, 184,
    185, 211
Hydroxypolyacetylenes, 84
Hydroxyproline, 50, *222*
5-Hydroxytryptamine, *see* Serotonin
3-Hydroxytyramine, *see* Dopamine
Hygrine, 256, 257, *258,* 259, 263, 267, 285
Hyoscine, 8, 257, *258, 259,* 260
Hyoscyamine, 8, 257, *258, 259,* 260
Hypericin, 100, *101,* 110
Hypoglycine A, *227,* 230
  glutamyl derivative, 227
Hypoxanthine, 293, 302, *303*

IAA, *see* Indolylacetic acid
Iboga-type alkaloids, *279*
IMP, 293
Indigotin, 9, *10*
Indole, *160*
Indole-3-acetic acid, *see* Indolylacetic acid

Indolylacetaldehyde, *250*
Indolylacetic acid, *4,* 28, 160, 249–50, *250*
Indolylacetonitrile, *250*
Indolylglucosinolates, 250
Indolyl-3-glycerol phosphate, *158,* 160
Indolylpyruvic acid, *250*
Inosine, 293
Inosinic acid, *294,* 296
Inositols, 30, 54, 74, 75
    see also *Myo*-inositol
Inulin, 43, *44,* 54
Iodine, 244
Ionones, *119*
Ipomeamarone, *5,* 149, *150*
Iridoids, 119–22, 153
Iron
    in flower pigmentation, 197
    in porphyrins, 289, 290, 291, 292
Isoamyl alcohol, 115, 116
Isoamylamine, *231,* 232
Isobutylamine, *231,* 232
Isobutyric acid, 95
Isocitrate, 57, 65
Isocoumarins, 181
Isodesmosine, 232
Isoflavanones, *200*
Isoflavones, 197–200, *200,* 203, 204
    C-glycosides, 198
Isoflavonoids, 184, 197–202, 203
Isofucosterol, 145
Isoleucine, *222, 227,* 238, 241, 260
Isomenthol, *120*
Isomenthone, *120*
Isomyristicin, *164*
Isopenicillin N, 245, 246
Isopentenol, 118
Isopentenyl pyrophosphate, *115, 117,* 118,
    122, 124, 149
Isoprene, 2, *112,* 113, 122
Isoprenoids, 96, 112–54, 189, 210–11
Isoquercitrin, 191, *192*
Isothiocyanates, *243,* 244
Isovaleric acid, 95
Isovincoside, 280, 283

Juglone, 212, *213,* 217
Juvabione, 149, *150*

Kaempferol, *190,* 191, *202*
Kaurenal, 126, *127*
Kaurene, *125, 126,* 127
Kaurenoic acid, *126,* 127
Kaurenol, *126,* 127
Kessyl alcohol, *123*
α-Kosin, *95*

Lanosterol, *130,* 133, 137
Lantoside C, 144
Lapachol, 214, *215,* 217
Lathyrine, 231
Lauric acid, 61, 62, *68,* 78
Lavandulol, *122*

Lawsone, 212, *213,* 214
Lecithin, 137
Leucanthemitol, 53
Leucine, *115,* 116, *222, 231,* 232, 238, 241
Leucoanthocyanins, 192, 204, 205
Leurocristine, 280, *282*
Levans, 43, 44
Lignans, 170–1, 179
Lignin, 9, 46, 162, 168–71, 180, 190, 193
Lignoceric acid, *68*
Limonene, *117, 118,* 119
Limonin, *132,* 133
Limonoids, 133, 153
Linalool, *117, 118*
Linamarin, 238, *239*
Linoleic acid, *61,* 62, *66, 67, 68,* 69, 77, 78
Linolenic acid, *61,* 62, *66,* 67, *68,* 69, 73,
    77, 78
α-Linolenic acid, *68,* 79, 80
γ-Linolenic acid, *68,* 79, 80
Linolenoyl CoA, *62*
Linoleoyl CoA, *62*
Lipids, 8, 73–81, 296
    chemosystematics, 79–81
    economic importance, 78–9
    function, 76–8
Lobeline, 8
Loganic acid, 120, *121*
Loganin, 119, 120, *121,* 279, *280,* 283, 284
Lotaustralin, 238, *239*
Lunularic acid, 206, 207
Lunularin, *207*
Lupanine, 264, *265, 266*
Lupinine, 264, *265*
Lupeol, *129*
Luteolin, 164, 188, 189
    7-glucoside, *188*
Lycopene, *146, 147*
Lycopodine, 284, *285*
Lycorine, 273, *274,* 275, 287
Lysergic acid, *278*
    derivatives, 277, 278
Lysine, 15, *222, 224,* 225, 255, 256, *263,*
    264, *265*

Magnesium
    in flower pigmentation, 196, 197
    in chlorophyll, 289
Malate, 57, 65
Malic acid, 20
Malonamide, *106,* 107
Malonate, *59,* 95, 207
Malonyl ACP, *58,* 59, 60, 62
Malonyl CoA, 24, *58,* 59, 62, 88, 183, *184,*
    185, 187, *207*
Maltose, *42,* 43, 47
Maltotriose, 43
Malvalic acid, *80*
Malvidin, 194, *195*
Mammein, *175*
Mangiferin, *36,* 197, *206,* 207
Mannans, 44–5, 49

Mannose, 37, 44, 45, 46, 48, 49
Mannuronic acid, *52*
Margaric acid, *68*
Marmesin, *174*
Meliacins, 133, 153
Melibiose, *46*
Melilotic acid glucoside, 172, *173*
Menaquinones, 153, 208, *215*, 217
Menthofuran, *120*
Menthol, *112*, *117*, 119, *120*
Menthone, 119, *120*
Mescaline, 235, 236
Messenger RNA, 301
Methacrylic acid, *248*
Methionine, 15, 74, 77, 95, 96, 97, 104, 106, 201, 211, 220, *222*, *228*, 229, *230*, 235, 236, 237, 243, *247*, 248, *251*, 255, 262, 296, 304, 305
Methoxycinnamic acid, 180
Methoxyisoflavones, 200, 201
Methoxymellein, *179*
Methylalkanes, 71
Methylanabasine, 268
Methylandrocymbine, *237*
Methylaspidinol, 94, *95*
Methylcysteine, *229*, *247*, 248, 251
  sulphoxide, *247*, 248
Methylcytosine, 300
Methylene-*bis*-aspidinol, *95*
Methylene-*bis*-4-hydroxycoumarin, *see* Dicoumarol
Methylene-*bis*-phloroglucinol derivatives, 95, 109
Methylenecholesterol, 145
Methylenecycloartenol, *134*, *135*
Methylenecyclopropylacetic acid, *227*
3-(Methylenecyclopropyl)alanine, *see* Hypoglycine A
Methylenelophenol, *135*
Methyleneoxindole, *250*
Methylenetetrahydrofolate, 295
Methyleugenol, 15, *16*
Methylflavonols, 205
Methylfucose, 50
Methyglucosamine, *29*
Methylglucuronic acid, 48
Methyljuglone, *98*, 212, *213*, 217
Methyllysine, 263
Methylmalonic acid, *107*
Methylnicotines, 268
Methylnorbelladine, 273, *274*
Methylornithine, 257
Methylorsellinic acid, *103*, 104
Methylphloroacetophenone, *96*
Methylpiperidinium chloride, 268
Methylpretetramide, *106*, 107
Methylputrescine, 257, *268*
Methylpyrrolinium salt, *267*, *268*
Methylsalicylic acid, 90, *91*, *92*, 98, *167*, *212*
Methylselenocysteine, *229*
Methylselenomethionine, *229*, *230*
Methyltyramine, *235*

Mevalonic acid, 24, 113, *114*, 118, 123, 127, 208, 209, 213, 216, *277*, 279
  phosphate, *114*, 277
  phosphorylation, *114*, *115*, 118
  pyrophosphate, 114
Mimosine, 224–5
Monoacylglycerol-3-phosphate, *74*
Monocrotalic acid, 262
Monocrotaline, *261*
Monoenoic acids, 60
Monogalactosyldiglyceride, *72*, 73, *74*, 76
Monosaccharides, 23–32, 33, 39, 53
Monoterpenoids, 2, 113, 117–22, 149, 152, 153
  acyclic, *117*
  bicyclic, *117*
  irregular structures, 122
  monocyclic, *117*
Morindone, *216*
Morphine, 8, 256, 257, 260, *270*, 271, 272, 283, 284, 287
  derivatives, 272
Mulberrin, *189*
Mundulone, 198, *199*
Mustard oils, 6, 38, 243, 244
Mycophenolic acid, 104–5, *104*
  halogenated, 105
Myo-inositol, 25–30, *26*
  methylethers, 28
  phosphate, 25, *26*
Myosmine, *267*, 268
Myricetin, 190, *191*
Myristic acid, *68*
Myristicin, *164*

NAD, 296, *297*, 306
NADH, *297*
NADP, 297
NADPH, 297
Naphthols, *116*
Naphthoquinone, 212, *213*
Naphthoquinones, 98, 110, 212–5, 217
Naringenin, *187*, *188*, *195*, 196, 203
Naringin, 187
NDP-glucose, 40
Necic acids, 260, 261, 262
Necine bases, 260, 261, 262
Neoflavonoids, 208
Neohesperidose, 187, 188
Neopinone, *270*, 271
Neral, *117*, 118, 119
Nerol, *117*, *118*, 119
Neryl pyrophosphate, *118*
Neurosporene, *146*, 147
Nicotianamine, *228*
Nicotinamide, *297*
  adenine dinucleotide, 296, *297*
  adenine dinucleotide phosphate, 297
Nicotinamide ribosyl monophosphate, 296
Nicotine, *8*, 9, 12, 255, 257, 266, *267*, 268, 269
Nicotinic acid, 14, *15*, 255, 256, 267, 268

Nitriles, 238, *239,* 243
Nona-2,4-dienal, *67*
Non-conjugated ethylenic acids, 68, 80
Non-2-enal, *67*
Non-protein amino acids, 221–231
  chemosystematics, 230–1
Noradrenaline, *234,* 236
Norbelladine, *273*
Norlaudanosoline, *269, 270, 271*
Normorphine, 271
Nornicotine, *267,* 268
Norpluviine, 273, *274*
Norpseudoephedrine, *see* Cathine
Norreticuline, *271*
Nucleic acids, 24, 57, 220, 232, 289, 296,
  300, 301, 302, 306
  catabolism, 302–6
  chemosystematics, 306–7
  methylation, 304
Nucleosides, 36, 293, 296, 302, 303, 306
Nucleotide-D-glucose, 40
Nucleotides, 36, 293, 296–7, 302, 303, 306
Nyctanthic acid, *132,* 133

Obtusifoliol, *135*
Octadec-6-ynoic acid, 70
Octopamine, *235*
Oenanthetol, *84*
Oenanthetoxin, *84*
Oestradiol, *145,* 151
Oestrogens, 145, 151
Oleandrose, *33, 34*
Oleanolic acid, *130*
Oleic acid, 59, 60, *61,* 62, *68,* 70, 78, 79,
  81, 82
Oleoyl ACP, 60
Orchinol, *179*
Organochlorine compounds, 83, 94, 98, 101,
  106, 109, 191, 215
Ornithine, 221, *222,* 226, 232, *233,* 255,
  256, 257, *258,* 260, *261, 268, 286*
Orotic acid, *295,* 296, 302
Orotidine monophosphate, *295,* 296
Orsellinic acid, 89, *90, 92,* 93, 109
Osthenol, *174*
Ouabagenin, 143, *144*
Ouabain, 8, 143, 144
Oxalacetate, 57, *59,* 65, 166
Oxalic acid, 31, *32, 225,* 226
Oxalyl CoA, *225*
Oxalyldiaminopropionic acid, 225–6, 231
Oxoglutarate, 57, 212, *213,* 220, 221
Oxogulonolactone, *30*
Oxo-*trans*-2-decenoic acid, 69
Oxyhaemoglobin, 289, 291, 292
Oxytetracycline, *106,* 107, 109

Pachybasin, *98*
Palmitic acid, *2,* 59, 60, 62, *68,* 73, 78, 145
Palmitoleic acid, *68,* 105, *106*
Palmityl ACP, *58,* 59, 60
Panose, *42*

Pantothenic acid, 298, *299*
Papaverine, 8, *271,* 272
Patulin, 91, *92,* 98
Pavine, 286, 287
Pectic acids, 50, 51
Pectic substances, 46, 50–1
Pectin, 46, 50–1, 197
Pectinic acids, 50, 51
Pelargonidin, 194, *195,* 196, 197
Pelletierine, *263,* 264, 267
Penicillic acid, 92, *93*
Penicillin
  G, *246*
  N, 245, *246*
  V, 246
Penicillins, 245–7
Pentahydroxyflavanone, *202*
Pentoses, 36
Peonidin, 194, *195*
PEP, *see* Phosphoenol pyruvate
Peptides, 245
Periplogenin, *143*
Peroxidase, 289, 291, 292
Petroselinic acid, 68, 80
Petunidin, 194, *195,* 196
Phaseollin, *201*
Phenols, 6, 32, 105–6, 152, 161–76, 178–81
  chemosystematics, 180–1
  function, 178–9
Phenoxyacetic acid, 246
Phenoxymethylpenicillin, *see* Penicillin V
Phenylacetic acid, 246, *249*
Phenylacetyl CoA, 246
Phenylalanine, *3,* 157, *158,* 160, 161, 162,
  164, 167, 176, *178,* 192, *206, 207,*
  211, 222, 225, 234, *236,* 237, 242,
  249, 255, 256, *257, 273*
  hydroxylation, 163
Phenylpropanes, 161, 162
Phenylpyruvic acid, *158, 178,* 190
Phloretin, *186*
Phloridzin, *39, 186*
Phloroglucinol, 94, 185
  derivatives, 94, 95, 96, 105
Phorbol, *125*
Phosphatidic acid, 73, *74,* 75
Phosphatidyl lipids, 72, 73, 75–7
Phosphatidylcholine, *72,* 73, 75
Phosphatidylethanolamine, *72,* 73, 75
Phosphatidylglycerol, *72,* 73, 75
Phosphatidylinositol, *72,* 73, 75
Phosphatidylserine, *72,* 73, 75
$3'$-Phosphoadenosine-$5'$-phosphosulphate,
  243
Phosphoenol pyruvate, 2, 24, 65, 157, *158*
Phosphoglycerate, *21*
Phosphoinositides, 29
Phosphopantetheine, 58, 298.
Phosphoribosyl
  anthranilic acid, 158, 160
  diphosphate, 296
  pyrophosphate, *294,* 295, 296, 297, 302

Phthalides (halogenated), *105*
Phylloquinones, 6, *7*, 125, 152, 208, *215*, 216
Physostigmine, 8, *276*
Phytane, *151*
Phytic acid, 27, *28*
Phytoecdysones, *137*
Phytoene, *146*, 147
Phytofluene, *146*, 147
Phytoglycogen, 43
Phytoglycolipid, 76
Phytol, *112*, 124, *125*, 151, 291
Phytosteroids, 134
Phytyl pyrophosphate, 114
α-Pinene, *117*
Pinitol, 54
Pinoresinol, 170, *171*
Pinosylvin, *206*, 207
Pipecolic acid, *223*
Piperideine, *233*, 267
Piperitenone, 119, *120*
Piperitone, 119, *120*
Pisatin, *5*, *201*
Piscodone, 198, *199*
Plastochromanol-8, 209, 210
Plastoquinones, 167, 208–12, *210*, 216
Plumbagin, *98*, 212, *213*, 214, 217
Podophyllotoxin, 170, *171*
Pollinastanol, 136, *137*
Polyacetylenes, 81–5
   chemosystematics, 84–5
   function, 84
Polyamines, 232–4, 236
Polygalacturonic acid, 51
Polyhydroxyfatty acids, 70, 76
Polyhydroxyphenols, 177–8
Polyisoprenes, 113, 148–9, 151
Polyketides, 56, 88–110
   biosynthesis, 88–9, 105–8
   chemosystematics, 109–10
   function, 108–9
Polyols, 53
Polysaccharides, 39–52, 54
Polysomes, 302
Polyuronide, 50
Porphobilinogen, *290*, 291
Porphyrins, 24, 289–92
Pregnanolone, *142*, 145
Pregnenolone, *140*, *141*, *142*, 143, 145
Prephenic acid, *158*, 160, *161*
Prephytoene, *146*, 147
Presqualene, 127, *128*, 147
Prestrychnine, *282*, 283
Pretazettine, 273, *274*
Pretyrosine, *161*
Primin, 208, *209*, 217
Pristane, *151*
Proanthocyanidin A2, *193*
Proanthocyanidins, 192–3, 194
Procyanidin B, *193*
Progesterone, *140*, *142*, 144, 145, 146
Progoitrin, *244*

Proline, *222*, *223*, 228, 256
Propenylcysteine, *248*
   sulphoxide, *248*, 249
Propenylsulphenic acid, *248*, 249
Propionaldehyde, *248*, 249
Propionic acid, *107*
Propionyl CoA, 63, 144
Proteins, 46, 48, 73, 77, 168, 170, 173, 175, 176, 179, 203, 220, 221, 222, 228, 251, 289, 301, 302
   synthesis, 108, 223, 301
Prothrombin, 173
Protoalkaloids, 140–1, 231, 234–6, 255
Protoanemonin, 38, *39*
Protoberberines, 287
Protocatechualdehyde, *273*
Protocatechuic acid, 157, *158*, 159, 168, 176
Protochlorophyllide *a*, *290*, 291
Protohaem, *290*, 291
Protohypericin, *101*
Protopectin, 50
Protopine alkaloids, 286
Protoporphyrin IX, *290*, 291
Prulaurasin, 238, *239*
Prunasin, 238, *239*, 240
Pseudoalkaloids, 255, 264, 304
Psilocin, *275*, 276
Psoralen, *116*, *174*
Pteridine, *177*, 304
   derivatives, 305
Pterocarpans, *200*, 201, 203
Puerarin, *198*
Pulegone, 119, *120*
Purines, 8, 23, 36–7, 254, 289, 293–6, 302, 304, 305
Putrescine, *232*, *233*, 234, 257, *258*, *261*, 267, *268*
Pyranose ring, *22*, 23, 49
3-Pyrazolylalanine, 231
Pyrethrins, *8*, 9
Pyridoxyl phosphate, 220, 224, 225, 263, 264
Pyrimidines, 36–7, 289, 293–6
Pyrocatechuic acid, 180
Pyrroles, 262, 289
Pyrroline, *233*
Pyrulic acid, 70
Pyruvate, 24, 166
Pyruvic acid, 225, 248

Quassin, *132*, 133
Quassins, 133, 153
Quercetagetin, *191*
Quercetin, 164, 190, *191*
   3,5-dimethylether, *see* Caryatin
   3-gentiobioside, 53
   7-glucoside, 205
   3-rutinoside, 205
   3-rutinoside-7-glucoside, 205
Quercitol, 53
Quinic acid, 157, *158*, 165, 176, 180
Quinidine, 8

Quinine, 8, *283,* 284, 285
Quinolinic acid, *297*
Quinones, 97–102, 109, 152, 179, 183, 207–17

Raffinose, *28,* 29, 40
 family of sugars, 40
Ranunculin, 38, *39*
Reserpine, 8
Resorcinol, 185
Reticuline, *270,* 271
Retinal, *148*
Retrochalcones, 185
Retronecine, 260, *261,* 262
Rhamnogalacturonans, 50
Rhamnose, *33,* 35, 36, 45, 50, 143, 188, 198
Rhombifoline, *266*
Riboflavin, 297, *298*
 phosphate, 211
Ribonucleic acids, 23, *300*
Ribonucleosides, 293
Ribonucleotides, 293, 299, *300*
Ribose, 20, 36, *37,* 293
 phosphate, *21,* 23, 294, 295, 296
Ribosomal RNA, 301
Ribosomes, 30, 108, 302
Ribulose, 20
 5-phosphate, *21*
 1,5-diphosphate, 20, *21*
Ricinoleic acid, *61,* 62, 63, *64,* 70, 78, 80
 oxidation, 64
Rifampicin, *108,* 109
Rifamycins, *108,* 109
RNA, 232, 245, 251, 300, 302
 clover leaf, 300, *301*
 m, 301
 r, 301
 replication, 301, 302
 secondary structures, 300, *301*
 synthesis, 124, 127, 301
 t, 301, 306
Rotenoids, 201–2, 203, 204
Rotenone, *8, 9, 201*
Rubber, 9, 148, *149,* 151
Rutin, 204
Rutinose, 34, 35, 37, 187, 188, 191

Salicin, *167,* 180
Salicyl alcohol, 167, 180
Salicylaldehyde, *167*
Salicylic acid, 91, 98, *163,* 167, 179, 180, 212
Saligenin, *see* Salicyl alcohol
Salutaridine, *270,* 271
Sambubiose, *34,* 35
Sambunigrin, 238, *239*
Santanolides, *154*
Santonin, 122, *123*
Sapogenins, 130–1, 137–8
 alkaloidal, 138, 153

Saponins, 6, 8, 35–6, 130–1, 137–8, 153
 steroidal, 130, 137–8, 150, 152
 triterpenoid, 130–1, 137–8, 150, 152
Sapotoxins, 137, 140
Sarmentose, *34*
Sarmentosigenin E, *143*
Schottenol, *150,* 151
Scillarenin, *144*
Scopolamine, *see* Hyoscine
Scopoletin, *171,* 172, 179
Scopolin, 172
Scyllitol, 29
Scyllo-inositol, 29
Secoiridoids, 120
Secologanin, 120, *121, 280*
Sedoheptulose, 20
 diphosphate, *21*
Selenium, 228–9
Selenoamino acids, 228–9
Selenocystathionine, 229, *230*
Selenocysteine, *229, 230*
Selenomethionine, 229, *230*
Senecic acids, 220, 262
Senecioic acid, *115,* 116
Senecionine, *261*
Sennosides, *100,* 110
Sequoyitol, 54
Serine, 74, 75, *160, 222,* 229, *242, 247,* 248, 305
Serotonin, *275,* 276
Sesamin, 170, *171*
Sesquiterpene lactones, 122, 153–4
Sesquiterpenoids, 2, 113, 122–4, 149, 153
Shikimate, 97, 116
Shikimic acid, 157, *158,* 159, 212, *213*
Simple indole alkaloids, 275–9
Sinalbin, *242*
Sinapic acid, *161,* 162, 169, 180, 183, 195
Sinapyl alcohol, 168, *169,* 180, 195
Sinigrin, *6, 242,* 244
Sitosterol, 134, *135,* 136, 145, 151
Skimmin, 37, *38*
Solanidine, *139*
Solanine, 256
Solasodine, 138, *139*
Sophorose, *34,* 35
Sorbitol, 53
Sorbose, *22,* 23
Soyasapogenol A, 131, *132*
Sparteine, 264, *265,* 266
Spermidine, *232,* 234
Spermine, *232,* 234
Sphingolipids, 65, 70, 76
Spiroketals, *82,* 85, 137
Sporopollenin, 152
Squalene, 2, *3, 14,* 113, 127, *128,* 147, 152
 epoxide, 127, *128, 129,* 133
Stachyose, 40
Starch, 9, 20, 39, 41–3, 47, 54
Stearic acid, 60, 62, *68,* 78
Stearidonic acid, 80

Stearolic acid, *70, 80*
Stearyl ACP, *60*
Sterculic acid, *71*
Sterigmatocystin, *102*, 103
Steroidal alkaloids, 138–40, 153
  glucoside lipids, 75
Steroids, 2, 6, 10, 13, 33, 56, 76, 113, 127,
  133–46, 150, 151, 152
  biosynthetic control, 114
  C_{12}, 140–1, 145
  chemosystematics, 151, 152
  degradation, 144–5
  function, 149–51
  in tissue cultures, 145–6
  pseudoalkaloids, 255
Stigmasterol, 135, 136, 145, 151
Stilbenes, 105, 183, 206–7
Stipitatic acid, *103*, 104
Stipitatonic acid, *103*, 104
Streptamine, *29*
Streptidine, *29*
Streptomycin, *29*, 30
Streptose, 29
Strictosidine, *282*, 283
Strophanthidin, *143*
Strophanthidol, *143*
Strophanthin G, 144
Strychnine, 8, 257, *282*, 283
Succinate, 57, 65
Succinic acid, 24, *290*, 291
Succinoylbenzoic acid, 212, 213, 216
Succinyl CoA, 291
Sucrose, 21, *22*, 23, 24, 28, 29, 39–41, 42,
  43, 44, 53
  phosphate, 40
Sugars, 20–54
  chemosystematics, 53–4
  sulphate esters, 52
Sulpholipid, *72, 73,* 75, 76
Sulphoquinovosyldiglyceride, *see* Sulpho-
  lipid
Sulphur, 228, 243, 247–9, 251
Sweroside, *121*
Synephrine, *235*
Syringic acid, *166*

Tabersonine, 280, *281*
Tannic acid, 175
Tannins, 6, 9, 27, 175–6, 179, 180, 193,
  194, 197, 202
Tartaric acid, *32*
Tasmanone, *95*
Taxiphyllin, 238, *239*
Terpenoids, 2, 5, 10, 13, 33, 35, 36, 56,
  112–54, 255
  biosynthetic control, 114
  chemosystematics, 151–4
  function, 149–54
  pseudoalkaloids, 255
α-Terpineol, *117*, 118
Testosterone, 144, *145*, 151
Tetracyclines, 107–8

Tetrahydrofolate, *177*
Tetrahydropapaverine, *271*
Tetrahydroxybenzophenone, *206,* 207
Tetrahydroxychalcone, *187*
Tetrammine copper hydroxide, 47
Theaflavins, *194*
Thearubigins, 194
Thebaine, *270,* 271, 272, 284
Theobromine, 8, *304*
Theophylline, 8, *304*
Thiocyanate, 240, 243, 244
Thioethers, 82, *83*
Thioethylamine, 298, *299*
Thiohydroxamates, *243*
Thiols, *82,* 83
Thiooxazolidones, 244
Thiophanic acid, *97*
Thiophenes, 81–5
  chemosystematics, 84–5
  function, 84
Threonine, *222*
Thrombocytin, *see* Serotonin
Thujone, *117*
Thymidine, 293
Thymidylic acid, *see* TMP
Thymine, *293,* 300, 301, 306
Thyroxine, 244
Tirucallol, *130*
TMP, 293
α-Tocopherol, 6, *7, 210, 211,* 212
Tocopherols, 125, 208–12, 216
Tocoquinone, 209, *210, 212*
Toluhydroquinone, 213, *214*
Tomatillidine, *139*
Trans-9-ethylenic acids, 69–70
Trans-12-ethylenic acids, 69–70
Trans-3-hexadecenoic acid, 68, 73
2-*Trans*,6-*trans*-farnesyl pyrophosphate, *122*
Transfer RNA, 301
Traumatic acid, *4, 66,* 67, 71, 77
Triacetic acid lactone, *91*
Trienoic acids, 61
Trihydroxychalcone, 185, 186, *198*
Trihydroxystearic acid, 70
Trisporic acid, *150*
Triterpenoids, 2, 113, 128–33, 150
  metabolism, 232–3
  pentacyclic, 129–32
  tetracyclic, 129, 130, 132
Tritium, 13
Tropic acid, 257, *258*
Tropine, 257, *258*
Tropolones, 103–4, 237
Tryptamine, *250, 275,* 280
  derivatives, 275–9
Tryptophan, *3,* 157, *158,* 160, *162, 222,*
  234, 238, 241, 242, *250,* 255, 256,
  275, 276, 277, 279, 283, 284, 285, 297
TTP, 293
Tubocurarine, 8
Tumerone, *11,* 112
Tumidulin, *97*

Turraeanthin, *132,* 133
Tyramine, *235, 273*
Tyrosine, *3,* 157, *158,* 160, 161, 162, *163,*
    211, 222, 225, *234, 235, 237,* 242, 255,
    256, *269,* 270, *273,* 284, 286

Ubiquinol, *211*
Ubiquinones, 152, 167, 208–11, *209,* 216
UDP, 293
UDP-apiose, 25
UDP-arabinose, 25
UDP-choline, 74
UDP-D-galactose, 25, 74, 75, 76
UDP-D-galacturonic acid, 25, 51
UDP-D-glucose, 24, 25, 26, 38, 40, 42
UDP-D-glucuronate, 25, 26
UDP-ethanolamine, 74
UDP-L-rhamnose, 25
UDP-sugars, 37, 48
UDP-xylose, 25
Umbelliferone, 37, *38, 116, 171, 174,* 180,
    181
UMP, 293, 295
Uracil, *293,* 298, 300
Urea, 226, *303,* 304
Ureidohomoserine, *226*
Uric acid, *303*
Uridine, 293
    diphosphate, 25, 296
    monophosphate, *295,* 296
Uridylic acid, *see* UMP
Uroporphyrinogen
    I, *290,* 291
    III, *290,* 291
Urushiol, 105, *106,* 110
Usnic acid, *96,* 97, 109
UTP, 293

Vaccenic acid, 79
Valine, *222, 231,* 232, 238, 241, 245, *246,*
    248, 260, 298, *299*
Vanillic acid, *166*
Vanillin, 9, *11*
Veramarine, *139*
Veramine, *139*
Verbascose, 40
Vernolic acid, *71,* 80
Versicolorin, *102,* 103
Viburnitol, 53

Vicianin, 35, 238, *239*
Vicianose, *34,* 35, 238
Vinblastine, 8, 282
Vincaleucoblastine, 275, 280, *282*
Vincoside, *280, 283,* 284
Vincristine, 8, 282
Vindoline, 280, *281*
Violanthin, *36*
Violutoside, 35
Vitamin
    A, 6, *7,* 78, 119, 148
        aldehyde, *148*
    B, 297, 305
    B$_c$, 177
    B$_{12}$, 290, *292,* 305
    C, 6, *7,* 30–2
    D, 6, *7,* 133
    E, 6, *7,* 125, 209
        see also Tocopherols
    K, 6, *7,* 125, 173, 215
        see also Phylloquinones and
            Menaquinones
    M, 177, 305
    P, 203
Vitamins, 6
Vitexin, *36,* 53, *189*

Waxes, 9, 60, 71, *76,* 78, 79
Withaferin A, *150*

Xanthine, *303,* 306
Xanthones, 105, 183, 197, 206–7
Xanthophylls, 148, 151
Xanthoxin, *123*
Ximenynic acid, *70*
Xylans, 48–9, 54
Xyloglucan, 50
Xylose, *33,* 35, 37, 46, 48, 49, 50, 188
    methyl, 50
Xylulose, 20
    phosphate, *21*

Yamogenin, 137, *138*

Zeatin, *4, 305,* 306
Zeaxanthin, *148*
Zingerone, *11*
Zingiberene, *123*

Botanical Index

Acacia, 45, 54, 231, 238, 241, 247
A. farnesiana, 247
A. senegal, 45
A. verek, 45
Acanthaceae, 184
Acer, 304
A. pseudoplatanus, 25
Aceraceae, 25, 304
Achras sapota, 148
Acocanthera, 7, 143
Aconite, 234
Aconitum, 150, 153, 234
A. napellus, 234
Adenostyles, 153
Aesculus californica, 227
Aestivum sativum, 27, 162, 229, 307
Agaricaceae, 276
Agavaceae, 138, 153, 227, 230
Agave sisalana, 138
Agrostemma githago, 132
Akee, 227
Aleurites montana, 70, 78
Alfalfa, 199
Algae, 52, 79, 160, 204, 206, 215, 216
   classification, 79, 80
Alkanna tinctoria, 214
Alliaceae, 247, 248
Allium, 248, 249
A. cepa, 248
A. sativum, 248
Almonds, 240
Aloe, 8, 36, 100
Amanita muscaria, 77
Amaryllidaceae, 153, 227, 230, 235, 273,
   285, 287
Amaryllis, 275
Ambrosia, 153
Ambrosiaceae, 153
Amorpha fruticosa, 201
Anabasis, 268
Anacardiaceae, 36, 71, 105, 106, 107, 110,
   175, 206
Anacardium occidentale, 105
Ananas comosus, 251
Anaphalis, 83
Angelica, 71, 174
Angelica archangelica, 71, 174
Angiosperms, 54, 169, 180, 204, 241, 259,
   286
Anise, 164
Annatto, 9
Annonaceae, 85
Anogeissus latifolia, 45
Anthemideae, 53, 85, 154
Anthemis, 82

Anthocercis, 285
Antiaris toxicaria, 7
Apium petroselinum, see Petroselinum
   crispum
Apocynaceae, 7, 8, 34, 53, 119, 140, 141,
   143, 148, 152, 279, 282, 285
Apples, 39, 77, 93, 251
Apricots, 45
Aquifoliaceae, 304
Arabis, 245
Araceae, 236
Arachis hypogaea, 78, 103
Araliaceae, 69, 80, 84, 85, 131, 152, 268
Araliales, *see* Umbellales
Araucaria, 54
Araucariaceae, 54
Arctotideae, 153, 154
Argemone, 286
Aristoliales, 270
Aristolochiaceae, 204
Armoracia lapathifolia, 292
Aronia, 53
Artemisia anthelmintica, 122
A. vulgares, 81
Artocarpus, 189
A. heterophyllus, 189
Asclepiadaceae, 35, 53
Aspergillus, 103
A. candidus, 191
A. flavus, 103
A. fumigatus, 173
A. parasiticus, 103
A. terreus, 101, 102
A. versicolor, 103
Astereae, 85, 153, 154
Astilbe, 180
Astragalus, 46, 228, 229
A. bisulcatus, 229
Atropa belladonna, 5, 9, 12, 257
Autumn crocus, 180, 237
Avena, 307
Aveneae, 54, 307
Avocado, 78

Bacillariophyta, 79
Bacillus leprae, 79
Bacteria, 30, 162, 177, 179, 204, 208, 212,
   215, 217, 226, 232, 247, 303
   gram negative, 108, 246
   gram positive, 108, 209, 246
Balsaminaceae, 212
Bamboo, 169
Bambusa, 169
Bananas, 179, 234
Baptisia, 205

B. alba x *B. tinctoria*, 205
B. leucantha x *B. sphaerocarpa*, 205
Barley, 235, 307
Basidiomycetes, 81, 85
Basil, 15, 164
Beech, 228
Beetroot, 286
Benniseed, 170
Berberidaceae, 264
Bergenia, 193
Beta vulgaris, 1, 39, 227, 286
Bignoniaceae, 70, 80, 217
Birdsfoot trefoil, 191
Bixa orellana, 9
Bixaceae, 9
Blackberry, 136
Blighia sapida, 227
Blue-green algae, 79, 152, 161, 209
Bombacaceae, 80
Boraginaceae, 80, 213, 214, 217, 260, 262, 303
Brassica, 242, 243, 244
B. campestris, 245
B. juncea, 245
B. napus, 78, 243, 244
B. nigra, 243, 245
B. oleracea, 244
B. rutabaga, 244
Broad beans, 40, 210
Bromeliaceae, 153, 251
Brown algae, 52, 79, 152
Bryophyta, 204
Buddleia, 153
Buddleiaceae, 153
Buellia canescens, 97
Buphane disticha, 275
Butia monosperma, 185
Buttercups, 5, 38
Butyrospermum paradoxum, 78
Buxaceae, 141
Buxus, 141

Cabbage, 244
Cactaceae, 46, 151, 235, 236
Caesalpinia coriaria, 176
Caesalpiniodeae, 230
Cajaninae, 231
Calabar bean, 9, 276, 277
Calenduleae, 85, 154
Caloplaca, 109
Camelia sinensis, 8, 192, 304
Campanulaceae, 84
Camphor, 124
Canavalia ensiformis, 226
Candida utilis, 220
Cannabinaceae, 95
Capparidaceae, 242, 244, 245
Caprifoliaceae, 236
Caricaceae, 245, 263
Carlina acaulis, 81
Carob bean, 44
Carrots, 147, 179, 205

Carthamus tinctorius, 9, 78
Caryophyllaceae, 36, 54, 80, 131, 132, 204, 286
Cashew, 105
Cassia, 8, 100, 110
C. angustifolia, 100
C. italica, 100
C. senna, 100
Castanea, 53
Castor oil, 8, 62, 63, 70, 78, 80
Casuarinaceae, 204
Catalpa ovata, 70
Catha edulis, 236
Catharanthus roseus, 8, 119, 279, 280, 281, 283, 284
Celastraceae, 236
Centaurea cyanus, 197
C. ruthenica, 83
Centella asiatica, 131
Centrospermae, 204, 285, 286, 287
Cephaelis ipecacuanha, 8
Cephalosporium, 245, 246
C. acremonium, 245, 246
Ceratonia siliqua, 44
Cestrum poeppigii, 165
Charophyceae, 204
Chenopodiaceae, 1, 39, 58, 119, 227, 234, 263, 264, 268, 286
Chenopodium ambrosioides, 119
Cherry, 45
Cherry laurel, 238
Chick peas, 190
Chicory, 172
Chimaphila umbellata, 213
Chlorophyceae, 54
Chlorophyta, 79, 152
Chondrodendron, 7, 8
Chrysanthemum, 53, 85
C. cinerariaefolium, 9
Cicer arietinum, 190
Cichorieae, 85, 154
Cichorium intybus, 172
Cicuta virosa, 84
Cinchona, 8, 284
C. ledgeriana, 284
Cinnamomum camphora, 124
C. zeylanicum, 10
Cinnamon, 10
Citrus, 9, 118, 187, 204, 205
C. paradisi, 187
Claviceps, 277
C. purpurea, 62
Clover, 77, 173, 199, 229
  red, 198
  subterranean, 198
  sweet, 163, 171
  white, 131, 199, 240, 241
Cloves, 10, 95
Cochlearia arctica, 285
Coca, *see* Cocaine plant
Cocaine plant, 257, 260
Cocoa, 78, 304

Coconut, 78
Cocus nucifera, 78
Coffea, 165, 304
Coffee, 165, 304
Cola, 304
Colchicum autumnale, 180, 237
Combretaceae, 45
Comfrey, 303
Compositae, 5, 9, 12, 43, 53, 54, 68,
    70, 71, 78, 80, 81, 82, 83, 84, 85, 118,
    122, 148, 153, 154, 162, 172, 180,
    184, 185, 186, 191, 196, 197, 204, 205,
    229, 234, 260, 268, 285, 286
  tribes, 85, 153
Coniferales, 49, 54
Conium maculatum, 5, 9, 264
Convallaria majalis, 143
Convolvulaceae, 43, 149, 166, 259, 285
Convolvulus, 285
Coreopsidineae, 205
Coreopsis, 185
Corn, 78
Corn cockle, 132
Cornaceae, 153, 236
Cornflowers, 197
Cotton, 78, 191
Cowslip, 180
Crassulaceae, 263, 264
Crataegus monogyna, 232
Crepis foetida, 70, 85
Crinum, 275
Crocus sativa, 9
Crotalaria, 5, 12, 225, 257, 260, 261, 262,
    264, 286
Croton tiglium, 303, 306
Cruciferae, 6, 9, 78, 229, 241, 242, 243,
    244, 245, 247, 250, 259, 285, 292
Cucumbers, 67, 151
Cucumis citrullus, 147
C. sativa, 67, 151
Cucurbita, 307
C. lundelliana, 307
C. maxima, 307
C. palmata, 307
C. pepo, 307
Cucurbitaceae, 67, 133, 147, 151, 153, 231,
    251, 307
Cupressaceae, 85
Curcuma longa, 9, 11, 105
Cyamopsis tetragonolobus, 44
Cyanophyta, 79, 152
Cymbopogon, 9, 119
Cynareae, 153, 154
Cyperaceae, 186, 208, 217, 235
Cyrtandroideae, 205

Daffodils, 273, 275
Dahlia, 43
Dalbergia, 208
Damson, 45
Daphne, 37, 181
D. odora, 172

Datisca cannabina, 191
Datiscaceae, 191
Datura, 257
D. innoxia, 258
D. metel, 257
Daucus carota, 147, 205
Deadly nightshade, 5, 9, 12, 257, 260
Delphinium, 5, 153
Dendrobium nobile, 149
Dermocybe sanguinea, 99
Derris, 9
D. elliptica, 201
Diatoms, 79
Dichapetalaceae, 8, 73
Dichapetalum, 80
D. toxicarium, 8, 73, 77, 80
Digitalis, 137, 141, 142, 143, 215
D. lanata, 8, 144, 145, 146
D. purpurea, 8, 144, 146, 190
Dimorphotheca sinuata, 70
Diocleinae, 231
Dionysia, 188
Dioscorea, 12, 137
D. deltoidea, 145
Dioscoreaceae, 12, 137, 153
Diploxylon, 205
Dipsaceae, 71
Divi-divi, 176
Djenkol bean, 247
Docynia, 180
Drosera, 98
Droseraceae, 98, 109, 110, 217
Duckweed, 27
Dyer's bugloss, 214

Ebenaceae, 98, 110, 204, 217
Eclipta erecta, 83
Elaeis guineensis, 78
Embelia ribes, 208
Ephedra, 8, 236
E. distachya, 236
Ergot, 5, 62, 277, 278
Ericaceae, 39, 163, 180, 205
Erythrina, 286, 287
Erythrophleum, 9
Erythroxylaceae, 257, 259, 285
Erythroxylum coca, 8, 257, 285
Escherichia coli, 157
Eucalyptus, 119
E. risdoni, 95
Eugenia caryophyllata, 10, 95
Euglenaceae, 79
Euglenoids, 79
Eupatorieae, 85, 154
Eupatorium rugosum, 5
Euphorbiaceae, 62, 70, 71, 78, 80, 148, 245,
    276, 285, 303

Fagaceae, 53, 228
Fagus, 53
F. silvatica, 228

Fasarium moniliforme, see Gibberella fujikuroi
Fennel, 164, 205
Ferns, 95, 96, 109, 152, 180, 207, 241, 256
Festuceae, 54
Flacourtiaceae, 71, 79, 80
Flax, 238, 240
Fly agaric, 77
Foeniculum vulgare, 164, 205
Fomes annosus, 179
Foxglove, 144, 146, 190
Franseria, 153
French beans, 201
Fritillaria, 139

Galanthus, 275
Gardenia lucida, 188
Garlic, 248
Garryaceae, 153
Gastrolobium, 80
Gaultheria procumbens, 163
Gelidium, 52
Gentiana, 40, 53, 120
G. lutea, 206, 207
Gentianaceae, 40, 53, 120, 206
Gentianales, 285
Genisteae, 264, 286
Geraniaceae, 9, 32, 176, 205, 207
Geranium, 205
G. pyrenaicum, 176
Geraniums, 32
Gesneriaceae, 184, 204, 205
Gesnerioideae, 205
Gibberella fujikuroi, 126, 127
Gloeosporium musarum, 179
Gloriosa, 238
G. superba, 180
Glycine max, 43, 78, 164, 169, 186, 190
Glycyrrhiza echinata, 185
G. glabra, 131
Gnetaceae, 236
Golden rod, 148
Gossypium, 78, 191
Gracilaria, 52
Gramineae, 5, 9, 27, 39, 41, 43, 54, 78, 119, 127, 162, 169, 176, 180, 203, 204, 215, 228, 229, 235, 238, 255, 277, 301, 307
Grapefruit, 187
Grapes, 32, 123
Grasses, 41, 43, 54, 173, 212, 277, 278
Green algae, 79, 84, 152, 204
Groundnuts, 78, 103
Guar, 44, 45
Gum arabic, 45
Gummiferae, 54, 231, 247
Guttiferae, 174, 207
Gymnospermae, 54, 151, 169, 180, 241, 284, 285
Gypsophila pacifica, 36, 131
Gyrostemonaceae, 245

Haemanthus, 275
Hagenia abyssinica, 95
Hamamelidaceae, 27
Hamamelis virginiana, 27
Haplopappus fremontii, 229
H. gracilis, 196
Haploxylon, 205
Hawkweed, 180
Hedera helix, 131
Helenieae, 85, 153, 154
Heliantheae, 85, 153, 154
Helianthus, 234
H. annuus, 78
Helichrysum, 83
Heliotropium, 262
Helleborus, 144, 219
Hemlock, 5, 9, 257, 264
Hevea brasiliensis, 148
Hieracium, 180, 181
Hippocastanaceae, 227, 230
Hippomane, 276
Holarrhena, 140
H. antidysenterica, 140, 145
H. floribunda, 140
Hops, 95
Hordeae, 54
Hordium sativum, 235
H. vulgare, 307
Horseradish, 257, 292
Humulus lupulus, 95
Hydrangea, 197
Hydrangeaceae, 197
Hymenanthes, 205
Hypericaceae, 100, 110
Hypericum, 100, 101, 110
H. perforatum, 100

Iberis, 245
Ilex, 304
Illiciaceae, 157
Illicium anisatum, 157
Indigofera tinctoria, 9
Inula, 43
Inuleae, 85, 154
Ipomoea batatas, 43, 149, 166
Iridaceae, 9, 207
Iris, 9
Isatis tinctoria, 9
Iva, 153
Ivy, 131

Jack bean, 226
Jack fruit, 189
Juglandaceae, 212, 217
Juglans, 212

Kale, 244
Kennediinae, 231
Kerria, 53
Kerrieae, 180
Khaya, 46

Koelreuteria paniculata, 241
Kola, 304

Labiatae, 9, 10, 40, 118, 119, 122, 164, 236
Laburnum, 266
L. anagyroides, 266
Larches, 49
Laris, 49
Lathyrus, 225, 231, 232, 236
L. hirsutus, 236
L. odoratus, 232, 236
L. roseus, 236
L. sativus, 225, 226
Lauraceae, 10, 78, 124, 268
Lavandula, 9
L. officinalis, 122
Lavender, 122
Lecanora, 97
Lecides carpathica, 97
Lecythidaceae, 229
Lecythis ollaria, 229
Leguminosae, 5, 9, 10, 12, 35, 40, 43, 44,
    45, 46, 54, 71, 78, 80, 85, 100, 103, 110,
    131, 160, 161, 163, 164, 169, 171, 174,
    176, 185, 186, 190, 191, 197, 198, 199,
    201, 203, 205, 208, 210, 215, 217, 224,
    225, 227, 228, 229, 230, 232, 238, 240,
    241, 260, 264, 266, 276, 285, 292
Lemna minor, 27
Lemnaceae, 27
Lemon grass, 119
Leucaena, 224
Lichens, 96, 97, 109, 215
Liliaceae, 8, 36, 100, 137–9, 142, 143,
    144, 153, 180, 184, 215, 223, 227, 228,
    230, 235, 237, 238, 285
Liliales, 285
Lily-of-the-valley, 143
Lima beans, 229
Limnanthaceae, 245
Linaceae, 78, 238
Linseed, 78
Linum usitatissimum, 78, 238
Liquorice, 131, 185
Liverworts, 152, 206
Lobelia, 8, 287
Lobeliaceae, 263, 287
Locust bean, 44, 45
Loganiaceae, 7, 153, 279, 283, 285
Lolium perenne, 215, 229
Lonchocarpus, 9
Lophocereus schottii, 151
Lophophora williamsii, 235
Lotus, 238, 241
L. corniculatus, 191
Lucerne, 131
Lunularia cruciata, 207
Lupin, 224, 264
Lupinus, 264
L. angustifolius, 224, 265
L. argenteus, 265
L. caudatus, 265

L. leucophyllus, 265
L. luteus, 265, 266
L. perennis, 265
L. sericeus, 265
Lycopersicum esculentum, 147, 165, 250
Lycopodiaceae, 284
Lycopodium, 151
Lycopods, 151

Macherium, 208
Madagascar periwinkle, 119, 279
Madder, 9, 215
Magnoliales, 241, 270, 285
Maize, 41, 43, 78, 162, 228, 301, 306
Malus, 39, 77
M. sylvestris, 251
Malvaceae, 78, 80, 191
Malvales, 71, 80
Mammea americana, 174
Mammy apple, 174
Mandragora officinalis, 9
Mandrake, 9
Mangifera indica, 36, 206
Mango, 206
Manihot esculenta, 238
Maté, *304*
Medicago sativa, 131, 199
Meliaceae, 46, 133, 153
Melilotus alba, 163, 171
Menispermaceae, 7, 53
Mentha aquatica, 119
M. piperita, 118, 119
Mescal, 235
Mimosa, 224
Mimosoideae, 230, 231, 247
Mimusops balata, 148
Molluginaceae, 204, 286
Monk's hood, 257
Moraceae, 7, 148, 189, 263
Morinda citrifolia, 216
M. reticulata, 229
Moringaceae, 244
Morus, 189
M. alba, 189
Mosses, 207, 255
Mulberry, 189
Mundulea sericea, 9, 198
Mung beans, 51, 198, 199
Musa, 179, 234
Musaceae, 179, 234
Mustard, 243, 245
Mutisieae, 85, 153, 154
Mycobacteria, 79
Myristica fragrans, 164
Myristicaceae, 164
Myrsinaceae, 208, 217
Myrtaceae, 10, 95, 119

Narcissus, 273
Neptunia amplexicaulis, 229
Nerine, 275
Nettles, 234, 275

Neviusia, 53
Nicotiana, 9, 15, 268
N. glauca, 268
N. tabacum, 9, 12, 25, 136, 160, 228, 268, 284, 306
Nigella damascena, 286
Nutmeg, 164

Oats, 307
Ocimum basilicum, 15, 164
Oenanthe crocata, 84
Oil palm, 78
Olacaceae, 70, 71, 80
Olea europaea, 78
Oleaceae, 78
Olive, 78
Onagraceae, 71
Onions, 248, 249, 273
Ophrestiinae, 231
Opium poppy, 271
Opuntia bulgida, 46
Orange, 118
  blossom, 119
Orchidaceae, 10, 44, 149, 179, 204, 259, 285
Orchidales, 285
Orchids, 179
Oryza sativa, 127
Oxalidaceae, 184
Oxylobium, 80

Palaquium gutta, 148
Palmae, 78, 204
Papaver, 8, 272, 287
P. paeoniflorum, 272
P. setigerum, 272, 287
P. somniferum, 8, 271, 287
Papaveraceae, 244, 245, 271, 285, 287
Papaverales, 270
Papilionoideae, 54, 226, 230, 231, 236, 241, 264, 266, 285, 286
Para rubber tree, 148
Parsley, 5, 27, 80, 183, 184, 187, 257
Passifloraceae, 238
Paullinia, 304
Peach, 238
Pear, 180
Peas, 160, 201, 225, 232, 266
Pedaliaceae, 78, 170
Pelargonium, 9, 32
P. crispum, 32
P. hortorum, 176
Penicillium, 93, 245
P. brevicompactum, 104, 105
P. chrysogenum, 246, 247
P. cyclopium, 93
P. patulum, 90, 91, 94, 97
P. puberulum, 103
P. stipatum, 103
P. urticae, 16, 17
Pentanthera, 205
Peppermint, 118, 119, 149
Periwinkles, 282

Persea americana, 78
Petroselinum crispum, 5, 27, 80, 183
Petunia hybrida, 183
Peyote, 235
Phaeophyceae, 52
Phaeophyta, 79, 152
Phaseoleae, 230
Phaseolus, 161
P. aureus, 198, 286
P. lunatus, 229, 238
P. vulgaris, 201
Phoma foveata, 98
Phyllanthus discoideus, 285
Phyllodoceae, 205
Physostigma, 276
P. venenosum, 8, 9, 276
Phytolaccaceae, 245
Picea, 170
P. pungens, 153
Picramnia, 70
Pilosella, 181
Pimpinella anisum, 164
Pinaceae, 124, 153, 170, 206
Pine, 188
Pineapple, 251
Pinus, 153, 170, 205, 206, 207
P. albicaulis, 124
Piperaceae, 263
Piscidia erythrina, 198
Pisum sativum, 160, 201
Pithecolobium lobatum, 247
Pittosporaceae, 84, 85, 152
Plantaginaceae, 245
Plumbaginaceae, 98, 109, 110, 205, 217
Plumbagineae, 205
Plumbago, 98
Podalyrieae, 264, 286
Podophyllaceae, 171
Podophyllum, 171
Poison ivy, 105
Polycarpicae, 241
Polygalaceae, 131
Polygonaceae, 100, 110, 215, 217
Polyporus hispidus, 166
P. tumulosus, 97
Pomegranate, 264
Pomoideae, 53
Potatoes, 5, 67, 165, 166, 234, 256, 257
Primula, 27, 188, 204
P. obconica, 208
P. sinensis, 35
P. veris, 180
Primulaceae, 27, 35, 180, 188, 204, 208, 217
Prosopis juliflora, 46
Proteaceae, 180
Prunoideae, 53
Prunus, 45, 238
P. amygdalis, 240
P. armeniaca, 45
P. cerasus, 45
P. insitia, 45

P. laurocerasus, 238
P. persica, 238
P. virginiana, 45
Psilocybe, 276
Psoralea corylifolia, 174
Pteridophyta, 284
Pueraria thunbergiana, 198
Punica granatum, 264
Punicaceae, 263, 264
Puriri, 189
Pyrethrum, 9, 53
Pyrolaceae, 213
Pyrus, 180

Quercus, 53
Quillaja, 53

Ragwort, 5, 261, 262
Ramalina ceruchis, 97
Ranunculaceae, 5, 38, 142, 144, 150, 153, 234, 285, 286
Ranunculales, 270, 285, 287
Ranunculus, 5, 38
Rape, 244
Rapeseed, 78, 244
Raspberry, 192
Ratsbane, 8, 73, 77
Rauvolfia, 8
Red algae, 52, 79, 152
Resedaceae, 244
Rhamnaceae, 100, 110, 215, 217
Rhamnus, 100
R. frangula, 100
Rheum, 100
Rhododendroideae, 205
Rhododendron, 205
Rhodophyceae, 52
Rhodophyta, 79, 152
Rhodoreae, 205
Rhodotypos, 53
Rhus semialata, 175
R. succedanea, 71
R. toxicodendron, 105
R. typhina, 176
Rice, 127
Ricinus communis, 8, 62, 78
Rohdea japonica, 228
Rosa, 9, 13, 118, 196, 197
Rosaceae, 9, 13, 35, 39, 45, 53, 77, 95, 118, 136, 180, 192, 232, 236, 238, 240, 241, 251
Rosales, 152, 180
Rose, 13, 118, 119
Rosoideae, 53, 180
Rubia tinctorium, 9, 215
Rubiaceae, 9, 110, 116, 165, 188, 215, 216, 217, 229, 279, 284, 285, 304
Rubus, 192
R. fruticosus, 136
Rue, 13, 174
Rumex, 100
Russian dandelion, 148

Ruta graveolens, 13, 174
Rutaceae, 9, 13, 118, 133, 153, 174, 180, 181, 187, 235, 285
Rye, 5, 277, 278, 307
Ryegrass, 215, 229

Saccharum officinarum, 39
Sacred bark, 100
Safflower, 78
St John's wort, 100
Salicaceae, 167, 180
Salix, 180
S. purpurea, 167
Salvadoraceae, 245
Sambucus, 238
Santalaceae, 70, 80
Santales, 70, 80
Sapindaceae, 85, 131, 227, 230, 238, 241, 304
Sapotaceae, 78, 131, 148
Saxifragaceae, 152, 180, 181, 193
Scilla maritima, 144
Scopolia lurida, 257
Scrophulariaceae, 40, 137, 153, 184, 190, 215, 285
Secale cereale, 5, 277, 307
Sedges, 208, 217
Sedum, 264
S. sarmentosum, 263, 264
Senecio, 5, 12, 257, 260, 261, 286
Senecioneae, 85, 153, 154
Senita cactus, 151
Sesame, 78
Sesamum, 170
S. indicum, 78
Shea, 78
Sikkimenses, 204
Simaroubaceae, 70, 133, 153
Sisal, 138
Snowdrops, 275
Solanaceae, 5, 8, 9, 12, 25, 136, 137, 138, 147, 150, 153, 160, 165, 183, 228, 234, 235, 257, 259, 263, 264, 268, 285, 306
Solanum, 12
S. tuberosum, 5, 165, 234
Solidago, 148
Sophora, 266
Sophoreae, 264, 286
Sorbus, 53
S. aucuparia, 232
Sorghum, 41, 176, 238
Sorghum, 176, 238
S. bicolor, 41
Soybeans, 43, 78, 164, 169, 186, 190
Sphaerophorus fragilis, 215
Spinach, 58, 75, 234
Spinacia oleracea, 58, 234
Spiraeoideae, 53
Stanleya pinnata, 229
Staphylococcus, 30, 108
Star-anise, 157
Staticeae, 205

Sterculia urens, 46
Sterculiaceae, 46, 78, 80, 304
Stoneworts, 204
Streptococcus, 177
Streptomyces, 107, 245, 302
S. erythreus, 107
S. griseus, 29
Strophanthus, 7, 34, 143
S. gratus, 144
Strychnos, 7, 8, 283
Sugar beet, 1, 39, 227
Sugar cane, 39
Sumach, 71, 175
Sunflower, 78
Swedes, 244
Sweet clover, 163, 171, 172
Sweet lupins, 265
Sweet pea, 232
Sweet potatoes, 43, 149, 166
Sycamore, 25, 50
Symphytum officinale, 303

Tanacetum vulgare, 118
Tansy, 118
Taraxacum kok-saghyz, 148
Taxaceae, 5
Taxus, 5, 238
T. baccata, 5
Tea, 192, 194, 304
Theaceae, 192, 304
Theobroma cacao, 8, 78, 304
Thermopsis, 205
Thymelaeaceae, 37, 172, 181
Tiliaceae, 80
Tobacco, 12, 25, 136, 160, 171, 228, 233, 267, 268, 284, 306
Tomato, 147, 165, 250
Tovariaceae, 245
Tribuleae, 53
Trifolium, 77
T. pratense, 198, 229
T. repens, 131, 199, 229, 238, 240
T. subterraneum, 198
Triticales, 307
Triticeae, 307
Tropaeolaceae, 245
Tubiflorales, 217
Tubuliflorae, 70
Tung, 69, 78
Turk's cap lily, 180
Turmeric, 11, 105
Turnip, 243

Ullmaria, 53
Umbellales, 80, 83, 151
Umbelliferae, 5, 9, 10, 15, 27, 69, 71, 80, 81, 83, 84, 131, 147, 164, 174, 180, 183, 205, 264

Upas tree, 7
Urginea maritima, 8
Urtica, 234, 275
Urticaceae, 234, 275
Usnea diffracta, 97

Valerianaceae, 71
Vanilla fragrans, 10
Veratriae, 153
Veratrum, 139, 153
Verbascum, 40
Verbenaceae, 36, 189, 217
Vernales, 204
Vernonia anthelmintica, 71
Vernonieae, 85, 154
Vetch, 224
Vicia, 224, 231, 238
V. angustifolia, 35
V. faba, 40, 210
V. sativa, 224
Vinca major, 282
V. minor, 282
V. rosea, see Catharanthus roseus
Viola, 9
V. cornuta, 35
V. tricolor, 36
Violaceae, 9, 35, 36
Violets, 119
Vitaceae, 32, 123
Vitex lucens, 36, 189
Vitis, 32, 123
Vulgares, 54

Walnuts, 212
Water dropwort, 84
Water hemlock, 84
Water melon, 147
Water mint, 119
Wheat, 27, 162, 229, 307
White cactus, 46
White clover, 131, 240
White squill, 144
Willow, 167, 180
Wintergreen, 163, 166
Witch hazel, 27
Withania somniferum, 150, 264
Woad, 9
Wormwood, 122

Xanthium, 153

Yams, 137
Yeast, 220, 295, 301
Yew, 5, 257

Zea mays, 41, 78, 162, 228, 301
Zingiber officinale, 11
Zingiberaceae, 9, 11, 105
Zygophyllaceae, 53